Lecture Notes in Computer Science   8633

Commenced Publication in 1973
Founding and Former Series Editors:
Gerhard Goos, Juris Hartmanis, and Jan van Leeuwen

Editorial Board

Arseny M. Shur   Mikhail V. Volkov (Eds.)

Developments in Language Theory

18th International Conference, DLT 2014
Ekaterinburg, Russia, August 26-29, 2014
Proceedings

Springer

Volume Editors

Arseny M. Shur
Mikhail V. Volkov
Ural Federal University
Institute of Mathematics and Computer Science
pr. Lenina, 51
620000 Ekaterinburg, Russia
E-mail:{arseny.shur, mikhail.volkov}@usu.ru

ISSN 0302-9743　　　　　　　　　　e-ISSN 1611-3349
ISBN 978-3-319-09697-1　　　　　　e-ISBN 978-3-319-09698-8
DOI 10.1007/978-3-319-09698-8
Springer Cham Heidelberg New York Dordrecht London

Library of Congress Control Number: 2014944651

LNCS Sublibrary: SL 1 – Theoretical Computer Science and General Issues

Typesetting: Camera-ready by author, data conversion by Scientific Publishing Services, Chennai, India

Printed on acid-free paper

Springer is part of Springer Science+Business Media (www.springer.com)

Preface

The 18th International Conference on Developments in Language Theory (DLT 2014) was held during August 26–29, 2014 in Ekaterinburg, Russia, hosted by Institute of Mathematics and Computer Science of the Ural Federal University.

The DLT conference series is one of the major international conference series in language theory and related areas. It started in Turku, Finland, in 1993. It was held initially once every two years. Since 2001, it has been held every year, odd years in Europe and even years on other continents. DLT 2014 continued this tradition: Ekaterinburg is situated in Asia, just a few kilometers away from the watershed of Ural mountains which serves as the border between Europe and Asia.

The scope of the conference includes, among others, the following topics and areas: grammars, acceptors, and transducers for words, trees, and graphs; algebraic theories of automata; algorithmic, combinatorial, and algebraic properties of words and languages; variable length codes; symbolic dynamics; cellular automata; polyominoes and multidimensional patterns; decidability questions; image manipulation and compression; efficient text algorithms; relationships to cryptography, concurrency, complexity theory, and logic; bio-inspired computing; quantum computing.

The papers submitted to DLT 2014 were from 17 countries including Brazil, Canada, Finland, France, Germany, India, Italy, Japan, Korea, The Netherlands, New Zealand, Poland, Portugal, Russia, Turkey, UK, and USA.

There were 38 qualified submissions. Each submission was reviewed by at least three referees and discussed by the Program Committee for presentation at the conference. The Committee decided to accept 22 regular and 5 short papers. There were 4 invited talks given by Kai Salomaa (Queen's University, Kingston, Ontario), Martin Kutrib (Universität Giessen), Pascal Weil (CNRS and Université de Bordeaux), and Yuri Gurevich (Microsoft Research). This volume of *Lecture Notes in Computer Science* contains the papers that were presented at DLT 2014 including the full versions of three invited lectures.

The reviewing process was organized using the EasyChair conference system created by Andrei Voronkov. We would like to acknowledge that this system helped greatly to improve the efficiency of the committee work.

Arto Salomaa, the co-founder of the DLT series and one of the major contributors to the area as a whole, celebrated his 80th birthday in June 2014. DLT 2014 included a special session in Arto's honor. The opening lecture of the session surveying Arto's achievements was delivered by Juhani Karhumäki (University of Turku, Finland).

The 7th School for students and young researchers "Computer Science Days in Ekaterinburg" (CSEdays 2014) was co-located with DLT 2014.

We are grateful to our sponsors:

- Russian Foundation for Basic Research
- SKB Kontur
- JetBrains

We greatly appreciate the assistance of the many departments of the Ural Federal University. Our special thanks to Vladimir Kruzhaev, the vice rector for science, and to Magaz Asanov, the director of the Institute of Mathematics and Computer Science. We also thank the local organizers: Alexandr Galperin, Alena Nevolina, Ekaterina Neznakhina, Elena Pribavkina, and Alexei Zverev.

August 2014   Arseny M. Shur
                                                          Mikhail V. Volkov

Organization

DLT 2014 was organized by the Ural Federal University in Ekaterinburg.

Program Committee

Jorge Almeida	University of Porto, Portugal
Marie-Pierre Beál	Université Paris-Est, France
Olivier Carton	Université Paris Diderot, France
Vesa Halava	University of Turku, Finland
Yo-Sub Han	Yonsei University, Korea
Oscar H. Ibarra	University of California, Santa Barbara, USA
Masami Ito	Kyoto Sangio University, Japan
Markus Lohrey	Universität Siegen, Germany
Dirk Nowotka	Universität Kiel, Germany
Giovanni Pighizzini	University of Milano, Italy
Igor Potapov	University of Liverpool, UK
Elena Pribavkina	Ural Federal University, Russia
Michel Rigo	University of Liege, Belgium
Marinella Sciortino	University of Palermo, Italy
Jeffrey Shallit	University of Waterloo, Canada
Arseny Shur (Co-chair)	Ural Federal University, Russia
Mikhail Volkov (Co-chair)	Ural Federal University, Russia
Hsu-Chun Yen	National Taiwan University, Taiwan

Organizing Committee

Alexandr Galperin	Ural Federal University, Russia
Alena Nevolina	Ural Federal University, Russia
Ekaterina Neznakhina	Ural Federal University, Russia
Elena Pribavkina	Ural Federal University, Russia
Arseny Shur	Ural Federal University, Russia
Mikhail Volkov (Chair)	Ural Federal University, Russia
Alexei Zverev	Ural Federal University, Russia

Steering Committee

Marie-Pierre Beál	Université Paris-Est, France
Veronique Bruyere	University of Mons, Belgium
Cristian S. Calude	University of Auckland, New Zealand
Volker Diekert	Stuttgart, Germany

Juraj Hromkovic	ETH Zurich, Switzerland
Oscar H. Ibarra	University of California, Santa Barbara, USA
Masami Ito	Kyoto Sangio University, Japan
Natasha Jonoska	University of South Florida, USA
Juhani Karhumäki (Chair)	University of Turku, Finland
Antonio Restivo	University of Palermo, Italy
Grzegorz Rozenberg	Leiden University, The Netherlands
Wojciech Rytter	University of Warsaw, Poland
Arto Salomaa	University of Turku, Finland
Kai Salomaa	Queen's University, Canada
Mikhail Volkov	Ural Federal University, Russia
Takashi Yokomori	Waseda University, Japan

External Reviewers

Sponsoring Institutions

Abstracts of Invited Talks

Privacy and Inverse Privacy

Yuri Gurevich

Microsoft Research, Redmond, WA, USA

"Civilization is the progress toward a society of privacy."

Ayn Rand

Abstract. Privacy is notoriously difficult to define. It is not even clear what language should be used to define it. Is privacy, quoting Helen Nissenbaum, "a claim, a right, an interest, a value, a preference, or merely a state of existence?"

The first part of the lecture is a quick sketch of the current privacy-research picture, admittedly and necessarily limited and imperfect.

In the second part of the lecture, we introduce inverse privacy. (It is not that what you might have thought, smile.) Roughly speaking, an item of your personal information is private if you have it but nobody else does. It is inversely private if somebody has it but you don't. The problem with inverse privacy is its mere existence. We analyze the provenance of inverse privacy and, limiting our attention to private sector, argue that the problem is solvable by and large.

The inverse-privacy part of the lecture is based on joint work with our Microsoft colleagues Efim Hudis and Jeannette Wing.

Measuring Communication
in Automata Systems

Martin Kutrib and Andreas Malcher

Institut für Informatik, Universität Giessen
Arndtstr. 2, 35392 Giessen, Germany
kutrib@informatik.uni-giessen.de

Abstract. We consider systems of interacting finite automata. On the
one hand, we look at automata systems consisting of a small constant
number of synchronous and autonomous finite automata that share a
common input and communicate with each other as weakly parallel
models. On the other hand, we consider cellular automata consisting
of a huge number of interacting automata as massively parallel systems.
The communication in both types of automata systems is quantitatively
measured by the number of messages sent by the components. In cellular
automata it is also qualitatively measured by the bandwidth of the communication links. We address several aspects concerning the complexity
of such systems. In particular, fundamental types of communication are
considered and the questions of how much communication is necessary
to accept a certain language and whether there are communication hierarchies are investigated. Since even for systems with few communication
many properties are undecidable, another question is to what extent the
communication has to be limited in order to obtain systems with decidable properties again. We present some selected results on these topics
and want to draw attention to the overall picture and to some of the
main ideas involved.

Input-Driven Pushdown Automata with Limited Nondeterminism

Alexander Okhotin[1,*] and Kai Salomaa[2,**]

[1] Department of Mathematics and Statistics,
University of Turku, FI-20014 Turku, Finland
alexander.okhotin@utu.fi
[2] School of Computing, Queen's University
Kingston, Ontario K7L 3N6, Canada
ksalomaa@cs.queensu.ca

Abstract. It is known that determinizing a nondeterministic input-driven pushdown automaton (NIDPDA) of size n results in the worst case in a machine of size $2^{\Theta(n^2)}$ (R. Alur, P. Madhusudan, "Adding nesting structure to words", J.ACM 56(3), 2009). This paper considers the special case of k-path NIDPDAs, which have at most k computations on any input. It is shown that the smallest deterministic IDPDA equivalent to a k-path NIDPDA of size n is of size $\Theta(n^k)$. The paper also gives an algorithm for deciding whether or not a given NIDPDA has the k-path property, for a given k; if k is fixed, the problem is P-complete.

* Supported by the Academy of Finland under grant 257857.
** Supported by NSERC under grant OGP0147224.

From Algebra to Logic: There and Back Again
The Story of A Hierarchy[*]

Pascal Weil[1,2]

[1] CNRS, LaBRI, UMR 5800, F-33400 Talence, France
pascal.weil@labri.fr
[2] Univ. Bordeaux, LaBRI, UMR 5800, F-33400 Talence, France

Abstract. This is a survey about a collection of results about a (double) hierarchy of classes of regular languages, which occurs in a natural fashion in a number of contexts. One of these occurrences is given by an alternated sequence of deterministic and co-deterministic closure operations, starting with the piecewise testable languages. Since these closure operations preserve varieties of languages, this defines a hierarchy of varieties, and through Eilenberg's variety theorem, a hierarchy of pseudo-varieties (classes of finite monoids that are defined by pseudo-identities). The point of this excursion through algebra is that it provides reasonably simple decision algorithms for the membership problem in the corresponding varieties of languages. Another interesting point is that the hierarchy of pseudo-varieties bears a formal resemblance with another hierarchy, the hierarchy of varieties of idempotent monoids, which was much studied in the 1970s and 1980s and is by now well understood. This resemblance provides keys to a combinatorial characterization of the different levels of our hierarchies, which turn out to be closely related with the so-called rankers, a specification mechanism which was introduced to investigate the two-variable fragment of the first-order theory of the linear order. And indeed the union of the varieties of languages which we consider coincides with the languages that can be defined in that fragment. Moreover, the quantifier alternation hierarchy within that logical fragment is exactly captured by our hierarchy of languages, thus establishing the decidability of the alternation hierarchy.

There are other combinatorial and algebraic approaches of the same logical hierarchy, and one recently introduced by Krebs and Straubing also establishes decidability. Yet the algebraic operations involved are seemingly very different, an intriguing problem. . .

---

[*] This work was partially supported by the ANR through ANR-2010-BLAN-0204

Table of Contents

Combinatorics and Algorithmics on Words

Algebraic, Decidability and Complexity Problems for Languages

On Automatic Transitive Graphs

Dmitry Berdinsky and Bakhadyr Khoussainov

Department of Computer Science, The University of Auckland,
Princes St., 38, 1010 Auckland, New Zealand
berdinsky@gmail.com, bmk@cs.auckland.ac.nz
http://www.cs.auckland.ac.nz

Abstract. We study infinite automatic transitive graphs. In particular we investigate automaticity of certain Cayley graphs. We provide examples of infinite automatic transitive graphs that are not Cayley graphs. We prove that Cayley graphs of Baumslag–Solitar groups and the restricted wreath products of automatic transitive graphs with $\mathbb{Z}$ are automatic.

Keywords: finite automata, automatic graphs, transitive graphs.

1   Introduction

This paper contributes to the field of automatic structures [11] with a particular emphasis on infinite automatic transitive graphs. Recall that a graph $G = (V, E)$ is **transitive** if for all vertices $u, v \in V$ there exists an automorphism of the graph that maps u into v. We postulate that the edges of our graphs are unordered pairs, and hence no loop exists from any vertex of the graph to itself.

Examples of transitive graphs are plentiful. The Rado graph, obtained as the Fraïssé limit of the class of all finite graphs, is transitive. So are Cayley graphs of finitely generated groups. Here we recall the definition of a Cayley graph. Let H be a finitely generated group with a finite set of generators X such that X does not contain the identity of the group. Define the Cayley graph of H with respect to X as follows: (a) the set of vertices of the Cayley graph is the set H, and (b) there is an edge between the vertices u and v if there exists an $x \in X$ such that $ux = v$ or $ux^{-1} = v$. Often Cayley graphs are directed and labeled graphs with labels from X, but for simplicity we omit the labels and directions on the edges.

We assume that the reader is familiar with the notion of finite automata and synchronous multi–head finite automata. Finite automata can be used to define the concept of an automatic graph. Call a graph $G = (V, E)$ **automatic** if it has an automatic presentation (FA–presentable), i.e., if both the set of vertices V and the set of edges E are recognized by finite automata. The automaton that recognizes the set of edges is a synchronous 2–head finite automaton. We recall that for two strings $u_1 \ldots u_k$ and $v_1 \ldots v_m$ representing two vertices of V, the shorter one is padded with a padding symbol $\diamond$. Then such an automaton reads

A.M. Shur and M.V. Volkov (Eds.): DLT 2014, LNCS 8633, pp. 1–12, 2014.

the string of paired symbols $(u_1, v_1) \ldots (u_n, v_n)$ for $n = \max\{k, m\}$. Here we give two sets of general examples of automatic transitive graphs.

Example 1. Let H be a group generated by a finite set X. Consider its labeled and directed Cayley graph $\Gamma(H, X)$. Call the group H **automatic** if there exists a regular subset $L \subseteq (X \cup X^{-1})^*$ such that the natural mapping from L into G is bijective, and for any $x \in X$ the set of directed edges of $\Gamma(H, X)$ labeled by x is recognized by a synchronous 2–head finite automaton. For the theory of automatic groups the reader is referred to [5]. It is proved that FA–presentable finitely generated groups are virtually Abelian [14]. This implies that the class of automatic groups properly contains FA–presentable finitely generated groups.

Example 2. As above, let H be a group generated by a finite set X. Consider its labeled and directed Cayley graph $\Gamma(H, X)$. We call the group H **Cayley automatic** [10] if there a regular subset $L \subset \Sigma^*$ for some finite alphabet Σ that uniquely represents elements of H, and for any $x \in X$ the set of directed edges of $\Gamma(H, X)$ labeled by x is recognized by a synchronous 2–head finite automaton, i.e., $\Gamma(H, X)$ is FA–presentable. The class of Cayley automatic groups properly contains automatic groups and retains many of their properties. Clearly a Cayley graph of a Cayley automatic group is automatic.

It is known that the Rado graph is not an automatic graph [12,4], while the examples above show the abundance of transitive automatic graphs. Taking into account that the Rado graph is not automatic and the two examples above, we postulate the following conditions on transitive graphs G that will be assumed for the rest of the paper: (1) the graph G is infinite and connected, and (2) the degree of every vertex v in G is bounded. Since the graph is transitive, the second condition implies that all vertices of the graph have the same degree.

2   Contributions of the Paper

- We give an example of a sequence of automatic infinite transitive graphs that are not Cayley graphs. We also show that the limit of this sequence is an automatic transitive graph such that no Cayley graph is quasi–isometric to it. These examples show that the class of automatic transitive graphs properly contains the class of Cayley graphs of all Cayley automatic groups.
- In [10] it is proved that Baumslag–Solitar groups $B(1, n)$ are Cayley automatic. We extend this result, and prove that all Baumslag–Solitar groups $B(m, n)$ are Cayley automatic. The proof is based on finding proper normal forms of group elements through HNN–extensions.
- In [10] it is proved that the wreath product of any finite group with the group $\mathbb{Z}$ of integers is Cayley automatic. These groups are examples of groups that are not automatic. We extend the result from [10] and prove that the wreath product of any Cayley automatic group with the group $\mathbb{Z}$ preserves Cayley automaticity. We also generalize this result to the class of locally finite automatic graphs. In addition, we make some relevant remarks on the Cayley automaticity of wreath products of finitely generated groups and infinite non–cyclic groups.

3   Examples of Infinite Transitive Automatic Graphs

We start this section with a simple example of an automatic infinite transitive graph which is not a Cayley graph. This example showcases that the class of transitive infinite automatic graphs properly contains the class of Cayley graphs of all Cayley automatic groups.

The example is presented in [15] but our description follows the beginning of § 2 in [3]. Consider the infinite 5–regular tree T_5 and the bipartite graph $K_{2,3}$. Define the graph $H_{2,3}$ as follows. First, replace vertices of T_5 by disjoint copies of $K_{2,3}$. Second, for each edge $\{u,v\}$ of T_5, identify a vertex of the $K_{2,3}$ corresponding to u with a vertex of the $K_{2,3}$ corresponding to v. No point in any $K_{2,3}$ is identified more than once, and a vertex in a class of size 2 is always identified with a vertex in a class of size 3 and vice versa. See Fig. 1.

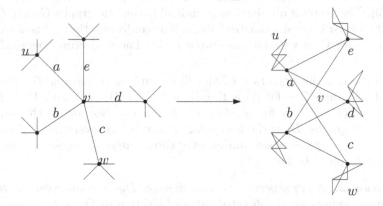

Fig. 1. Constructing the non–Cayley graph $H_{2,3}$ from T_5

Proposition 1. *The graph $H_{2,3}$ is a transitive infinite automatic graph that is not a Cayley graph for any finitely generated group.*

Proof. It is clear that $H_{2,3}$ is transitive. We prove that $H_{2,3}$ is not a Cayley graph for any finitely generated group. Recall that a graph H is a Cayley graph of some group if and only if there exists a subgroup S of automorphisms of H that acts freely and transitively on H. Here S acts transitively on H if for every pair u,v of vertices of H there exists an action of S that maps u to v; also, S acts freely on H if every nontrivial action of S moves every vertex of H.

Suppose that there exists a subgroup S of automorphisms of $H_{2,3}$ that acts freely and transitively on $H_{2,3}$. Let K be one of the $K_{2,3}$ making up $H_{2,3}$, and let $\{a,b\}$ and $\{c,d,e\}$ be the vertices in a class of size 2 and 3, respectively. It is not hard to see that any automorphism that sends an element of $\{c,d,e\}$ back into $\{c,d,e\}$ must fix K. Now let $\theta \in S$ be such that $\theta(c) = d$; therefore, it must swap a and b. Let $\theta' \in S$ be such that $\theta'(c) = e$; again, it must swap a and b.

Then $\theta'\theta^{-1} \in S$ sends d to e and fixes a and b. This gives us a contradiction. Hence $H_{2,3}$ is not a Cayley graph for any finitely generated group.

To prove automaticity of $H_{2,3}$, we note that T_5 is an automatic graph. The definition of $H_{2,3}$ shows that we can identify the elements of $H_{2,3}$ and a quotient set $T_5 \times K_{2,3}/ \equiv$, where $\equiv$ is recognizable by a finite automaton equivalence relation on $T_5 \times K_{2,3}$. An easy analysis can then show that $H_{2,3}$ under the representation induced by $T_5 \times K_{2,3}/ \equiv$ is automatic.   □

The construction above can easily be generalized to build automatic transitive graphs $H_{n,m}$ from the regular tree T_{n+m} and the bipartite graph $K_{n,m}$, where $n \neq m$ and $n \geqslant 2$ and $m \geqslant 3$:

Corollary 1. *The graphs $H_{n,m}$ are transitive infinite automatic graphs that are not Cayley graphs.*   □

We now construct another transitive automatic graph $G^\star$. Our description follows [3]. The graph is obtained as a limit of automatic graphs $G_0, G_1, G_2, \ldots$ such that each G_i is quasi–isometric[1] to T_5. The conjecture that $G^\star$ is not quasi–isometric to any Cayley graph was stated in [3]. The conjecture was confirmed in [7].

First we recall the definition of the line graph of a digraph D. The **line graph** of a digraph $D = (V, E)$ is the digraph $D' = (V', E')$ with $V' = E$ and $E' = \{((u,v), (u',v')) \mid (u,v), (u',v') \in V' \land v = u'\}$. Now define the sequence $D_0, D_1, \ldots$ of graphs: D_0 is the 5–regular tree such that every node of the tree has exactly two ingoing edges and exactly three outgoing edges, and set D_{i+1} be the line graph of D_i for $i \geq 0$.

Proposition 2. *[3, Proposition 3] The digraph D_n is isomorphic to the digraph whose vertices are the directed paths of length n in D_0, with an edge from $x_1 x_2 \ldots x_{n+1}$ to $y_1 y_2 \ldots y_{n+1}$ if $y_i = x_{i+1}$ for all $1 \leqslant i \leqslant n$.*   □

Since D_0 is automatic, Proposition 2 implies that each digraph D_n is automatic.

Let G_n be the graph obtained from D_n by removing the directions from the edges of D_n. For instance, G_1 is isomorphic to $H_{2,3}$. Since each D_n is automatic, the graph G_n is also automatic. All the graphs $G_1, G_2, \ldots$ are transitive and are not Cayley graphs. Moreover, each G_i is quasi–isometric to T_5. In [3] it is proved that the sequence $G_0, G_1, \ldots$ converges to $G^\star$, and that the graph $G^\star$ can be described as follows.

Let X be a 3–regular tree where each node has in–degree 2 and out–degree 1. Let Y be a 4–regular tree in which each node has in–degree 1 and out–degree 3. Fix a vertex $O_1 \in V(X)$ and a vertex $O_2 \in V(Y)$. For each $x \in X$, set $r(x)$ to be

---

[1] We recall that two graphs G_1 and G_2 are quasi–isometric if there exist a map $\theta : V(G_1) \to V(G_2)$ and some $\lambda \geqslant 1$ such that $\frac{1}{\lambda} d_{G_1}(x,y) - \lambda \leqslant d_{G_2}(\theta(x),\theta(y)) \leqslant \lambda d_{G_1}(x,y) + \lambda$ for all $x, y \in V(G_1)$, and for any point $y \in V(G_2)$ there is some $x \in V(G_1)$ such that $d_{G_2}(\theta(x),y) \leqslant \lambda$; where $d_{G_1}(x,y)$ and $d_{G_2}(\theta(x),\theta(y))$ are the graph distances between x, y and $\theta(x), \theta(y)$ in G_1 and G_2, respectively. For more on quasi-isometries we refer to [8].

the signed distance from O_1 to x, i.e., if the unique undirected path from O_1 to x in X has s forward edges and t backward edges then $r(x) = s - t$. Define $r(y)$ similarly for each $y \in Y$. The set of vertices of G^* is the set $\{(x, y) \in X \times Y : r(x) = r(y)\}$, and G^* has an arc from (x, y) to (x', y') if $(x, x') \in E(X)$ and $(y, y') \in E(Y)$. Thus, the results in [3] and [7] imply:

Theorem 1. *The structure G^* is a transitive automatic graph such that no Cayley graph is quasi–isometric to it.* $\qquad\qquad\square$

4 Baumslag–Solitar Groups

The Baumslag–Solitar groups are defined as a particular class of two-generator one-relator groups [1]. They play an important role in combinatorial and geometric group theory.

Definition 1. *For given non-negative integers m, n, the Baumslag–Solitar group $BS(m, n)$ is a finitely generated one–relator group defined as follows:*

$$BS(m, n) = \langle a, t \mid t^{-1}a^m t = a^n \rangle.$$

We will further suppose that $m, n \neq 0$; if $m = 0$ or $n = 0$ then $BS(m, n)$ is isomorphic to a free product of a cyclic group and $\mathbb{Z}$. The groups $BS(m, n)$ are automatic if and only if $m = n$ [5]. It is also known that $BS(m, n)$ are all asynchronously automatic [5]. In this section we prove that the Baumslag–Solitar groups $BS(m, n)$ are Cayley automatic groups for all $n, m \in \mathbb{N}$.

In [10] it is proved that some special cases of Baumslag–Solitar group, namely $BS(1, n), n \in \mathbb{N}$, are Cayley automatic. The proof is based on representation of groups elements as special type of linear functions acting on the real line. We extend this theorem to the class of all Baumslag–Solitar groups. Our proof is based on a normal form for elements of $BS(m, n)$ that comes from representing $BS(m, n)$ as Higman–Neumann–Neumann (HNN) extension.

We now recall the general construction for HNN extension and the normal form theorem for HNN extension; we will follow [13, Chapter IV, § 2].

The HNN extension of G relative to the subgroups $A, B \leqslant G$ and an isomorphism $\phi : A \to B$ is the group

$$G^* = \langle G, t; t^{-1}at = \phi(a), a \in A \rangle,$$

where t is usually called a stable letter. The normal form theorem for HNN extension, as presented in [13], is the following. For the proof see statement (II) of Theorem 2.1 in [13]:

Theorem 2. *Suppose we have fixed all the representatives for right cosets of subgroups A and B in G, where the identity $1 \in G$ represents the subgroups A and B. Then every element w of G^* has a unique representation as $w = g_0 t^{\epsilon_1} \cdots t^{\epsilon_\ell} g_\ell$ where*

- g_0 is an arbitrary element of G,
- if $\epsilon_i = -1$, then g_i is a representative of a right coset of A in G,
- if $\epsilon_i = +1$, then g_i is a representative of a right coset of B in G, and
- there is no consecutive subsequence $t^\epsilon, 1, t^{-\epsilon}$. $\square$

By replacing the representatives to left cosets of subgroups A and B, with 1 representing both A and B, one obtains the following corollary of Theorem 2.

Corollary 2. *Suppose that we have fixed all the representatives for left cosets of A and B in G with the identity $1 \in G$ representing A and B. Then every element w of G^* has a unique representation as $w = g_\ell t^{\epsilon_\ell} \cdots g_1 t^{\epsilon_1} g_0$, where*

- g_0 is an arbitrary element of G,
- if $\epsilon_i = -1$, then g_i is a representative of a left coset of B in G,
- if $\epsilon_i = +1$, then g_i is a representative of a left coset of A in G,
- there is no consecutive subsequence $t^\epsilon, 1, t^{-\epsilon}$. $\square$

Note that the Baumslag- $\mathbb{Z} = \langle a \rangle$ relative to subgroups $m\mathbb{Z}$ and $n\mathbb{Z}$, and the isomorphism $\phi : m\mathbb{Z} \to n\mathbb{Z}$ that maps a^m to a^n. By using Corollary 2, our goal is to show that $BS(m, n)$ is a Cayley automatic group.

Theorem 3. *The group $BS(m, n)$ is a Cayley automatic group.*

Proof: We put $1, \ldots, a^{m-1}$ and $1, \ldots, a^{n-1}$ to be the representatives for left cosets of subgroups $m\mathbb{Z}$ and $n\mathbb{Z}$ in $\mathbb{Z}$ respectively. By Corollary 2, every element $w \in BS(m, n)$ has a unique representation as

$$w = g_\ell t^{\epsilon_\ell} \cdots g_1 t^{\epsilon_1} g_0, \tag{1}$$

where $g_0 = a^k$ for some $k \in \mathbb{Z}$, and if $\epsilon_i = -1$ then $g_i \in \{1, a, \ldots, a^{n-1}\}$, if $\epsilon_i = +1$ then $g_i \in \{1, a, \ldots, a^{m-1}\}$, and there is no consecutive subsequence $t^\epsilon, 1, t^{-\epsilon}$.

The right–multiplication by the generator a transforms the normal form of w as follows:

$$g_\ell t^{\epsilon_\ell} \cdots g_1 t^{\epsilon_1} a^k \xrightarrow{\times a} g_\ell t^{\epsilon_\ell} \cdots g_1 t^{\epsilon_1} a^{k+1}. \tag{2}$$

Let $k = mp + r$ where $p \in \mathbb{Z}$ and $r \in \{0, \ldots, m - 1\}$. The right–multiplication by the generator t transforms a normal form of w as follows

- if $r \neq 0$ then

$$g_\ell t^{\epsilon_\ell} \cdots g_1 t^{\epsilon_1} a^k \xrightarrow{\times t} g_\ell t^{\epsilon_\ell} \cdots g_1 t^{\epsilon_1} a^r t a^{np}, \tag{3}$$

- if $r = 0$, and $\ell \geqslant 1, \epsilon_1 = -1$ then

$$g_\ell t^{\epsilon_\ell} \cdots g_2 t^{\epsilon_2} g_1 t^{-1} a^k \xrightarrow{\times t} g_\ell t^{\epsilon_\ell} \cdots g_2 t^{\epsilon_2} (g_1 a^{np}), \tag{4}$$

– if $r = 0$ and $\ell \geqslant 1, \epsilon_1 = 1$ then

$$g_\ell t^{\epsilon_\ell} \cdots g_1 t a^k \xrightarrow{\times t} g_\ell t^{\epsilon_\ell} \cdots g_1 t 1 t a^{np}, \tag{5}$$

– if $r = 0$ and $\ell = 0$ then

$$a^k \xrightarrow{\times t} a^r t a^{np}. \tag{6}$$

Combining (1) with (2)–(6), we see that $BS(m,n)$ is Cayley automatic. We note that in order to construct relevant finite automata, the m–ary representation of $k \in \mathbb{Z}$ (for which the map $k = mp + r \to np$ is recognizable by a finite automaton) should be used. □

Remark 1. We note that the map $k = mp + r \to mp$ is recognizable by a finite automaton for a unary representation of k. It follows from the proof of Theorem 3 that if $m = n$ then $BS(m,n)$ is automatic.

Remark 2. We mentioned that $BS(m,n)$ is not automatic if $m \neq n$. Indeed, a group that is automatic satisfies a quadratic isoperimetric inequality (for the proof see, e.g., [5, Theorem 2.3.12]). However, by choosing a proper family of cycles in the Cayley graph, it can be shown that $BS(m,n)$ does not satisfy a polynomial isoperimetric inequality (see the explanation in [5, § 7.4]).

The next remark concerns asynchronously automatic groups (see [5, § 7]). This class of groups extends the class of groups that are automatic.

Remark 3. It is known that $BS(m,n)$ is an asynchronously automatic group (see the explanation in [5, § 7.4]). This fact can be alternatively derived from Theorem 3. To show it, we observe that the map $k = mp + r \to np$ is recognizable by an asynchronous automaton for the unary representation of k.

5   Wreath Products

We start by recalling the definition of the restricted wreath products of two groups A and B; we will follow [9, § 6.2].

Let A and B be groups. We denote by $A^{(B)}$ the group of all functions $B \to A$ having finite supports and the usual multiplication rule; recall that a function $f : B \to A$ has finite support if $f(x) \neq e$ for only finite number of $x \in B$, where e is the identity of A. Let us define the homomorphism $\tau : B \to \mathrm{Aut}(A^{(B)})$ as follows: for a given $f \in A^{(B)}$, the automorphism $\tau(b)$ maps f to $f^b \in A^{(B)}$, where $f^b(x) = f(bx)$ for all $x \in B$. For given groups A and B, the restricted wreath product $A \wr B$ is defined to be the semidirect product $A^{(B)} \rtimes_\tau B$. Thus, $A \wr B$ is the set product $B \times A^{(B)}$ with multiplication given by

$$(b, f) \cdot (b', f') = (bb', f^{b'} f'), \tag{7}$$

where $f^{b'}(x) = f(b'x)$.

For our purposes we will use the converse order for representing elements of a wreath product. Namely, we will represent an element of $A \wr B$ as a pair (f, b), where $f \in A^{(B)}$ and $b \in B$. For such a representation, the group multiplication is given by

$$(f, b) \cdot (f', b') = (ff'^{b^{-1}}, bb'), \tag{8}$$

where $f'^{b^{-1}}(x) = f'(b^{-1}x)$. There exist natural embeddings $B \to A \wr B$ and $A^{(B)} \to A \wr B$ mapping b to $(\mathbf{e}, b)$ and f to (f, e) respectively, where $\mathbf{e}$ is the identity of $A^{(B)}$ and e is the identity of B. For the sake of simplicity, we will identify B and $A^{(B)}$ with the corresponding subgroups of $A \wr B$.

5.1   Wreath Products of Finitely Generated Groups and $\mathbb{Z}$

For a given finitely generated group G, let us consider the wreath product $G \wr \mathbb{Z}$. We represent an element of $G \wr \mathbb{Z}$ as a pair (f, z), where $f \in G^{(\mathbb{Z})}$ and $z \in \mathbb{Z}$. By (8), the multiplication in $G \wr \mathbb{Z}$ on pairs (f, z) and (f', z') is given by:

$$(f, z)(f', z') = (ff'^{-z}, z + z'),$$

where $f'^{-z}(x) = f'(x - z)$.

Let $g_1, \ldots, g_n$ be generators of G and $\mathbb{Z} = \langle t \rangle$; let us identify t with $1 \in \mathbb{Z}$. We denote by $f_1, \ldots, f_n$ the functions from $G^{(\mathbb{Z})}$ such that $f_i(0) = g_i$ and $f_i(x) = e$ for all $x \neq 0 \in \mathbb{Z}$, $i = 1 \ldots n$, where e is the identity of G. The pairs $(f_1, 0), \ldots, (f_n, 0)$ and $(\mathbf{e}, t)$ are the generators of $G \wr \mathbb{Z}$. The right–multiplications of a pair (f, z) by these generators are given by:

$$(f, z)(\mathbf{e}, t) = (f, z + 1), \tag{9}$$

$$(f, z)(f_i, 0) = (ff_i^{-z}, z), \ i = 1, \ldots, n, \tag{10}$$

where $f_i^{-z}(z) = g_i$, $f_i^{-z}(x) = e$ for all $x \neq z$ and $\mathbf{e}$ is the identity of $G^{(\mathbb{Z})}$; one can obtain ff_i^{-z} from f by changing the value of the function f at the point z to the one equal to $f(z)g_i$.

It is useful to imagine an element $(f, z) \in G \wr \mathbb{Z}$ as a bi–infinite string of elements of G, such that only a finite number of elements are not equal to $e \in G$, with the tape head in position z pointing at an element of this string. Then the identity $(\mathbf{e}, 0)$ of $G \wr \mathbb{Z}$ corresponds to the string where all elements are equal to $e \in G$ with the tape head in the origin $0 \in \mathbb{Z}$.

For the sake of convenience, let t and f_i denote the corresponding elements $(\mathbf{e}, t)$ and (f_i, e) in $G \wr \mathbb{Z}$. Then, in the wreath product $G \wr \mathbb{Z}$, the right–multiplication by t shifts the tape head to the right by one whilst that of by t^{-1} shifts the tape head to the left by one; the right–multiplication by f_i changes the value of the element that the tape head is pointed to multiplying it by g_i. Therefore, a word $t^k f_{i_1} \cdots f_{i_\ell} t^{-k}$ can be interpreted as follows: the tape head makes k steps from the origin either to the right or left (for $k > 0$ or $k < 0$ respectively) then the kth element is changed from e to $g_{i_1} \cdots g_{i_\ell}$, and then the tape head returns back to the origin.

Suppose that we chose minimal length representatives (in terms of generators $g_1, \ldots, g_n$) for all elements of G, i.e., for each $u \in G$ we have its minimal length representative $v = g_{i_1}^{\pm 1} \cdots g_{i_\ell}^{\pm 1} \in (\{g_1, \ldots, g_n\} \cup \{g_1^{-1}, \ldots, g_n^{-1}\})^*$. There exist two natural ways to represent uniquely elements of $G \wr \mathbb{Z}$ [2]. For a given $w \in G \wr \mathbb{Z}$ let the right–first normal form $rf(w)$ be the following:

$$rf(w) = (t^{i_1} v_{i_1} t^{-i_1}) \cdots (t^{i_k} v_{i_k} t^{-i_k})(t^{-j_1} v_{j_1} t^{j_1}) \cdots (t^{-j_\ell} v_{j_\ell} t^{j_\ell}) t^m, \qquad (11)$$

where $i_k > \ldots i_1 \geqslant 0$, $j_l > \ldots j_1 > 0$, $m \in \mathbb{Z}$, and $v_{i_1}, \ldots, v_{i_k}, v_{j_1}, \ldots, v_{j_\ell}$ are the corresponding minimal representatives in G. The left–first normal form $lf(w)$ is given by:

$$lf(w) = (t^{-j_1} v_{j_1} t^{j_1}) \cdots (t^{-j_\ell} v_{j_\ell} t^{j_\ell})(t^{i_1} v_{i_1} t^{-i_1}) \cdots (t^{i_k} v_{i_k} t^{-i_k}) t^m. \qquad (12)$$

In the right–first normal form the tape head first moves to the right from the origin whilst in the left–first form the tape head first moves to left from the origin. By [2], either $rf(w)$ or $lf(w)$ (taken in the reduced form) provide the minimal length representative of w.

The normal forms (11) and (12) give an immediate prompt about how to show that the wreath product $G \wr \mathbb{Z}$ is Cayley automatic if G itself is Cayley automatic.

Theorem 4. *For a given Cayley automatic finitely generated group G the wreath product $G \wr \mathbb{Z}$ is Cayley automatic.*

Proof. Let G be a Cayley automatic group with a finite set of generators $S = \{g_1, \ldots, g_n\}$. Let P be a regular language that gives a graph automatic representation of the labeled and directed Cayley graph $\Gamma(G, S)$. Without loss of generality we will suppose that each element $w \in G$ has a unique word $v \in P$ as a representative. In order to prove that $G \wr \mathbb{Z}$ is Cayley automatic, we need to construct a regular language Q representing elements of $G \wr \mathbb{Z}$ and describe the automata recognizing right–multiplications by the generators $(\mathbf{e}, t)$ and $(f_i, 0)$ (see (9) and (10) respectively).

To obtain the alphabet of Q, let us add to the alphabet of P four additional symbols: A, B, C and $\#$. For a given element of $G \wr \mathbb{Z}$ we use the symbol A to specify the position of the origin $0 \in \mathbb{Z}$, the symbol C to specify the position of the tape head $m \in \mathbb{Z}$, and the symbol B will be used instead of A and C in case the positions of the origin and the tape head coincide: $m = 0$. The symbol $\#$ is used to separate subwords representing elements of G; at the positions corresponding to the origin and the tape head, the symbols A, C or B should be used instead of $\#$. Given an element of $G \wr \mathbb{Z}$, we represent it as a finite string of the following form:

$$v_{-j} \# \ldots \# v_{-1} A v_0 \# v_1 \# \ldots \# v_{m-1} C v_m \# v_{m+1} \ldots \# v_i, \qquad (13)$$

where $v_{-j}, \ldots, v_i$ are the words of P and v_{-j} and v_i are the representatives for the leftmost and rightmost nontrivial elements of G. The language of such finite strings form the regular language Q.

The right–multiplication by $(\mathbf{e}, t)$ corresponds to the following relation:

$$\begin{pmatrix} \cdots \, \# \, v_{m-1} \; C \; v_m \; \# \; v_{m+1} \; \# \cdots \\ \cdots \, \# \, v_{m-1} \; \# \; v_m \; C \; v_{m+1} \; \# \cdots \end{pmatrix},$$

which is clearly recognizable by a finite automaton.

The right–multiplication by $(f_i, 0)$ corresponds to the following relation:

$$\begin{pmatrix} \cdots \, \# \, v_{m-1} \; C \; v_m \; \# \; v_{m+1} \; \# \cdots \\ \cdots \, \# \, v_{m-1} \; C \; u_m \; \# \; v_{m+1} \; \# \cdots \end{pmatrix}. \tag{14}$$

The relation $\begin{pmatrix} v_m \\ u_m \end{pmatrix}$ is recognizable by a finite automaton by the initial assumption that G is Cayley automatic. The difference between the lengths of the words v_m and u_m is bounded by some constant C for all generators $g_i \in S$. Thus, the relation (14) is recognizable by a finite automaton.   □

The next remark concerns asynchronously automatic groups.

Remark 4. The wreath product $G \wr \mathbb{Z}$ is not an asynchronously automatic group for any nontrivial group G. Indeed, if a group is asynchronous automatic then it is finitely presented (see [5, Theorem 7.3.4]). But, the wreath product $G \wr \mathbb{Z}$ is finitely presented only if G is trivial.

5.2   Generalization

The wreath product on groups can naturally be defined for graphs. Indeed, let B be a graph and A be a graph with a distinguished vertex $a_0 \in V(A)$. For a function $f : V(B) \to V(A)$, the support of f is the set $\{b \in V(B) : f(b) \neq a_0\}$.

Definition 2. *([6, Definition 2.1]) The wreath product $A \wr B$ of graphs A and B is the graph the vertices of which are pairs (f, b), where $b \in V(B)$ and $f : V(B) \to V(A)$ is a function with a finite support. Two vertices (f_1, b_1) and (f_2, b_2) are joined by an edge if either*

- $b_1 = b_2$ *and* $f_1(x) = f_2(x)$ *for all* $x \neq b_1$ *and there is an edge in A between* $f_1(b_1)$ *and* $f_2(b_2)$, *or*
- $f_1(x) = f_2(x)$ *for all* $x \in V(B)$ *and there is an edge in B between* b_1 *and* b_2.

Note that in this definition, we do not require that the graphs A and B are transitive. However, it is not hard to see that if A and B are transitive then so is their wreath product $A \wr B$.

Let G_1 and G_2 be Cayley graphs of two finitely generated groups H_1 and H_2, and the distinguished vertex of G_1 is the identity of H_1. Then it is easy to see that the wreath product of graphs $G_1 \wr G_2$ is the Cayley graph of the wreath product of groups $H_1 \wr H_2$.

We denote by $\Gamma(\mathbb{Z})$ the Cayley graph of $\mathbb{Z}$ with respect to the standard generator $t = 1 \in \mathbb{Z}$. Recall that a graph is locally finite if the degree of every vertex of the graph is finite. Theorem 4 can be straightforwardly generalized as follows.

Theorem 5. *For any given locally finite automatic graph Γ the wreath product $\Gamma \wr \Gamma(\mathbb{Z})$ is automatic.*   □

5.3 Is the Wreath Product $\mathbb{Z}_2 \wr \mathbb{Z}^2$ Cayley Automatic?

The analysis of Cayley automaticity of a wreath product $G \wr H$ becomes more complex if H is an not a cyclic group. For example, we do not know whether the groups $\mathbb{Z}_2 \wr \mathbb{F}_2$ and $\mathbb{Z}_2 \wr \mathbb{Z}^2$ are Cayley automatic or not. In this subsection we show that if $\mathbb{Z}_2 \wr \mathbb{Z}^2$ is Cayley automatic then its automatic presentation is unnatural in the sense described below.

Consider the wreath product $G \wr H$, where G is a Cayley automatic group and H is an infinite Cayley automatic non–cyclic group, say $H = \mathbb{Z}^2$. It perfectly seems reasonable, at first, to construct a Cayley automatic representation following the approach used for $G \wr \mathbb{Z}$. Indeed, for any bijective map $\tau : \mathbb{Z} \to H$, a regular language representing the elements of $G \wr H$ can be constructed composing the representation of $G \wr \mathbb{Z}$ given in the proof of Theorem 4 and the one–to–one correspondence between H and $\mathbb{Z}$. Moreover, if G is a finite group, then such a representation gives the regular representation for the subgroup $G^{(H)}$ for which the group operation is recognizable by a finite automaton. Assume that $\tau(0)$ is the identity of H. Then the regular language representing $G^{(H)} \trianglelefteq G \wr H$ consists of all words of Q having the symbol B as a subword that corresponds to the configurations when the tape head is pointing to the identity of H. It is then not surprising, as implied by Proposition 3 below, that such representations fail to be Cayley automatic.

Assume that P is a regular language that gives a Cayley automatic representation of $\mathbb{Z}_2 \wr \mathbb{Z}^2$. We suppose that each element of $\mathbb{Z}_2 \wr \mathbb{Z}^2$ has a unique representative in P. Let $f_0 \in \mathbb{Z}_2^{(\mathbb{Z}^2)}$ be the function such that $f_0(0,0) = 1$ and $f_0(z_1, z_2) = 0$ if $(z_1, z_2) \neq (0,0)$; let $r = (1,0) \in \mathbb{Z}^2$ and $u = (0,1) \in \mathbb{Z}^2$. These are the generators of $\mathbb{Z}_2 \wr \mathbb{Z}^2$. Cayley automaticity of $\mathbb{Z}_2 \wr \mathbb{Z}^2$ implies that the right–multiplications by f_0, r and u are recognizable by finite automata.

Proposition 3. *Assume that $\mathbb{Z}_2 \wr \mathbb{Z}^2$ is Cayley automatic with respect to P such that the subset $P' \subset P$ of representatives of the subgroup $\mathbb{Z}_2^{(\mathbb{Z}^2)}$ is a regular language. Then the group operation in $\mathbb{Z}_2^{(\mathbb{Z}^2)}$ is not recognizable by a finite automaton.*

Proof. We will prove the proposition by contradiction. Suppose that the group operation in $\mathbb{Z}_2^{(\mathbb{Z}^2)}$ is recognizable by a finite automaton.

For a given $n \in \mathbb{N}$, we denote by $H_n \leqslant \mathbb{Z}_2^{(\mathbb{Z}^2)}$ the subgroup of functions $f \in \mathbb{Z}_2^{(\mathbb{Z}^2)}$ having supp $f \subset \{(i,j) \mid -n \leqslant i, j \leqslant n\}$. Since the group operation in $\mathbb{Z}_2^{(\mathbb{Z}^2)}$ and right multiplications by u, r and f_0 are recognizable by finite automata, it directly follows from the Constant Growth lemma (see, e.g., [10, Lemma 14.1]) that there exists a constant C such that $|f|_{P'} \leqslant Cn$ for all $f \in H_n$, where $|f|_{P'}$

denotes the length of the representative of f in P'. On the other hand, the number of elements in H_n is equal to $2^{(2n+1)^2}$. Thus, we get a contradiction and, therefore, the group operation in $\mathbb{Z}_2^{(\mathbb{Z}^2)}$ is not recognizable by a finite automaton.

$\square$

References

1. Baumslag, G., Solitar, D.: Some two-generator one-relator non-Hopfian groups. Bulletin of American Mathematical Society 68, 199–201 (1962)
2. Cleary, S., Taback, J.: Dead end words in lamplighter groups and other wreath products. Quarterly Journal of Mathematics 56, 165–178 (2005)
3. Deistel, R., Leader, I.: A conjecture concerning a limit of non–Cayley graphs. Journal of Algebraic Combinatorics 14(1), 17–25 (2001)
4. Delhommé, C.: Automaticité des ordinaux et des graphes homogènes. Comptes Rendus Mathematique 339(1), 5–10 (2004)
5. Epstein, D.B.A., Cannon, J.W., Holt, D.F., Levy, S.V.F., Paterson, M.S., Thurston, W.P.: Word Processing in Groups. Jones and Barlett Publishers, Boston (1992)
6. Erschler, A.: Generalized wreath products. International Mathematics Research Notices Article ID 57835, 1–14 (2006)
7. Eskin, A., Fisher, D., Whyte, K.: Quasi-isometries and rigidity of solvable groups. arXiv:math/0511647v3 [math.GR] (2006)
8. Gromov, M.: Asymptotic invariants of infinite groups. In: Niblo, A., Roller, M.A. (eds.) Geometric Group Theory, Vol. 2. London Math. Soc. Lecture Note Ser., vol. 182, Cambridge Univ. Press, Cambridge (1993)
9. Kargapolov, M.I., Merzljakov, J.I.: Fundamentals of the theory of groups. Springer, Heidelberg (1979)
10. Kharlampovich, O., Khoussainov, B., Miasnikov, A.: From automatic structures to automatic groups. arXiv:1107.3645v2 [math.GR] (2011)
11. Khoussainov, B., Nerode, A.: Automatic presentations of structures. In: Leivant, D. (ed.) LCC 1994. LNCS, vol. 960, pp. 367–392. Springer, Heidelberg (1995), http://dx.doi.org/10.1007/3-540-60178-3_93
12. Khoussainov, B., Nies, A., Rubin, S., Stephan, F.: Automatic structures: richness and limitations. Logical Methods in Computer Science 3(2:2), 1–18 (2007)
13. Lyndon, R.C., Schupp, P.E.: Combinatorial group theory. Springer, Heidelberg (1977)
14. Oliver, G.P., Thomas, R.M.: Automatic presentations for finitely generated groups. In: Diekert, V., Durand, B. (eds.) STACS 2005. LNCS, vol. 3404, pp. 693–704. Springer, Heidelberg (2005), http://dx.doi.org/10.1007/978-3-540-31856-9_57
15. Thomassen, C., Watkins, M.E.: Infinite vertex-transitive, edge-transitive, non-1-transitive graphs. Proceedings of American Mathematical Society 105(1), 258–261 (1989)

Upper Bounds on Syntactic Complexity
of Left and Two-Sided Ideals*

Janusz Brzozowski[1] and Marek Szykuła[2]

[1] David R. Cheriton School of Computer Science, University of Waterloo,
Waterloo, ON, Canada N2L 3G1
brzozo@uwaterloo.ca
[2] Institute of Computer Science, University of Wrocław,
Joliot-Curie 15, PL-50-383 Wrocław, Poland
msz@cs.uni.wroc.pl

Abstract. We solve two open problems concerning syntactic complexity. We prove that the cardinality of the syntactic semigroup of a left ideal or a suffix-closed language with n left quotients (that is, with state complexity n) is at most $n^{n-1} + n - 1$, and that of a two-sided ideal or a factor-closed language is at most $n^{n-2} + (n - 2)2^{n-2} + 1$. Since these bounds are known to be reachable, this settles the problems.

Keywords: factor-closed, left ideal, regular language, suffix-closed, syntactic complexity, transition semigroup, two-sided ideal, upper bound.

1 Introduction

The *syntactic complexity* [4] of a regular language is the size of its syntactic semigroup [5]. The *transition semigroup* T of a deterministic finite automaton (DFA) $\mathcal{D}$ is the semigroup of transformations of the state set of $\mathcal{D}$ generated by the transformations induced by the input letters of $\mathcal{D}$. The transition semigroup of a minimal DFA of a language L is isomorphic to the syntactic semigroup of L [5]; hence syntactic complexity is equal to the cardinality of T.

The number n of states of $\mathcal{D}$ is known as the *state complexity* of the language [1,6], and it is the same as the number of left quotients of the language. The *syntactic complexity of a class* of regular languages is the maximal syntactic complexity of languages in that class expressed as a function of n.

A *right ideal* (respectively, *left ideal, two-sided ideal*) is a non-empty language L over an alphabet Σ such that $L = L\Sigma^*$ (respectively, $L = \Sigma^*L$, $L = \Sigma^*L\Sigma^*$). We are interested only in regular ideals; for reasons why they deserve to be studied see [2, Section 1]. Ideals appear in pattern matching. For example, if a *text* is a word w over some alphabet Σ, and a *pattern* is an arbitrary language L over Σ, then an occurrence of a pattern represented by L in text w is a triple (u, x, v) such that $w = uxv$ and x is in L. Searching text w for words in L is

* This work was supported by the Natural Sciences and Engineering Research Council of Canada grant No. OGP000087, and by Polish NCN grant DEC-2013/09/N/ST6/01194.

A.M. Shur and M.V. Volkov (Eds.): DLT 2014, LNCS 8633, pp. 13–24, 2014.
© Springer International Publishing Switzerland 2014

equivalent to looking for prefixes of w that belong to the language $\Sigma^* L$, which is the left ideal generated by L.

The syntactic complexity of right ideals was proved to be n^{n-1} in [4]. The syntactic complexities of left and two-sided ideals were also examined in [4], where it was shown that $n^{n-1} + n - 1$ and $n^{n-2} + (n-2)2^{n-2}$, respectively, are lower bounds on these complexities, and it was conjectured that they are also upper bounds. In this paper we prove these conjectures.

If $w = uxv$ for some $u, v, x \in \Sigma^*$, then v is a *suffix* of w and x is a *factor* of w. A suffix of w is also a factor of w. A language L is *suffix-closed* (respectively, *factor-closed*) if $w \in L$ implies that every suffix (respectively, factor) of w is also in L. We are interested only in regular suffix- and factor-closed languages. Since every left (respectively, two-sided) ideal is the complement of a suffix-closed (respectively, factor-closed) language, and syntactic complexity is preserved by complementation, our theorems also apply to suffix- and factor-closed languages, but our proofs are given for left and two-sided ideals only.

2   Preliminaries

The *left quotient* or simply *quotient* of a regular language L by a word w is denoted by Lw and defined by $Lw = \{x \mid wx \in L\}$. A language is regular if and only if it has a finite number of quotients. The number of quotients of L is called its *quotient complexity*. We denote the set of quotients by $K = \{K_0, \ldots, K_{n-1}\}$, where $K_0 = L = L\varepsilon$ by convention. Each quotient K_i can be represented also as Lw_i, where $w_i \in \Sigma^*$ is such that $Lw_i = K_i$.

A *deterministic finite automaton (DFA)* is a quintuple $\mathcal{D} = (Q, \Sigma, \delta, q_0, F)$, where Q is a finite non-empty set of *states*, Σ is a finite non-empty *alphabet* $\delta \colon Q \times \Sigma \to Q$ is the *transition function*, $q_0 \in Q$ is the *initial* state, and $F \subseteq Q$ is the set of *final* states.

The *quotient DFA* of a regular language L with n quotients is defined by $\mathcal{D} = (K, \Sigma, \delta, K_0, F)$, where $\delta(K_i, w) = K_j$ if and only if $K_i w = K_j$, and $F = \{K_i \mid \varepsilon \in K_i\}$. To simplify the notation, we use the set $Q = \{0, \ldots, n-1\}$ of subscripts of quotients to denote the states of $\mathcal{D}$; then $\mathcal{D}$ is denoted by $\mathcal{D} = (Q, \Sigma, \delta, 0, F)$. The quotient corresponding to $q \in Q$ is then $K_q = \{w \mid \delta(q, w) \in F\}$. The quotient $K_0 = L$ is the *initial* quotient. A quotient is *final* if it contains ε. A state q is *empty* if its quotient K_q is empty.

The quotient DFA of L is isomorphic to each complete minimal DFA of L. The number of states in the quotient DFA of L (the quotient complexity of L) is therefore equal to the state complexity of L.

In any DFA, each letter $a \in \Sigma$ defines a transformation of the set Q of n states. Let $\mathcal{T}_Q$ be the set of all n^n transformations of Q; then $\mathcal{T}_Q$ is a monoid under composition. The *identity* transformation $\mathbf{1}$ maps each element to itself. For $k \geq 2$, a transformation (permutation) t of a set $P = \{q_0, q_1, \ldots, q_{k-1}\} \subseteq Q$ is a *k-cycle* if $q_0 t = q_1, q_1 t = q_2, \ldots, q_{k-2} t = q_{k-1}, q_{k-1} t = q_0$. A k-cycle is denoted by $(q_0, q_1, \ldots, q_{k-1})$. If a transformation t of Q acts like a k-cycle on some $P \subseteq Q$, we say that t has a *k-cycle*. A transformation has a *cycle* if it has a k-cycle for some $k \geq 2$. A 2-cycle (q_0, q_1) is called a *transposition*. A transformation is *constant* if it maps all states to a single state q; it is denoted by $(Q \to q)$. If w is a word of Σ^*, the fact that w induces transformation t is denoted by $w \colon t$. A transformation mapping i to q_i for $i = 0, \ldots, n-1$ is sometimes denoted by $[q_0, \ldots, q_{n-1}]$.

3 Left Ideals

3.1 Basic Properties

Let $Q = \{0, \ldots, n-1\}$, let $\mathcal{D}_n = (Q, \Sigma_{\mathcal{D}}, \delta_{\mathcal{D}}, 0, F)$ be a minimal DFA, and let T_n be its transition semigroup. Consider the sequence $(0, 0t, 0t^2, \ldots)$ of states obtained by applying transformation $t \in T_n$ repeatedly, starting with the initial state. Since Q is finite, there must eventually be a repeated state, that is, there must exist i and j such that $0, 0t, \ldots, 0t^i, 0t^{i+1}, \ldots, 0t^{j-1}$ are distinct, but $0t^j = 0t^i$; the integer $j - i$ is the *period* of t. If the period is 1, t is said to be *initially aperiodic*; then the sequence is $0, 0t, \ldots, 0t^{j-1} = 0t^j$.

Lemma 1 ([4]). *If $\mathcal{D}_n$ is a DFA of a left ideal, all the transformations in T_n are initially aperiodic, and no state of $\mathcal{D}_n$ is empty.*

Remark 1 ([2]). A language $L \subseteq \Sigma^*$ is a left ideal if and only if for all $x, y \in \Sigma^*$, $Ly \subseteq Lxy$. Hence, if $Lx \neq L$, then $L \subset Lx$ for any $x \in \Sigma^+$.

It is useful to restate this observation it terms of the states of $\mathcal{D}_n$. For DFA $\mathcal{D}_n$ and states $p, q \in Q$, we write $p \prec q$ if $K_p \subset K_q$.

Remark 2. A DFA $\mathcal{D}_n$ is a minimal DFA of a left ideal if and only if for all $s, t \in T_n \cup \{1\}$, $0t \preceq 0st$. If $0t \neq 0$, then $0 \prec 0t$ for any $t \in T_n$. Also, if $r \in Q$ has a t-predecessor, that is, if there exists $q \in Q$ such that $qt = r$, then $0t \preceq r$. (This follows because $q = 0s$ for some transformation s since q is reachable from 0; hence $0 \preceq q$ and $0t \preceq qt = r$.) In particular, if r appears in a cycle of t or is a fixed point of t, then $0t \preceq r$.

We consider chains of the form $K_{i_1} \subset K_{i_2} \subset \cdots \subset K_{i_h}$, where the K_{i_j} are quotients of L. If L is a left ideal, the smallest element of any maximal-length chain is always L. Alternatively, we consider chains of states starting from 0 and strictly ordered by $\prec$.

Proposition 1. *For $t \in T_n$ and $p, q \in Q$, $p \prec q$ implies $pt \preceq qt$. If $p \prec pt$, then $p \prec pt \prec \cdots \prec pt^k = pt^{k+1}$ for some $k \geq 1$. Similarly, $p \succ q$ implies $pt \succeq qt$, and $p \succ pt$ implies $p \succ pt \succ \cdots \succ pt^k = pt^{k+1}$ for some $k \geq 1$.*

It was proved in [4, Theorem 4, p. 124] that the transition semigroup of the following DFA of a left ideal meets the bound $n^{n-1} + n - 1$.

Definition 1 (Witness: Left Ideals). *For $n \geq 3$, we define the DFA $\mathcal{W}_n = (Q, \Sigma_{\mathcal{W}}, \delta_{\mathcal{W}}, 0, \{n-1\})$, where $Q = \{0, \ldots, n-1\}$, $\Sigma_{\mathcal{W}} = \{a, b, c, d, e\}$, and $\delta_{\mathcal{W}}$ is defined by $a \colon (1, \ldots, n-1)$, $b \colon (1, 2)$, $c \colon (n-1 \to 1)$, $d \colon (n-1 \to 0)$, and $e \colon (Q \to 1)$. For $n = 3$, a and b coincide, and we can use $\Sigma = \{b, c, d, e\}$.*

Remark 3. In $\mathcal{W}_n$, the transformations induced by a, b, and c restricted to $Q \setminus \{0\}$ generate all the transformations of the last $n - 1$ states. Together with the transformation of d, they generate all transformations of Q that fix 0. To see this, consider any transformation t that fixes 0. If some states from $\{1, \ldots, n-1\}$ are mapped to 0 by t, we can map them first to $n - 1$ and $n - 1$ to one of them by the transformations of a, b, and c, and then map $n - 1$ to 0 by the transformation of d. Also the words of the form ea^i for $i \in \{0, \ldots, n-2\}$ induce constant transformations $(Q \to i + 1)$. Hence the transition semigroup of $\mathcal{W}_n$ contains all the constant transformations.

Example 1. One verifies that the maximal-length chains of quotients in $\mathcal{W}_n$ have length 2. On the other hand, for $n \geq 2$, let $\Sigma = \{a, b\}$ and let $L = \Sigma^* a^{n-1}$. Then L has n quotients and the maximal-length chains are of length n.

3.2   Upper Bound

Our main result of this section shows that the lower bound $n^{n-1} + n - 1$ is also an upper bound. Our approach is as follows: We consider a minimal DFA $\mathcal{D}_n = (Q, \Sigma_\mathcal{D}, \delta_\mathcal{D}, 0, F)$, where $Q = \{0, \ldots, n-1\}$, of an arbitrary left ideal with n quotients and let T_n be the transition semigroup of $\mathcal{D}_n$. We also deal with the witness DFA $\mathcal{W}_n = (Q, \Sigma_\mathcal{W}, \delta_\mathcal{W}, 0, \{n-1\})$ of Definition 1 that has the same state set as $\mathcal{D}_n$ and whose transition semigroup is S_n. We shall show that there is an injective mapping $f \colon T_n \to S_n$, and this will prove that $|T_n| \leq |S_n|$.

Remark 4. If $n = 1$, the only left ideal is Σ^* and the transition semigroup of its minimal DFA satisfies the bound $1^0 + 1 - 1 = 1$. If $n = 2$, there are only three allowed transformations, since the transposition $(0, 1)$ is not initially aperiodic and so is ruled out by Lemma 1. Thus the bound $2^1 + 2 - 1 = 3$ holds.

Lemma 2. *If $n \geq 3$ and a maximal-length chain in $\mathcal{D}_n$ strictly ordered by $\prec$ has length 2, then $|T_n| \leq n^{n-1} + n - 1$ and T_n is a subsemigroup of S_n.*

Proof. Consider an arbitrary transformation $t \in T_n$ and let $p = 0t$. If $p = 0$, then any state other than 0 can possibly be mapped by t to any one of the n states; hence there are at most n^{n-1} such transformations. All of these transformations are in S_n by Remark 3.

If $p \neq 0$, then $0 \prec p$. Consider any state $q \notin \{0, p\}$; by Remark 2, $p \preceq qt$. If $p \neq qt$, then $p \prec qt$. But then we have the chain $0 \prec p \prec qt$ of length 3, contradicting our assumption. Hence we must have $p = qt$, and so t is the constant transformation $t = (Q \to p)$. Since p can be any one of the $n - 1$ states other than 0, we have at most $n - 1$ such transformations. Since all of these transformations are in S_n by Remark 3, T_n is a subsemigroup of S_n.   □

Theorem 1 (Left Ideals, Suffix-Closed Languages). *If $n \geq 3$ and L is a left ideal or a suffix-closed language with n quotients, then its syntactic complexity is less than or equal to $n^{n-1} + n - 1$.*

Proof. It suffices to prove the result for left ideals. For a transformation $t \in T_n$, consider the following cases:

Case 1: $t \in S_n$.

Let $f(t) = t$; obviously $f(t)$ is injective.

Case 2: $t \notin S_n$ and $0t^2 \neq 0t$.

Note that $t \notin S_n$ implies $0t \neq 0$ by Remark 3. Let $0t = p$. We have $p = 0t \prec 0tt = pt$ by Remark 2. Let $p \prec \cdots \prec pt^k = pt^{k+1}$ be the chain defined from p; this chain is of length at least 2. Let $f(t) = s$, where s is the transformation defined by

$$0s = 0, \quad pt^k s = p, \quad qs = qt \text{ for the other states } q \in Q.$$

Transformation s is shown in Figure 1, where the dashed transitions show how s differs from t.

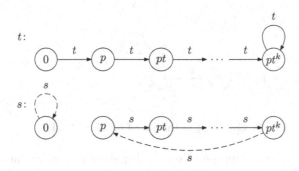

Fig. 1. Case 2 in the proof of Theorem 1

By Remark 3, $s \in S_n$. However, $s \notin T_n$, as it contains the cycle $(p, \ldots, pt^k)$ with states strictly ordered by $\prec$ in DFA $\mathcal{D}_n$, which contradicts Proposition 1. Since $s \notin T_n$, it is distinct from the transformations defined in Case 1.

In going from t to s, we have added one transition ($0s = 0$) that is a fixed point, and one ($pt^k s = p$) that is not. Since only one non-fixed-point transition has been added, there can be only one cycle in s with states strictly ordered by $\prec$. Since 0 can't appear in this cycle, p is its smallest element with respect to $\prec$.

Suppose now that $t' \neq t$ is another transformation that satisfies Case 2, that is, $0t' = p' \neq 0$ and $p't' \neq p'$; we shall show that $f(t) \neq f(t')$. Define s' for t' as s was defined for t. For a contradiction, assume $s = f(t) = f(t') = s'$.

Like s, s' contains only one cycle strictly ordered by $\prec$, and p' is its smallest element. Since we have assumed that $s = s'$, we must have $p = 0t = 0t' = p'$ and the cycles in s and s' must be identical. In particular, $pt^k t = pt^k = p(t')^k t' = p(t')^k$. For q of $Q \setminus \{0, pt^k\}$, we have $qt = qs = qs' = qt'$. Hence $t = t'$—a contradiction. Therefore $t \neq t'$ implies $f(t) \neq f(t')$.

Case 3: $t \notin S_n$ and $0t^2 = 0t$.

As before, let $0t = p$. Consider any state $q \notin \{0, p\}$; then $0 \prec q$ by Remark 2 and $0t \preceq qt$ by Proposition 1. Thus either $p \prec qt$, or $p = qt$. We consider the following sub-cases:

• **(a):** t has a cycle.

Since t has a cycle, take a state r from the cycle; then r and rt are not comparable under $\preceq$ by Proposition 1, and $p \prec r$ by Remark 2. Let $f(t) = s$, where s is the transformation shown in Figure 2 and defined by

$$0s = 0, \quad ps = r, \quad qs = qt \text{ for the other states } q \in Q.$$

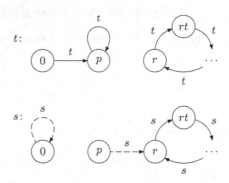

Fig. 2. Case 3(a) in the proof of Theorem 1

By Remark 3, $s \in S_n$. Suppose that $s \in T_n$; since $p \prec r$, we have $r = ps \preceq rs = rt$ by the definition of s and Proposition 1; this contradicts that r and rt are not comparable. Hence $s \notin T_n$, and so s is distinct from the transformations of Case 1.

We claim that p is not in a cycle of s; this cycle would have to be

$$p \xrightarrow{s} r \xrightarrow{s} rt \xrightarrow{s} \cdots \xrightarrow{s} rt^{k-1} \xrightarrow{s} p, \text{ that is, } p \xrightarrow{s} r \xrightarrow{t} rt \xrightarrow{t} \cdots \xrightarrow{t} rt^{k-1} \xrightarrow{t} p,$$

for some $k \geq 2$ because $r \neq p = pt$ and $rt \neq p$. Since $p \prec r$ we have $p \prec rt$; but then we have a chain $p \prec rt \prec \cdots \prec rt^k = p$, contradicting Proposition 1.

Since p is not in a cycle of s, it follows that s does not contain a cycle with states strictly ordered by $\prec$, as such a cycle would also be in t. So s is distinct from the transformations of Case 2.

We claim there is a unique state q such that (a) $0 \prec q \prec qs$, (b) $qs \npreceq qs^2$. First we show that p satisfies these conditions: (a) holds because $ps = r$ and $p \prec r$; (b) holds because $ps = r$, $ps^2 = rt$ and r and rt are not comparable. Now suppose that q satisfies the two conditions, but $q \neq p$. Note that $qs \neq p$, because $qs = p$ implies $qs = p \prec r = qs^2$, contradicting (b). Since $q, qs \notin \{0, p\}$, we have $qt = qs \npreceq qs^2 = qt^2$. But Proposition 1 for $q \prec qt$ implies that $qt \preceq qt^2$—a contradiction. Thus p is the only state satisfying these conditions.

If $t' \neq t$ is another transformation satisfying the conditions of this case, we define s' like s. Suppose that $s = f(t) = f(t') = s'$. Since both s and s' contain a unique state p satisfying the two conditions above, we have $0t = 0t' = p$ and $pt = pt' = p$. Since the other states are mapped by s exactly as by t and t', we have $t = t'$.

• **(b):** t has no cycles and has a fixed point $r \neq p$.

Because $0 \prec r$ by Remark 2, $0t \preceq rt$ by Proposition 1. If r is a fixed point of t, then $p = 0t \preceq rt = r$. Since $r \neq p$, we have $p \prec r$. Let $f(t) = s$, where s is the transformation shown in Figure 3 and defined by

$$0s = 0, \quad qs = 0 \text{ for each fixed point } q \neq p,$$
$$qs = qt \text{ for the other states } q \in Q.$$

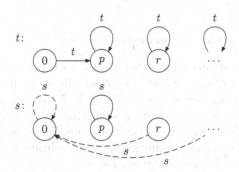

Fig. 3. Case 3(b) in the proof of Theorem 1

By Remark 3, $s \in S_n$. Suppose that $s \in T_n$; because $p \prec r$, $ps = p$, $rs = 0$, and $ps \preceq rs$ by Proposition 1, we have $p \prec 0$, which is a contradiction. Hence s is not in T_n and so is distinct from the transformations of Case 1. Also, s maps at least one state other than 0 to 0, and so is distinct from the transformations of Case 2 and also from the transformations of Case 3(a).

If $t' \neq t$ is another transformation satisfying the conditions of this case, we define s' like s. Now suppose that $s = f(t) = f(t') = s'$. There is only one fixed point of s other than 0 ($ps = p$), and only one fixed point of s' other than 0 ($p's' = p'$); hence $0t = p = p' = 0t'$. By the definition of s, for each state $q \neq 0$ such that $qs = 0$, we have $qt = q$. Similarly, for each state $q \neq 0$ such that $qs' = 0$, we have $qt' = q$. Hence t and t' agree on these states. Since the remaining states are mapped by s exactly as they are mapped by t and t', we have $t = t'$. Thus we have proved that $t \neq t'$ implies $f(t) \neq f(t')$.

• **(c):** t has no cycles, has no fixed point $r \neq p$ and there is a state r such that $p \prec r$ with $rt = p$.

Let $f(t) = s$, where s is the transformation shown in Figure 4 and defined by

$$0s = 0, \ ps = r, \ qs = 0 \text{ for each } q \succ p \text{ such that } qt = p,$$
$$qs = qt \text{ for the other states } q \in Q.$$

By Remark 3, $s \in S_n$. Suppose that $s \in T_n$; because $p \prec r$, $ps = r$, $rs = 0$, and $r = ps \preceq rs = 0$ by Proposition 1, we have $r \prec 0$—a contradiction. Hence $s \notin T_n$ and s is distinct from the transformations of Case 1.

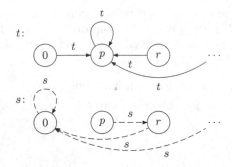

Fig. 4. Case 3(c) in the proof of Theorem 1

Because s maps at least one state other than 0 to 0 ($rs = 0$), it is distinct from the transformations of Case 2 and 3(a). Also s does not have a fixed point other than 0, while the transformations of Case 3(b) have such a fixed point.

We claim that there is a unique state q such that (a) $0 \prec q \prec qs$ and (b) $qs^2 = 0$. First we show that p satisfies these conditions. By assumption $0 \prec p \prec r$ and $rt = p$; also $rs = 0$ by the definition of s. Condition (a) holds because $0 \prec p \prec r = ps$, and (b) holds because $0 = rs = ps^2$.

Now suppose that $0 \prec q \prec qs$, $qs^2 = 0$ and $q \neq p$. Since $qs \neq 0$, we have $qs = qt$ by the definition of s. Because qt has a t-predecessor, $p \preceq qt$ by Remark 2. Also $qt = qs \neq p$, for $qs = p$ implies $0 = qs^2 = ps = r$—a contradiction. Hence $p \prec qt$. From $qt = qs$ and $q \prec qs$, we have $q \prec qt$. Since $qs^2 = 0$ we have $(qt)s = 0$ and so $(qt)t = p$, by the definition of s. By Proposition 1, from $q \prec qt$ we have $qt \preceq (qt)t = p$, contradicting $p \prec qt$. So $q = p$.

If $t' \neq t$ is another transformation satisfying the conditions of this case, we define s' like s. Suppose that $s = f(t) = t(t') = s'$. Since s and s' contain a unique state p satisfying the two conditions above, we have $0t = 0t' = p$ and $pt = pt' = p$. Then r and the states $q \succ p$ with $qt = p$ are determined by p, since they are precisely the states $q \succ p$ with $qs = 0$. Since the other states are mapped by s exactly as by t and t', we have $t = t'$, and f is again injective.

- **All Cases Are Covered**

Now we need to ensure that any transformation t fits in at least one case. It is clear that t fits in Case 1 or 2 or 3. For Case 3, it is sufficient to show that if (i) $t \notin S_n$ does not contain a fixed point $r \neq p$, and (ii) there is no state r with $p \prec r$ and $rt = p$, then t contains a cycle.

First, if there is no r such that $p \prec r$, we claim that t is the constant transformation $(Q \to p)$. Consider any state $q \in Q$ such that $qt \neq p$. Then $p \prec qt$ by Remark 2, contradicting that there is no state r such that $p \prec r$.

So let r be some state such that $p \prec r$. Consider the sequence $r, rt, rt^2, \ldots$. By Remark 2, $p \preceq rt^i$ for all $i \geq 0$. If $rt^k = p$ for some $k \geq 1$, let i be the smallest such k; we have $(rt^{i-1})t = p$, contradicting (ii). Since p is the only fixed point by (i), we have $rt^i \neq rt^{i-1}$. Since there are finitely many states, $rt^i = rt^j$ for some i and j such that $0 \leq i < j - 1$, and so the states $rt^i, rt^{i+1}, \ldots, rt^j = rt^i$ form a cycle.

We have shown that for every transformation t in T_n there is a corresponding transformation $f(t)$ in S_n, and f is injective. So $|T_n| \leq |S_n| = n^{n-1} + n - 1$.   □

Next we prove that S_n is the only transition semigroup meeting the bound. It follows that minimal DFAs of left ideals with the maximal syntactic complexity have maximal-length chains of length 2.

Theorem 2. *If T_n has size $n^{n-1} + n - 1$, then $T_n = S_n$.*

Proof. Consider a maximal-length chain of states strictly ordered by $\prec$ in $\mathcal{D}_n$. If its length is 2, then by Lemma 2, T_n is a subsemigroup of S_n. Thus only $T_n = S_n$ reaches the bound in this case.

Assume now that the length of a maximal-length chain is at least 3. Then there are states p and r such that $0 \prec p \prec r$. Let $R = \{q \mid p \prec q\}$, and let $X = Q \setminus (R \cup \{0, p\})$. We shall show that there exists a transformation s that is in S_n but not in $f(T_n)$. To define s we use the constant transformation $u = (Q \to p)$ as an auxiliary transformation. Let $0s = 0$, $ps = r$, $rs = 0$ for all $r \in R$, and $qs = qu = p$ for $q \in X$; these are precisely the rules we used in Case 3(c) in the proof of Theorem 1. By Remark 3, $s \in S_n$.

It remains to be shown that there is no transformation $t \in T_n$ such that $s = f(t)$. The proof that s is different from the transformations $f(t)$ of Cases 1, 2, 3(a) and 3(b) is exactly the same as the corresponding proof in Case 3(c) following the definition of s.

It remains to verify that there is no $u' \in T_n$ in Case 3(c) such that $f(u') = s$. Suppose there is such a u'. Recall that states p and r satisfying $0 \prec p \prec r$ have been fixed by assumption. By the definition of s, state p satisfies the conditions (a) $0 \prec p \prec ps$ and (b) $ps^2 = 0$. We claim that p is the only state satisfying these conditions. Indeed, if $q \neq p$ then either $qs = 0$, $q \not\prec qs = 0$ and (a) is violated, or $qs = p$, $qs^2 = ps = r \neq 0$ and (b) is violated. This observation is used in the proof of Case 3(c) to prove the claim below.

Both u and u' satisfy the conditions of Case 3(c), except that u fails the condition $u \notin S_n$. However, that latter condition is not used in the proof that if $u \neq u'$ and u' satisfy the other conditions of Case 3(c), then $s' \neq s$, where s' is the transformation obtained from u' by the rules of s. Thus s is also different from the transformations in $f(T_n)$ from Case 3(c).

Because $s \notin f(T_n)$, $s \in S_n$ and $f(T_n) \subseteq S_n$, the bound $n^{n-1} + n - 1$ cannot be reached if the length of the maximal-length chains is not 2.   □

4   Two-Sided Ideals

If a language L is a right ideal, then $L = L\Sigma^*$ and L has exactly one final quotient, namely Σ^*; hence this also holds for two-sided ideals. For $n \geq 3$, in a two-sided ideal every maximal chain is of length at least 3: it starts with L, every quotient contains L and is contained in Σ^*.

It was proved in [4, Theorem 6, p. 125] that the transition semigroup of the following DFA of a two-sided ideal meets the bound $n^{n-2} + (n-2)2^{n-2} + 1$.

Definition 2 (Witness: Two-Sided Ideals). *For* $n \geq 4$, *define the DFA* $\mathcal{W}_n = (Q, \Sigma_\mathcal{W}, \delta_\mathcal{W}, 0, \{n-1\})$, *where* $a\colon (1, 2, \ldots, n-2)$, $b\colon (1, 2)$, $c\colon (n-2 \to 1)$, $d\colon (n-2 \to 0)$, *for* $q = 0, \ldots, n-2$, $\delta(q, e) = 1$ *and* $\delta(n-1, e) = n-1$, *and* $f\colon (1 \to n-1)$. *For* $n = 4$, *inputs* a *and* b *coincide.*

Remark 5. If $n = 1$, the only two-sided ideal is Σ^*, its syntactic complexity is 1, and the bound above is not tight. If $n = 2$, each two-sided ideal is of the form $L = \Sigma^* \Gamma \Sigma^*$, where $\emptyset \subsetneq \Gamma \subseteq \Sigma$, its syntactic complexity is 2, and the bound is tight. If $n = 3$, there are eight transformations that are initially aperiodic and such that $(n-1)t = t$ (the property of a right-ideal transformation). We have verified that the DFA having all eight or any seven of the eight transformations is not a two-sided ideal. Hence 6 is an upper bound, and we know from [4] that the transformations $[1, 2, 2]$, $[0, 0, 2]$, and $[0, 1, 2]$ generate a 6-element semigroup. From now on we may assume that $n \geq 4$.

We consider a minimal DFA $\mathcal{D}_n = (Q, \Sigma_\mathcal{D}, \delta_\mathcal{D}, 0, \{n-1\})$, where $Q = \{0, \ldots, n-1\}$, of an arbitrary two-sided ideal with n quotients, and let T_n be the transition semigroup of $\mathcal{D}_n$. We also deal with the witness DFA $\mathcal{W}_n = (Q, \Sigma_\mathcal{W}, \delta_\mathcal{W}, 0, \{n-1\})$ of Definition 2 with transition semigroup S_n.

Remark 6. In $\mathcal{W}_n$, the transformations induced by a, b, and c restricted to $Q \setminus \{0, n-1\}$ generate all the transformations of the states $1, \ldots, n-2$. Together with the transformations of d and f, they generate all transformations of Q that fix 0 and $n-1$. For any subset $S \subseteq \{1, \ldots, n-2\}$, there is a transformation— induced by a word w_S, say—that maps S to $n-1$ and fixes $Q \setminus S$. Then the words of the form $w_S e a^i$, for $i \in \{0, \ldots, n-3\}$, induce all transformations that maps $S \cup \{n-1\}$ to $n-1$ and $Q \setminus (S \cup \{n-1\})$ to $i+1$. In $\mathcal{W}_n$, there is also the constant transformation $ef\colon (Q \to n-1)$.

Lemma 3. *If* $n \geq 4$ *and a maximal-length chain in* $\mathcal{D}_n$ *strictly ordered by* $\prec$ *has length 3, then* $|T_n| \leq n^{n-2} + (n-2)2^{n-2} + 1$, *and* T_n *is a subsemigroup of* S_n.

Proof. Consider an arbitrary transformation $t \in T_n$; then $(n-1)t = n-1$. If $0t = 0$, then any state not in $\{0, n-1\}$ can possibly be mapped by t to any one of the n states; hence there are at most n^{n-2} such transformations.

If $0t \neq 0$, then $0 \prec 0t$. Consider any state $q \notin \{0, 0t\}$; since $\mathcal{D}_n$ is minimal, q must be reachable from 0 by some transformation s, that is, $q = 0s$. If $0st \notin \{0t, n-1\}$, then $0t \prec 0st$ by Remark 2. But then we have the chain $0 \prec 0t \prec 0st \prec n-1$ of length 4, contradicting our assumption. Hence we must have either $0st = 0t$, or $0st = n-1$. For a fixed $0t$, a subset of the states in $Q \setminus \{0, n-1\}$ can be mapped to $0t$ and the remaining states in $Q \setminus \{0, n-1\}$ to $n-1$, thus giving 2^{n-2} transformations. Since there are $n-2$ possibilities for $0t$, we obtain the second part of the bound. Finally, all states can be mapped to $n-1$.

By Remark 6 all of the above-mentioned transformations are in S_n. $\square$

Theorem 3 (Two-Sided Ideals, Factor-Closed Languages). *If* L *is a two-sided ideal or a factor-closed language with* $n \geq 4$ *quotients, then its syntactic complexity is less than or equal to* $n^{n-2} + (n-2)2^{n-2} + 1$.

Proof. It suffices to prove the result for two-sided ideals. As we did for left ideals, we show that $|T_n| \leq |S_n|$, by constructing an injective function $f \colon T_n \to S_n$.

We have $q \preceq n - 1$ for any $q \in Q$, and $n - 1$ is a fixed point of every transformation in T_n and S_n.

We omit here the detailed proof of injectivity of f. The complete proof can be found in [3].

For a transformation $t \in T_n$, consider the following cases:

Case 1: $t \in S_n$.
The proof is the same as that of Case 1 of Theorem 1.

Case 2: $t \notin S_n$, and $0t^2 \neq 0t$.
Let $0t = p \prec \cdots \prec pt^k = pt^{k+1}$ be the chain defined from p.

- **(a):** $pt^k \neq n - 1$.
The proof is the same as that of Case 2 of Theorem 1.

- **(b):** $pt^k = n - 1$ and $k \geq 2$.
Let $f(t) = s$, where s is the following transformation:

$$0s = 0, \quad pt^i s = pt^{i-1} \text{ for } 1 \leq i \leq k - 1, \quad ps = n - 1,$$
$$qs = qt \text{ for the other states } q \in Q.$$

- **(c):** $pt = n - 1$.
Let $P = \{0, p, n - 1\}$. Since $n \geq 4$, there must be a state $r \notin P$. If $p \prec r$ for all $r \notin P$, then $n - 1 = pt \preceq rt$; hence $rt = n - 1$ for all such r, and $qt \in \{p, n - 1\}$ for all $q \in Q$. By Remark 6, there is a transformation in S_n that maps $S \cup \{n-1\}$ to $n - 1$, and $Q \setminus (S \cup \{n - 1\})$ to p for any $S \subseteq \{1, \ldots, n - 2\}$. Thus $t \in S_n$—a contradiction.

In view of the above, there must exist a state $r \notin P$ such that $p \npreceq r$. By Remark 2, we have $p \preceq rt$ and of course $rt \preceq n - 1$. If rt is p or $n - 1$ for all $r \notin P$, we again have the situation described above, showing that $t \in S_n$. Hence there must exist an $r \notin P$ such that $p \npreceq r$ and $p \prec rt \prec n - 1$.

Let $f(t) = s$, where s is the following transformation:

$$0s = 0, \quad ps = rt, \quad (rt)s = p, \quad rs = 0,$$
$$qs = qt \text{ for the other states } q \in Q.$$

Case 3: $t \notin S_n$, $0t = p \neq 0$ and $pt = p$.
- **(a):** t has a cycle.
The proof is analogous to that of Case 3(a) in Theorem 1, but we need to ensure that s is different from the s of Cases 2(b) and 2(c).
- **(b):** t has no cycles and has a fixed point $r \notin \{p, n - 1\}$.
The proof is analogous to that of Case 3(b) in Theorem 1, but we need to ensure that s is different from the s of Cases 2(b) and 2(c).
- **(c):** t has no cycles and no fixed point $r \notin \{p, n-1\}$, but has a state $r \succ p$ mapped to p.
The proof is analogous to that of Case 3(c) in Theorem 1, but we need to ensure that s is different from the s of Cases 2(b) and 2(c).

- **(d):** t has no cycles, no fixed point $r \notin \{p, n-1\}$, and no state $r \succ p$ mapped to p, but has a state r such that $p \prec r \prec n-1$, mapped to $n-1$. Let $f(t) = s$, where s is the following transformation:

$$0s = 0, \quad qs = q \text{ for states } q \text{ such that } qt = n-1, \quad ps = n-1$$
$$qs = qt \text{ for the other states } q \in Q.$$

- **All Cases Are Covered**

We need to ensure that any transformation t fits in at least one case. It is clear that t fits in Case 1 or 2 or 3. Any transformation from Case 2 fits in Case 2(a) or 2(b) or 2(c). For Case 3, it is sufficient to show that if (i) $t \notin S_n$ does not contain a fixed point $r \notin \{p, n-1\}$, and (ii) there is no state r, $p \prec r \prec n-1$, mapped to p or $n-1$, then t has a cycle.

If there is no state r such that $p \prec r \prec n-1$, then $qt \in \{p, n-1\}$ for any $q \in Q$, since $qt \succeq p$; by Remark 6, $t \in S_n$—a contradiction.

So let r be some state such that $p \prec r \prec n-1$. Consider the sequence $r, rt, rt^2, \ldots$. By Remark 2, $p \preceq rt^i$ for all $i \geq 0$. If $rt^k \in \{p, n-1\}$ for some $k \geq 1$, then let i be the smallest such k. Then we have $(rt^{i-1})t \in p$, contradicting (ii). Since p and $n-1$ are the only fixed points by (i), we have $rt^i \neq rt^{i-1}$. Since there are finitely many states, $rt^i = rt^j$ for some i and j such that $0 \leq i < j-1$, and so the states $rt^i, rt^{i+1} \ldots, rt^j = rt^i$ form a cycle. □

Theorem 4. *If T_n has size $n^{n-2} + (n-2)2^{n-2} + 1$, then $T_n = S_n$.*

Proof. The proof is very similar to that of Theorem 2. It can be found in [3]. □

References

1. Brzozowski, J.: Quotient complexity of regular languages. J. Autom. Lang. Comb. 15(1/2), 71–89 (2010)
2. Brzozowski, J., Jirásková, G., Li, B.: Quotient complexity of ideal languages. Theoret. Comput. Sci. 470, 36–52 (2013)
3. Brzozowski, J., Szykuła, M.: Upper bounds on syntactic complexity of left and two-sided ideals (2014), http://arxiv.org/abs/1403.2090
4. Brzozowski, J., Ye, Y.: Syntactic complexity of ideal and closed languages. In: Mauri, G., Leporati, A. (eds.) DLT 2011. LNCS, vol. 6795, pp. 117–128. Springer, Heidelberg (2011)
5. Pin, J.E.: Syntactic semigroups. In: Handbook of Formal Languages: Word, Language, Grammar, vol. 1, pp. 679–746. Springer, New York (1997)
6. Yu, S.: State complexity of regular languages. J. Autom. Lang. Comb. 6, 221–234 (2001)

On the Average Complexity of Brzozowski's Algorithm for Deterministic Automata with a Small Number of Final States*

Sven De Felice[1] and Cyril Nicaud[2]

[1] LIAFA, Université Paris Diderot - Paris 7 & CNRS UMR 7089, France
Sven.De-Felice@liafa.univ-paris-diderot.fr
[2] LIGM, Université Paris-Est & CNRS UMR 8049, France
nicaud@univ-mlv.fr

Abstract. We analyze the average complexity of Brzozowski's minimization algorithm for distributions of deterministic automata with a small number of final states. We show that, as in the case of the uniform distribution, the average complexity is super-polynomial even if we consider random deterministic automata with only one final state. We therefore go beyond the previous study where the number of final states was linear in the number of states. Our result holds for alphabets with at least 3 letters.

1 Introduction

In this article we continue our investigation of the average complexity of Brzozowski's algorithm [4] that was started in [10]. Recall that Brzozowski's method is based on the fact that determinizing a trim co-deterministic automaton that recognizes a language $\mathcal{L}$ yields the minimal automaton for $\mathcal{L}$. Hence, starting from an automaton $\mathcal{A}$ that recognizes the language $\mathcal{L}$, one can compute its minimal automaton by first determinizing its reversal, then by determinizing the reversal of the resulting automaton.

This elegant method is not efficient in the worst case, since the first determinization can produce an automaton that has exponentially many states, even if one starts with a deterministic automaton (see [10] for a classical example). We are therefore far from the efficient solutions available to minimize deterministic automata, such as Hopcroft's algorithm [13], which runs in $\mathcal{O}(n \log n)$ time.

In [10] we proved that for the uniform distribution on deterministic and complete automata with n states, or for distributions where each state is final with (fixed) probability $b \in (0, 1)$, the running time of Brzozowski's algorithm is super-polynomial[1] with high probability. One limitation of this result is that under such a distribution, an automaton with n states has around bn final states, for fixed

---

* This work is supported by the French National Agency (ANR) through ANR-10-LABX-58 and through ANR-2010-BLAN-0204.

[1] Grows quicker than n^d for any positive d.

A.M. Shur and M.V. Volkov (Eds.): DLT 2014, LNCS 8633, pp. 25–36, 2014.

b, which therefore grows linearly with the number of states. However, in many situations the automata that are built do not have that many final states (see, for instance, Aho-Corasick automaton [1], which is used for pattern matching). A natural question is whether this result still holds for automata with, for instance, a fixed number of final states. This is the question we investigate in this article.

The precise definition of a *distribution of automata with a small number of final states* is given in Section 4, but it covers the cases of random size-n automata with just one final state, with $\log n$ final states, or where each state is final with probability $\frac{3}{n}$ or $\frac{2}{\sqrt{n}}$, and so on. It therefore differs significantly from the cases studied in [10].

Notice that analyzing distributions of automata with a small number of final states is an up-to-date question in the statistical study of automata. The main results in this field, the average complexity of Moore's algorithm and the asymptotic number of minimal automata, only hold for distributions of automata with "sufficiently many" final states [2,7,3]. Some effort have been undertaken to extend them to, say, automata with only one final state, but with no success so far. To our knowledge, we present in this article the first result of this kind.

We will see that the proof of our main result is not just simply an adaptation of the proof proposed in [10] and we will need some deeper understanding of the typical properties of a random automaton. In return, we will establish some new facts that are interesting on their own, and that may be reused for further work on statistical properties of random automata.

The paper is organized as follows. After recalling some basic definitions in Section 2, we briefly revisit the article [10] in Section 3 to point out the difficulties encountered when trying to reduce the number of final states. In Section 4 we state our main result and prove it for automata with only one final state in Section 5. In Section 6, we explain how to extend it to get the full proof.

2   Definitions

Let $[n]$ denote the set of integers between 1 and n. If x, y are two real numbers, let $[\![x, y]\!]$ denote the set of integers i such that $x \leq i \leq y$. For any positive integer n, let $\mathfrak{S}_n$ denote the set of all permutations on $[n]$.

Automata. Let A be a finite alphabet, an *automaton* $\mathcal{A}$ is a tuple (Q, δ, I, F), where Q is its finite set of *states*, $I \subseteq Q$ is its set of *initial states* and $F \subseteq Q$ is its set of *final states*. Its *transition function* δ is a (partial) map from $Q \times A$ to 2^Q. A *transition* of $\mathcal{A}$ is a tuple $(p, a, q) \in Q \times A \times Q$, which we write $p \xrightarrow{a} q$, such that $q \in \delta(p, a)$. The map δ is classically extended by morphism to $Q \times A^*$. We denote by $\mathcal{L}(\mathcal{A})$ the set of words recognized by $\mathcal{A}$. A *deterministic and complete automaton* is an automaton such that $|I| = 1$ and for every $p \in Q$ and $a \in A$, $|\delta(p, a)| = 1$; for such an automaton we consider that δ is a (total) map from $Q \times A^*$ to Q to simplify the notations. A state q is *accessible* when there exists a path an initial state to q. It is *co-accessible* when there exists a path from

q to a final state. If $\mathcal{A}$ is an automaton, we let $\mathrm{Trim}(\mathcal{A})$ denote the automaton obtained after removing states that are not accessible or not co-accessible.

For any automaton $\mathcal{A} = (Q, \delta, I, F)$, we denote by $\tilde{\mathcal{A}}$ the *reverse* of $\mathcal{A}$, which is the automaton $\tilde{\mathcal{A}} = (Q, \tilde{\delta}, F, I)$, where $p \xrightarrow{a} q$ is a transition of $\tilde{\mathcal{A}}$ if and only if $q \xrightarrow{a} p$ is a transition of $\mathcal{A}$. The automaton $\tilde{\mathcal{A}}$ recognizes the reverse[2] of $\mathcal{L}(\mathcal{A})$. An automaton is *co-deterministic* when its reverse is deterministic.

Recall that the *minimal automaton* of a rational language $\mathcal{L}$ is the smallest deterministic and complete automaton[3] that recognizes $\mathcal{L}$. To each rational language $\mathcal{L}$ corresponds a minimal automaton, which is unique up to isomorphism.

Subset Construction and Brzozowski's Algorithm. If $\mathcal{A} = (Q, \delta, I, F)$ is a non-deterministic automaton, it is classical that the subset automaton of $\mathcal{A}$ defined by

$$\mathcal{B} = \left(2^Q, \gamma, \{I\}, \{X \in 2^Q \mid F \cap X \neq \emptyset\}\right)$$

is a deterministic automaton that recognizes the same language, where for every $X \in 2^Q$ and every $a \in A$, $\gamma(X, a) = \cup_{p \in X} \delta(p, a)$. This is of course still true if we only take the accessible part of $\mathcal{B}$, and this is not a difficulty when implementing it, since the accessible part of $\mathcal{B}$ can be built on the fly, using the rule for γ in a depth-first traversal of $\mathcal{B}$ starting from I. We denote by $\mathrm{Subset}(\mathcal{A})$ the accessible part of the subset automaton of $\mathcal{A}$.

In [4], Brzozowski established the following result:

Theorem 1 (Brzozowski). *If $\mathcal{A}$ is a trim co-deterministic automaton then $\mathrm{Subset}(\mathcal{A})$ is the minimal automaton of $\mathcal{L}(\mathcal{A})$.*

This theorem readily yields an algorithm to compute the minimal automaton of the language recognized by an automaton $\mathcal{A}$, based on the subset construction: since $\mathcal{B} = \mathrm{Subset}(\mathrm{Trim}(\tilde{\mathcal{A}}))$ is a deterministic automaton recognizing the reverse of $\mathcal{L}(\mathcal{A})$, then $\mathrm{Subset}(\mathrm{Trim}(\tilde{\mathcal{B}}))$ is the minimal automaton of $\mathcal{L}(\mathcal{A})$.

Mappings. A *mapping* of size n is a total function from $[n]$ to $[n]$. A mapping f can be seen as a directed graph with an edge $i \rightarrow j$ whenever $f(i) = j$. Such a graph is a union of cycles of Cayley trees (i.e., rooted labelled trees), as depicted in Fig. 1 (see [11] for more information on this graph description). Let f be a size-n mapping. An element $x \in [n]$ is a *cyclic point* of f when there exists an integer $i > 0$ such that $f^i(x) = x$. The *cyclic part* of a mapping f is the permutation obtained when restricting f to its set of cyclic points. The *normalized cyclic part* of f is obtained by relabelling the c cyclic points of f with the elements of $[c]$, while keeping their relative order[4].

Automata as Combinatorial Structures. In the sequel, A is always a fixed alphabet with $k \geq 2$ letters. Let $\mathfrak{A}_n$ (or $\mathfrak{A}_n(A)$ when we want to specify the alphabet) denote the set of all deterministic and complete automata with input

---

[2] If $u = u_0 \cdots u_{n-1}$ is a word of length n, the *reverse* of u is the word $\tilde{u} = u_{n-1} \cdots u_0$.

[3] Minimal automata are not always required to be complete in the literature.

[4] The notion of normalization will be used for other substructures, always for relabelling the atoms with an initial segment of the positive integers, while keeping their relative order.

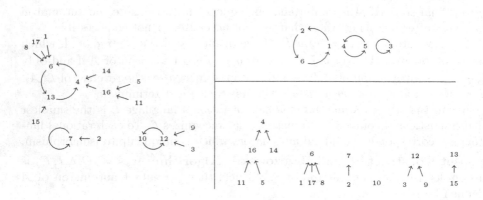

Fig. 1. A mapping of size 17, on the left. On the upper right we have its normalized cyclic part, and on the lower right its Cayley trees (not normalized).

alphabet A whose set of states is $[n]$ and whose initial state is 1. Such an automaton $\mathcal{A}$ is characterized by the tuple (n, δ, F). A *transition structure* is an automaton without final states, and we denote by $\mathfrak{T}_n$ the set of n-state transition structures with the same label restrictions as for $\mathfrak{A}_n$. If $\mathcal{A} \in \mathfrak{A}_n$, an a-cycle of $\mathcal{A}$ is a cycle of the mapping induced by a, i.e., $p \mapsto \delta(p, a)$. A state of an a-cycle is called an *a-cyclic state*.

Distributions of Combinatorial Structures. Let E be a set of combinatorial objects with a notion of size such that the set E_n of elements of size n of E is finite for every $n \geq 0$. The *uniform distribution* (which is a slight abuse of notation since there is one distribution for each n) on the set E is defined for any $e \in E_n$ by $\mathbb{P}_n(\{e\}) = \frac{1}{|E_n|}$. The reader is referred to [12] for more information on combinatorial probabilistic models.

Probabilities on Automata. Let A be an alphabet. We consider two kinds of distribution on size-n deterministic and complete automata. The first one is the *fixed-size distribution* on $\mathfrak{A}_n$ of parameter m. It is the uniform distribution on size-n automata with exactly m states. The parameter m may depend[5] on n; one can for instance consider the fixed-size distribution of parameter $\lfloor \sqrt{n} \rfloor$. The second one is the *p-distribution* on $\mathfrak{A}_n$, where the transition structure of the automaton is chosen uniformly at random and where each state is final with probability p independently; in this model also, p may depend on n, for instance $p = \frac{2}{n}$ yields automata with two final states on average.

Note that the Bernoulli model of parameter b of [10] is the same as the p-distribution for $p = b$: it is the case where p does not depend on n.

Some Terminology. We consider a (sequence of) distributions on $E = \bigcup_n E_n$. Let P be a property defined on E. We say that P holds *generically* (or *with high probability*) when the probability it holds tends to 1 as n tends to infinity. We say that P is *visible* (or holds with *visible probability*) when there exists a

---

[5] The term "fixed" stands for: for any given n, the number of final states is fixed.

positive constant C and an integer n_0 such that for every $n \geq n_0$ the probability that P holds on size-n elements is at least C.

In the sequel we will implicitly use that if P and Q are generic then $P \wedge Q$ is generic; if P is generic and Q is visible, then $P \wedge Q$ is visible; if P and Q are visible and independent, then $P \wedge Q$ is visible, and so on.

3 Result for a Large Number of Final States

In [10] we proved that the complexity of Brzozowski's algorithm is generically super-polynomial for the uniform distribution on deterministic complete automata with n states. For this distribution, every state is final with probability $\frac{1}{2}$. Thus if one take such an automaton uniformly at random, with high probability it has around $\frac{n}{2}$ final states. The article also considers the case where the probability of being final is some fixed $b \in (0, 1)$.

In this paper we consider distributions on $\mathfrak{A}_n$ where the typical number of final states can be small, for instance in $o(n)$, and we will try to reuse some ideas from [10] to do the analysis. We therefore recall in this section, very briefly, the proof of the following result:

Theorem 2 (De Felice, Nicaud [10]). *Let A be an alphabet with at least 2 letters. If $\mathcal{L}$ is the language recognized by a deterministic and complete n-state automaton over A taken uniformly at random, then generically the minimal automaton of the reverse of $\mathcal{L}$ has a super-polynomial number of states.*

The first observation is the following lemma, which will be used in a slightly modified version in this paper. It is the only result needed from automata theory; the remainder of the proof consists in analyzing the combinatorial structure of the underlying graph of a random automaton. A cycle is *primitive* when the sequence of types of states (final and non-final) in the cycle forms a primitive word.

Lemma 1 ([10]). *Let $\mathcal{A} \in \mathfrak{A}_n$ be a deterministic automaton that contains m primitive a-cycles $\mathcal{C}_1, \ldots, \mathcal{C}_m$ of lengths at least two that are all accessible. The minimal automaton of $\mathcal{L}(\tilde{\mathcal{A}})$ has at least $\mathrm{lcm}(|\mathcal{C}_1|, \ldots, |\mathcal{C}_m|)$ states.*

The proof of Theorem 2 is organized as follows:

1. If we look just at the action of a in a random automaton, it is a random mapping from the set of states to itself. Using the classical properties of random mappings [14] we get that, with high probability, there are at least $n^{1/3}$ a-cyclic states.
2. The cyclic part of a uniform random mapping behaves likes a uniform random permutation. We therefore want to use a celebrated result of Erdős and Turán [9] which states that the lcm of the lengths of the cycles of a random permutation[6] is super-polynomial with high probability.

---

[6] It is exactly the order of the permutation.

3. We prove that the a-cycles of lengths at least $\log n$ are generically primitive and accessible in a random automaton, using properties of random automata established in [5].
4. We conclude by proving that even if we remove the cycles of lengths smaller than $\log n$ in Erdős and Turán's result, we still have a super-polynomial lcm with high probability.

Now assume that we are considering the uniform distribution on automata with just one final state. It is no longer true that the large cycles are generically primitive: with high probability the final state is not in any a-cycle, which is therefore not primitive. In the sequel we will show how to get around this problem, which has consequences at every step of the proof (in particular we can no longer use the result of Erdős and Turán).

4   Main Result

A distribution on automata is said to have *a small number of final states* when it is either a fixed size distribution or a p-distribution on size-n automata such that the number of final states is in $[\![1, \frac{n}{2}]\!]$ with visible probability. Our main result is the following:

Theorem 3. *Let A be an alphabet with at least 3 letters. If $\mathcal{L}$ is the language recognized by a random n-state deterministic and complete automaton following a distribution with a small number of final states, then for any $d > 0$, the minimal automaton of the reverse of $\mathcal{L}(\mathcal{A})$ has, with visible probability, more than n^d states.*

Compared to the main result of [10] we capture many more distributions on automata, by weakening the statement a bit: it holds for an alphabet of 3 or more letters and it does not hold generically but with positive probability. The latter is unavoidable: as proved in [5] there is a linear number of states that are not accessible in a typical random automaton. Thus for the fixed size distribution with one final state, the final state has a positive probability of not being accessible.

The average complexity of Brzozowski's algorithm is a direct consequence of Theorem 3.

Corollary 1. *Let A be an alphabet with at least 3 letters. The average complexity of Brzozowski's algorithm is super-polynomial for distributions with a small number of final states.*

Proof. For any $d > 0$, the expected number of states after the first determinization is at least n^d times the probability that an automaton has at least n^d states after the first determinization. This probability is greater than some positive constant C for n sufficiently large by Theorem 3, concluding the proof.   □

Our results hold, for instance, for the p-distributions with $p = \frac{\alpha}{n}$ for some positive real α: there are α final states on average, and it is straightforward to check that it has a small number of final states. They also hold for the fixed-size distribution with $\lfloor \sqrt{n} \rfloor$ final states, since $1 \leq \lfloor \sqrt{n} \rfloor \leq \frac{n}{2}$ for n sufficiently large.

5   Proof of Theorem 3 for Automata with One Final State

In this section we prove our main theorem for the fixed-size distribution of parameter 1, that is, for the uniform distribution on the sets $\mathfrak{A}_n^1$ of automata with exactly one final state. From now on we are working over the alphabet $A = \{a, b, c\}$, as adding more letters just makes the problem easier.

We start with a generalization of Lemma 1 from [10]. Let $\mathcal{A}$ be an automaton with transition function δ and set of final states F, let $\mathcal{C}$ be an a-cycle of $\mathcal{A}$ of length ℓ and let $u \in A^*$. The u-word of $\mathcal{C}$ is the word $v = v_0 \ldots v_{\ell-1}$ of length ℓ on $\{0, 1\}$ defined as follows: let x be the smallest element of $\mathcal{C}$, we set $v_i = 1$ if and only if $\delta(x, a^i u) \in F$, for $i \in \{0, \ldots, \ell - 1\}$. This is a generalization of [10] where the word associated with $\mathcal{C}$ is exactly the ε-word of $\mathcal{C}$. The cycle $\mathcal{C}$ is u-primitive when its u-word is a primitive word.

Lemma 2. *Let u be a word of A^* and let $\mathcal{A} \in \mathfrak{A}_n$ be a deterministic automaton that contains m u-primitive a-cycles $\mathcal{C}_1, \ldots, \mathcal{C}_m$ of lengths at least two that are all accessible. The minimal automaton of $\mathcal{L}(\tilde{\mathcal{A}})$ has at least $\mathrm{lcm}(|\mathcal{C}_1|, \ldots, |\mathcal{C}_m|)$ states.*

Proof. (sketch) This is the same proof as in [10], except that we are considering the sets $\delta^{-1}(F, u) \cap \mathcal{C}$ instead of $F \cap \mathcal{C}$, where $\mathcal{C} = \cup_{i=1}^m \mathcal{C}_i$. □

In the sequel we first find a suitable word u, then a collection of a-cycles with good properties, in order to apply Lemma 2.

5.1   Finding $u \in \{b, c\}^*$ Such That $\delta_{\mathcal{A}}^{-1}(f, u)$ Is Sufficiently Large

For the first step, we consider letters b and c only and try to build a sufficiently large set of states in the determinization of the reverse of an automaton with one final state. In this section we prove the following result.

Proposition 1. *There exists a constant $w > 0$ such that if we draw an element $\mathcal{A}$ of $\mathfrak{A}_n^1(\{b, c\})$ uniformly at random, there exists a word $u \in \{b, c\}^*$ such that $\delta_{\mathcal{A}}^{-1}(f, u)$ has size between $w\sqrt{n}$ and $w\sqrt{n}\log n$ with visible probability, where f denotes the final state of $\mathcal{A}$.*

Of course, in Proposition 1 the word u depends on $\mathcal{A}$. We need some preliminary results on random mappings to establish the proposition.

Lemma 3. *Generically, a random mapping of $[n]$ has no element with more than $\log n$ preimages.*

Lemma 3 is used the following way. If we find a word v such that $\delta_{\mathcal{A}}^{-1}(f, v)$ has size greater than $w\sqrt{n}\log n$, then there exists a prefix u of v such that $w\sqrt{n} \leq \delta_{\mathcal{A}}^{-1}(f, u) \leq w\sqrt{n}\log n$ since the image of a set of states X by a letter in $\tilde{\mathcal{A}}$ generically has size at most $|X|\log n$. Hence, we just have to find a word v such that $\delta_{\mathcal{A}}^{-1}(f, v) \geq w\sqrt{n}$ to conclude the proof.

A set of vertices X of a digraph is *stable* when there is no edge $x \to y$ for $x \in X$ and $y \notin X$.

Lemma 4. *Let $\mathcal{A}$ be an element of $\mathfrak{A}_n^1$ taken uniformly at random, and let G be the digraph induced on $[n]$ by the actions of b and c (there is an edge $x \to y$ if and only if $\delta_{\mathcal{A}}(x,b) = y$ or $\delta_{\mathcal{A}}(x,c) = y$). Generically, G has a unique stable strongly connected component, which has size greater than $\frac{1}{2}n$.*

Proof. (sketch) We first prove that generically there is no stable set of states of size smaller than $\frac{1}{4}n$: we overcount the number of transition structures having a stable subset X of size ℓ by choosing the ℓ states, their images by both letters in X and the images of the other states. This yields an upper bound of $\binom{n}{\ell}\ell^{2\ell}n^{2n-2\ell}$ for the number of such transition structures. Summing for ℓ from 1 to $n/4$ this upper bound is sufficient to prove that it generically does not happen.

It is proven in [5] that in a random transition structure with n states on a two-letter alphabet, the accessible part is generically of size greater than $\frac{1}{2}n$. If we have a transition structure with a stable strongly connected component $\mathcal{C}$ of size between $\frac{1}{4}n$ and $\frac{1}{2}n$, then by symmetry, the initial state is in $\mathcal{C}$ with probability at least $\frac{1}{4}$. But in such a case, the accessible part has size at most $\frac{1}{2}n$, which generically cannot happen according to [5]. This concludes the proof, as there can be at most one component of size greater $\frac{1}{2}n$.   □

Remark 1. If an automaton has a unique stable strongly connected component $\mathcal{C}$, then for every state q there exists a path from q to any state of $\mathcal{C}$, as one can see on the acyclic graph of the strongly connected components. In particular, $\mathcal{C}$ is necessarily accessible.

Recall that a mapping f on $[n]$ can be seen as a union of cycles of Cayley trees. Define the *largest tree* of f as its largest Cayley tree, taking the tree with the smallest root label if there are several trees with the maximum number of nodes. In a transition structure or in an automaton, the largest b-tree is the largest tree for the mapping associated with the action of b. Our next lemma states that the largest b-tree of a random structure behaves like a uniform random tree (when there is only one tree of maximum size). Thus we can use classical results on random mappings and Cayley trees to estimate the typical width of such a tree.

Lemma 5. *Let t and n be two integers such that $1 \leq \frac{n}{2} < t \leq n$ and let $\mathfrak{M}_n^{(t)}$ denote the set of mapping on $[n]$ whose largest tree has t nodes. The normalized largest tree of a uniform element of $\mathfrak{M}_n^{(t)}$ is distributed as a uniform random Cayley tree with t nodes.*

The following result is the mix between a classical result on the largest tree in a random mapping [14] and the analysis of the width of a random Cayley tree done in [6].

Theorem 4 (Kolčin, Chassaing and Marckert). *There exist two positive constants w and C such that the probability that the largest tree of a random mapping on $[n]$ has width at least $w\sqrt{n}$ is greater than C, for n sufficiently large.*

We can now give the proof of Proposition 1.

Proof. (of Proposition 1) By Theorem 4, there is a b-tree T of width at least $w\sqrt{n}$ in a random size-n transition structure with input alphabet $A' = \{b, c\}$ with positive probability. Moreover, by Lemma 4 such a transition structure generically has only one stable strongly connected component $\mathcal{C}$, which contains more than $\frac{1}{2}n$ states. Hence if we add a final state uniformly at random, it is in $\mathcal{C}$ with probability at least $\frac{1}{2}$. As stated in Remark 1, if the final state is in the unique stable strongly connected component, then there exists a word v that labels a path from the root of T to f. Consider the word $v' = \tilde{v}b^i$, where i is the layer of T with the maximal number of nodes (the level that gives its width). Then $\delta^{-1}(f, v')$ contains all the states of the ith layer of T, and it therefore contains at least $w\sqrt{n}$ elements.

By Lemma 3, every state has generically less than $\log n$ preimages. Thus if $|\delta^{-1}(f, v')| \geq w\sqrt{n}$, then there exists a prefix u of v' such that $\delta^{-1}(f, u)$ contains between $w\sqrt{n}$ and $w\sqrt{n}\log n$ elements, concluding the proof. $\square$

5.2   Finding a Good Collection of a-cycles

We now switch to a-cycles and consider only the action of the letter a (the actions of the three letters are independent). Recall that conditioned by its number of cyclic points m, the cyclic permutation of a uniform random mapping is a uniform permutation of $\mathfrak{S}_m$. We first establish some properties of random permutations.

If σ is a permutation of $[n]$, its *sequence of cycles* is the ordered sequence of its cycles $(\mathcal{C}_1, \mathcal{C}_2, \ldots, \mathcal{C}_m)$, where the cycles are ordered by their smallest element. If $(\mathcal{C}_1, \mathcal{C}_2, \ldots, \mathcal{C}_m)$ is the sequence of cycles of σ and $d \leq m$, the d *first cycles of σ* are the cycles $\mathcal{C}_1, \mathcal{C}_2, \ldots, \mathcal{C}_d$. Let $L_d(\sigma) = (|\mathcal{C}_1|, \ldots, |\mathcal{C}_d|)$ denote the d first cycles of σ, when σ has at least d cycles and let $L_d(\sigma) = \bot$ otherwise.

Lemma 6. *Let d be a positive integer. For any $(\ell_1, \ldots, \ell_d) \in [\![\frac{n}{3d}, \frac{n}{2d}]\!]^d$, the following lower bound holds for n sufficiently large:*

$$\mathbb{P}\big(L_d = (\ell_1, \ldots, \ell_d)\big) \geq \frac{1}{n^d}.$$

We now turn our attention to the lcm of the first d cycles of a random permutation, and establish the following proposition.

Proposition 2. *Let $(x_1, \ldots, x_d)$ be a uniform element of $[\![\frac{n}{3d}, \frac{n}{2d}]\!]^d$. There exists a constant $\lambda > 0$ such that $\mathrm{lcm}(x_1, \ldots, x_d) \geq \lambda n^d$ with visible probability.*

Let $\mathrm{Cycle}_d(n)$ be the set of permutations σ of $[n]$ such that $L_d(\sigma) \in [\![\frac{n}{3d}, \frac{n}{2d}]\!]^d$ and $\mathrm{lcm}(\ell_1, \ldots, \ell_d) \geq \lambda n^d$, with $L_d(\sigma) = (\ell_1, \ldots, \ell_d)$. We use the λ of Proposition 2.

If we take a permutation σ uniformly at random, conditioned by $L_d(\sigma) \in [\![\frac{n}{3d}, \frac{n}{2d}]\!]^d$, the vector $L_d(\sigma)$ is not uniformly distributed in $[\![\frac{n}{3d}, \frac{n}{2d}]\!]^d$. However, we can control the lack of uniformity and use Proposition 2 to obtain sufficiently many permutations such that the lcm of their d first cycles is large enough.

Lemma 7. *For any positive integer d, a uniform random permutation of $[n]$ is in $\mathrm{Cycle}_d(n)$ with visible probability.*

In [10], having generically more than $n^{1/3}$ a-cyclic states was enough to implies the desired result for the uniform distribution. Here we need something more precise. In the next lemma we show that with positive probability, the number of a-cyclic states is in $\Theta(\sqrt{n})$.

Lemma 8. *The cyclic part of a uniform random mapping of size n has size in $[\![\sqrt{n}, 2\sqrt{n}]\!]$ with visible probability.*

Proof. (sketch) We rely on tools from analytic combinatorics [12] applied to the decomposition of a mapping into a union of cycles of Cayley trees, as in [11]. We obtain that the expected number of cyclic points is asymptotically equivalent to $\sqrt{\frac{\pi n}{2}}$, which is already in [11], and that the standard deviation is asymptotically equivalent to $\sqrt{\frac{(4-\pi)n}{2}}$. The result follows by Chebyshev's inequality. $\qquad\square$

At this point, we know that with visible probability, the cyclic permutation of the action of a is in $\mathrm{Cycle}_d(x)$ for $x \in [\![\sqrt{n}, 2\sqrt{n}]\!]$. To complete the proof, we need to verify that they are accessible (which is easy since large a-cycles are generically accessible) and that they are sufficiently often u-primitive for the u of Proposition 1.

5.3 Completing the Proof

We will use the following lemma to establish the primitivity of the first d a-cycles:

Lemma 9. *Let $n \geq 2$ be an integer and let $i \in [\![1, n-1]\!]$. For the uniform distribution on binary words of length n having i occurrences of the letter 0, the probability that a word is not primitive is smaller than $\frac{2}{n}$.*

We therefore have two independent random sets, $\delta^{-1}(f, u)$ and the union of the first d a-cycles, and are interested in their intersection. The two following lemmas establish that this intersection is not trivial with visible probability. Together with Lemma 9, this will ensure that these a-cycles are u-primitive with positive probability.

Lemma 10. *Let α and β be two positive real numbers. Let X be a subset of $[n]$ of size $\lceil \alpha\sqrt{n} \rceil$ and let Y be a uniform random subset of $[n]$ of size $\lceil \beta\sqrt{n} \rceil$. For every integer $j \geq 0$, there exists a positive constant M_j such that $|X \cap Y| = j$ with probability at least M_j, for n sufficiently large.*

Lemma 11. *Let α be a positive real number. Let X be a subset of $[n]$ of size $m = \lceil \alpha\sqrt{n} \rceil$ and let Y be a uniform random subset of $[n]$ of size m' with $1 \leq m' < \frac{n}{2}$. The probability that $X \subseteq Y$ is smaller than $B\sqrt{n}2^{-\alpha\sqrt{n}}$ for some positive constant B and for n sufficiently large.*

We can now establish the proof of Theorem 3 for automata with one final state as follows. For every $x \in [\![\sqrt{n}, 2\sqrt{n}]\!]$, let $E(x)$ denote the set of mappings of size n whose cyclic part σ has size x and belongs to $\mathrm{Cycle}_d(x)$. By Lemma 8 and Lemma 7, a random mapping is in $\cup_{x \in [\![\sqrt{n}, 2\sqrt{n}]\!]} E(x)$ with visible probability.

Let us fix some mapping f_a of $E(x)$ for the action of a, and let σ_a be its cyclic part. Let S_1, S_2, ..., S_d be arbitrary subsets of size $m = \lceil \frac{\sqrt{n}}{3d} \rceil$ of the first d cycles of σ_a, and let $S = \cup_{i=1}^d S_i$. Since the actions of b and c are independent of the action of a, by Proposition 2 there exists a word u such that $Y = \delta^{-1}(f, u)$ has size in $[\![w\sqrt{n}, w\sqrt{n} \log n]\!]$ with positive probability. Let Y' be a uniform subset of Y of size $y = \lceil w\sqrt{n} \rceil$. By symmetry, the set Y' is a uniform random subset of size y of $[n]$. Therefore by Lemma 10, with positive probability we have $|S \cap Y'| = d$, and a direct computation shows that this implies that, with positive probability, $|S_i \cap Y'| = 1$ for every $i \in [d]$. Moreover, since $|Y| \leq \frac{n}{2}$, by Lemma 11, the probability that at least one S_i is a subset of Y is smaller than $dB\sqrt{n}2^{-\sqrt{n}}$. Hence, with visible probability, the intersection of Y and S_i is non-trivial for every $i \in [d]$, and so is the intersection of Y and the first d cycles of σ_a (since they contain S_i). Hence, by Lemma 9, there exists a constant $M > 0$ such that the first d cycles are u-primitive with probability at least M for n sufficiently large; and importantly, the value of M is the same for any $x \in [\![\sqrt{n}, 2\sqrt{n}]\!]$. Therefore, if we sum the contributions for all x in $[\![\sqrt{n}, 2\sqrt{n}]\!]$, we get that with visible probability the first d a-cycles are u-primitive (for some word u) and therefore that the lcm of their lengths is at least $\lambda n^{d/2}$.

But the first d a-cycle have lengths greater than $\frac{\sqrt{n}}{2d}$ and are therefore generically accessible (this is Proposition 1 of [10]). This concludes the proof by Lemma 2: by choosing $d = 2d' + 1$ there are more than $n^{d'}$ states in the first determinization step of Brzozowski's algorithm with visible probability. $\qquad \square$

6 General Case

The proof for a general distribution with a small number of final states is not difficult once we have establish the result for the uniform distribution on automata with one final state. We consider two cases depending on whether the automaton has between 1 and $w\sqrt{n}$ final states or between $w\sqrt{n}$ and $\frac{n}{2}$ final states.

For the first case, we select one of the final states $\hat{f}$ and apply the same construction as in Section 5. With visible probability, we therefore obtain a word u such that $\delta^{-1}(\hat{f}, u)$ has size at least $w\sqrt{n}$, and therefore $\delta^{-1}(f, u)$ also has size at least $w\sqrt{n}$. Hence, by Lemma 3, there generically exists a prefix u' of u such that $\delta^{-1}(f, u) \in [\![w\sqrt{n}, w\sqrt{n} \log n]\!]$, and we can continue the proof as in Section 5.

The second case is easier. We do not need to build the word u since F is already large enough to apply Lemma 10 and still small enough to apply Lemma 11. The general statement of Theorem 3 follows. $\qquad \square$

A natural question is whether the average super-polynomial complexity of Brzozowski's algorithm still holds for alphabets with two letters. The proof of this paper relies on the fact that we built $u \in \{b, c\}^*$ and the a-cycles independently, so that we can apply Lemma 9, Lemma 10 and Lemma 11. If u uses the letter a, we need a more complicated proof that takes the dependency into account, which is usually difficult. Therefore, the best way to obtain the result is probably to find a completely different approach.

Acknowledgment. We would like to thanks Jean-François Marckert for his patient explanation of his result on the width of random Cayley trees [6]. We also thank the referees for providing helpful comments, which helped to improve the quality of this article.

References

1. Aho, A.V., Corasick, M.J.: Efficient string matching: An aid to bibliographic search. Commun. ACM 18(6), 333–340 (1975)
2. Bassino, F., David, J., Nicaud, C.: Average case analysis of Moore's state minimization algorithm. Algorithmica 63(1-2), 509–531 (2012)
3. Bassino, F., David, J., Sportiello, A.: Asymptotic enumeration of minimal automata. In: Dürr, Wilke (eds.) [8], pp. 88–99
4. Brzozowski, J.A.: Canonical regular expressions and minimal state graphs for definite events. In: Mathematical theory of Automata. MRI Symposia Series, vol. 12, pp. 529–561. Polytechnic Press, Polytechnic Institute of Brooklyn, N.Y (1962)
5. Carayol, A., Nicaud, C.: Distribution of the number of accessible states in a random deterministic automaton. In: Dürr, Wilke (eds.) [8], pp. 194–205
6. Chassaing, P., Marckert, J.F.: Parking functions, empirical processes, and the width of rooted labeled trees. Electron. J. Combin. 8(1) (2001)
7. David, J.: Average complexity of Moore's and Hopcroft's algorithms. Theor. Comput. Sci. 417, 50–65 (2012)
8. Dürr, C., Wilke, T. (eds.): 29th International Symposium on Theoretical Aspects of Computer Science, STACS 2012, Paris, France, February 29-March 3. LIPIcs, vol. 14. Schloss Dagstuhl - Leibniz-Zentrum fuer Informatik (2012)
9. Erdos, P., Turán, P.: On some problems of a statistical group theory, III. Acta Math. Acad. Sci. Hungar. 18(3-4), 309–320 (1967)
10. De Felice, S., Nicaud, C.: Brzozowski algorithm is generically super-polynomial for deterministic automata. In: Béal, M.-P., Carton, O. (eds.) DLT 2013. LNCS, vol. 7907, pp. 179–190. Springer, Heidelberg (2013)
11. Flajolet, P., Odlyzko, A.M.: Random mapping statistics. In: Quisquater, J.-J., Vandewalle, J. (eds.) EUROCRYPT 1989. LNCS, vol. 434, pp. 329–354. Springer, Heidelberg (1990)
12. Flajolet, P., Sedgewick, R.: Analytic Combinatorics. Cambridge University Press (2009)
13. Hopcroft, J.E.: An $n \log n$ algorithm for minimizing the states in a finite automaton. In: Kohavi, Z. (ed.) The Theory of Machines and Computations, pp. 189–196. Academic Press (1971)
14. Kolčin, V.: Random Mappings: Translation Series in Mathematics and Engineering. Translations series in mathematics and engineering. Springer London, Limited (1986), http://books.google.fr/books?id=roOiPwAACAAJ

State Complexity of Deletion

Yo-Sub Han[1,*], Sang-Ki Ko[1,*], and Kai Salomaa[2,**]

[1] Department of Computer Science, Yonsei University, 50, Yonsei-Ro,
Seodaemun-Gu, Seoul 120-749, Korea
{emmous,narame7}@cs.yonsei.ac.kr
[2] School of Computing, Queen's University, Kingston, Ontario K7L 3N6, Canada
ksalomaa@cs.queensu.ca

Abstract. It is well known that the language obtained by deleting an arbitrary language from a regular language is regular. We give an upper bound for the state complexity of deleting an arbitrary language from a regular language and a matching lower bound. We show that the state complexity of deletion is $n \cdot 2^{n-1}$ (respectively, $(n + \frac{1}{2}) \cdot 2^n - 2$) when using complete (respectively, incomplete) deterministic finite automata.

1 Introduction

The descriptional complexity of finite automata has been studied for over half a century [21–24] and there has been much renewed interest since the early 90's [11, 13, 20, 29]. Operational state complexity investigates the size of a DFA (deterministic finite automaton) needed to recognize the language obtained by applying a regularity preserving operation to given DFAs. The precise worst case state complexity of many basic language operations has been established; see [1, 4, 5, 7, 12, 14, 22, 25, 30] where further references can be found. Also there has been much work on the state complexity of combinations of basic language operations [2, 6, 8, 9, 15, 26].

Deletion is one of the basic operations in formal language theory [17, 18]. The deletion of a string v from a string u consists of erasing a contiguous substring v from u. We denote the result of deleting a language L_2 from a language L_1 by $L_1 \rightsquigarrow L_2$.[1]

Deletion is the simplest and most natural generalization of the left/right quotient [18]. It is known that for L_1 recognized by a DFA with n states and an arbitrary language L_2, the worst case state complexity of the left-quotient $L_2 \backslash L_1$ is $2^n - 1$ and the state complexity of the right-quotient L_1 / L_2 is n [29]. Recently, the state complexity of insertion which, using the terminology of [16], is the left inverse of deletion, was investigated in [10].

---

* Han and Ko were supported by the Basic Science Research Program through NRF funded by MEST (2012R1A1A2044562).
** Salomaa was supported by the Natural Sciences and Engineering Research Council of Canada Grant OGP0147224.
[1] See section 2 for a formal definition.

A.M. Shur and M.V. Volkov (Eds.): DLT 2014, LNCS 8633, pp. 37–48, 2014.

It is well known that $L_1 \leadsto L_2$ is always regular for a regular language L_1 and an arbitrary language L_2 [18]. However, in spite of deletion being a fundamental language operation its precise state complexity has not been studied in the literature. When L_1 is recognized by a DFA with n states, the proof of Theorem 1 of [18] yields an upper bound 2^{2n} for the size of the DFA needed to recognize $L_1 \leadsto L_2$. The proof works for an arbitrary language L_2 and is not, in general, effective.

More general types of deletion operations, called *deletion along trajectories* have been considered by Domaratzki [3] and Kari and Sosik [19]. In this context a set of trajectories is a set $T \subseteq \{i, d\}^*$ where i stands for "insert" and d stands for "delete". The result of deleting a language L_2 from a language L_1 along a set of trajectories T is denoted $L_1 \leadsto_T L_2$. Deletion along a regular set of trajectories preserves regularity [3], that is, for regular languages L_1, L_2 and T, also $L_1 \leadsto_T L_2$ is regular. The ordinary deletion operation we consider here is defined by the set of trajectories $i^* d^* i^*$, and the construction used in the proof of Lemma 3.1 of [3] would yield an upper bound 2^{3mn} for the state complexity of $L_1 \leadsto_{i^* d^* i^*} L_2$ when L_1 (respectively, L_2) is recognized by a DFA of size m (respectively, n). Naturally, Lemma 3.1 of [3] deals with deletion along arbitrary regular sets of trajectories and the result cannot be expected to yield a good bound in the very special case we are considering here.

Most of the literature uses complete DFAs to measure the state complexity of a regular language, but state complexity based on incomplete DFAs also has been considered. Câmpeanu et al. [1] give the state complexity of shuffle in terms of incomplete DFAs while the precise state complexity of shuffle in terms of complete DFAs remains still open. For a given regular language the sizes of the minimal complete and the minimal incomplete DFA differ by at most one state, however, there can be a more significant difference in the state complexity functions when the measure is based on complete and incomplete DFAs, respectively.

In this paper we give a tight state complexity bound for the language obtained from a regular language by deleting an arbitrary language. We show that if L_1 is recognized by a complete DFA with n states and L_2 is an arbitrary language, the complete DFA for the language $L_1 \leadsto L_2$ needs $n \cdot 2^{n-1}$ states in the worst case. The corresponding state complexity function based on incomplete DFAs is shown to be $(n + \frac{1}{2}) \cdot 2^n - 2$. While the upper bounds hold for arbitrary languages L_2 (that need not be even recursively enumerable) we show that matching lower bound constructions can be found where L_2 consists of a single string of length one. We give conditions based on L_2 and the DFA for L_1 that are necessary for the state complexity of deletion to reach the worst case bound.

2 Preliminaries

We assume that the reader is familiar with the basics of finite automata and formal languages and recall here just some definitions and notation. For more information on the topic the reader may consult the monographs [27, 28] or the survey [29].

In the following Σ always stands for a finite alphabet and the set of strings over Σ is Σ^*. A language is a subset of Σ^*. The cardinality of a finite set S is denoted $|S|$.

The set of strings obtained from $u \in \Sigma^*$ by deleting a string $v \in \Sigma^*$ is

$$u \rightsquigarrow v = \{w \in \Sigma^* \mid (\exists u_1, u_2 \in \Sigma^*) \; w = u_1 u_2 \text{ and } u = u_1 v u_2\}.$$

For example, $bababa \rightsquigarrow aba = \{bba, bab\}$. The *deletion operation* is extended in the natural way for languages $L_1, L_2 \subseteq \Sigma^*$ by setting

$$L_1 \rightsquigarrow L_2 = \bigcup_{u \in L_1, v \in L_2} u \rightsquigarrow v.$$

An *incomplete deterministic finite automaton* (incomplete DFA) is a five-tuple $A = (Q, \Sigma, \delta, q_0, F)$ where Q is a finite set of states, Σ is an alphabet, δ is a partial function $Q \times \Sigma \to Q$, $q_0 \in Q$ is the initial state and $F \subseteq Q$ is a set of final (or accepting) states.

The transition function δ is in the usual way extended as a partial function $Q \times \Sigma^* \to Q$ and the language recognized by A is $L(A) = \{w \in \Sigma^* \mid \delta(q_0, w) \in F\}$. A language is regular if it is recognized by some DFA.

For $q \in Q$, $P \subseteq Q$, $b \in \Sigma$ and $L \subseteq \Sigma^*$ we also denote

$$\delta(P, b) = \{\delta(p, b) \mid p \in P\} \quad \text{and} \quad \delta(q, L) = \{\delta(q, w) \mid w \in L\}.$$

A DFA $A = (Q, \Sigma, \delta, q_0, F)$ is said to be *complete* if δ is a total function $Q \times \Sigma \to Q$. We will use both complete and incomplete DFAs and, when not separately mentioned, by a DFA we mean an incomplete DFA.[2]

A DFA $A = (Q, \Sigma, \delta, q_0, F)$ is *minimal* if each state $q \in Q$ is reachable from the initial state q_0 (that is, $\delta(q_0, w) = q$ for some string w) and no two states $q_1, q_2 \in Q$, $q_1 \neq q_2$, are equivalent. States $q_1, q_2 \in Q$ are said to be equivalent if

$$(\forall w \in \Sigma^*) \; \delta(q_1, w) \in F \text{ iff } \delta(q_2, w) \in F.$$

The minimal (complete or incomplete) DFA for a given regular language L is unique and the sizes of the minimal complete and incomplete DFAs for L differ by at most one state. The minimal complete DFA may have a dead state (or sink state). In the minimal incomplete DFA the dead state can always be omitted.

The *state complexity* of L, sc(L), is the size of the minimal complete DFA recognizing L. Similarly, the *incomplete state complexity* of L, isc(L), is the size of the minimal incomplete DFA recognizing L. For each regular language L either sc$(L) = $ isc$(L) + 1$ or sc$(L) = $ isc(L).

3 Upper Bound for Deletion

It is known that the result of deleting an arbitrary language from a regular language is regular [18]. Hence in the lemmas establishing the upper bound for

---

[2] Naturally, a complete DFA is just a special case of an incomplete DFA.

deletion (for complete or incomplete DFAs) we do not need to assume that the deleted language L_2 is regular. However, in the case of an arbitrary L_2 finding a DFA for the language $L_1 \rightsquigarrow L_2$ is not, in general, effective.

First we give an upper bound construction for complete DFAs.

Lemma 1. *Consider $L_1, L_2 \subseteq \Sigma^*$ where L_1 is recognized by a complete DFA with n states. Then*

$$\mathrm{sc}(L_1 \rightsquigarrow L_2) \leq n \cdot 2^{n-1}.$$

Proof. Let $A = (Q, \Sigma, \delta, q_0, F_A)$ be a complete DFA for L_1 where $|Q| = n$. To recognize the language $L_1 \rightsquigarrow L_2$ we define a DFA

$$B = (P, \Sigma, \gamma, p_0, F_B),$$

where $P = \{(r, R) \mid r \in Q, R \subseteq Q, \delta(r, L_2) \subseteq R\}$, $p_0 = (q_0, \delta(q_0, L_2))$ and

$$F_B = \{(r, R) \mid r \in Q, R \subseteq Q, \delta(r, L_2) \subseteq R \text{ and } R \cap F_A \neq \emptyset\}.$$

It remains to define the transitions of γ. For $(r, R) \in P$ and $b \in \Sigma$ we set

$$\gamma((r, R), b) = (\delta(r, b), \ \delta(R, b) \cup \delta(\delta(r, b), L_2)). \tag{1}$$

The transition relation always adds the elements of $\delta(\delta(r, b), L_2)$ to the second component and, consequently, the state $\gamma((r, R), b)$, as defined above, is an element of P.

The intuitive idea of the construction is as follows. In order to recognize the language $L_1 \rightsquigarrow L_2$, the DFA B must check that the input string w can be completed to a string of L_1 by inserting a string $u \in L_2$ in some position, that is, for some decomposition $w = w_1 w_2$ we have $w_1 u w_2 \in L_1$. Since we do not know at which position the string $u \in L_2$ is to be inserted and B has to be deterministic, roughly speaking, B has to keep track of all computations of A on strings where a string of L_2 was deleted from some earlier position.

The first component of the states of B simply simulates the computation of A, i.e., it keeps track of the state of A, assuming that up to the current position in the input a string of L_2 was not yet deleted. The second component of the states of B keeps track of all states that A could be in assuming that at some point in the preceding computation a string of L_2 was deleted from the input.

We need to verify that the transitions of B (as defined in (1)) preserve these properties. For the first component it is clear that the simulation works as claimed. To verify the claim for the second component, assume that the input is ubv, $u, v \in \Sigma^*$, $b \in \Sigma$ and after reading the prefix u the DFA B has reached a state (r, R). In the following discussion b refers to the particular symbol occurrence just after the prefix u. After reading the symbol b, the second component of the state of B will be $\delta(R, b) \cup \delta(\delta(r, b), L_2))$, where R consists of states that A could be in, assuming a string of L_2 was deleted somewhere before the symbol b and the states of $\delta(R, b)$ are then the states A could be in after reading b assuming a string of L_2 was deleted before symbol occurrence b. On the other hand, $r = \delta(q_0, u)$, i.e., r is the state A reaches after reading the prefix

u where no deletion has occurred, and hence $\delta(\delta(r,b), L_2)$ consists of exactly all states A can be in the simulated computation, assuming a string of L_2 was deleted directly after the symbol occurrence b. This means that the transition relation γ correctly preserves the property that the second component of the state of B consists of all states that A could be in assuming a string of L_2 was deleted some time previously.

The choice of final states guarantees that B accepts exactly the strings obtained from strings of $L(A)$ by deleting a string of L_2 at some position.

We still need to verify that the number of states of B is as claimed. If $L_2 = \emptyset$, then $L_1 \rightsquigarrow L_2 = \emptyset$ and $L_1 \rightsquigarrow L_2$ has a DFA of size one. Hence in what follows we can assume that $L_2 \neq \emptyset$.

Since A is a complete DFA and $L_2 \neq \emptyset$, for each $r \in Q$ we have $|\delta(r, L_2)| \geq 1$. This means that for a given $r \in Q$, there exist at most $2^{|Q|-1}$ sets R such that (r, R) is a state of B. Thus, the number of states of B is at most $|Q| \cdot 2^{|Q|-1}$. $\quad\square$

Next we consider the case of incomplete DFAs. The upper bound construction uses similar ideas as the above proof of Lemma 1, and we just need to modify the construction to allow the possibility of undefined transitions.

Lemma 2. Let $L_1, L_2 \subseteq \Sigma^*$ where L_1 is recognized by an incomplete DFA A with n states. Then

$$\mathrm{isc}(L_1 \rightsquigarrow L_2) \leq (n+1) \cdot 2^n - (2^{n-1} + 2).$$

Proof. Let $A = (Q, \Sigma, \delta, q_0, F_A)$ be an incomplete DFA for L_1, $|Q| = n$. We define the completion of δ as a function $\delta' : (Q \cup \{\text{dead}\}) \times \Sigma \to Q \cup \{\text{dead}\}$ by setting for $r \in Q \cup \{\text{dead}\}$ and $b \in \Sigma$,

$$\delta'(r,b) = \begin{cases} \delta(r,b), & \text{if } r \in Q \text{ and } \delta(r,b) \text{ is defined;} \\ \text{dead}, & \text{otherwise.} \end{cases}$$

To recognize the language $L_1 \rightsquigarrow L_2$ we define a DFA

$$B = (P, \Sigma, \gamma, p_0, F_B),$$

where $P = (Q \cup \{\text{dead}\}) \times 2^Q - \{(\text{dead}, \emptyset), (\text{dead}, Q)\}$, $p_0 = (q_0, \delta(q_0, L_2))$ and

$$F_B = \{(r, R) \mid r \in Q \cup \{\text{dead}\}, R \subseteq Q \text{ and } R \cap F_A \neq \emptyset\}.$$

(Note that $|P| = (n+1) \cdot 2^n - 2$. However, as will be seen below at least 2^{n-1} elements of P will be unreachable as states of B.)

The transitions of γ are defined by setting, for $(r, R) \in P$ and $b \in \Sigma$,

$$\gamma((r,R),b) = \begin{cases} (\delta'(r,b),\ \delta(R,b) \cup \delta(\delta(r,b), L_2)), & \text{if } r \in Q \text{ and } [\delta'(r,b) \neq \text{dead} \\ & \text{or } \delta(R,b) \cup \delta(\delta(r,b), L_2) \neq \emptyset]; \\ (\text{dead}, \delta(R,b)), & \text{if } r = \text{dead and } \delta(R,b) \neq \emptyset; \\ \text{undefined}, & \text{otherwise.} \end{cases}$$

As in the proof of the previous lemma, the idea is that the first component simulates the computation of the original DFA A, assuming a string of L_2 has so far not been deleted (and using the new state "dead" to indicate that the simulated computation of A has failed), and the second component keeps track of the set of all possible states that A can be in, assuming a string of L_2 was deleted somewhere previously (this set can now be empty because A is incomplete). We leave the details of verifying that B recognizes $L(A) \rightsquigarrow L_2$ to the reader.

Below we explain why the states $(\text{dead}, \emptyset)$ and (dead, Q) can be omitted from the state set of B, and that, furthermore at least 2^{n-1} elements of P must be unreachable as states of B.

The state $(\text{dead}, \emptyset)$ would be a sink state of the DFA B and, according to the definition of γ, it is never entered. Also, we note that when the computation, for the first time, reaches a state where the first component is "dead" this has to occur on an alphabet symbol that has at least one undefined transition in A. This means that when the computation initially reaches a state with first component "dead", the cardinality of the second component is at most $|Q| - 1$, and after that point the transitions of γ do not add new states to the second component because if the deletion of the string of L_2 did not occur previously, the computation has already failed. Thus, the state (dead, Q) is always unreachable.

To verify the unreachability of 2^{n-1} further states, without loss of generality, we can assume that for some $q_1 \in Q$ and $w_1 \in L_2$, $\delta(q_1, w_1)$ is defined. Note that in the opposite case, no string of L_2 can occur as a substring of a string of $L(A) = L_1$ and, hence, $L_1 \rightsquigarrow L_2 = \emptyset$. Now all transitions of B that enter a state with the first component q_1 add the element $\delta(q_1, w_1)$ to the second components. This means that all elements (q_1, R), $\delta(q_1, w_1) \notin R$, are unreachable.   □

In the above proof we noted that 2^{n-1} states of the constructed DFA B are unreachable for each state q of A such that $\delta(q, w)$ is defined for some $w \in L_2$. Thus, the worst case state complexity blow-up can occur only when transitions spelling out a string in L_2 originate only from one state of A and, slightly more precisely, we get the following necessary condition for languages that can reach the worst case state complexity of the deletion operation.

Corollary 1. *Let $A = (Q, \Sigma, \delta, q_0, F)$ be an incomplete DFA with n states, $L_1 = L(A)$ and L_2 is an arbitrary language. Then a necessary condition for $\mathrm{isc}(L_1 \rightsquigarrow L_2)$ to reach the upper bound $(n+1) \cdot 2^n - (2^{n-1} + 2)$ given by Lemma 2 is that*

$$(\exists q \in Q) \; [\; |\delta(q, L_2)| = 1 \; and \; (\forall p \in Q, p \neq q) \; \delta(p, L_2) = \emptyset \;].$$

4   Lower Bound Constructions

As our main result we show here that the bounds given in the previous section are optimal. We begin with the case of complete DFAs where the construction is somewhat simpler.

4.1   Lower Bound for Complete DFAs

From the construction of the complete DFA B for $L_1 \rightsquigarrow L_2$ in Lemma 1 we know that the possible states of B are pairs (r, R) where r (respectively, R) is a state (respectively, a set of states) of the DFA A recognizing L_1 and R has the property that it contains all states that are reachable from r on a string of L_2 (and possibly other states). Hence a possible worst case construction should use a singleton language L_2 or a language L_2 such that for any given state r of A, all strings of L_2 take r to the same state. In the proof of Lemma 3 we choose L_2 to be a singleton language consisting of a string of length one.

Lemma 3. *Let $\Sigma = \{a, b, c, d, e\}$. For every $n \geq 3$ there exists a complete DFA A over Σ with n states such that*

$$\mathrm{sc}(L(A) \rightsquigarrow \{c\}) = n \cdot 2^{n-1}.$$

Proof. Choose $A = (Q, \Sigma, \delta, 0, \{0\})$ where $Q = \{0, 1, \ldots, n-1\}$ and the transitions of δ are defined by setting

- $\delta(i, a) = i + 1$ for $0 \leq i \leq n - 2$, $\delta(n - 1, a) = 0$;
- $\delta(0, b) = 0$, $\delta(i, b) = i + 1$ for $1 \leq i \leq n - 2$, $\delta(n - 1, b) = 0$;
- $\delta(0, c) = 1$, $\delta(i, c) = i$ for $1 \leq i \leq n - 1$;
- $\delta(0, d) = 0$, $\delta(1, d) = 1$, $\delta(i, d) = i + 1$ for $2 \leq i \leq n - 2$, $\delta(n - 1, d) = 0$;
- $\delta(0, e) = \delta(1, e) = 1$, $\delta(i, e) = i + 1$ for $2 \leq i \leq n - 2$, $\delta(n - 1, e) = 0$.

The DFA A is depicted in Figure 1.

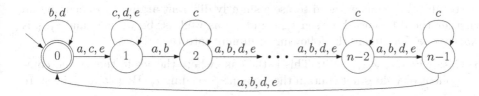

Fig. 1. The complete DFA A used in the proof of Lemma 3

Let $B = (P, \Sigma, \gamma, p_0, F_B)$ be the complete DFA recognizing the language $L(A) \rightsquigarrow \{c\}$ that is constructed as in the proof of Lemma 1. Since the deleted language $\{c\}$ consists of only one string, B has $n \cdot 2^{n-1}$ states and in order to prove the lemma it is sufficient to show that all states of B are reachable from the initial state $p_0 = (0, \{1\})$ and all states of B are pairwise inequivalent.

Claim 1. All states $(0, R) \in P$ where $R \supseteq \delta(0, \{c\}) = \{1\}$ are reachable from $(0, \{1\})$.

We prove the claim by induction on the cardinality of R. In the base case $|R| = 1$ there is nothing to prove because $R = \{1\}$ is the only set satisfying the required condition on R.

Inductively, now assume that the claim holds for all sets R of cardinality $1 \leq k < n$, that is, all states $(0, R)$ where $1 \in R$ and $|R| \leq k$ are reachable. Now consider a state of P,

$$\mathbf{u} = (0, \{1, i_1, \ldots i_k\}), \quad 1 < i_1 < i_2 < \cdots < i_{k-1} \text{ and } (i_{k-1} < i_k \text{ or } i_k = 0).$$

Above the elements $i_1, \ldots, i_k$ are listed in increasing order, except that 0 is considered the largest element. By the inductive assumption the state

$$\mathbf{u}' = (0, \{1, i_2 - i_1 + 1, i_3 - i_1 + 1, \ldots, i_k - i_1 + 1\})$$

is reachable. Here the arithmetic operations are done modulo n. If $i_k = 0$, above $i_k - i_1 + 1$ stands for $n - i_1 + 1$. Now

$$\gamma(\mathbf{u}', b) = (0, \{1, 2, i_2 - i_1 + 2, i_3 - i_1 + 2, \ldots, i_k - i_1 + 2\}),$$

because the transition where the first component enters 0 adds 1 to the second component and otherwise the transition on b cycles "upwards" the states of $\{1, i_2 - i_1 + 1, i_3 - i_1 + 1, \ldots, i_k - i_1 + 1\}$. (Note that this set does not contain the element 0.) Next applying to the state $\gamma(\mathbf{u}', b)$ $i_1 - 2$ times a transition on d we reach $\mathbf{u}$. Note that transitions on d "cycle upwards" the states in $\{2, 3, \ldots, n-1\}$, and keep the states 0 and 1 stationary, and do not add (in the transition relation of B), new elements to the second component of the state.

Next we show that all states (r, R), $R \supseteq \delta(r, \{c\})$, are reachable. We need to consider only cases where $r \neq 0$ (because the case $r = 0$ was handled in Claim 1 above) and, hence, $\delta(r, \{c\}) = \{r\}$. Consider an arbitrary state

$$\mathbf{v} = (i_j, \{i_1, i_2, \ldots, i_{j-1}, i_j, i_{j+1}, \ldots, i_k\}), \quad 0 \leq i_1 < i_2 < \cdots < i_k \leq n - 1.$$

For technical reasons we need to use a slightly different argument depending on whether the difference between i_{j+1} and i_j, as well as, between i_j and i_{j-1} is exactly one. We divide the following argument into three cases.

(i) Case where $i_j \neq i_{j-1} + 1$: This is the case where the set in the second component of $\mathbf{v}$ does not contain the element preceding i_j. By Claim 1 the state

$$\mathbf{v}' = (0, \{1, i_1 - i_j + 1, i_2 - i_j + 1, \ldots, i_{j-1} - i_j + 1, i_{j+1} - i_j + 1, \ldots, i_k - i_j + 1\})$$

is reachable. In the preceding line all quantities are computed modulo n. Now

$$\gamma(\mathbf{v}', c) = (1, \{1, i_1 - i_j + 1, i_2 - i_j + 1, \ldots, i_{j-1} - i_j + 1, i_{j+1} - i_j + 1, \ldots, i_k - i_j + 1\}).$$

Note that, because $i_j \neq i_{j-1} + 1$, the sequence $i_x - i_j + 1$, $x = 1, \ldots, j - 1, j + 1, \ldots, k$, does not contain the element 0, and hence a transition on c is the identity on these elements. In the DFA A the a-transitions just cycle through the states and hence applying $i_j - 1$ times the a-transition to state $\gamma(\mathbf{v}', c)$ we get the state $\mathbf{v}$.

(ii) Case where $i_j = i_{j-1} + 1$ and $i_{j+1} = i_j + 1$: This is the case where the set in the second component of $\mathbf{v}$ contains both the element preceding i_j and the element following i_j. By Claim 1 the state

$$\mathbf{v_1} = (0, \{i_1 - i_j, i_2 - i_j, \ldots, i_k - i_j\})$$

is reachable. Note that the second component contains the element 1 and hence $\mathbf{v_1}$ is a legal state of B. Now $\gamma(\mathbf{v_1}, a^{i_j}) = \mathbf{v}$.

(iii) Case where $i_j = i_{j-1} + 1$ and $i_{j+1} \neq i_j + 1$: This corresponds to the situation where the second component of $\mathbf{v}$ contains the element preceding i_j but does not contain the element following i_j. Here we cannot use a-transitions alone, because a state where the first component is 0 must contain 1 in the the second component. By Claim 1 the state

$$\mathbf{v_2} = (0, \{1, i_1 - i_j, i_2 - i_j, \ldots, i_{j-1} - i_j, i_{j+1} - i_j, \ldots, i_k - i_j\})$$

is reachable and

$$\gamma(\mathbf{v_2}, e) = (1, \{1, i_1 - i_j + 1, i_2 - i_j + 1, \ldots, i_k - i_j + 1\}).$$

Here we need the fact that $i_{j+1} \neq i_j + 1$ and hence an e-transition adds one to the state $i_{j+1} - i_j$. (Note also that $i_{j-1} - i_j = n - 1$ and, consequently, beginning with a c-transition as in case (i) above would not work, because the second component at the end would not contain i_{j-1}.)

Applying $i_j - 1$ a-transitions to the state $\gamma(\mathbf{v_2}, e)$ we get $\mathbf{v}$.

We have shown that all states of B are reachable and it remains to show that they are all pairwise inequivalent.

First consider two states $(r_1, R_1), (r_2, R_2) \in P$ where $R_1 \neq R_2$. Without loss of generality we can find $s \in R_1 - R_2$ since the other possibility is completely symmetric. If $s = 0$ then (r_1, R_1) is a final state and (r_2, R_2) is a nonfinal state of B. Thus it is sufficient to consider cases $s \in \{1, 2, \ldots, n - 1\}$. Now $\gamma((r_1, R_1), a^{n-s}) \in F_B$ because the string a^{n-s} takes the state $s \in R_1$ to the element 0. We show that $\gamma((r_2, R_2), a^{n-s}) \notin F_B$. Since $s \notin R_2$, the string a^{n-s} does not take any element of R_2 to the element 0 (which is the only final state of A). We note that $r_2 \neq s$ because from the definition of legal states of B we know that $\delta(r_2, c)$ must be an element of R_2 (and $\delta(s, c) = s$ when $1 \leq s \leq n - 1$). Also, if during the computation on a^{n-s} starting from (r_2, R_2), the transitions of γ add the element 1 to the second component when the first component becomes 0, then the added element 1 cannot cycle through all states to reach the final state 0 because after adding the element 1 there remains at most $n - s - 1$ input symbols and $s \geq 1$.

Second, consider two states $(r_1, R_1), (r_2, R_2) \in P$, where $r_1 \neq r_2$. Due to symmetry between r_1 and r_2 we can assume that $r_2 \neq 0$.

(i) Case where $r_1 = 0$ and $r_2 \neq n - 1$: Since $r_2 \neq r_1$, we have $r_2 \neq 0$ and we note that $\gamma((0, R_1), b) = (0, R_1')$ where $1 \in R_1'$ (because the self-loop on state 0 adds the element 1 to the second component) and $\gamma((r_2, R_2), b) = (r_2 + 1, R_2')$. Here $1 \notin R_2'$ because no transition of A labeled by b reaches the state 1 and also since $r_2 + 1 \neq 0$ the transition cannot add the element 1 to the second component. Since the second components are distinct sets we know that the states $(0, R_1')$ and $(r_2 + 1, R_2')$ are distinguishable.

(ii) Case where $r_1 = 0$ and $r_2 = n - 1$: We note that $\gamma((0, R_1), a) = (1, R_1')$ and $\gamma((n - 1, R_2), a) = (0, R_2')$. The states $(1, R_1')$ and $(0, R_2')$ are inequivalent by case (i) above.

(iii) Case where $r_1 \neq 0$ and $r_2 \neq 0$: By cycling with $n - r_1$ a-transitions we get states $(0, R_1')$ and $(n - r_1 + r_2, R_2')$. If $r_1 - r_2 \neq 1$, this case was covered in (i) above and, if $r_1 - r_2 = 1$, this case was covered in (ii) above.

Above (i)–(iii) cover all cases where $r_1 \neq r_2$ and $r_2 \neq 0$. This concludes the proof of the lemma.   □

Now by combining Lemmas 1 and 3 we get a tight state complexity bound for deletion.

Theorem 1. *For languages $L_1, L_2 \subseteq \Sigma^*$ where L_1 is regular,*

$$\mathrm{sc}(L_1 \rightsquigarrow L_2) \leq \mathrm{sc}(L_1) \cdot 2^{\mathrm{sc}(L_1)-1}.$$

For every $n \geq 3$ there exists a regular language L_1 over a five-letter alphabet with $\mathrm{sc}(L_1) = n$ and a singleton language L_2 such that in the above inequality we have equality.

4.2   Lower Bound for Incomplete DFAs

We show that the state complexity upper bound for incomplete DFAs from Lemma 2 can be reached by DFAs defined over a five-letter alphabet. Based on the observations made in Corollary 1, as the language of deleted strings, we use a singleton set $\{c\}$ where, furthermore, a c-transition is defined only for one state of the DFA recognizing L_1. The conditions of Corollary 1 do not force L_2 to be a singleton set, however, the conditions indicate that a construction may be simpler to achieve using a singleton set. (The proof of Lemma 4 is omitted due to the limitation on the number of pages.)

Lemma 4. *Let $\Sigma = \{a, b, c, d, e\}$. For every $n \geq 4$ there exists a regular language $L_1 \subseteq \Sigma^*$ recognized by an incomplete DFA with n states such that*

$$\mathrm{isc}(L_1 \rightsquigarrow \{c\}) = (n+1) \cdot 2^n - (2^{n-1} + 2).$$

As a result of Lemma 4 we conclude that also the upper bound for the size of an incomplete DFA for the language $L_1 \rightsquigarrow L_2$ given in Lemma 2 is tight.

Theorem 2. *For languages $L_1, L_2 \subseteq \Sigma^*$ where L_1 is regular,*

$$\mathrm{isc}(L_1 \rightsquigarrow L_2) \leq (\mathrm{isc}(L_1) + 1) \cdot 2^{\mathrm{isc}(L_1)} - (2^{\mathrm{isc}(L_1)-1} + 2).$$

For every $n \geq 4$ there exists a language L_1 over a five-letter alphabet recognized by an incomplete DFA with n states and a singleton language L_2 such that in the above inequality we have an equality.

5   Conclusion and Further Work

We have established tight state complexity bounds for the deletion of an arbitrary language L_2 from a regular language L_1 both in the case where L_1 and $L_1 \rightsquigarrow L_2$ are represented by complete DFAs and when they are represented by incomplete DFAs. Furthermore, in the lower bound construction the deleted language can be chosen to be a singleton set consisting of a string of length one. This result is in some sense the strongest possible because deleting the empty string from L_1 yields just L_1.

Roughly speaking, in the upper bound constructions given in Section 3, the DFA B for $L_1 \rightsquigarrow L_2$ is based only on the DFA A for L_1 and B depends on L_2 only by way of the transitions the strings of L_2 define on A. This causes that the transitions in parts of B that, respectively, simulate the original DFA A and the computation of A after a string was deleted are closely related and, perhaps partly because of this reason, the lower bound constructions that match the upper bound (respectively, for complete and for incomplete DFAs) are fairly involved. For the constructions we used a five-letter alphabet. The alphabet size could likely be reduced, but this would lead to considerably more complicated proofs of correctness. Furthermore, it does not seem clear whether the general upper bound can be reached using a binary alphabet. Note that for a unary alphabet, deletion coincides with right-quotient and the state complexity is known to be n [29].

More general types of deletion operations have been considered within the context of deletion along trajectories [3, 19]. Our "ordinary" deletion operation is defined by the set of trajectories $i^* d^* i^*$ and the left-quotient and right-quotient operations are defined, respectively, by the sets of trajectories $d^* i^*$ and $i^* d^*$. The set of trajectories $d^* i^* d^*$ defines the *bipolar deletion* operation [3, 16, 18]. The language $L_1 \rightsquigarrow_{d^* i^* d^*} L_2$ consists of all strings v such that for some string $u = u_1 u_2 \in L_2$, the string $u_1 v u_2$ is in L_1. From [3, 18] it is known that bipolar deletion preserves regularity but the state complexity bound given by these results is not optimal. Differing from deletion, the state complexity of bipolar deletion would need to depend on the size of DFAs for both of the argument languages.

References

1. Câmpeanu, C., Salomaa, K., Yu, S.: Tight lower bound for the state complexity of shuffle of regular languages. J. Autom. Lang. Comb. 7, 303–310 (2002)
2. Cui, B., Gao, Y., Kari, L., Yu, S.: State complexity of combined operations with two basic operations. Theoret. Comput. Sci. 437, 98–107 (2012)
3. Domaratzki, M.: Deletion along trajectories. Theoret. Comput. Sci. 320, 293–313 (2004)
4. Domaratzki, M., Okhotin, A.: State complexity of power. Theoret. Comput. Sci. 410, 2377–2392 (2009)
5. Domaratzki, M., Salomaa, K.: State complexity of shuffle on trajectories. J. Automata, Languages and Combinatorics 9, 217–232 (2004)
6. Eom, H.-S., Han, Y.-S.: State complexity of combined oeprations for suffix-free regular languages. Theoret. Comput. Sci. 510, 87–93 (2013)

7. Eom, H.-S., Han, Y.-S., Jirásková, G.: State complexity of basic operations on non-returning regular languages. In: Jurgensen, H., Reis, R. (eds.) DCFS 2013. LNCS, vol. 8031, pp. 54–65. Springer, Heidelberg (2013)

8. Eom, H.-S., Han, Y.-S., Salomaa, K.: State complexity of k-union and k-intersection for prefix-free regular languages. In: Jurgensen, H., Reis, R. (eds.) DCFS 2013. LNCS, vol. 8031, pp. 78–89. Springer, Heidelberg (2013)

9. Gao, Y., Kari, L.: State complexity of star of union and square of union on k regular languages. Theoret. Comput. Sci. 499, 38–50 (2013)

10. Gao, Y., Piao, X.: State complexity of insertion, manuscript in preparation (2014)

11. Gao, Y., Yu, S.: State complexity and approximation. Int. J. Found. Comput. Sci. 23(5), 1085–1098 (2012)

12. Han, Y.-S., Salomaa, K.: State complexity of basic operations on suffix-free regular languages. Theoret. Comput. Sci. 410, 2537–2548 (2009)

13. Holzer, M., Kutrib, M.: Descriptional and computational complexity of finite automata — A survey. Inf. Comput. 209, 456–470 (2011)

14. Jirásek, J., Jiráskova, G., Szabari, A.: State complexity of concatenation and complementation. Internat. J. Foundations Comput. Sci. 16, 511–529 (2005)

15. Jirásková, G., Shallit, J.: The state complexity of star-complement-star. In: Yen, H.-C., Ibarra, O.H. (eds.) DLT 2012. LNCS, vol. 7410, pp. 380–391. Springer, Heidelberg (2012)

16. Kari, L.: On language equations with invertible operations. Theoret. Comput. Sci. 132, 129–150 (1994)

17. Kari, L.: On insertion and deletion in formal languages. PhD thesis. University of Turku (1991)

18. Kari, L.: Deletion operations: Closure properties. Internat. J. Comput. Math. 52, 23–42 (1994)

19. Kari, L., Sosik, P.: Aspects of shuffle and deletion on trajectories. Theoret. Comput. Sci. 332, 47–61 (2005)

20. Kutrib, M., Pighizzini, G.: Recent trends in descriptional complexity of formal languages. Bulletin of the EATCS 111, 70–86 (2013)

21. Lupanov, O.B.: A comparison of two types of finite sources. Problemy Kibernetiki 9, 328–335 (1963)

22. Maslov, A.N.: Estimates on the number of states of finite automata. Soviet Math. Dokl. 11, 1373–1375 (1970)

23. Meyer, A.R., Fischer, M.J.: Economy of description by automata, grammars and formal systems. In: Proc. SWAT (FOCS), pp. 188–191. IEEE Computer Society (1971)

24. Moore, F.R.: On the bounds for state-set size in the proofs of equivalence between deterministic, nondeterministic, and two-way finite automata. IEEE Transactions on Computers C-20, 1211–1214 (1971)

25. Rampersad, N.: The state complexity of L^2 and L^k. Inform. Proc. Letters 98, 231–234 (2006)

26. Salomaa, A., Salomaa, K., Yu, S.: State complexity of combined operations. Theoret. Comput. Sci. 383, 140–152 (2007)

27. Shallit, J.: A Second Course in Formal Languages and Automata Theory. Cambridge University Press (2009)

28. Wood, D.: Theory of Computation. John Wiley & Sons, Inc., New York (1987)

29. Yu, S.: Regular languages. In: Rozenberg, G., Salomaa, A. (eds.) Handbook of Formal Languages, vol. I, pp. 41–110. Springer (1997)

30. Yu, S., Zhuang, Q., Salomaa, K.: The state complexities of some basic operations on regular languages. Theoret. Comput. Sci. 125, 315–328 (1994)

Semisimple Synchronizing Automata and the Wedderburn-Artin Theory

Jorge Almeida and Emanuele Rodaro

Centro de Matemática, Faculdade de Ciências
Universidade do Porto, 4169-007 Porto, Portugal
{jalmeida,emanuele.rodaro}@fc.up.pt

Abstract. We approach Černý's conjecture using the Wedderburn-Artin theory. We first introduce the radical ideal of a synchronizing automaton, and then the natural notion of semisimple synchronizing automata. This is a rather broad class since it contains simple synchronizing automata like those in Černý's series. Furthermore, semisimplicity gives the advantage of "factorizing" the problem of finding a synchronizing word into the sub-problems of finding words that are zeros in the projections into the simple components in the Wedderburn-Artin decomposition. This situation is applied to prove that Černý's conjecture holds for the class of strongly semisimple synchronizing automata. These are automata whose sets of synchronizing words are cyclic ideals, or equivalently are ideal regular languages which are closed by takings roots.

1 Introduction

A deterministic finite automaton (DFA) $\mathscr{A} = \langle Q, \Sigma, \delta \rangle$ is called synchronizing if there exists a word $w \in \Sigma^*$ "sending" all the states into a single state, i.e. $\delta(q, w) = \delta(q', w)$ for all $q, q' \in Q$. Any such word is said to be synchronizing (or reset) for the DFA $\mathscr{A}$. These automata have been widely studied since the work of Černý in 1964 [8] and his well known conjecture regarding an upper bound for the length of the shortest reset word. This conjecture states that any synchronizing automaton $\mathscr{A}$ with n states admits at least a reset word w with $|w| \le (n-1)^2$. In [8] it is shown that this bound is tight by exhibiting an infinite series of synchronizing automata $\mathscr{C}_n$ having a shortest synchronizing word of length $(n-1)^2$. For more information on synchronizing automata we refer the reader to Volkov's survey [20]. In this paper we follow a representation theoretic approach to the Černý conjecture and synchronizing automata initially pursued in [1,2,4,18]. Our approach has a more ring theoretic flavor making use of the well-known Wedderburn-Artin theory of semisimple rings.

The paper is organized as follows. In Section 2 we introduce the notion of radical of a synchronizing automaton. In Section 3 we characterize this ideal and we introduce the natural notion of semisimple synchronizing automaton. We show that synchronizing automata which are simple, or equivalently they do not have non-trivial congruences, are also semisimple. Finally we exhibit some classes of semisimple and simple synchronizing automata. Section 4 shows how

A.M. Shur and M.V. Volkov (Eds.): DLT 2014, LNCS 8633, pp. 49–60, 2014.
© Springer International Publishing Switzerland 2014

the Wedderburn-Artin Theory can be used to factorize the problem of finding
a synchronizing words into the sub-problem of finding words which are zero in
the projections into the simple components. This approach works in case the
0-minimal ideals in the factor monoids do not contain 0-$\mathcal{H}$-classes. In Section 5
we introduce cyclic ideal languages and characterize them. These are in particular cyclic languages as introduced in [6]. Finally, using the results of Section
4, we show that Černý's conjecture holds for a particular class of semisimple
synchronizing automata: the strongly semisimple automata. This is the class of
synchronizing automata whose set of reset words is a cyclic ideal language.

2   The Radical of a Synchronizing Automaton

In this section we fix some notations used throughout the paper, and we introduce the central notion of radical of a synchronizing automaton. Hence-
forth, we consider a synchronizing automaton $\mathscr{A} = \langle Q, \Sigma, \delta \rangle$ with set of states
$Q = \{q_1, \ldots, q_n\}$, and by $\mathcal{S}$ we denote the set of the synchronizing (reset) words
of $\mathscr{A}$. It is an easy exercise to check that the set $\mathcal{S}$ is a regular language which is a
two-sided ideal of Σ^*, i.e. $\Sigma^* \mathcal{S} \Sigma^* \subseteq \mathcal{S}$. Let $\mathrm{M}(\mathscr{A})$ be the transition monoid of $\mathscr{A}$
and let $\pi : \Sigma^* \to \mathrm{M}(\mathscr{A})$ be the natural epimorphism and put $\mathscr{A}^* = \mathrm{M}(\mathscr{A})/\pi(\mathcal{S})$.
There is a natural action of $\mathrm{M}(\mathscr{A})$ on the set Q given by $q \cdot \pi(u) = \delta(q, u)$; we
often omit the map π and we use the simpler notation $q \cdot u$. This action is extended to subsets of Q in the obvious way. It is a well known fact that $\mathrm{M}(\mathscr{A})$
embeds into the ring $\mathbb{M}_n(\mathbb{C})$ of $n \times n$ matrices with entries in $\mathbb{C}$ and with a slight
abuse of notation we still denote by $\pi : \Sigma^* \to \mathbb{M}_n(\mathbb{C})$ the representation induced
by this embedding. This representation induces an action of Σ^* on the vector
space $\mathbb{C}Q$ defined by $v \cdot u = v\pi(u)$. Consider the vector $Q = q_1 + \ldots + q_n$ formed
by summing all the elements of the canonical basis. Using the fact that the $\pi(a)$,
$a \in \Sigma$, are functions it is not difficult to see that Σ^* acts on the orthogonal
space $Q^\perp = \{u \in \mathbb{C}Q : \langle u|Q \rangle = 0\}$ where $\langle \cdot|\cdot \rangle$ is the usual scalar product (see for
instance [4]). This fact can be easily verified on the base of $Q^\perp$ formed by the
vectors $q_1 - q_i$ for $i = 2, \ldots, n$. Furthermore, it is an easy exercise to check that
$u \in \mathcal{S}$ if and only if for every $v \in Q^\perp$ we get $v \cdot u = 0$. This induces a representa-
tion $\varphi : \Sigma^*/\mathcal{S} \to \mathrm{End}(Q^\perp) \simeq \mathbb{M}_{n-1}(\mathbb{C})$ with $\varphi(\Sigma^*/\mathcal{S}) \simeq \mathscr{A}^*$. Therefore, we see
$\mathscr{A}^*$ as a finite multiplicative submonoid of $\mathbb{M}_{n-1}(\mathbb{C})$. Let $\mathcal{R}$ be the $\mathbb{C}$-subalgebra
of $\mathbb{M}_{n-1}(\mathbb{C})$ generated by $\mathscr{A}^*$. Clearly $\mathcal{R}$ is a finitely generated $\mathbb{C}$-algebra called
the *synchronized $\mathbb{C}$-algebra associated to* the synchronizing DFA $\mathscr{A}$ where $\mathscr{A}^*$
embeds into $\mathcal{R}$, and with a slight abuse of notation we identify $\mathscr{A}^*$ with the
image of this embedding $\mathscr{A}^* \hookrightarrow \mathcal{R}$. Therefore, we define the *radical* $\mathrm{Rad}(\mathscr{A}^*)$ of
$\mathscr{A}^*$ as the restriction of the radical $\mathrm{Rad}(\mathcal{R})$ (see [10]) of the $\mathbb{C}$-subalgebra $\mathcal{R}$ to
$\mathscr{A}^*$, i.e. $\mathrm{Rad}(\mathscr{A}^*) = \mathrm{Rad}(\mathcal{R}) \cap \mathscr{A}^*$.

Let $\theta : \Sigma^* \to \Sigma^*/\mathcal{S}$ be the Rees morphism. Throughout the paper we consider
the morphism $\rho : \Sigma^* \to \mathscr{A}^*$ defined by $\rho = \varphi \circ \theta$. Since $\mathrm{Rad}(\mathcal{R})$ is an ideal of $\mathcal{R}$
[10, Corollary 4.2] we have that $\mathrm{Rad}(\mathscr{A}^*)$ is also an ideal of the (finite) monoid
$\mathscr{A}^*$. We have the following definition of radical of a synchronizing automaton.

Definition 1. *The set* $\mathrm{Rad}(\mathscr{A}) = \rho^{-1}(\mathrm{Rad}(\mathscr{A}^*)) \subseteq \Sigma^*$ *is a two-sided ideal which is clearly a regular language called the radical of the synchronizing automaton* $\mathscr{A}$.

Note that $\mathcal{S} \subseteq \mathrm{Rad}(\mathscr{A})$. The elements of $\mathrm{Rad}(\mathscr{A})$ are called the *radical words* of $\mathscr{A}$.

3 Semisimple Synchronizing Automata

With the definition of radical of Section 2 it is natural to call a synchronizing DFA $\mathscr{A}$ *semisimple* whenever $\mathrm{Rad}(\mathscr{A}^*) = \{0\}$. Note that $\mathscr{A}$ is semisimple if and only if $\mathrm{Rad}(\mathscr{A}) = \mathcal{S}$. This last fact shows that the search for synchronizing words in a semisimple synchronizing automaton is reduced to the search of words $u \in \Sigma^*$ for which $\psi(\rho(u)) = 0$, where $\psi : \mathcal{R} \to \overline{\mathcal{R}} := \mathcal{R}/\mathrm{Rad}(\mathcal{R})$ is the canonical epimorphism. Besides the semisimple case, in general radical words can still give some information on synchronizing words. We now make some general considerations on $\mathrm{Rad}(\mathscr{A})$. We recall that an ideal I in a monoid with zero (or in a ring) is *nilpotent* whenever there is an integer m such that $I^m = 0$. The following proposition characterizes the radical words of a synchronizing automaton.

Proposition 1. $\mathrm{Rad}(\mathscr{A})$ *is an ideal containing* $\mathcal{S}$, *moreover* $\mathrm{Rad}(\mathscr{A})/\mathcal{S}$ *is the largest nilpotent left (right) ideal of* $\Sigma^*/\mathcal{S}$.

Proof. Since $\mathcal{R}$ is both noetherian and artinian, by [10, Theorem 4.12] $\mathrm{Rad}(\mathcal{R})$ is the largest nilpotent left (right) ideal of $\mathcal{R}$. We claim that $\mathrm{Rad}(\mathscr{A}^*)$ is the largest nilpotent left (right) ideal of $\mathscr{A}^*$. Indeed, assume that H is a nilpotent left ideal of $\mathscr{A}^*$, and let $\mathcal{H}$ be the $\mathbb{C}$-algebra generated by H. Since $\mathcal{R}$ is generated by $\mathscr{A}^*$, then $\mathcal{H}$ is also a left (right) ideal of $\mathcal{R}$. Moreover it is nilpotent: if $H^m = 0$, then it is straightforward to check that also $\mathcal{H}^m = 0$. Thus, $\mathcal{H} \subseteq \mathrm{Rad}(\mathcal{R})$ and so in particular we have $H \subseteq \mathcal{H} \cap \mathscr{A}^* \subseteq \mathrm{Rad}(\mathcal{R}) \cap \mathscr{A}^* = \mathrm{Rad}(\mathscr{A}^*)$. If $\varphi : \Sigma^*/\mathcal{S} \to \mathscr{A}^*$ is the representation map, then it is routine to check that $\mathrm{Rad}(\mathscr{A})/\mathcal{S} = \varphi^{-1}(\mathrm{Rad}(\mathscr{A}^*))$ is also the largest nilpotent left (right) ideal of $\Sigma^*/\mathcal{S}$. $\square$

From this proposition it is evident that if one is able to find a radical word u, then a synchronizing word can be obtained by considering a suitable power of u. Therefore, for $u \in \mathrm{Rad}(\mathscr{A})$ it is important defining the index $\mathrm{Depth}(u)$ as the smallest positive integer $n \geq 1$ such that $u^n \in \mathcal{S}$. We can extend this parameter to the whole automaton by putting $\mathrm{Depth}(\mathscr{A}) = \min\{m : \mathrm{Rad}(\mathscr{A})^m = \mathcal{S}\}$. It is obvious that $\mathrm{Depth}(u) \leq \mathrm{Depth}(\mathscr{A})$ and the study of bounds for such quantities can be interesting in finding bounds for short synchronizing word assuming that finding a short radical word can be easier. In this way we can split the task of finding bounds for the shortest synchronizing words into the problem of bounding the shortest radical word u and to bound one of the quantities $\mathrm{Depth}(\mathscr{A})$, $\mathrm{Depth}(u)$. Note that $\mathrm{Depth}(\mathscr{A}) = 1$ iff $\mathscr{A}$ is semisimple.

We now frame the combinatorial class of simple synchronizing automata into the class of semisimple synchronizing automata. Given an automaton $\mathscr{A} = \langle Q, \Sigma, \delta \rangle$, we recall that an equivalence relation σ on the set of states Q is an (automaton) *congruence* if $x\sigma y$ implies $(x \cdot u)\sigma(y \cdot u)$ for any $u \in \Sigma^*$. We denote the lattice of congruences of $\mathscr{A}$ by $\mathrm{Cong}(\mathscr{A})$ with maximum given by the universal relation $\omega_{\mathscr{A}}$ and minimum the identity relation $1_{\mathscr{A}}$. For a given equivalence relation σ we denote by $\mathrm{Cong}_\sigma(\mathscr{A})$ the sub-lattice of the congruences of $\mathscr{A}$ contained in σ. We have the following lemma.

Lemma 1. *Let $\mathscr{A} = \langle Q, \Sigma, \delta \rangle$ be a synchronizing automaton which is not semisimple. Let $g \in \mathrm{Rad}(\mathscr{A}) \setminus \mathcal{S}$ and let $\mathrm{Ker}(g)$ be the kernel of the transformation induced by the word g. Then*

$$\mathrm{Cong}_{\mathrm{Ker}(g)}(\mathscr{A}) \neq \{1_{\mathscr{A}}\}$$

Proof. We have to show that $\mathrm{Cong}_{\mathrm{Ker}(g)}(\mathscr{A})$ is non-trivial. Since $g \in \mathrm{Rad}(\mathscr{A}) \setminus \mathcal{S}$, by Proposition 1 there is a minimum integer $m > 1$ such that:

$$\underbrace{g\Sigma^* g\Sigma^* g \ldots \Sigma^* g}_{m\text{--times}} = 0 \text{ in } \Sigma^*/\mathcal{S} \tag{1}$$

By the minimality of m there are words $u_1, \ldots u_{m-2} \in \Sigma^*$ such that

$$gu_1gu_2 \ldots u_{m-2}g$$

is not synchronizing. Using the usual characterization of synchronizing automata this is equivalent to saying that there are two different states $q_1, q_2 \in Q$ such that

$$\bar{q}_1 = q_1 \cdot (gu_1gu_2 \ldots u_{m-2}g) \neq q_2 \cdot (gu_1gu_2 \ldots u_{m-2}g) = \bar{q}_2$$

Define the relation σ by putting $x\sigma y$ if there is a $u \in \Sigma^*$ such that $\{\bar{q}_1, \bar{q}_2\} \cdot u = \{x, y\}$. It is evident that σ is symmetric. Let σ^t be the reflexive and transitive closure of σ. This is an equivalence relation which is also a congruence, in fact it is the smallest congruence generated by identifying $\bar{q}_1$ with $\bar{q}_2$. We claim that $\sigma^t \subseteq \mathrm{Ker}(g)$ or equivalently $|[q]_{\sigma^t} \cdot g| = 1$ for any $q \in Q$, where $[q]_{\sigma^t}$ is the equivalence class of σ^t containing q. Indeed, assume, contrary to our claim, that there is some non-trivial class $[q]_{\sigma^t}$ such that $|[q]_{\sigma^t} \cdot g| > 1$. Thus, there are two distinct states $h, h' \in [q]_{\sigma^t} \cdot g$ and a sequence $x_1, \ldots, x_n$ of states of $[q]_{\sigma^t}$ such that $x_i\sigma x_{i+1}$ for $i = 1, \ldots, n-1$ and $x_1 \cdot g = h$, $x_n \cdot g = h'$. It is straightforward to check that this implies the existence of an index $i \in \{1, \ldots, n-1\}$ such that $x_i \cdot g \neq x_{i+1} \cdot g$. Hence there is some word $u \in \Sigma^*$ such that $\{\bar{q}_1, \bar{q}_2\} \cdot u = \{x_i, x_{i+1}\}$ which implies

$$q_1 \cdot (gu_1gu_2 \ldots u_{m-2}gug) \neq q_2 \cdot (gu_1gu_2 \ldots u_{m-2}gug)$$

which contradicts (1). Hence $|[q]_{\sigma^t} \cdot g| = 1$ for every $q \in Q$, and $\sigma^t \subseteq \mathrm{Ker}(g)$. □

As an immediate consequence we have the following corollary.

Corollary 1. *Let $\mathscr{A}$ be a synchronizing automaton which is not semisimple, then $\mathrm{Cong}(\mathscr{A}) \neq \{\omega_{\mathscr{A}}, 1_{\mathscr{A}}\}$.*

Proof. Since $\mathscr{A}$ is not semisimple, we take any $g \in \mathrm{Rad}(\mathscr{A}) \setminus \mathcal{S}$. Then by Lemma 1 there is a congruence $\sigma \in \mathrm{Cong}_{\mathrm{Ker}(g)}(\mathscr{A})$ with $\sigma \neq 1_{\mathscr{A}}$. Moreover, since g is not synchronizing $\mathrm{Ker}(g) \neq \omega_{\mathscr{A}}$, whence $\sigma \neq \omega_{\mathscr{A}}$ as well. □

We recall that an automaton $\mathscr{A}$ is called *simple* whenever $\mathrm{Cong}(\mathscr{A}) = \{\omega_{\mathscr{A}}, 1_{\mathscr{A}}\}$ (see [5,19]). Therefore, by the previous lemma we have the following immediate theorem:

Theorem 1. *A synchronizing simple automaton is also semisimple.*

Using Corollary 1 we can find another class of semisimple automata as the following proposition shows.

Proposition 2. *Let $\mathscr{A} = \langle Q, \Sigma, \delta \rangle$ be a synchronizing automaton with $|Q|$ prime and having a subset $P \subseteq \Sigma$ such that P^* acts transitively on Q like a permutation group. Then $\mathscr{A} = \langle Q, \Sigma, \delta \rangle$ is semisimple.*

Proof. If $\mathscr{A}$ is not semisimple, then by Proposition 1 there is a congruence $\sigma \in \mathrm{Cong}(\mathscr{A})$ with $\sigma \neq 1_{\mathscr{A}}, \omega_{\mathscr{A}}$. Thus there is a class $[q]_{\sigma}$ of Q/σ with $1 < |[q]_{\sigma}| < |Q|$. Since σ is a congruence, and P^* acts like a permutation group transitively on Q, then $|[q']_{\sigma} \cdot u| = |[q']_{\sigma}|$ for any $q' \in Q$ and $u \in P^*$. Thus by the transitivity we can factorize $|Q| = |Q/\sigma| |[q]_{\sigma}|$ with $1 < |[q]_{\sigma}| < |Q|$, a contradiction. Hence $\mathscr{A}$ is semisimple. □

The previous proposition holds in the more general context of groups acting on a set that are primitive. We recall that a group G acting on a set Q is called *primitive* whenever there are no non-trivial equivalence relations on Q preserved by G. Thus, any automaton having a subset $P \subseteq \Sigma$ acting primitively on Q is simple and so semisimple.

Another example of simple, and therefore semisimple, automaton is given by the well known series of Černý. We recall that this series is formed by the automatata (see Fig. 1) $\mathscr{C}_n = \langle \{0, \ldots, n-1\}, \{a, b\}, \delta \rangle$ where $\delta(i, a) = i + 1$ mod n, $\delta(i, b) = i$ for all $0 \leq i \leq n - 2$ and $\delta(n-1, b) = 0$. We have the following proposition.

Proposition 3. *For all $n \geq 1$ the automata $\mathscr{C}_n$ are simple.*

Proof. Any non-trivial congruence ρ on the set of states $\mathbb{Z}_n = \{0, \ldots, n-1\}$ is a non-trivial congruence for the (regular) action of the group $\mathbb{Z}_n$ on itself. Hence, the congruence classes are the cosets of a non-trivial subgroup H of $\mathbb{Z}_n$. We claim that $H = \mathbb{Z}_n$. Indeed, assume, contrary to the claim, that H is a proper subgroup of $\mathbb{Z}_n$. It is not difficult to check that $n - 1 \notin H$, and so $n - 1$ and 0 belong to different classes. Since H is proper, the class $[n-1]_{\rho}$ contains also an element $i \neq 0, n - 1$. Since $0 \cdot b = (n - 1) \cdot b = 0$ and $i \cdot b = i$ we get $[0]_{\rho} = [n - 1]_{\rho} \cdot b = [i]_{\rho} \cdot b = [n - 1]_{\rho}$, a contradiction. Whence, $H = \mathbb{Z}_n$ and $\rho = \omega_{\mathscr{A}}$. □

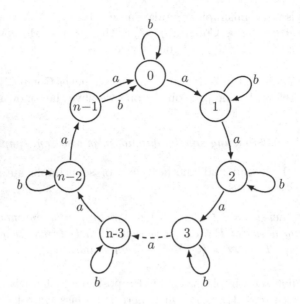

Fig. 1. Černý's series $\mathscr{C}_n$

4   Factoring the Problem via the Wedderburn-Artin Theorem

Since $\overline{\mathcal{R}} = \mathcal{R}/\operatorname{Rad}(\mathcal{R})$ is left artinian and $\operatorname{Rad}(\overline{\mathcal{R}}) = \{0\}$, by [10, Theorem 4.14] $\overline{\mathcal{R}}$ is actually semisimple. Therefore, the Wedderburn-Artin Theorem [10, Theorem 3.5] gives us the advantage of having the following nice decomposition of $\overline{\mathcal{R}}$:

$$\overline{\mathcal{R}} \simeq \mathbb{M}_{n_1}(D_1) \times \ldots \times \mathbb{M}_{n_k}(D_k)$$

for some (uniquely determined) positive integers $n_1, \ldots, n_k$, and $D_1, \ldots, D_k$ (finite dimensional) $\mathbb{C}$-division algebras. Since $\mathbb{C}$ is algebraically closed field we have that actually

$$D_1 = \ldots = D_k = \mathbb{C}$$

for, if $d \in D_i$, then $\mathbb{C}[d]$ is a finite field extension of $\mathbb{C}$, whence $d \in \mathbb{C}$. Let $\varphi_i : \overline{\mathcal{R}} \to \mathbb{M}_{n_i}(\mathbb{C})$ for $i = 1, \ldots, k$ be the projection map onto the i-th simple component. We recall that $\psi : \mathcal{R} \to \overline{\mathcal{R}}$ is the canonical epimorphism, and ρ is composition of the Rees morphism $\theta : \Sigma^* \to \Sigma^*/\mathcal{S}$ with the representation φ. Henceforth, we consider the morphism $\overline{\varphi}_i : \Sigma^* \to \mathbb{M}_{n_i}(\mathbb{C})$ for $i = 1, \ldots, k$ defined by:

$$\overline{\varphi}_i = \varphi_i \circ \psi \circ \rho$$

and let $\mathcal{M}_i = \overline{\varphi}_i(\Sigma^*)$ be the subsemigroup of $\mathbb{M}_{n_i}(\mathbb{C})$ generated by Σ^*, for $i = 1, \ldots, k$. We call $\mathcal{M}_i$ the i-th factor monoid. Thus, in case $\mathscr{A}$ is semisimple we factorize the problem of finding a synchronizing word into the problems of finding words u_i, $i = 1, \ldots, k$ with $\overline{\varphi}_i(u_i) = 0$. Indeed, a synchronizing word can be obtained by the concatenation of these words u_i, $i = 1, \ldots, k$. We have the following lemma.

Lemma 2. *The i-th factor monoid $\mathcal{M}_i$ has a unique 0-minimal ideal $\mathcal{I}_i$ which is a 0-simple semigroup. Furthermore, $\mathcal{M}_i$ acts faithfully on the right and left of $\mathcal{I}_i$.*

Proof. Note that $\mathcal{M}_i$ being the image of a finite semigroup is a finite semigroup with 0. Therefore there is a 0-minimal ideal $\mathcal{I}_i$. By [9, Proposition 3.1.3] either $\mathcal{I}_i^2 = 0$ or $\mathcal{I}_i$ is a 0-simple semigroup. Since $\mathbb{M}_{n_i}(\mathbb{C})$ is a simple ring, for any $r \in \mathcal{I}_i$ we have:

$$\mathbb{M}_{n_i}(\mathbb{C}) r \mathbb{M}_{n_i}(\mathbb{C}) = \mathbb{M}_{n_i}(\mathbb{C})$$

Therefore, if 1_i is the unit of $\mathbb{M}_{n_i}(\mathbb{C})$, since $\mathcal{I}_i$ is an ideal we have $\sum_j \lambda_j r_j = 1_i$ for some $r_j \in \mathcal{I}_i$, $\lambda_i \in \mathbb{C}$. Suppose that $\mathcal{I}_i^2 = 0$ and take any non-zero element $s \in \mathcal{I}_i$. Hence $\sum_j \lambda_j r_j s = s$ and since $r_j s \in \mathcal{I}_i^2 = 0$, we have $s = 0$, a contradiction. Therefore $\mathcal{I}_i$ is a 0-simple semigroup. Since $\sum_j \lambda_j r_j = 1_i$ it is also immediate to see that $\mathcal{M}_i$ acts faithfully on both the right and left of $\mathcal{I}_i$.

The proof of the uniqueness follows the same argument. Indeed, assume, contrary to our statement, that $\mathcal{I}_i'$ is another 0-minimal ideal, then for all $b \in \mathcal{I}_i'$ we have $\sum_j \lambda_j r_j b = b$. Since $r_j b \in \mathcal{I}_i' \cap \mathcal{I}_i = \{0\}$, then $b = 0$ for all $b \in \mathcal{I}_i'$, a contradiction. $\square$

We recall that given $u \in \Sigma^*$, the *rank* of u is the cardinality of the rank of the function associated to u, equivalently $\mathrm{rk}(u) = |Q \cdot u|$. By the usual laws of composition of functions, it is straightforward to check that the following holds:

$$\mathrm{rk}(uv) \leq \min\{\mathrm{rk}(u), \mathrm{rk}(v)\}, \quad \forall u, v \in \Sigma^* \tag{2}$$

The following definition extends this index to elements of $\mathcal{I}_i \setminus \{0\}$.

Definition 2. *For any $g \in \mathcal{I}_i \setminus \{0\}$ the i-th rank of g is given by:*

$$\mathrm{Rk}_i(g) = \min\{\mathrm{rk}(u) : \overline{\varphi}_i(u) = g\}$$

The rank of $\mathcal{I}_i$ is defined as:

$$\mathrm{Rk}_i(\mathcal{I}_i) = \min\{\mathrm{Rk}_i(g) : g \in \mathcal{I}_i \setminus \{0\}\}$$

Note that for a non-zero element g, we have $\mathrm{Rk}_i(g) > 1$. Indeed if for some $u \in \Sigma^*$ with $\overline{\varphi}_i(u) = g$, we have $\delta(Q, u) = 1$, then $u \in \mathcal{S}$, whence $\overline{\varphi}_i(u) = 0$. Consequently we have $\mathrm{Rk}_i(\mathcal{I}_i) > 1$. The following lemma shows that this index is equal for any element of $\mathcal{I}_i \setminus \{0\}$.

Lemma 3. *For any $g \in \mathcal{I}_i \setminus \{0\}$ we have $\mathrm{Rk}_i(g) = \mathrm{Rk}_i(\mathcal{I}_i)$.*

Proof. Let $g \in \mathcal{I}_i \setminus \{0\}$, by the definition we have $\mathrm{Rk}_i(g) \geq \mathrm{Rk}_i(\mathcal{I}_i)$. On the other hand, let $s \in \mathcal{I}_i \setminus \{0\}$ with $\mathrm{Rk}_i(s) = \mathrm{Rk}_i(\mathcal{I}_i)$, and let $u \in \Sigma^*$ such that

$\overline{\varphi}_i(u) = s$, $\mathrm{Rk}_i(s) = \mathrm{rk}(u)$. Since by Lemma 2 $\mathcal{I}_i$ is 0-simple, $\mathcal{I}_i \setminus \{0\}$ is a $\mathcal{J}$-class. Thus, there are $x, y \in \Sigma^*$ such that $g = \overline{\varphi}_i(xuy)$. Hence, by (2) we get

$$\mathrm{Rk}_i(g) \leq \mathrm{rk}(xuy) \leq \mathrm{rk}(u) = \mathrm{Rk}_i(s) = \mathrm{Rk}_i(\mathcal{I}_i)$$

from which it follows the statement $\mathrm{Rk}_i(g) = \mathrm{Rk}_i(\mathcal{I}_i)$.   □

We recall that *the deficiency* of a word $w \in \Sigma^*$ with respect to $\mathscr{A}$ is the (positive) integer $\mathrm{df}(w) = |Q| - |Q \cdot w|$. We make use of the following result which is a consequence of [11, Corollary 3.4].

Proposition 4. *Given a synchronizing automaton* $\mathscr{A} = \langle Q, \Sigma, \delta \rangle$ *and the words* $v', v \in \Sigma^+$ *such that* $\mathrm{df}(v') = \mathrm{df}(v) = k > 1$, *there exists a word* $u \in \Sigma^*$, *with* $|u| \leq k + 1$, *such that* $\mathrm{df}(v'uv) > k$.

An interesting situation, similarly to the case explored in [1,2], occurs when $\mathcal{I}_i \setminus \{0\}$ is a semigroup.

Lemma 4. *If* $\mathcal{I}_i \setminus \{0\}$ *is a semigroup, then for any* $g \in \mathcal{I}_i \setminus \{0\}$ *there is a word* $u \in \Sigma^*$ *with* $|u| \leq n - \mathrm{Rk}_i(\mathcal{I}_i) + 1$ *such that* $g\overline{\varphi}_i(u) = 0$.

We have the following proposition.

Proposition 5. *If* $\mathcal{I}_i \setminus \{0\}$ *is a semigroup, then there is a word* $u \in \Sigma^*$ *with* $|u| \leq n_i(n-1)$ *with* $\overline{\varphi}_i(u) = 0$.

We have the following theorem.

Theorem 2. *With the above notation, consider an ideal* I *of* $\overline{\mathcal{R}}$ *of the form*

$$I = \mathbb{M}_{n_{i_1}}(\mathbb{C}) \times \ldots \times \mathbb{M}_{n_{i_m}}(\mathbb{C})$$

for some choices $i_1, \ldots, i_m$ *of* $\{1, \ldots, k\}$, *such that the associated 0-minimal ideals* $\mathcal{I}_{i_j} \setminus \{0\}$, $j = 1, \ldots, m$, *are all semigroups. Let* $J = \psi^{-1}(I)$, *then there is a word* $u \in \Sigma^*$ *with* $|u| \leq (n-1)^2$ *such that*

$$\rho(u)J = 0$$

5   Černý's Conjecture for Strongly Semisimple Synchronizing Automata

In this section we use Theorem 2 to prove that Černý's conjecture holds for a particular class of semisimple synchronizing automata. Note that if $\mathcal{S}$ is closed under taking roots, then from Section 3 and Proposition 1 we have that $\mathrm{Rad}(\mathscr{A}) = \mathcal{S}$. This condition can be expressed using the *root operator* on a regular language. For any regular language L on an alphabet Σ, this operator is defined by:

$$\mathrm{root}(L) = \{u \in \Sigma^* : \exists m \geq 1 \text{ such that } u^m \in L\}$$

This is an operator that returns a regular language (see for instance [17]). Henceforth, we call an *ideal language* any regular language $I \subseteq \Sigma^*$ which is also a two-sided ideal of Σ^*. We say that a synchronizing automaton $\mathscr{A}$ with the ideal language of the reset words S is *strongly semisimple* if root$(S) = S$. The approach of studying Černý's conjecture from the language theoretic point of view of the ideal language of the synchronizing words is relatively recent and can be drawn back to [13]. In general, given an ideal language I it is easy to see that the minimal DFA recognizing I is actually a synchronizing automaton whose language of reset words is exactly I. However, this automaton has a sink state, and Černý's conjecture has been verified for such automata [16]. On the other hand, it is well known that if the Černý conjecture is solved for the class of strongly connected synchronizing automaton, then this conjecture holds in general. Thus, the approach of studying synchronizing automata via their languages of synchronizing words is supported by the following result presented in a first version in [14] and then improved in [15] which we partially report here in the following theorem.

Theorem 3. *Let I be an ideal language on a non-unary alphabet, then there is a strongly connected synchronizing automaton having I as the set of synchronizing words.*

We now characterize the ideal languages I satisfying root$(I) = I$ but first we need some definitions. We recall that for a regular language $L \subseteq \Sigma^*$, the language $\sqrt{L} = \{u \in \Sigma^* : u^2 \in L\}$ is also regular. Given two words $x, y \in \Sigma^*$ with $x = vx'$ and $y = y'v$, where $v \in \Sigma^*$ is the maximal prefix of x which is also a suffix of y, we define the *concatenation with overlap* as $y \circ x = y'vx'$. An ideal language $I \subseteq \Sigma^*$ is called a *cyclic ideal language* whenever root$(I) = I$. We have the following characterization.

Proposition 6. *Given an ideal language I, the following are equivalent.*

  i) $\sqrt{I} \subseteq I$;
  ii) for any $u \in I$ and any factorization $u = xy$, for some $x, y \in \Sigma^*$, then
      $y \circ x \in I$;
 iii) root$(I) = I$;
 iv) $I = \eta^{-1}(0)$ where $\eta : \Sigma^* \to S$ is a morphism onto a finite monoid with 0
      satisfying the condition $x^2 = 0 \Rightarrow x = 0$.

Proof. i) $\Rightarrow$ *ii)*. Assume $\sqrt{I} \subseteq I$ and let $u \in I$ with $u = xy$. Suppose that $x = vx'$ and $y = y'v$ for some $v \in \Sigma^*$ which is the maximal prefix of x which is also a suffix of y. Let $h = y \circ x = y'vx'$. Since I is an ideal we have $h^2 = y'vx'y'vx' = y'ux' \in I$. Hence, since $\sqrt{I} \subseteq I$, we deduce $h \in I$.

   ii) $\Rightarrow$ *iii)*. Assume I closed by concatenation with overlap. Since $I \subseteq$ root(I), we have to prove the other inclusion. Let $u \in$ root(I) and let $n > 1$ be the integer such that $u^n \in I$. Since $u^n = uu^{n-1}$ and I is closed by concatenation with overlap, then we have $u^{n-1} = u^{n-1} \circ u \in I$. Thus using induction we get $u \in I$.

   iii) $\Rightarrow$ *iv)*. Since I is regular there is a morphism $\chi : \Sigma^* \to T$, for some finite monoid with $I = \chi^{-1}(\chi(I))$. Since $J = \chi(I)$ is a two-sided ideal of T, we can consider the Rees quotient semigroup $S = T/J$. Thus, the morphism

$\eta : \Sigma^* \to S$, which is the composition of χ with the Rees morphism $T \to T/J$, satisfies $\eta^{-1}(0) = I$. Furthermore, if $x^2 = 0$ in T/J, then $x^2 \in J$ in T. Hence, if $u \in \Sigma^*$ such that $\chi(u) = x$, then $u \in \text{root}(I) = I$, and so $\chi(u) \in J$, i.e. $x = 0$ in T/J.

$iv) \Rightarrow i)$. If $u^2 \in I$, then $\eta(u)^2 = 0$, whence $\eta(u) = 0$, i.e. $u \in I$.   □

This proposition also justifies the name cyclic since these are ideal languages that are also cyclic languages in the sense of [6]. Indeed the first condition: $u \in I$ iff $u^n \in I$ is satisfied since I is an ideal and $\text{root}(I) = I$. The second condition $uv \in I$ iff $vu \in I$ is also satisfied. Consider the direct implication and let $uv \in I$. Since I is an ideal, then $v(uv)u \in I$. Hence $vu \in I$ since $\text{root}(I) = I$. It is worth noting that the syntactic algebras of cyclic languages are semisimple [7], and in general languages whose syntactic algebras of their characteristic series are semisimple have been studied recently in [12].

By Proposition 1 and the definition it is clear that a strongly semisimple synchronizing automaton is semisimple, whence $\overline{\mathcal{R}} \simeq \mathcal{R}$. We have the following main theorem.

Theorem 4. *A strongly semisimple synchronizing automaton $\mathscr{A}$ satisfies the Černý conjecture.*

Proof. Keeping the notation of Section 4, let $\mathcal{M}_i$ for $i = 1, \ldots, k$ be the factor monoids. We say that $T \subseteq \{1, \ldots, k\}$ is a core whenever $\overline{\varphi}_i(u) = 0$ for all $i \in T$, implies $\overline{\varphi}_i(u) = 0$ for all $i \in \{1, \ldots, k\}$. Let $C \subseteq \{1, \ldots, k\}$ be a minimal core with respect to inclusion and let

$$I = \prod_{j \in C} \mathbb{M}_{n_j}(\mathbb{C})$$

be the corresponding ideal in $\overline{\mathcal{R}} \simeq \mathcal{R}$. For a fixed $j \in C$, we claim that the associated 0-minimal ideal $\mathcal{I}_j \setminus \{0\}$ is a semigroup. Since C is a minimal core, there is a word $w \in \Sigma^*$ such that $\overline{\varphi}_j(w) \neq 0$ and $\overline{\varphi}_r(w) = 0$ for all $r \in C \setminus \{j\}$. By Lemma 2 $\mathcal{M}_j$ acts faithfully on the right of $\mathcal{I}_j$, whence there is an element $t \in \mathcal{I}_j \setminus \{0\}$ such that:

$$t\overline{\varphi}_j(w) \neq 0. \tag{3}$$

We have two possibilities: either $\overline{\varphi}_j(vwvw) = 0$ or $\overline{\varphi}_j(vwvw) \neq 0$. In the former case $vwvw \in \mathcal{S}$ (being $\mathscr{A}$ semisimple). Since $\mathscr{A}$ is also strongly semisimple we get $vw \in \text{root}(\mathcal{S}) = \mathcal{S}$, which implies $\overline{\varphi}_j(vw) = 0$. However, this contradicts (3). Hence, we must have $\overline{\varphi}_j(vwvw) \neq 0$ which implies by [9, Lemma 3.2.7] that $\overline{\varphi}_j(vw)$ belong to some $\mathcal{H}$-class containing an idempotent e. In particular, since $\mathcal{M}_j$ is finite, we can assume $\overline{\varphi}_j((vw)^m) = e$ for some integer $m \geq 0$. Assume, contrary to our claim, that $\mathcal{I}_j \setminus \{0\}$ is not a semigroup. Hence by [9, Lemma 3.2.7], there is a 0-$\mathcal{H}$-class H of $\mathcal{I}_j \setminus \{0\}$. Let $h \in H$. Suppose that $h \in R_e$ and let $s \in \Sigma^*$ such that $\overline{\varphi}_j(s) = h$. Consider the word $(vw)^m s$. Since e is a left identity we have $\overline{\varphi}_j((vw)^m s) = h \neq 0$. However, $\overline{\varphi}_j(((vw)^m s)^2) = h^2 = 0$, which with $\overline{\varphi}_r(w) = 0$ for all $r \in C \setminus \{j\}$, implies $(vw)^m s \in \text{root}(\mathcal{S}) = \mathcal{S}$, and so $h = 0$, a contradiction. On the other hand, we can assume that $R_e \cap L_h$ is an

$\mathcal{H}$-class containing an idempotent g (otherwise I could have chosen as 0-$\mathcal{H}$-class $H = R_e \cap L_h$). Let $u \in \Sigma^*$ such that $\overline{\varphi}_j(u) = g$. Using the fact that e is a left identity for g and g is a right identity for h we get $\overline{\varphi}_j(s(vw)^m u) = h \neq 0$. However, $\overline{\varphi}_j((s(vw)^m u)^2) = h^2 = 0$ which implies $s(vw)^m u \in \text{root}(\mathcal{S}) = \mathcal{S}$, and so $h = 0$, a contradiction. Therefore $\mathcal{I}_j \setminus \{0\}$ is a semigroup. Applying Theorem 2 to I we deduce that there is a word u with $|u| \leq (n-1)^2$ such that $\overline{\varphi}_i(u) = 0$ for all $i \in C$. Hence, since C is a core, $\overline{\varphi}_i(u) = 0$ for all $i = 1, \ldots k$, i.e. $u \in \text{Rad}(\mathcal{A}) = \mathcal{S}$ since $\mathcal{A}$ is semisimple. $\square$

6   Open Problems

We list some directions of research originated from this paper.

1) The main open problem is certainly to find quadratic upper bounds for the length of the shortest radical word. Having Lemma 4 without the condition that $\mathcal{I}_i \setminus \{0\}$ does not contain 0-$\mathcal{H}$-classes would be a major breakthrough to a solution of the previous point. Actually, without this condition we obtain the following weaker version:

Lemma 5. *For any $g \in \mathcal{I}_i \setminus \{0\}$ there is a word $u \in \Sigma^*$ with $|u| \leq n - \text{Rk}_i(\mathcal{I}_i) + 1$ such that either $g\overline{\varphi}_i(u) = 0$, or $g\overline{\varphi}_i(u)$ is contained in some 0-$\mathcal{H}$-class.*

which may be useful in dealing with the general case.

2) If point 1) is solved, then the next step to find a quadratic upper bound for the length of the shortest synchronizing words in the general case could be to reduce this problem to the case of simple automata, and then use Theorem 1. Another reduction from the semisimple case to the general one could be the study of upper bounds of the two indices $\text{Depth}(u), \text{Depth}(\mathcal{A})$ introduced in Section 1.

3) Proposition 3 shows that Černý's automata are simple. This is probably not a coincidence, and, in general, synchronizing automata which are "difficult" to synchronize (with "long" shortest reset words) may be simple or semisimple. In particular, is it always the case that a circular synchronizing automaton on n states with letters having rank at least $n - 1$ is simple (semisimple)? What about when it is considered one-cluster automata? Some interesting cases in which one can prove (or disprove) simplicity (semisimplicity) could be the series of slowly synchronizing automata found in [3].

Acknowledgements. The authors acknowledge precious suggestions and comments from the anonymous referees. This work has been partially supported by the European Regional Development Fund through the program COMPETE and by the Portuguese Government through the FCT – Fundação para a Ciência e a Tecnologia under the project PEst-C/MAT/UI0144/2013. The second author also acknowledges the support of the FCT project SFRH/BPD/65428/2009.

References

1. Almeida, J., Margolis, S., Steinberg, B., Volkov, M.: Representation theory of finite semigroups, semigroup radicals and formal language theory. Trans. Amer. Math. Soc. 361(3), 1429–1461 (2009)
2. Almeida, J., Steinberg, B.: Matrix Mortality and the Černý-Pin Conjecture. In: Diekert, V., Nowotka, D. (eds.) DLT 2009. LNCS, vol. 5583, pp. 67–80. Springer, Heidelberg (2009)
3. Ananichev, D.S., Volkov, M.V., Gusev, V.V.: Primitive digraphs with large exponents and slowly synchronizing automata. Zap. Nauchn. Sem. S.-Peterburg. Otdel. Mat. Inst. Steklov (POMI) 402, no. Kombinatorika i Teoriya Grafov. IV, pp. 9–39 (2012)
4. Arnold, F., Steinberg, B.: Synchronizing groups and automata. Theoretical Computer Science 359(1-3), 101–110 (2006)
5. Babcsànyi, I.: Automata with Finite Congruence Lattices. Acta Cybernetica 18(1), 155–165 (2007)
6. Béal, M.P., Carton, O., Reutenauer, C.: Cyclic languages and Strongly cyclic languages. In: Puech, C., Reischuk, R. (eds.) STACS 1996. LNCS, vol. 1046, pp. 49–59. Springer, Heidelberg (1996)
7. Berstel, J., Perrin, D., Reutenauer, C.: Codes and Automata. Encyclopedia of Mathematics and its Applications. Cambridge University Press (2009)
8. Černý, J.: Poznámka k homogénnym eksperimentom s konečnými automatami. Mat.-Fyz. Čas. Slovensk. Akad. Vied. 14, 208–216 (1964) (in Slovak)
9. Howie, J.M.: Fundamentals of Semigroup Theory. Clarendon Press, Oxford (1995)
10. Lam, T.Y.: A first course in noncommutative rings. Springer
11. Margolis, S., Pin, J.E., Volkov, M.: Words guaranteeing minimum image. Int. J. Found. Comput. Sci. 15, 259–276 (2004)
12. Perrin, D.: Completely Reducible Sets. Int. J. Alg. Comp. 23(4), 915–942 (2013)
13. Pribavkina, E., Rodaro, E.: Synchronizing automata with finitely many minimal synchronizing words. Information and Computation 209(3), 568–579 (2011)
14. Reis, R., Rodaro, E.: Regular Ideal Languages and Synchronizing Automata. In: Karhumäki, J., Lepistö, A., Zamboni, L. (eds.) WORDS 2013. LNCS, vol. 8079, pp. 205–216. Springer, Heidelberg (2013)
15. Reis, R., Rodaro, E.: Ideal Regular Languages and Strongly Connected Synchronizing Automata (preprint, 2014)
16. Rystov, I.: Reset words for commutative and solvable automata. Theor. Comp. Sci. 172(1-2, 10), 273–279 (1997)
17. Shallit, J.: A Second Course in Formal Languages and Automata Theory. Cambridge University Press
18. Steinberg, B.: The Černý conjecture for one-cluster automata with prime length cycle. Theor. Comp. Sci. 412(39, 9), 5487–5491 (2011)
19. Thierrin, G.: Simple automata. Kybernetika 6(5), 343–350 (1970)
20. Volkov, M.V.: Synchronizing automata and the Černý conjecture. In: Martín-Vide, C., Otto, F., Fernau, H. (eds.) LATA 2008. LNCS, vol. 5196, pp. 11–27. Springer, Heidelberg (2008)

On Two Algorithmic Problems
about Synchronizing Automata
(Short Paper)

Mikhail V. Berlinkov[*]

Institute of Mathematics and Computer Science,
Ural Federal University, 620000 Ekaterinburg, Russia
berlm@mail.ru

Abstract. Under the assumption $P \neq NP$, we prove that two natural problems from the theory of synchronizing automata cannot be solved in polynomial time. The first problem is to decide whether a given reachable partial automaton is synchronizing. The second one is, given an n-state binary complete synchronizing automaton, to compute its reset threshold within performance ratio less than $d \ln(n)$ for a specific constant $d > 0$.

1 Testing for Synchronization

A *deterministic finite automaton* (DFA) $\mathscr{A}$ is a triple $\langle Q, \Sigma, \delta \rangle$ where Q is the state set, Σ is the input alphabet and $\delta : Q \times \Sigma \to Q$ is the *transition function*. If δ is totally defined on $Q \times \Sigma$ then $\mathscr{A}$ is called *complete*, otherwise $\mathscr{A}$ is called *partial*. The transition function can be naturally extended to Σ^* as follows. For every state $q \in Q$ we let $\delta(q, \lambda) = q$ where λ is the empty word, and for every $u \in \Sigma^*$ we inductively define $\delta(q, ua) = \delta(\delta(q, u), a)$ for each $a \in \Sigma$ whenever both $\delta(q, u)$ and $\delta(\delta(q, u), a)$ are defined. Sometimes we simplify the notation by writing $S.w$ instead of $\{\delta(q, w) \mid q \in S\}$ for a subset $S \subseteq Q$ and a word $w \in \Sigma^*$.

A DFA $\mathscr{A} = \langle Q, \Sigma, \delta \rangle$ is called *synchronizing* if there exists a word $w \in \Sigma^*$ such that $|Q.w| = 1$. Notice that here, in contrast to some other versions of synchronizability studied in the realm of partial automata (see e.g. [7]), w is not assumed to be defined at every state. Each word w with this property is said to be a *reset* word for $\mathscr{A}$. The minimum length of reset words is called the *reset threshold* of $\mathscr{A}$ and denoted by $rt(\mathscr{A})$. Analogously, a word w *synchronizes* a subset $S \subseteq Q$ if $|S.w| = 1$.

Recall that a DFA $\mathscr{A} = \langle Q, \Sigma, \delta \rangle$ is called *reachable* if one can choose an *initial* state $q_0 \in Q$ and a set of *final* states $F \subseteq Q$ such that each state $q \in Q$ is accessible from q_0 and co-accessible from F, i.e., there are words $u, v \in \Sigma^*$ such that $q_0.u = q$ and $q.v \in F$. This case is of certain interest due to its applications in DNA-computing, namely, reset words for partial reachable DFAs serve as *constants* for the corresponding *splicing* systems (see e.g. [3]). It is known that the

---

[*] Supported by the Presidential Programme for young researchers, grant MK-3160.2014.1 and by the Russian Foundation for Basic Research, grant 13-01-00852.

A.M. Shur and M.V. Volkov (Eds.): DLT 2014, LNCS 8633, pp. 61–67, 2014.

problem of testing whether or not a given strongly connected partial DFA is synchronizing can be solved in polynomial time (see [11, Algorithm 3]). In contrast, we show here that the problem becomes PSPACE-complete for reachable DFAs.

We use the following results from [12] about subset synchronization in complete strongly connected DFAs.

Theorem 1 ([12, Theorem 7]). *There exists a series $\mathscr{A}_n$ of complete strongly connected 2-letter DFAs such that the minimum length of synchronizing words for a certain set of states in $\mathscr{A}_n$ is of magnitude $2^{\Omega(n)}$.*

Theorem 2 ([12, Theorem 10]). *The following problem is PSPACE-complete: given a complete strongly connected 2-letter DFA $\mathscr{A}$ and a set S of states in $\mathscr{A}$, to decide whether or not S can be synchronized in $\mathscr{A}$.*

Now we reduce subset synchronization in complete strongly connected DFAs to synchronization of partial reachable DFAs.

Lemma 1. *For each complete strongly connected 2-letter DFA $\mathscr{A} = \langle Q, \{a, b\}, \delta \rangle$ and each subset $S \subseteq Q$ one can construct in $O(|Q|)$ time a reachable partial 3-letter DFA $\mathscr{B} = \langle Q', \{a, b, c\}, \delta' \rangle$ with at most $2|Q|$ states with the following properties.*

1. *If $u \in \{a, b\}^*$ synchronizes S in $\mathscr{A}$, then the word cu synchronizes $\mathscr{B}$.*
2. *If $w \in \{a, b, c\}^*$ synchronizes $\mathscr{B}$ then w has a suffix $u \in \{a, b\}^*$ that synchronizes S in $\mathscr{A}$.*
3. *S can be synchronized in $\mathscr{A}$ if and only if $\mathscr{B}$ is synchronizing.*
4. *If S cannot be synchronized in $\mathscr{A}$ by words of length less than R, then the reset threshold of $\mathscr{B}$ is at least R.*

Proof. Denote $|S|$ by k and let $S = \{s_0, s_1, \ldots, s_{k-1}\}$. Take a k-element set $Z = \{z_0, z_1, \ldots, z_{k-1}\}$ and let $Q' = Q \cup Z$. Now define δ' as follows:

$$
\delta'(q, x) = \begin{cases} \delta(q, x) & \text{if } q \in Q,\ x \in \{a, b\}; \\ s_i & \text{if } q = z_i,\ x = c; \\ z_{i+1 \,(\mathrm{mod}\,k)} & \text{if } q = z_i,\ x \in \{a, b\}, \\ \text{undefined} & \text{if } q \in Q,\ x = c. \end{cases}
$$

Since the automaton $\mathscr{A}$ is strongly connected and the letter a acts as a cyclic permutation on Z, the automaton $\mathscr{B} = \langle Q', \{a, b, c\}, \delta' \rangle$ is reachable: one can choose z_0 as its initial state and Q as its set of final states.

Since $Q'.c = S \subseteq Q$, Claim 1 follows from the fact that c is undefined on Q and the letters a, b act on Q in $\mathscr{A}$ and $\mathscr{B}$ in the same way. For Claim 2, let w synchronize $\mathscr{B}$. Since $Z.a = Z.b = Z$ and c is undefined on Q, we conclude that w should be of the form vcu where $u, v \in \{a, b\}^*$. Since $Q'.vc = S$, the word u should synchronize S and Claim 2 follows. Claims 3 and 4 immediately follow from Claims 1 and 2.

Theorem 3. *Testing a given reachable partial 3-letter DFA for synchronization is PSPACE-complete. There is a series of reachable partial 3-letter n-state DFAs with reset threshold of magnitude $2^{\Omega(n)}$.*

Proof. The existence of a series of reachable partial 3-letter n-state DFAs with reset threshold of magnitude $2^{\Omega(n)}$ follows from Theorem 1 and Claim 4 of Lemma 1.

The problem of testing synchronization for reachable partial 3-letter DFAs is PSPACE-hard by Theorem 2 and Claim 3 of Lemma 1. It remains to prove that the problem is in PSPACE. Given a DFA $\langle Q, \Sigma, \delta \rangle$, we can check whether it is synchronizing storing only a current subset P initialized by Q. For this, in an endless loop, we nondeterministically choose a letter $a \in \Sigma$ and let $P := P.a$. If at some step we have $|P| = 1$, we return "yes", otherwise we continue the iteration. Since this algorithm uses only $O(|Q|)$ memory, the problem is in NPSPACE, but NPSPACE = PSPACE by Savitch's theorem [9].

The following lemma relies on the usual technique of encoding letters in states (see e.g. [2]).

Lemma 2. *For each reachable d-letter partial DFA $\mathscr{A} = \langle Q, \Sigma, \delta \rangle$, one can construct in polynomial time a 2-letter reachable partial DFA $\mathscr{B} = \langle Q', \{a, b\}, \delta' \rangle$ such that $\mathscr{A}$ is synchronizing if and only if $\mathscr{B}$ is synchronizing, $|Q'| = d|Q|$ and $rt(\mathscr{A}) \leq rt(\mathscr{B}) \leq d * rt(\mathscr{A}) + 1$ if $\mathscr{A}$ is synchronizing.*

Proof. Let $\Sigma = \{a_1, a_2, \ldots, a_d\}$. We construct $\mathscr{B}$ by letting $Q' = Q \times \Sigma$ and defining $\delta' : Q' \times \{a, b\} \to Q'$ as follows:

$$\delta'((q, a_i), x) = \begin{cases} (q, a_{\min(i+1,d)}) & \text{if } x = a, \\ (\delta(q, a_i), a_1) & \text{if } x = b \text{ and } a_i \text{ is defined on } q, \\ \text{undefined} & \text{if } x = b \text{ and } a_i \text{ is undefined on } q. \end{cases}$$

Thus, the action of a on a state $q' \in Q'$ substitutes an appropriate letter from the alphabet Σ of $\mathscr{A}$ for the second component of q' while the action of b imitates the action of the second component of q' on its first component and resets the second component to a_1.

Given a word $w = a^{i_1} b a^{i_2} b \ldots a^{i_k} b \in \{a, b\}^*$, let

$$r(w) = a^{\min(i_1,d)} b a^{\min(i_2,d)} b \ldots a^{\min(i_k,d)} b.$$

Define the map $f : \{a, b\}^* \mapsto \Sigma^*$ by $f(w) = a_{\min(i_1,d)} a_{\min(i_2,d)} \cdots a_{\min(i_k,d)}$. Given $w \in \{a, b\}^*$, by the definition of f we get that if $f(w)$ resets $\mathscr{A}$ then bw resets $\mathscr{B}$, and if w resets $\mathscr{B}$ then $f^{-1}(r(w))$ resets $\mathscr{A}$. The lemma follows.

As a straightforward corollary of Theorem 3 and Lemma 2 (for $d = 3$) we get the main result of this section.

Corollary 1. *Testing a given reachable partial 2-letter DFA for synchronization is PSPACE-complete. There is a series of reachable partial 2-letter n-state DFAs with reset threshold of magnitude $2^{\Omega(n)}$.*

2   Approximation of Reset Thresholds

In this section we restrict ourself to the case of complete DFAs. For this case, testing for synchronization is polynomial. When a DFA is synchronizing, the next natural problem is to calculate its reset threshold. It is known that a precise calculation of the reset threshold is computationally hard (see, e.g., [4,8]). There are polynomial time algorithms that, given a synchronizing DFA, find a reset word for it, see, e.g., [4]. These algorithms can be used as approximation algorithms for calculating the reset threshold, and it is quite natural to ask how good such a polynomial approximation can be. The quality of an approximation algorithm is measured by its performance ratio, which for our problem can be defined as follows. Let K be a class of synchronizing DFAs. We say that an algorithm M *approximates the reset threshold in K* if, for an arbitrary DFA $\mathscr{A} \in K$, the algorithm calculates an integer $M(\mathscr{A})$ such that $M(\mathscr{A}) \geq rt(\mathscr{A})$. The *performance ratio* of M at $\mathscr{A}$ is $R_M(\mathscr{A}) = \dfrac{M(\mathscr{A})}{rt(\mathscr{A})}$. The author [2] proved that, unless $P = NP$, for no constant r, a polynomial time algorithm can approximate the reset threshold in the class of all 2-letter synchronizing DFAs with performance ratio less than r.

When no polynomial time approximation within a constant factor is possible, the next natural question is whether or not one can approximate within a logarithmic factor. Gerbush and Heeringa [5] conjectured that if $P \neq NP$, then there exists $\alpha > 0$ such that no polynomial time algorithm approximating the reset threshold in the class of all synchronizing DFAs with a fixed number $k > 1$ of input letters achieves the performance ratio $\alpha \log n$ at all n-state DFAs. Using a reduction from the problem Set-Cover and a powerful non-approximation result from [1], Gerbush and Heeringa proved a weaker form of this conjecture when the number of input letters is allowed to grow with the state number.

Here we prove the conjecture from [5] in its full generality, for each fixed size $k > 1$ of the input alphabet. Though we depart from the same reduction from Set-Cover as in [5], we use not only the result from [1], but also some ingredients from its proof, along with an appropriate encoding of letters in states.

Given a universe $\mathcal{U} = \{u_1, \ldots, u_n\}$ and a family of its subsets $\mathcal{S} = \{S_1, \ldots, S_m\} \subseteq 2^{\mathcal{U}}$ such that $\bigcup_{S_j \in \mathcal{S}} S_j = \mathcal{U}$, Set-Cover is the problem of finding there a minimal sub-family $\mathsf{C} \subseteq \mathcal{S}$ that covers the whole universe in the sense that $\bigcup_{S_j \in \mathsf{C}} S_j = \mathcal{U}$. Denote the size of the minimal sub-family by $OPT(\mathcal{U}, \mathcal{S})$. Set-Cover is a classic NP-hard combinatorial optimization problem, and it is known that it can be approximated in polynomial time to within $\ln(n) - \ln(\ln(n)) + \Theta(1)$ (see [6,10]).

The following transparent reduction from Set-Cover is presented in [5]. Given a Set-Cover instance $(\mathcal{U}, \mathcal{S})$, define the automaton

$$\mathscr{A}(\mathcal{U}, \mathcal{S}) = \langle \mathcal{U} \cup \{\hat{q}\}, \Sigma = \{a_1, \ldots a_m\}, \delta \rangle$$

where the transition function is defined as follows:

$$\delta(u, a_i) = \begin{cases} \hat{q}, & u \in S_i \\ u, & u \notin S_i. \end{cases}$$

Remark 1. Let $\mathscr{A} = \langle Q, \Sigma, \delta \rangle$ be the automaton defined by $(\mathcal{U}, \mathcal{S})$ as above. Then $rt(\mathscr{A}) = OPT(\mathcal{U}, \mathcal{S})$, $|Q| = |\mathcal{U}| + 1$, $|\Sigma| = |\mathcal{S}|$.

The following powerful result has been obtained in [1].

Theorem 4 ([1, Theorem 7]). Unless P = NP, no polynomial time algorithm can approximate Set-Cover within performance ratio less than $c_{sc} \ln n$, where n is the size of the universe and $c_{sc} > 0.2267$ is a constant.

Here we prove the aforementioned conjecture from [5] by encoding a binary representation of letters in states and using some properties from the proof of Theorem 4.

Lemma 3. For every m-letter synchronizing DFA $\mathscr{A} = \langle Q, \Sigma, \delta \rangle$, there is a 2-letter synchronizing DFA $\mathscr{B} = \mathscr{B}(\mathscr{A}) = \langle Q', \{0, 1\}, \delta' \rangle$ such that

$$rt(\mathscr{A})\lceil \log_2 m + 1 \rceil \leq rt(\mathscr{B}) \leq \lceil \log_2 m + 1 \rceil (1 + rt(\mathscr{A})).$$

$\mathscr{B}$ has at most $4m|Q|$ states and can be constructed in time polynomial of m and $|Q|$.

Proof. Let $\Sigma = \{a_1, \ldots, a_m\}$ and for simplicity assume that m is a power of 2, i.e., $m = 2^k$ (otherwise we can add at most $m - 1$ letters with trivial action without impact on the bounds). Let $\ell : \{0, 1\}^k \mapsto \Sigma$ be a bijective function. Set $Q' = Q \times \{0, 1\}^{\leq k}$ and define the transition function $\delta' : Q' \times \{0, 1\} \to Q'$ as follows. For each $q \in Q$, each binary sequence $w \in \{0, 1\}^{\leq k}$ and each bit $x \in \{0, 1\}$, we let

$$\delta'((q, w), x) = \begin{cases} (q, wx) & \text{if } |w| < k; \\ (q.\ell(w), \lambda) & \text{if } |w| = k, x = 1; \\ (q, w) & \text{if } |w| = k, x = 0. \end{cases} \tag{1}$$

Let $u = a_{j_1} a_{j_2} \ldots a_{j_t}$ be a reset word for $\mathscr{A}$. Then the word

$$1^{k+1} \ell^{-1}(a_{j_1}) 1 \ldots \ell^{-1}(a_{j_t}) 1$$

is reset for $\mathscr{B}$ and its length equals $(k + 1)(t + 1)$. The upper bound follows.

In order to prove the lower bound it is enough to consider the shortest binary word u which synchronizes the subset (Q, λ) in $\mathscr{B}$. Since u is chosen shortest, $u = w_1 1 w_2 1 \ldots w_r 1$ where $|w_j| = k$ for each $j \in \{1, \ldots r\}$. Indeed, after applying a word $w \in \{0, 1\}^k$ to the state of the form (q, λ) it make no sense to apply 0 in view of the third choice of definition 1. Then the word $\ell(w_1)\ell(w_2)\ldots\ell(w_r)$ resets $\mathscr{A}$ and the lower bound follows.   □

Now, suppose that for some constant $d > 0$, there is a polynomial time algorithm f_2 such that

$$rt(\mathscr{B}) \leq f_2(\mathscr{B}) \leq d \ln(n) rt(\mathscr{B})$$

for every 2-letter n-state synchronizing DFA $\mathscr{B}$. Then Lemma 3 implies that for each $m \geq 2$ there is also a polynomial time algorithm f_m such that

$$rt(\mathscr{A}) \leq f_m(\mathscr{A}) \leq d \ln(4nm)(1 + rt(\mathscr{A}))$$

for every m-letter n-state synchronizing DFA $\mathscr{A}$. Indeed, such algorithm first constructs $\mathscr{B}(\mathscr{A})$ with at most $4nm$ states as in Lemma 3, and then runs f_2 on $\mathscr{B}(\mathscr{A})$:

$$rt(\mathscr{A}) \leq f_m(\mathscr{A}) = \frac{f_2(\mathscr{B}(\mathscr{A}))}{\lceil \log_2 m + 1 \rceil} \leq d \ln(4nm)(rt(\mathscr{A}) + 1).$$

Combining this with Theorem 4 and Remark 1 we immediately get the following corollary.

Corollary 2. *Let $g(n)$ be an upper bound on the cardinality of the family S as a function of the size of the universe $\mathcal{U}$ in the instances of Set-Cover witnessing Theorem 4. Then, unless* $\mathrm{P} = \mathrm{NP}$, *no polynomial time algorithm approximates reset threshold with performance ratio $\frac{d}{\log_n g(n)+1} \ln(n)$ for any $d < c_{sc}$ in the class of all 2-letter synchronizing automata.*

Thus it suffices to find a lower bound on the size of the universe $\mathcal{U}$ and an upper bound on the size of the family of subsets S in the instances of Set-Cover that witness Theorem 4. Namely, we need to find a polynomial upper bound for $g(n)$.

Due to the space constraints, we use some notions from [1] without going into details. Recall that a PCP instance is a triple $(X, \mathbb{F}, \Phi)$ where $X = \{x_1, \ldots, x_n\}$ is a set of formal variables that range over a domain $\mathbb{F}$ and $\Phi = \{\phi_1, \ldots, \phi_m\}$ is a so-called uniform collection of local tests, $m = poly(n)$. Essentially in the proof of non-approximability of Set-Cover in [1], a PCP instance $(X, \mathbb{F}, \Phi)$ is duplicated $D = \Theta(\frac{|\Phi|}{|X|})$ times and then polynomially reduced to Set-Cover. The universe of Set-Cover instance $\mathcal{U}$ is defined as $\{1, \ldots, D\} \times \Phi \times B$ where B is a so-called universal set and does not matter for us. It follows that the rough lower bound for the size of the universe $\mathcal{U}$ is $|\Phi|$.

The size of the family of subsets S is equal to $D|X||F| + |\Phi||F|^d$ where F is a field of cardinality at most $|X|$ and $d \geq 2$ is a positive integer which can be taken equal 3. Hence the upper bound for $|S|$ is $\Theta(1)|\Phi||X|^d$. Thus we get that

$$\log_{|\mathcal{U}|} |S| \leq \frac{d + \log_{|X|} |\Phi|}{\log_{|X|} |\Phi|}.$$

Notice that $|\Phi|$ is only restricted to be some polynomial of $|X|$, i.e., it can be chosen to be $|X|^r$ for an arbitrary large constant r. As a conclusion, we get the following lemma, which gives a nice property of Set-Cover itself.

Lemma 4. *Given any $\gamma > 0$, unless P = NP, no polynomial time algorithm approximates the Set-Cover with performance ratio $d \ln n$ for any $d < c_{sc}$ in the class of all Set-Cover instances $(\mathcal{U}, \mathcal{S})$ satisfying $\log_{|\mathcal{U}|} |\mathcal{S}| \leq 1 + \gamma$.*

Combining this with Corollary 2 gives us the second main result.

Theorem 5. *Unless P = NP, no polynomial time algorithm approximates the reset threshold with performance ratio less than $0.5c_{sc} \ln n$ in the class of all n-state synchronizing DFAs with 2 input letters.*

Of course, the same bound holds for any fixed non-singleton alphabet.

It is known (see, e.g., [6,10]) that the greedy algorithm for Set-Cover has a logarithmic performance ratio so the bound of Theorem 4 is tight. A natural question is whether or not the bound in Theorem 5 is tight. In this connection we mention that Ananichev (unpublished) has recently constructed a series of synchronizing DFAs for which the greedy algorithm from [4] cannot approximate reset threshold with a logarithmic performance ratio.

Acknowledgement. The author thanks the anonymous referees for their useful remarks and suggestions.

References

1. Alon, N., Moshkovitz, D., Safra, S.: Algorithmic construction of sets for k-restrictions. ACM Trans. Algorithms 2(2), 153–177 (2006)
2. Berlinkov, M.: Approximating the minimum length of synchronizing words is hard. Theory Comput. Syst. 54(2), 211–223 (2014)
3. Bonizzoni, P., Jonoska, N.: Regular splicing languages must have a constant. In: Mauri, G., Leporati, A. (eds.) DLT 2011. LNCS, vol. 6795, pp. 82–92. Springer, Heidelberg (2011)
4. Eppstein, D.: Reset sequences for monotonic automata. SIAM J. Comput. 19, 500–510 (1990)
5. Gerbush, M., Heeringa, B.: Approximating minimum reset sequences. In: Domaratzki, M., Salomaa, K. (eds.) CIAA 2010. LNCS, vol. 6482, pp. 154–162. Springer, Heidelberg (2011)
6. Lovász, L.: On the ratio of optimal integral and fractional covers. Discr. Math. 13, 383–390 (1975)
7. Martyugin, P.: Computational complexity of certain problems related to carefully synchronizing words for partial automata and directing words for nondeterministic automata. Theory Comput. Syst. 54(2), 293–304 (2014)
8. Olschewski, J., Ummels, M.: The complexity of finding reset words in finite automata. In: Hliněný, P., Kučera, A. (eds.) MFCS 2010. LNCS, vol. 6281, pp. 568–579. Springer, Heidelberg (2010)
9. Savitch, W.: Relationships between nondeterministic and deterministic tape complexities, J. Comp. System Sci. 4(2), 177–192 (1970)
10. Slavik, P.: A tight analysis of the greedy algorithm for set cover. In: Proc. 28th ACM Symp. on Theory of Computing, pp. 435–441 (1996)
11. Travers, N., Crutchfield, J.: Exact synchronization for finite-state sources. J. Stat. Phys. 145(5), 1181–1201 (2011)
12. Vojtěch, V.: Subset synchronization of transitive automata, arXiv:1403.3972 (accepted to AFL 2014)

Synchronizing Automata with Random Inputs*
(Short Paper)

Vladimir V. Gusev

Institute of Mathematics and Computer Science,
Ural Federal University, Ekaterinburg, Russia
vl.gusev@gmail.com

Abstract. We study the problem of synchronization of automata with
random inputs. We present a series of automata such that the expected
number of steps until synchronization is exponential in the number of
states. At the same time, we show that the expected number of letters
to synchronize any pair of the famous Černý automata is at most cubic
in the number of states.

1 Introduction

A *complete deterministic finite automaton* $\mathscr{A}$, or simply *automaton*, is a triple
$\langle Q, \Sigma, \delta \rangle$, where Q is a finite *set of states*, Σ is a finite *input alphabet*, and
$\delta : Q \times \Sigma \mapsto Q$ is a totally defined *transition function*. Following standard
notation, by Σ^* we mean the set of all finite words over the alphabet Σ, including
the empty word ε. The function δ naturally extends to the free monoid Σ^*; this
extension is still denoted by δ. Thus, via δ, every word $w \in \Sigma^*$ acts on the set Q.

An automaton $\mathscr{A}$ is called *synchronizing*, if there is a word $w \in \Sigma^*$ which
brings all states of the automaton $\mathscr{A}$ to a particular one, i.e. there exists a state
$t \in Q$ such that $\delta(s, w) = t$ for every $s \in Q$. Any such word w is said to be
a *reset* (or *synchronizing*) *word* for the automaton $\mathscr{A}$. The minimum length of
reset words for $\mathscr{A}$ is called the *reset threshold* of $\mathscr{A}$. Note, that the language $\mathscr{L}$ of
synchronizing words of the automaton $\mathscr{A}$ is a *two-sided ideal*, i.e. $\Sigma^* \mathscr{L} \Sigma^* = \mathscr{L}$.
We say that that the word w synchronizes a pair $\{s, t\}$ if $\delta(s, w) = \delta(t, w)$.

Synchronizing automata serve as transparent and natural models of error-
resistant systems in many applied areas such as robotics, coding theory, and
bioinformatics. At the same time, synchronizing automata surprisingly arise in
some parts of pure mathematics: algebra, symbolic dynamics, and combinatorics
on words. See recent surveys by Sandberg [9] and Volkov [12] for a general
introduction to the theory of synchronizing automata.

The interest to the field is heated also by the famous *Černý conjecture*. In
1964 Černý exhibited a series $\mathscr{C}_n$ of automata with n states whose reset threshold
equals $(n-1)^2$ [3]. Soon after he conjectured, that this series represents the worst
possible case, i.e. the reset threshold of every n-state synchronizing automaton
is at most $(n-1)^2$. In spite of its simple formulation and intensive researchers'

---

* Supported by the Presidential Program for Young Researchers, grant MK-
3160.2014.1 and by the Russian Foundation for Basic Research, grant 13-01-00852.

A.M. Shur and M.V. Volkov (Eds.): DLT 2014, LNCS 8633, pp. 68–75, 2014.

efforts, the Černý conjecture remains unresolved for fifty years. The best known upper bound on the reset threshold of a synchronizing n-state automaton is $\frac{n^3-n}{6}$ by Pin [7].

The focus of this paper is on probabilistic aspects of synchronization. One general question that was actively studied in the literature is the following: what are synchronizing properties of a *random automaton*? Skvortsov and Zaks have shown that a random automaton with sufficiently large number of letters is synchronizing with high probability [10]. Later on, they proved that a random 4-letter automaton is synchronizing with a positive probability that is independent of the number of states [14]. The last step in this direction seems to be done by Berlinkov [2]. He has shown that a random automaton over a binary alphabet is synchronizing with high probability. Another direction within this setting is devoted to reset thresholds of random synchronizing automata. It was shown in [10] that a random automaton with large number of letters satisfies the Černý conjecture with high probability. Furthermore, computational experiments performed in [11,6] suggest that expected reset threshold of a random synchronizing automaton is sub-linear.

In [5] probability distributions over states of an automaton $\mathscr{A}$ were used to define the synchronizing probability function of $\mathscr{A}$. This function encompasses various synchronizing properties of $\mathscr{A}$.

The setting of the present paper is different. In our considerations we investigate how *random input* acts on a *fixed* automaton. Assume that several copies of a synchronizing automaton $\mathscr{A}$ simultaneously read a common input from a fixed source of random letters. Initially these automata may be in different states. What is the expected number of steps E until all copies will be in the same state? We can give the following illustration of this approach. Let $\mathscr{D}$ be a decoder of a code. Due to data transmission errors the decoder $\mathscr{D}$ may be in a different state compared to a correct decoder $\mathscr{D}_c$. Then the number E computed for decoders $\mathscr{D}$ and $\mathscr{D}_c$ represents an average number of steps before recovery of the decoder $\mathscr{D}$ after an error.

Our setting heavily depends on a model of a random input. In the present paper we restrict ourselves to a binary alphabet $\Sigma = \{a, b\}$ and the *Bernoulli model*, i.e. every succeeding letter is drawn independently with probability p for the letter a and probability $q = 1 - p$ for the letter b. In section 2 we present a series of n-state automata $\mathscr{U}_n$ over Σ and a pair S such that the expected number of steps to synchronize S is exponential in n. At the same time, in section 3 we show that the expected number of steps to synchronize any pair of the famous example $\mathscr{C}_n$ by Černý is at most cubic in n. These results reveal that despite the fact that synchronization of $\mathscr{C}_n$ is hard in the deterministic case, it is relatively easy in the random setting.

2   Automata $\mathscr{U}_n$ with the Sink State

Let Σ be a binary alphabet $\{a, b\}$. Let $\mathscr{U}_n$ be the minimal automaton recognizing the language L_n, where L_n is equal to $\Sigma^* a^{\frac{n+1}{2}} b^{\frac{n-1}{2}} \Sigma^*$ if n is odd, and to

$\Sigma^* a^{\frac{n}{2}} b^{\frac{n}{2}} \Sigma^*$ if n is even. Note, that the automaton $\mathscr{U}_n$ is synchronizing, and its language of synchronizing words coincides with L_n.

First, we will consider the case when n is odd. Let us define $\mathscr{U}_n$ more formally, see fig. 1. The set of states of $\mathscr{U}_n$ is equal to $\{1, 2, \ldots n+1\}$. The transition function δ of $\mathscr{U}_n$ is defined as follows:

$$\delta(i,a) = \begin{cases} i+1, & \text{if } i < \frac{n+3}{2} \\ i, & \text{if } i = \frac{n+3}{2} \\ 2, & \text{if } \frac{n+3}{2} < i < n+1 \\ n+1, & \text{if } i = n+1; \end{cases} \qquad \delta(i,b) = \begin{cases} 1, & \text{if } i < \frac{n+3}{2} \\ i+1, & \text{if } \frac{n+3}{2} \leq i < n+1 \\ n+1, & \text{if } i = n+1. \end{cases}$$

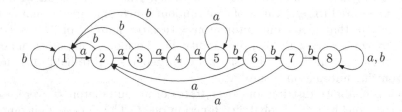

Fig. 1. Automaton $\mathscr{U}_7$

Let $\mathcal{B}(p,q)$ be the source of random letters such that each letter is drawn independently with probability p for the letter a and probability $q = 1 - p$ for the letter b. Let $\mathscr{A} = \langle Q, \Sigma, \delta \rangle$ be a synchronizing automaton. We consider the following random process:

1. $S := Q$
2. Until $|S| = 1$ do
3. $x \leftarrow \mathcal{B}(p,q)$
4. $S := \delta(S, x)$

We start with the set S equal to the state set Q. On each step we draw a random letter x from the source $\mathcal{B}(p,q)$ and apply it to S. We stop when S is a singleton.

In general, we are interested in the average number of steps that this process takes for a given automaton $\mathscr{A}$. In particular, we have the following theorem.

Theorem 1. *The expected number of letters, that are drawn from $\mathcal{B}(p,q)$, until $\mathscr{U}_n$ is synchronized, is equal to $\dfrac{1}{p^{\frac{n+1}{2}} q^{\frac{n-1}{2}}}$ if n is odd, and is equal to $\dfrac{1}{p^{\frac{n}{2}} q^{\frac{n}{2}}}$ if n is even.*

Proof. We will only prove the theorem for the case when n is odd. Proof of the other case is similar. It is rather easy to see that the word w synchronizes the automaton $\mathscr{U}_n$ if and only if $\delta(1, w) = n + 1$. Thus, the average number of steps in our random process equals the average length of a random walk that brings the state 1 to the state $n + 1$, where the probability of the transition labeled by a is p, and the probability of the transition labeled by b is q. It is well-known how to compute the latter quantity[1] [8, section 6.2]. For $1 \leq i \leq n+1$ let μ_i be the expected length of a random walk that brings the state i to the state $n+1$. These quantities necessarily satisfy the following system of equations:

$$
\begin{cases}
\mu_1 = p\mu_2 + q\mu_1 + 1 & (1) \\
\mu_i = p\mu_{i+1} + q\mu_1 + 1, \text{ if } 1 \leq i \leq \frac{n+1}{2} & (2) \\
\mu_{\frac{n+3}{2}} = p\mu_{\frac{n+3}{2}} + q\mu_{\frac{n+5}{2}} + 1 & (3) \\
\mu_i = p\mu_2 + q\mu_{i+1} + 1, \text{ if } \frac{n+5}{2} \leq i \leq n-1 & (4) \\
\mu_n = p\mu_2 + q\mu_{n+1} + 1 & (5) \\
\mu_{n+1} = 0 & (6)
\end{cases}
$$

We will solve this system in several steps:

1. Let us show that $\mu_i = \mu_1 - \frac{p^{i-1}-1}{p^i - p^{i-1}}$ for $2 \leq i \leq \frac{n+3}{2}$. Equation (1) implies that this statement is true for $i = 2$. Suppose now that the statement is true for μ_i. Let us show that it is true for μ_{i+1}. From equation (2) we get $\mu_1 - \frac{p^{i-1}-1}{p^i-p^{i-1}} = p\mu_{i+1} + q\mu_1 + 1$. Therefore, $\mu_1 - \frac{p^i-1}{p^{i+1}-p^i} = \mu_{i+1}$.

We will denote $\frac{p^{\frac{n+1}{2}}-1}{p^{\frac{n+3}{2}}-p^{\frac{n+1}{2}}}$ as C in order to simplify notation. Therefore, $\mu_{\frac{n+3}{2}} = \mu_1 - C$.

2. Equation (3) immediately implies $\mu_{\frac{n+5}{2}} = \mu_{\frac{n+3}{2}} - \frac{1}{q}$. Therefore, we have $\mu_{\frac{n+5}{2}} = \mu_1 - C - \frac{1}{q}$.

3. Now we will show that $\mu_i = \mu_1 - \frac{C}{q^{i-\frac{n+5}{2}}} - \frac{1}{q^{i-\frac{n+3}{2}}}$ for $\frac{n+5}{2} \leq i \leq n$. This statement is true for $i = \frac{n+5}{2}$. Let us show that it is true for every succeeding $i \leq n$. Since $\mu_2 \overset{(1)}{=} \mu_1 - \frac{1}{p}$ we can rewrite equation (4) in the following way: $\mu_i = p\mu_1 + q\mu_{i+1}$. Our assumption states that $\mu_i = \mu_1 - \frac{C}{q^{i-\frac{n+5}{2}}} - \frac{1}{q^{i-\frac{n+3}{2}}}$. Therefore, $q\mu_1 - \frac{C}{q^{i-\frac{n+5}{2}}} - \frac{1}{q^{i-\frac{n+3}{2}}} = q\mu_{i+1}$. Finally, $\mu_{i+1} = \mu_1 - \frac{C}{q^{i-\frac{n+3}{2}}} - \frac{1}{q^{i-\frac{n+1}{2}}}$.

Note, we have $\mu_n = \mu_1 - \frac{C}{q^{\frac{n-5}{2}}} - \frac{1}{q^{\frac{n-3}{2}}}$.

4. Equation (5) and (6) imply $\mu_n = p\mu_1$.
Therefore, $\mu_1 = \frac{C}{q^{\frac{n-3}{2}}} + \frac{1}{q^{\frac{n-1}{2}}} = \frac{qC+1}{q^{\frac{n-1}{2}}} = \frac{1}{p^{\frac{n+1}{2}}q^{\frac{n-1}{2}}}$.

We also want to mention the following simple fact. If an automaton $\mathscr{A}$ has a synchronizing word w of length ℓ then the expected number of letters, that are drawn from $\mathcal{B}(p,q)$, until $\mathscr{A}$ is synchronized, is at most $\frac{\ell}{p^{\ell_a}q^{\ell-\ell_a}}$, where ℓ_a is the number of letters a in the word w.

3   The Černý Automata $\mathscr{C}_n$

Now we study a classical example introduced by Černý in 1964 [3]. Recall the definition of the Černý automaton $\mathscr{C}_n$. The state set of $\mathscr{C}_n$ is $Q = \{0, 1, \ldots, n-1\}$, and the letters a and b act on Q as follows.

---

[1] It is also called the mean absorption time of a Markov chain.

$$\delta(i,a) = \begin{cases} 1 & \text{if } i = 0, \\ i & \text{if } i > 0; \end{cases} \qquad \delta(i,b) = \begin{cases} i+1 & \text{if } i < n-1, \\ 0 & \text{if } i = n-1. \end{cases}$$

The reset threshold of $\mathscr{C}_n$ is equal to $(n-1)^2$, see [1,4,3].

The goal of the present section is to find the expected number of letters, that are drawn from $\mathcal{B}(p,q)$, until the pair of states $\{1, \frac{n+1}{2}\}$, when n is odd, and the pair $\{1, \frac{n+2}{2}\}$, when n is even, is synchronized. At the same time, we will see that the expectation for these pairs is the largest among other pairs.

Let $\mathscr{A} = \langle Q, \Sigma, \delta \rangle$ be an automaton. The *pair automaton* $\mathcal{P}(\mathscr{A})$ is defined as follows. The set of states of $\mathcal{P}(\mathscr{A})$ is equal to $\{\{s,t\} \mid s \neq t\} \cup \{\mathbf{z}\}$. The transition function $\delta_{\mathcal{P}}$ of $\mathcal{P}(\mathscr{A})$ for each $x \in \Sigma$, $s,t \in Q$ is defined by the following rules:

$$\delta_{\mathcal{P}}(\{s,t\}, x) = \begin{cases} \{\delta(s,x), \delta(t,x)\}, \text{if } \delta(s,x) \neq \delta(t,x) \\ \mathbf{z}, \text{if } \delta(s,x) = \delta(t,x); \end{cases} \qquad \delta_{\mathcal{P}}(\mathbf{z}, x) = \mathbf{z}.$$

Note, that a word w that synchronize a pair $\{s,t\}$ labels a path in $\mathcal{P}(\mathscr{A})$ from $\{s,t\}$ to $\mathbf{z}$. Furthermore, a word w is synchronizing for $\mathscr{A}$ if and only if w is synchronizing for $\mathcal{P}(\mathscr{A})$. The proof of this easy fact can be found for instance in [12].

First, let n be a positive odd integer. In order to prove the main result of this section we will require another representation of the pair automaton of $\mathscr{C}_n$. We will denote it by $\mathscr{P}_n$. The state set of $\mathscr{P}_n$ is the set of ordered pairs

$$\{(i, \ell) \mid 0 \leq i \leq n-1,\ 1 \leq \ell \leq \frac{n-1}{2}\} \cup \{\mathbf{z}\}.$$

The transition function δ is defined as follows.

$\delta(\mathbf{z}, x) = \mathbf{z}$ for every $x \in \Sigma$,
$\delta((i, \ell), b) = ((i+1) \mod n, \ell)$ for every admissible i and ℓ.
$\delta((i, \ell), a) = (i, \ell)$ for every admissible i and ℓ with the exception of the following cases:
$\delta((0,1), a) = \mathbf{z}$,
$\delta((0, \ell), a) = (1, \ell-1)$ if $2 \leq \ell \leq \frac{n-1}{2}$,
$\delta((n-\ell, \ell), a) = (n-\ell, \ell+1)$ if $1 \leq \ell \leq \frac{n-3}{2}$,
$\delta((\frac{n+1}{2}, \frac{n-1}{2}), a) = (1, \frac{n-1}{2})$.

Lemma 1. *Let n be a positive odd integer. The automaton $\mathscr{P}_n$ is isomorphic to the pair automaton of $\mathscr{C}_n$.*

Proof. We will construct the desired isomorphism. The sink state $\mathbf{z}$ of the pair automaton is mapped to the sink state $\mathbf{z}$ of $\mathscr{P}_n$. Let $\{s,t\}$ be an arbitrary pair of states. Let $\delta_{\mathscr{C}}$ be the transition function of the automaton $\mathscr{C}_n$. There is a positive integer m that satisfies equations $\delta_{\mathscr{C}}(s, b^m) = t$ and $\delta_{\mathscr{C}}(t, b^{n-m}) = s$. Let ℓ be the minimum of m and $n-m$. Since n is odd $m \neq n-m$. Let

$$i = \begin{cases} s, & \text{if } \delta_{\mathscr{C}}(s, b^\ell) = t \\ t, & \text{if } \delta_{\mathscr{C}}(t, b^\ell) = s. \end{cases}$$

Then the pair $\{s,t\}$ of the pair automaton is mapped to the state (i,ℓ) of the automaton $\mathscr{P}_n$. It is easy to check that the presented mapping is an isomorphism.

Now we are ready to formulate the main result of this section.

Theorem 2. *Let n be a positive odd integer. The expected number of letters, that are drawn from $\mathcal{B}(p,q)$, until the pair $\{1, \frac{n+1}{2}\}$ of $\mathscr{C}_n$ is synchronized, is equal to $\frac{(n-1)((n-1)^2+q(3n-5)+4q^2)}{8pq^2}$.*

Proof. It is not hard to see that a word w labels a path from (i,ℓ) to $\mathbf{z}$ in the automaton $\mathscr{P}_n$ if and only if the word w synchronizes the pair $\{i,(i+\ell) \bmod n\}$ of the automaton $\mathscr{C}_n$. Thus, the expected number of letters until the pair $\{1, \frac{n+1}{2}\}$ is synchronized is equal to the expected length of a random walk in automaton $\mathscr{P}_n$ from the state $(1, \frac{n-1}{2})$ to the state $\mathbf{z}$, where the probability of the transition labeled by a is p, and the probability of the transition labeled by b is q. For $0 \leq i \leq n-1$ and $1 \leq \ell \leq \frac{n-1}{2}$ let $\mu_{i,\ell}$ be the expected length of a random walk that brings the state (i,ℓ) of $\mathscr{P}_n$ to the state $\mathbf{z}$. As in the proof of the theorem 1 these values have to satisfy a particular system of linear equations, see [8, section 6.2]. For convenience, we will split this system into three parts. The first part:

$$
\begin{cases}
\mu_{0,1} = q\mu_{1,1} + 1 & (1) \\
\mu_{i,1} = p\mu_{i,1} + q\mu_{i+1,1} + 1, & \text{if } 1 \leq i \leq n-2 \quad (2) \\
\mu_{n-1,1} = p\mu_{n-1,2} + q\mu_{0,1} + 1 & (3) \\
\mu_{\mathbf{z}} = 0
\end{cases}
$$

The second part, $2 \leq \ell \leq \frac{n-3}{2}$:

$$
\begin{cases}
\mu_{0,\ell} = p\mu_{1,\ell-1} + q\mu_{1,\ell} + 1 & (4) \\
\mu_{i,\ell} = p\mu_{i,\ell} + q\mu_{i+1,\ell} + 1, & \text{if } 1 \leq i \leq n-\ell-1 \quad (5) \\
\mu_{n-\ell,\ell} = p\mu_{n-\ell,\ell+1} + q\mu_{n-\ell+1,\ell} + 1, & (6) \\
\mu_{i,\ell} = p\mu_{i,\ell} + q\mu_{i+1,\ell} + 1, & \text{if } n-\ell+1 \leq i \leq n-2 \quad (7) \\
\mu_{n-1,\ell} = p\mu_{n-1,\ell} + q\mu_{0,\ell} + 1, & (8)
\end{cases}
$$

And the third part:

$$
\begin{cases}
\mu_{0,\frac{n-1}{2}} = p\mu_{1,\frac{n-3}{2}} + q\mu_{1,\frac{n-1}{2}} + 1 & (9) \\
\mu_{i,\frac{n-1}{2}} = p\mu_{i,\frac{n-1}{2}} + q\mu_{i+1,\frac{n-1}{2}} + 1, & \text{if } 1 \leq i \leq \frac{n-1}{2} \quad (10) \\
\mu_{\frac{n+1}{2},\frac{n-1}{2}} = p\mu_{1,\frac{n-1}{2}} + q\mu_{\frac{n+3}{2},\frac{n-1}{2}} + 1, & (11) \\
\mu_{i,\frac{n-1}{2}} = p\mu_{i,\frac{n-1}{2}} + q\mu_{i+1,\frac{n-1}{2}} + 1, & \text{if } \frac{n+3}{2} \leq i \leq n-2 \quad (12) \\
\mu_{n-1,\frac{n-1}{2}} = p\mu_{n-1,\frac{n-1}{2}} + q\mu_{0,\frac{n-1}{2}} + 1, & (13)
\end{cases}
$$

Let us resolve the first part. Applying equations (2) in successive order we get
$\mu_{1,1} \overset{(2)}{=} \mu_{n-1,1} + \frac{n-2}{q} \overset{(3)}{=} p\mu_{n-1,2} + q\mu_{0,1} + 1 + \frac{n-2}{q}$. Since $\mu_{n-1,2} \overset{(8)}{=} \mu_{0,2} + \frac{1}{q} \overset{(4)}{=} p\mu_{1,1} + q\mu_{1,2} + 1 + \frac{1}{q}$ and $\mu_{0,1} \overset{(1)}{=} q\mu_{1,1} + 1$ we have $\mu_{1,1} = p(p\mu_{1,1} + q\mu_{1,2} +$

$1 + \frac{1}{q}) + q(q\mu_{1,1} + 1) + 1 + \frac{n-2}{q}$. After trivial simplification, using the fact that $1 - p^2 - q^2 = 2pq$, we obtain

$$2\mu_{1,1} = \mu_{1,2} + \frac{n-p}{pq^2} \qquad (14)$$

Let us focus on the second part. Let $2 \leq \ell \leq \frac{n-3}{2}$. Applying equations (5) several times in successive order we get $\mu_{1,\ell} \overset{(5)}{=} \mu_{n-\ell,\ell} + \frac{n-\ell-1}{q} \overset{(6)}{=} p\mu_{n-\ell,\ell+1} + q\mu_{n-\ell+1,\ell} + 1 + \frac{n-\ell-1}{q}$. Since $\mu_{n-\ell,\ell+1} \overset{(7\,\text{or}\,12)}{=} \mu_{n-1,\ell+1} + \frac{\ell-1}{q} \overset{(8\,\text{or}\,13)}{=} \mu_{0,\ell+1} + \frac{\ell}{q} \overset{(4\,\text{or}\,9)}{=} p\mu_{1,\ell} + q\mu_{1,\ell+1} + 1 + \frac{\ell}{q}$ and $\mu_{n-\ell+1,\ell} \overset{(7)}{=} \mu_{n-1,\ell} + \frac{\ell-2}{q} \overset{(8)}{=} \mu_{0,\ell} + \frac{\ell-1}{q} \overset{(4)}{=} p\mu_{1,\ell-1} + q\mu_{1,\ell} + 1 + \frac{\ell-1}{q}$ we have $\mu_{1,\ell} = p(p\mu_{1,\ell} + q\mu_{1,\ell+1} + 1 + \frac{\ell}{q}) + q(p\mu_{1,\ell-1} + q\mu_{1,\ell} + 1 + \frac{\ell-1}{q}) + 1 + \frac{n-\ell-1}{q}$. After simplification we obtain the following equation:

$$2\mu_{1,\ell} = \mu_{1,\ell+1} + \mu_{1,\ell-1} + \frac{n-p}{pq^2} \qquad (15)$$

Let us resolve the third part. Applying equations (10) in successive order we get $\mu_{1,\frac{n-1}{2}} \overset{(10)}{=} \mu_{\frac{n+1}{2},\frac{n-1}{2}} + \frac{n-1}{2q} \overset{(11)}{=} p\mu_{1,\frac{n-1}{2}} + q\mu_{\frac{n+3}{2},\frac{n-1}{2}} + 1 + \frac{n-1}{2q}$. Since $\mu_{\frac{n+3}{2},\frac{n-1}{2}} \overset{(12)}{=} \mu_{n-1,\frac{n-1}{2}} + \frac{n-5}{2q} \overset{(13)}{=} \mu_{0,\frac{n-1}{2}} + \frac{n-3}{2q} \overset{(9)}{=} p\mu_{1,\frac{n-3}{2}} + q\mu_{1,\frac{n-1}{2}} + 1 + \frac{n-3}{2q}$ we have $\mu_{1,\frac{n-1}{2}} = p\mu_{1,\frac{n-1}{2}} + q(p\mu_{1,\frac{n-3}{2}} + q\mu_{1,\frac{n-1}{2}} + 1 + \frac{n-3}{2q}) + 1 + \frac{n-1}{2q}$. After an easy simplification we obtain the following equation:

$$\mu_{1,\frac{n-1}{2}} = \mu_{1,\frac{n-3}{2}} + \frac{q^2 + \frac{n-1}{2}q + \frac{n-1}{2}}{pq^2} \qquad (16)$$

Summing up equations (14),(15) for $2 \leq \ell \leq \frac{n-3}{2}$, and (16) we obtain the following equation:

$$\mu_{1,1} = \frac{n-3}{2} \cdot \frac{n-p}{pq^2} + \frac{q^2 + \frac{n-1}{2}q + \frac{n-1}{2}}{pq^2} \qquad (17)$$

Now we can show that

$$\mu_{1,\ell} = \ell\mu_{1,1} - \frac{\ell(\ell-1)}{2} \cdot \frac{n-p}{pq^2} \qquad (18)$$

Equation (14) serves as the induction base. Using equation (15) we make the induction step.

From equation (18) for $\ell = \frac{n-1}{2}$ we get $\mu_{1,\frac{n-1}{2}} = \frac{n-1}{2}\mu_{1,1} - \frac{(n-1)(n-3)}{8} \cdot \frac{n-p}{pq^2}$. Using (17) after tedious simplification we get the final result:

$$\mu_{1,\frac{n-1}{2}} = \frac{(n-1)((n-1)^2 + q(3n-5) + 4q^2)}{8pq^2} \qquad (19)$$

Note, that the leading term of $\mu_{1,\frac{n-1}{2}}$ is equal to $\frac{n^3}{8pq^2}$. It is easy to see, that the minimum of $\frac{1}{8pq^2}$ is reached at $p = \frac{1}{3}$. Therefore, the expected number of letters until the pair $\{1, \frac{n+1}{2}\}$ of $\mathscr{C}_n$ is synchronized is close to the minimum for the source of random letters $\mathcal{B}(\frac{1}{3}, \frac{2}{3})$. In this case we have

$$\mu_{1,\frac{n-1}{2}} = \frac{27n^3}{32} - \frac{27n^2}{32} - \frac{15n}{32} + \frac{15}{32}$$

In a similar way one can also prove the following theorem.

Theorem 3. *Let n be a positive even integer. The expected number of letters, that are drawn from $\mathcal{B}(p,q)$, until the pair $\{1, \frac{n+2}{2}\}$ of $\mathscr{C}_n$ is synchronized, is equal to $\frac{n((n-1)(n-2)+q(3n-6)+4q^2)}{8pq^2}$.*

References

1. Ananichev, D.S., Gusev, V.V., Volkov, M.V.: Primitive digraphs with large exponents and slowly synchronizing automata. Journal of Mathematical Sciences (US) 192(3), 263–278 (2013)
2. Berlinkov, M.V.: On the probability of being synchronizable (2013), ArXiv: http://arxiv.org/abs/1304.5774
3. Černý, J.: Poznámka k homogénnym eksperimentom s konečnými automatami. Matematicko-fyzikalny Časopis Slovensk. Akad. Vied 14(3), 208–216 (1964) (in Slovak)
4. Gusev, V.V.: Lower bounds for the length of reset words in eulerian automata. Int. J. Found. Comput. Sci. 24(2), 251–262 (2013)
5. Jungers, R.M.: The synchronizing probability function of an automaton. SIAM J. Discret. Math. 26(1), 177–192 (2012)
6. Kisielewicz, A., Kowalski, J., Szykuła, M.: A Fast Algorithm Finding the Shortest Reset Words. In: Du, D.-Z., Zhang, G. (eds.) COCOON 2013. LNCS, vol. 7936, pp. 182–196. Springer, Heidelberg (2013)
7. Pin, J.-E.: On two combinatorial problems arising from automata theory. Ann. Discrete Math. 17, 535–548 (1983)
8. Privault, N.: Understanding Markov Chains. Springer (2013)
9. Sandberg, S.: Homing and synchronizing sequences. In: Broy, M., Jonsson, B., Katoen, J.-P., Leucker, M., Pretschner, A. (eds.) Model-Based Testing of Reactive Systems. LNCS, vol. 3472, pp. 5–33. Springer, Heidelberg (2005)
10. Skvortsov, E.S., Zaks, Y.: Synchronizing random automata. Discr. Math. and Theor. Comp. Sci. 12(4), 95–108 (2010)
11. Skvortsov, E., Tipikin, E.: Experimental study of the shortest reset word of random automata. In: Bouchou-Markhoff, B., Caron, P., Champarnaud, J.-M., Maurel, D. (eds.) CIAA 2011. LNCS, vol. 6807, pp. 290–298. Springer, Heidelberg (2011)
12. Volkov, M.V.: Synchronizing automata and the Černý conjecture. In: Martín-Vide, C., Otto, F., Fernau, H. (eds.) LATA 2008. LNCS, vol. 5196, pp. 11–27. Springer, Heidelberg (2008)
13. Volkov, M.V.: Synchronizing automata preserving a chain of partial orders. Theoret. Comput. Sci. 410, 2992–2998 (2009)
14. Zaks, Y.I., Skvortsov, E.S.: Synchronizing random automata on a 4-letter alphabet. Journal of Mathematical Sciences 192(3), 303–306 (2013)

Graph Spectral Properties
of Deterministic Finite Automata[*]
(Short Paper)

Ryoma Sin'ya

Department of Mathematical and Computing Sciences,
Tokyo Institute of Technology
shinya.r.aa@m.titech.ac.jp

Abstract. We prove that a minimal automaton has the minimal adjacency matrix rank and the minimal adjacency matrix nullity among all equivalent deterministic automata. Our proof uses equitable partition (from graph spectra theory) and Nerode partition (from automata theory). This result leads to the notion of rank of a regular language L, which is the minimal adjacency matrix rank of a deterministic automaton that recognises L. We then define and focus on rank-one languages. We also define the expanded canonical automaton of a rank-one language.

1 Introduction

The *counting function*[1] $C_L : \mathbb{N} \to \mathbb{N}$ of a language L over a finite alphabet maps a natural number n into the number of words in L of length n defined as $C_L(n) := |\{w \in L \mid |w| = n\}|$.

The counting function is a fundamental object in formal language theory and has been studied extensively (cf. [8,9]). If L is a regular language, we can represent its counting function $C_L(n)$ using the n-th power of an adjacency matrix of a deterministic automaton that recognises L as: $C_L(n) = IM^nF$, where M is an adjacency matrix, I is an initial vector and F is a final vector of any deterministic automaton which recognises L (cf. Lemma 1 in [11]).

Our interest is in an *"easily countable"* class of languages, in the intuitive sense of the word.

Ranking and Its Applications. *Ranking* is one of the variants of counting. The ranking function of L over a finite alphabet A is a bijective function $R_L : L \to \mathbb{N}$ that maps a word w in L to its index in the lexicographic ordering $\prec$ over A^* defined as $R_L(w) := |\{v \in L \mid v \prec w\}|$.

In 1985, Goldberg and Sipser introduced a ranking-based string compression in [3]. Recently, the author studied a ranking-based compression on a regular language to analyse its compression ratio and improve a ranking algorithm in [11]. We show an example of a ranking-based compression on a regular language.

---

[*] The full paper of this work with more detailed description and many useful examples is avalable at [12].

[1] Also called *growth function* or *combinatorial complexity*.

A.M. Shur and M.V. Volkov (Eds.): DLT 2014, LNCS 8633, pp. 76–83, 2014.
© Springer International Publishing Switzerland 2014

Example 1. The formal grammar of *Uniform Resource Identifier* (URI) is defined in RFC 3986[2], and is known to be regular (cf. [10]). Because the language of all URIs U is regular, we can apply a ranking-based compression on a regular language. For example, the index of the URI $w_1 =$ `http://dlt2014.sciencesconf.org/` is:

$$R_U(w_1) = 72855229650479606638211370075845591039390765035063493.$$

The word w_1 is 32 bytes long ($|w_1| = 32$), whereas its index $R_U(w_1)$ is 23 bytes long ($\lfloor \log_{256} R_U(w_1) \rfloor = 23$). w_1 is compressed up to 72% and, clearly, we can decompress it by the inverse of R_U since ranking is bijective.   ◇

In the case of a ranking on a regular language, the ranking function and its inverse (*unranking*) of L can be calculated using the adjacency matrix of the deterministic automaton L (cf. [2,11])[3]. The computational complexity of an unranking function is higher than a ranking function because the former requires matrix multiplication but the latter does not (cf. Table 1 in [11]). In Example 1, the calculation of ranking (compression) was performed in less than one second; however, the calculation of unranking (decompression) took about two minutes due to the cost of matrix multiplication (the naive algorithm for it has cubic complexity). The minimal automaton of U has 180 states, and its multiplication cost is high in practice.

Rank-one Languages and Our Results. There are several classes of matrices that have a matrix power that can be computed efficiently (*e.g.* diagonalisable matrices and low-rank matrices). We focus on *rank-one matrices* from these classes. As we describe in Section 4, the power of a rank-one matrix has constant time complexity with linear-time preprocessing. We investigate *rank-one languages*: the class of languages for which the rank of minimal automaton is one. We define an automaton as *rank-n* if its adjacency matrix is rank-n. Next, we introduce the definition of the rank of a language.

Definition 1. A regular language L is *rank-n* if there exists a rank-n deterministic automaton that recognises L, and there does not exist a rank-m deterministic automaton that recognises L for any m less than n.   ◇

However, Definition 1 raises the question of how to find a minimal rank. It is a classical theorem in automata theory that for any regular language L, there is a unique automaton $\mathcal{A}$ that recognises L that has a minimal number of states, and $\mathcal{A}$ is called the *minimal automaton* of L. We intend to refine Definition 1 as the following definition.

Definition 1 (refined). A regular language L is *rank-n* if its minimal automaton is rank-n.   ◇

---

[2] `http://www.ietf.org/rfc/rfc3986.txt`
[3] Example 1 uses RANS [10], which is open source software implemented by the author based on the algorithms in [11].

Nevertheless, to achieve this we have to show that a minimal automaton has the minimal rank for consistency of the above two definitions. Hence in Section 3, we prove the following theorem, which has a more general statement.

Theorem 1. *An automaton $\mathcal{A}$ is minimal if and only if both the rank and the nullity of its adjacency matrix are minimal.*   ◇

Theorem 1 provides a necessary and sufficient condition for the minimality of an automaton and is a purely algebraic characterisation of minimal automata. This theorem is not obvious because, in general, for an automaton $\mathcal{A}$, the number of states of $\mathcal{A}$ and the rank (nullity) of $\mathcal{A}$ are not related. This is illustrated in Figure 1, where the deterministic automaton $\mathcal{B}_1$ has three states and its rank is two, whereas $\mathcal{C}_1$ has four states and its rank is one, which equals the rank of the minimal automaton $\mathcal{A}_1$. Therefore, we cannot argue naively that "any minimal automaton has the minimal rank (nullity)" by its minimality of states.

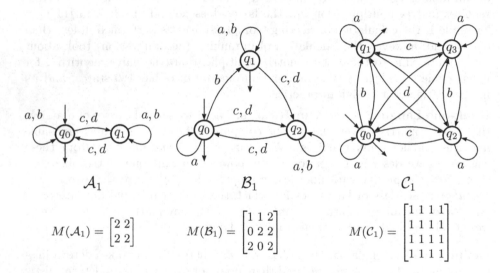

$$M(\mathcal{A}_1) = \begin{bmatrix} 2 & 2 \\ 2 & 2 \end{bmatrix} \qquad M(\mathcal{B}_1) = \begin{bmatrix} 1 & 1 & 2 \\ 0 & 2 & 2 \\ 2 & 0 & 2 \end{bmatrix} \qquad M(\mathcal{C}_1) = \begin{bmatrix} 1 & 1 & 1 & 1 \\ 1 & 1 & 1 & 1 \\ 1 & 1 & 1 & 1 \\ 1 & 1 & 1 & 1 \end{bmatrix}$$

Fig. 1. Three equivalent deterministic trim automata and their adjacency matrices

The proof consists of the use of two fundamental tools: *equitable partition* from graph spectra theory and *Nerode partition* from automata theory.

2   Nerode Partition and Equitable Partition

We assume that the reader has a basic knowledge of automata, graphs and linear algebra. All results in this section are well known, and for more details, we refer the reader to [7] for automata theory and [4] for graph spectra theory.

2.1 Automata and Languages

A *deterministic finite automaton* $\mathcal{A}$ is a quintuple $\mathcal{A} = \langle Q, A, \delta, q_0, F \rangle$; the finite set of *states* Q, the finite set A called *alphabet*, the *transition function* $\delta : Q \times A \to Q$, the *initial state* q_0, and the set of *final states* $F \subseteq Q$. If $\delta(q, w) = p$ is a transition of automaton $\mathcal{A}$, w is said to be the *label* of the transition. We call a transition $\delta(p, w) = q$ *successful* if its destination is in the final states F of $\mathcal{A}$. A word w in A^* is *accepted* by $\mathcal{A}$ if it is the label of a successful transition from the initial state of $\mathcal{A}$: $\delta(q_0, w) \in F$. The symbol $|\mathcal{A}|$ denotes the number of states of an automaton $\mathcal{A}$. The set of all accepted words of $\mathcal{A}$, or the *language* of $\mathcal{A}$, is denoted by $L(\mathcal{A})$. We call two automata *equivalent* if their languages are identical. A deterministic automaton $\mathcal{A} = \langle Q, A, \delta, q_0, F \rangle$ is *trim* if, for each state $q \in Q$, there exist two words v and w such that $\delta(q_0, v) = q$ (*accessible*) and $\delta(q, w) \in F$ (*co-accessible*).

2.2 Graphs and Adjacency Matrices

A *multidigraph* G is a pair $G = \langle N, E \rangle$; the set of *nodes* N, the multiset of *edges* E. The *adjacency matrix* $M(G)$ of G is an $N \times N$ matrix defined by $M(G)_{ij} :=$ the number of edges from node i to node j. The *spectrum* of a matrix M is the multiset of the eigenvalues of M and is denoted by $\lambda(M)$. The *kernel* of a matrix M is the subspace defined by $\ker(M) = \{v \mid Mv = 0\}$. We denote the dimension, rank and nullity (the dimension of the kernel) of M by $\dim(M), \operatorname{rank}(M)$ and $\operatorname{null}(M)$ respectively. The following dimension formula is known as the *rank-nullity theorem*:

$$\dim(M) = \operatorname{rank}(M) + \operatorname{null}(M).$$

A *partition* π of a multidigraph $G = \langle N, E \rangle$ is a set of nodal sets $\pi = \{C_1, C_2, \ldots, C_k\}$ that satisfies the following three conditions:

$$\emptyset \notin \pi \quad \text{and} \quad \bigcup_{C \in \pi} C = N \quad \text{and} \quad \forall i, j \in [1, k], i \neq j \Rightarrow C_i \cap C_j = \emptyset.$$

Let $M_{i,j}$ be the submatrix of M formed by the rows in C_i and the columns in C_j. We call the matrix

$$\pi M = \begin{bmatrix} M_{1,1} & \cdots & M_{1,k} \\ \vdots & & \vdots \\ M_{k,1} & \cdots & M_{k,k} \end{bmatrix} \tag{2.1}$$

the *partitioned matrix* induced on M by π. The *characteristic matrix* S of π is the $|N| \times k$ matrix defined $S_{ij} = 1$ if $i \in C_j$ and 0 otherwise. In general, S is a full rank matrix ($\operatorname{rank}(S) = \min(|N|, k) = k$) and $S^T S = \operatorname{diag}(|C_1|, |C_2|, \ldots, |C_k|)$ where S^T is the transpose of S. The *quotient matrix* M^π of M by π is defined as the $k \times k$ matrix:

$$M^\pi = (S^T S)^{-1} S^T M S \tag{2.2}$$

where $(S^T S)^{-1} = \operatorname{diag}(\frac{1}{|C_1|}, \frac{1}{|C_2|}, \ldots, \frac{1}{|C_k|})$. That is, $(M^\pi)_{ij}$ denotes the average row sum of the block matrix $(\pi M)_{i,j}$, in the intuitive sense of the word.

2.3   Nerode Partition

Let q be a state of $\mathcal{A}$. We denote by $F(q)$ the set of words w that are labels of a successful transition starting from q. It is called the future of the state q. Two states p and q are said to be *Nerode equivalent* if and only if $F(p) = F(q)$. *Nerode partition* is the partition induced by Nerode equivalence.

Nerode's theorem states that states of minimal automaton are blocks of Nerode partition, edges and terminal states are defined accordingly (cf. [5,1]). That is, note that the adjacency matrix of a minimal automaton equals the quotient matrix of the adjacency matrix of an equivalent automaton by its Nerode partition.

2.4   Equitable Partition

If the row sum of each block matrix $M_{i,j}$ in Equation (2.1) induced by π is constant, then the partition π is called *equitable*. In this case the characteristic matrix S of π satisfies the following equation (cf. Article 15 in [4]):

$$MS = SM^\pi \tag{2.3}$$

If v is an eigenvector of M^π belonging to the eigenvalue λ, then Sv is an eigenvector of M belonging to the same eigenvalue λ. Indeed, left-multiplication of the eigenvalue equation $M^\pi v = \lambda v$ by S yields:

$$\lambda Sv = (SM^\pi)v = (MS)v = M(Sv).$$

We conclude the the following lemma.

Lemma 1. *Let π be an equitable partition of a matrix M and M^π be its induced quotient matrix, then $\lambda(M^\pi) \subseteq \lambda(M)$ holds.* $\square$

3   Minimality Properties of Minimal Automata

The "if" direction of the Theorem 1 is obvious from the rank-nullity theorem. The proof of the "only if" direction is based on the following two propositions.

Proposition 3.1. *Let π be an equitable partition of a matrix M and M^π be its induced quotient matrix, then the following inequalities hold.*

$$\dim(M^\pi) \leq \dim(M), \quad \mathrm{rank}(M^\pi) \leq \mathrm{rank}(M), \quad \mathrm{null}(M^\pi) \leq \mathrm{null}(M).$$

Proof. Since $\dim(M^\pi) \leq \dim(M)$ is obvious, we prove only the rest two inequalities.

Let v be a vector in the kernel of M^π and w be a vector not in the kernel of M^π, then the following equations hold.

$$MSv = S(M^\pi v) = \mathbf{0}, \tag{3.1}$$
$$MSw = S(M^\pi w) \neq \mathbf{0}. \tag{3.2}$$

Equation (3.2) is induced by $M^\pi w \neq 0$ and $Su \neq 0$ for any $u \neq 0$ since S has full rank. Equation (3.1) and (3.2) imply

$$v \in \ker(M^\pi) \Rightarrow Sv \in \ker(M) \quad \text{and} \quad w \notin \ker(M^\pi) \Rightarrow Sw \notin \ker(M).$$

For any linearly independent vectors u and u', the vectors Su and Su' are also linearly independent since S has full rank. This gives the rest two inequalities.

$\square$

Proposition 3.2. *Nerode partition is equitable.*

Proof. Let $\mathcal{A} = \langle Q, A, \delta, q_0, F \rangle$ be a deterministic automaton and $\pi = (C_1, C_2, \ldots, C_k)$ be its Nerode partition. The proof is by contradiction.

Assume π is not equitable, then there exist C_i and C_j in π and p and q in C_i such that p and q have different number of transition rules into C_j. We assume without loss of generality that the number of transition rules into C_j of p is larger than q's. Then there exists at least one letter a in A such that $\delta(p, a) \in C_j$ and $\delta(q, a) \notin C_j$. Let $p_a = \delta(p, a)$ and $q_a = \delta(q, a)$, then p_a and q_a are not Nerode equivalent since p_a belongs to another partition of q_a's. Hence $F(p_a) \neq F(q_a)$ holds and either $F(p_a)$ or $F(q_a)$ is not empty. We assume without loss of generality that $F(p_a)$ is not empty. Because $F(p_a) \neq F(q_a)$ and $F(p_a) \neq \emptyset$, there exists w in $F(p_a)$ such that $w \notin F(q_a)$ and then w satisfies:

$$\delta(p, aw) = \delta(p_a, w) \in F \quad \text{and} \quad \delta(q, aw) = \delta(q_a, w) \notin F.$$

This leads that p and q are not Nerode equivalent even though p and q belong to the same Nerode equivalent class C_i. This is a contradiction. $\square$

4 Rank-one Languages and Expanded Canonical Automata

In this section, we focus on rank-one languages and introduce expanded canonical automata. Firstly, we introduce the well-known general properties of rank-one matrices (*cf.* Proposition 1 in [6]).

Property 1 (characterization of a rank one matrix). Let M, $n \geq 2$, be a n-dimensional real matrix of rank one. Then

1. There exist vectors $x, y \in \mathbb{C}^n \setminus \{0\}$ such that $M = xy^T$;
2. M has at most one non-zero eigenvalue with algebraic multiplicity 1;
3. This eigenvalue is $y^T x$. $\diamond$

Property 1 shows that, for any rank-one language L, its counting function can be represented as a monomial: $C_L(n) = \alpha\lambda^n$ for $n > 0$ and natural numbers α and λ. In addition, rank-one matrices have beneficial property that their power can be computable in constant time with linear-time preprocessing. Indeed, for any m-dimensional rank-one matrix M, there exists x, y such that $xy^T = M$ hence, the following equation holds for $\lambda = y^T x$:

$$M^n = (xy^T)^n = x(y^T x)^{n-1} y^T = \lambda^{n-1} xy^T = \lambda^{n-1} M.$$

This shows that $(M^n)_{ij}$ equals $\lambda^{n-1}(xy^T)_{ij}$, and the inner product of x and y has linear-time complexity with respect to its dimension m.

4.1   In-vector and Out-vector

For any rank-one matrix M, we can construct $\boldsymbol{x}$ and $\boldsymbol{y}$ such that $M = \boldsymbol{x}\boldsymbol{y}^T$ from the ratio of the number of incoming edges and outgoing edges, respectively.

Definition 2. Let M be an n-dimensional rank-one matrix. The *in-vector* of M is a non-zero row vector having minimum length in M and denoted by in(M). Because M is rank-one, each row vector $\boldsymbol{v}_i$ in M can be represented as $\boldsymbol{v}_i = \alpha_i \cdot$ in(M) for some natural number $\alpha_i \geq 1$. The *out-vector* of M is the n-dimensional column vector that has an i-th element defined as the above coefficient α_i and denoted by out(M).

By this construction, it is clear that an in-vector and out-vector satisfy out$(M) \cdot$ in$(M) = M$.

4.2   Expanded Canonical Automata

First, we define a normal form of a rank-one automaton.

Definition 3. A rank-one automaton $\mathcal{A}$ is *expanded normal* if for its adjacency matrix M, each element of the in-vector of M equals to zero or one.   ◇

Secondly, we propose the operation *expansion* that expands the given matrix (graph or automaton) algebraically.

Definition 4. Let $^\pi M$ and M be two matrices of dimension m and n, respectively. We define $^\pi M$ as an expansion of M if there exists a partition $\pi = \{C_1, C_2, \ldots, C_n\}$ of $^\pi M$ such that the characteristic matrix S of π satisfies:

$$^\pi M = SM(S^T S)^{-1} S^T.$$ ◇

Expansion is an algebraic transformation that increases the dimension of the given matrix. Intuitively, expansion can be regarded as an inverse operation of quotient. Indeed, for any expanded matrix πM of some n-dimensional M by π and its characteristic matrix S, we have the following equation:

$$(^\pi M)^\pi = (S^T S)^{-1} S^T (^\pi M) S = (S^T S)^{-1} S^T \left(SM(S^T S)^{-1} S^T \right) S = M$$

which holds by Equation (2.2) and Definition 4. If M is rank-one, then for any expanded matrix $^\pi M$ of M, the out-vector of $^\pi M$ consists of same elements as the out-vector of M. This reflects the invariance of the number of outgoing transition rules of the Nerode equivalent states.

Finally, we define a canonical automaton of a rank-one language: *expanded canonical automaton*. The minimal automaton of a regular language K is uniquely determined by K, whereas the expanded canonical automaton of a rank-one language L is not uniquely determined, but its graph structure is uniquely determined by L.

Definition 5. Let L be a rank-one language, then we define its expanded canonical automaton $^\pi\mathcal{A}_L$ as the expanded automaton of the minimal automaton $\mathcal{A}_L$ of L by a partition $\pi = \{C_1, C_2, \ldots C_{|\mathcal{A}_L|}\}$ such that, for all $C_i \in \pi$, $|C_i| = \text{in}(\mathcal{A}_L)_i$ if $\text{in}(\mathcal{A}_L)_i \neq 0$ and 1 otherwise.   ◇

By the definition, it is clear that for any rank-one language L, its expanded canonical automaton is expanded normal (cf. $\mathcal{A}_1$ and its expanded canonical automaton C_1 in Example 1). We have introduced expanded canonical automata for analysis and evaluation of the closure properties of rank-one languages. Because of the limitations of space, a detailed discussion of expanded canonical automata is not possible here.

References

1. Béal, M.P., Crochemore, M.: Minimizing incomplete automata. In: Finite-State Methods and Natural Language Processing (FSMNLP 2008), pp. 9–16. Joint Research Center (2008), http://igm.univ-mlv.fr/~beal/Recherche/Publications/minimizingIncomplete.pdf
2. Choffrut, C., Goldwurm, M.: Rational transductions and complexity of counting problems. Mathematical Systems Theory 28(5), 437–450 (1995), http://dblp.uni-trier.de/db/journals/mst/mst28.html#ChoffrutG95a
3. Goldberg, A., Sipser, M.: Compression and ranking. In: Proceedings of the Seventeenth Annual ACM Symposium on Theory of Computing, STOC 1985, pp. 440–448. ACM, New York (1985), http://doi.acm.org/10.1145/22145.22194
4. Mieghem, P.V.: Graph Spectra for Complex Networks. Cambridge University Press, New York (2011)
5. Nerode, A.: Linear automaton transformations. Proceedings of the American Mathematical Society 9(4), 541–544 (1958)
6. Osnaga, S.M.: On rank one matrices and invariant subspaces. Balkan Journal of Geometry and its Applications (BJGA) 10(1), 145–148 (2005), http://eudml.org/doc/126283
7. Sakarovitch, J.: Elements of Automata Theory. Cambridge University Press, New York (2009)
8. Shur, A.M.: Combinatorial complexity of regular languages. In: Hirsch, E.A., Razborov, A.A., Semenov, A., Slissenko, A. (eds.) CSR 2008. LNCS, vol. 5010, pp. 289–301. Springer, Heidelberg (2008)
9. Shur, A.M.: Combinatorial characterization of formal languages. CoRR abs/1010.5456 (2010), http://dblp.uni-trier.de/db/journals/corr/corr1010.html#abs-1010-5456
10. Sin'ya, R.: Rans: More advanced usage of regular expressions, http://sinya8282.github.io/RANS/
11. Sin'ya, R.: Text compression using abstract numeration system on a regular language. Computer Software 30(3), 163–179 (2013), English extended abstract is available at http://arxiv.org/abs/1308.0267 (in Japanese)
12. Sin'ya, R.: Graph spectral properties of deterministic finite au tomata (full paper, 12 pages) (2014), http://www.shudo.is.titech.ac.jp/members/sinya

Input-Driven Pushdown Automata
with Limited Nondeterminism
(Invited Paper)

Alexander Okhotin[1],* and Kai Salomaa[2],**

[1] Department of Mathematics and Statistics,
University of Turku, FI-20014 Turku, Finland
`alexander.okhotin@utu.fi`
[2] School of Computing, Queen's University,
Kingston, Ontario K7L 3N6, Canada
`ksalomaa@cs.queensu.ca`

Abstract. It is known that determinizing a nondeterministic input-driven pushdown automaton (NIDPDA) of size n results in the worst case in a machine of size $2^{\Theta(n^2)}$ (R. Alur, P. Madhusudan, "Adding nesting structure to words", J.ACM 56(3), 2009). This paper considers the special case of k-path NIDPDAs, which have at most k computations on any input. It is shown that the smallest deterministic IDPDA equivalent to a k-path NIDPDA of size n is of size $\Theta(n^k)$. The paper also gives an algorithm for deciding whether or not a given NIDPDA has the k-path property, for a given k; if k is fixed, the problem is P-complete.

1 Introduction

In computations of an *input-driven pushdown automaton* (IDPDA), the current input symbol determines whether the automaton pushes to the stack, pops from the stack, or does not touch the stack. These automata were originally introduced by Mehlhorn [18], and the early research on the model, carried out in the 1980s, was mostly concerned with the computational complexity of the languages they recognize [4,9]. Alur and Madhusudan [1] reintroduced the model in 2004 under the name of *visibly pushdown automata*, and considerably more work has been done on the model since then [2,5,7]. Further equivalent automaton models studied in recent literature include *nested word automata* [2,6,8,26] and *pushdown forest automata* [10]. Nested words provide a natural model for applications such as XML document processing, where data has a dual linear-hierarchical structure.

In the 1980s, von Braunmühl and Verbeek [4] established that for input-driven automata, nondeterminism is equivalent to determinism: this was done by presenting a simulation of an n-state nondetermistic input-driven automaton (NIDPDA) by a deterministic input-driven automaton (DIDPDA) with 2^{n^2} states. Later, Alur and Madhusudan [1,2] proved that $2^{\Omega(n^2)}$ states are necessary

---

* Supported by the Academy of Finland under grant 257857.
** Supported by NSERC under grant OGP0147224.

A.M. Shur and M.V. Volkov (Eds.): DLT 2014, LNCS 8633, pp. 84–102, 2014.

in the worst case. Thus, determinization of general NIDPDAs incurs a $2^{\Theta(n^2)}$ size blow-up. These and other complexity issues of input-driven automata are reported in the authors' [22] recent survey paper.

This paper investigates succinctness questions for a subclass of input-driven automata, in which nondeterminism is regarded as a limited resource. In general, various ways of quantifying the nondeterminism of automaton models have been considered in the literature [11,12,14]. Possibly the most widely studied approach is to measure the number of accepting computations on each input string. *Unambiguous* automata have a unique accepting computation on every accepted input, whereas the number of rejecting computations is irrelevant. Similarly, in *finitely ambiguous* automata, the number of accepting computations on each input is bounded by a constant. When the number of accepting computations is unbounded, one can consider its dependence on the length of the input string, and accordingly distinguish between *polynomially ambiguous* and *exponentially ambiguous* automata, etc. [16].

In the case of finite automata, it is well known that simulating an n-state unambiguous finite automaton (UFA) by a deterministic finite automaton (DFA) requires in the worst case 2^n states, and converting a general nondeterministic finite automaton (NFA) to a UFA requires $2^n - 1$ states [17]. The authors [21] introduced and studied *unambiguous input-driven pushdown automata* (UIDPDA) and established analogous descriptional complexity trade-offs: determinizing an n-state UIDPDA requires $2^{\Theta(n^2)}$ states and the succinctness trade-off between NIDPDA and UIDPDA is more or less the same.

Another approach to quantifying the nondeterminism is to limit the total number of computations, whether accepting or rejecting. The simplest example of this approach is a deterministic automaton with k initial states (a *k-entry automaton*) where all allowed nondeterminism is in the choice of the initial state. More generally, a nondeterministic automaton is said to be *k-path*, with $k \geqslant 1$, if, for every input, there are at most k nondeterministic computations on that input. An automaton that is k-path for some k is called *finite-path*. In the case of finite automata, this measure has been studied under the name 'leaf size' [3,14] or 'tree width' [23,24]. Multiple-entry DFAs also received some attention [13].

This paper defines finite-path input-driven automata, along with a more restricted multiple-entry variant, and considers the descriptional complexity of determinizing these models. A simulation of multiple-entry input-driven automata (MDIDPDA) by equivalent deterministic automata (DIDPDA) is described in Section 3, where it is also proved to be optimal with respect to the number of states. This result is fairly similar to the case of finite automata. On the other hand, as shown in Section 4, transforming a finite-path NIDPDA to a DIDPDA recognizing the same language incurs some complications specific for input-driven automata. Nevertheless, the resulting complexity trade-off is asymptotically comparable to the one in the case of finite automata.

A related question considered in this paper is the descriptional complexity of converting a general NIDPDA to a finite-path automaton. In Section 5, it is proved that in the worst case, a finite-path NIDPDA equivalent to an NIDPDA of the general form is not more succinct than a DIDPDA. The proof adapts the ideas of Goldstine et al. [12], who investigated the succinctness of NFAs with limited nondeterminism.

The last subject, investigated in Section 6, is the computational complexity of deciding whether a given NIDPDA has the k-path property. If the number of paths $k \geqslant 1$ is given, this problem is shown to be decidable. Furthermore, if k is treated as a constant, then the problem is P-complete. However, the decidability of testing whether a given NIDPDA has the finite path property (that is, is k-path for some unknown k) remains open.

2   Definitions

This section recalls the basic definitions and notation of input-driven pushdown automata. For more details on the applications of this model, the reader is referred to the seminal paper by Alur and Madhusudan [2], whereas its complexity is surveyed in the authors' [22] recent paper. General references on automata theory and on descriptional complexity include a book by Shallit [27], a handbook chapter by Yu [28] and a survey by Goldstine et al. [11].

In the following, Σ denotes a finite alphabet and Σ^* is the set of strings over Σ, the length of a string $w \in \Sigma^*$ is $|w|$, the reversal of w is w^R, and the empty string (of length 0) is ε. The cardinality of a finite set S is $|S|$.

In an input-driven pushdown automaton, the input alphabet Σ is split into three disjoint parts as $\Sigma = \Sigma_{+1} \cup \Sigma_{-1} \cup \Sigma_0$, where the symbols in Σ_{+1} and in Σ_{-1} are called *left brackets* and *right brackets*, respectively, and the elements of Σ_0 are known as *neutral symbols*. The type of the current input symbol determines whether the automaton pushes onto the stack (Σ_{+1}), pops from the stack (Σ_{-1}), or does not touch the stack (Σ_0). Throughout this paper, this partition is assumed whenever Σ is an input alphabet of an input-driven automaton.

Input-driven automata normally operate over *well-nested* strings, in which every left bracket has a matching right bracket and vice versa. Well-nested strings over Σ are formally defined by induction: the empty string is well-nested; every symbol from Σ_0 is a well-nested string; a concatenation of well-nested strings is well-nested; for every well-nested string u and for every left bracket $< \, \in \Sigma_{+1}$ and right bracket $> \, \in \Sigma_{-1}$, the string $<u>$ is also well-nested.

The definitions of input-driven automata given below assume that the input string is always well-nested. For extended definitions of automata that can handle ill-nested inputs, the reader is referred to Alur and Madhusudan [1], as well as to the authors' survey [22]. All results in this paper can be reformulated for not necessarily well-nested strings without any difficulties. However, the well-nestedness assumption leads to a clearer presentation.

2.1   Deterministic Automata

Definition 1. *A deterministic input-driven pushdown automaton (DIDPDA) is a 6-tuple $A = (\Sigma, Q, \Gamma, q_0, [\delta_a]_{a \in \Sigma}, F)$, where*

- $\Sigma = \Sigma_{+1} \cup \Sigma_{-1} \cup \Sigma_0$ is an input alphabet split into three disjoint parts;
- Q is a finite set of (internal) states of the automaton, with an initial state $q_0 \in Q$ and with a subset of accepting states $F \subseteq Q$;
- Γ is a finite pushdown alphabet;
- the transition function by each left bracket symbol $< \, \in \Sigma_{+1}$ is a partial function $\delta_< : Q \to Q \times \Gamma$, which, for a given current state, provides the next state and the symbol to be pushed onto the stack;
- for every right bracket symbol $> \, \in \Sigma_{-1}$, there is a partial function $\delta_> : Q \times \Gamma \to Q$ specifying the next state, assuming that the given stack symbol is popped from the stack;
- for a neutral symbol $c \in \Sigma_0$, the state change is described by a partial function $\delta_c : Q \to Q$.

A configuration of A is a triple (q, w, x), where $q \in Q$ is the state, $w \in \Sigma^*$ is the remaining input and $x \in \Gamma^*$ is the stack contents. For a well-nested input string $w_0 \in \Sigma^*$, the computation of A on w_0 is a uniquely determined sequence of configurations, which begins with the initial configuration (q_0, w_0, ε), and proceeds as follows. For each configuration with at least one remaining input symbol, the next configuration is given by a single step transition function:

- for each left bracket $< \, \in \Sigma_{+1}$, let $(q, <w, x) \vdash_A (q', w, \gamma x)$, where $\delta_<(q) = (q', \gamma)$;
- for every right bracket $> \, \in \Sigma_{-1}$, let $(q, >w, \gamma x) \vdash_A (\delta_>(q, \gamma), w, x)$;
- for a neutral symbol $c \in \Sigma_0$, define $(q, cw, x) \vdash_A (\delta_c(q), w, x)$.

Normally, the computation continues until the input string is exhausted, at which point the stack shall always be empty; then the last configuration $(q, \varepsilon, \varepsilon)$ is accepting if $q \in F$. If q is not in F, or if the computation ends prematurely by an undefined transition, this is a rejecting computation. The language $L(A)$ recognized by the automaton is the set of all strings $w \in \Sigma^*$, on which the computation from (q_0, w, ε) is accepting.

Note that the next computation step depends on the top-of-stack symbol only when the input symbol is a right bracket and the automaton accordingly pops from the stack. When reading a left bracket or a neutral symbol, the computation does not depend on the stack contents.

If the alphabet contains only neutral symbols, that is, if $\Sigma_{+1} = \Sigma_{-1} = \varnothing$, then a DIDPDA becomes a deterministic finite automaton (DFA).

2.2   Nondeterministic Automata

In nondeterministic automata, the transition function is multi-valued, and accordingly, there may be multiple computations on the same input. A string is considered accepted, if at least one of these computations is accepting.

Definition 2. A nondeterministic input-driven pushdown automaton (NIDPDA) is defined as a 6-tuple $B = (\Sigma, Q, \Gamma, Q_0, [\delta_a]_{a \in \Sigma}, F)$, in which the input alphabet $\Sigma = \Sigma_{+1} \cup \Sigma_{-1} \cup \Sigma_0$, the set of states Q, the pushdown alphabet Γ, and the set of accepting states $F \subseteq Q$ are as in Definition 1, and

- there is a set of initial states $Q_0 \in Q$, and a computation may begin from any of them;
- for each left bracket symbol $< \in \Sigma_{+1}$, the transition function $\delta_< : Q \to 2^{Q \times \Gamma}$ provides, for a given current state, a set of possible outcomes, which are pairs of the next state and the stack symbol to be pushed;
- for every right bracket symbol $> \in \Sigma_{-1}$, there is a function $\delta_> : Q \times \Gamma \to 2^Q$ that lists all possible next states, if the given stack symbol is popped from the stack;
- for a neutral symbol $c \in \Sigma_0$, there is a function $\delta_c : Q \to 2^Q$.

A configuration of B is again a triple (q, w, x), with $q \in Q$, $w \in \Sigma^*$ and $x \in \Gamma^*$. On an input string $w_0 \in \Sigma^*$, each configuration (q_0, w_0, ε), with $q_0 \in Q_0$, is called an initial configuration. For every configuration, all possible next configurations are defined by a transition relation as follows:

- for each left bracket $< \in \Sigma_{+1}$, and for all pairs (q', γ) in $\delta_<(q)$, let $(q, <w, x) \vdash_B (q', w, \gamma x)$;
- for every right bracket $> \in \Sigma_{-1}$, and for all $q' \in \delta_>(q, \gamma)$, let $(q, >w, \gamma x) \vdash_B (q', w, x)$;
- for a neutral symbol $c \in \Sigma_0$, for every $q' \in \delta_c(q)$, let $(q, cw, x) \vdash_B (q', w, x)$.

Now, a computation of B on an input w_0 is a sequence of configurations C_0, C_1, ..., C_m, with $m \geqslant 0$, where C_0 is an initial configuration on w_0, $C_i \vdash_B C_{i+1}$ for each $i \in \{0, \ldots, m-1\}$, and C_m has no next configuration. A computation is accepting if it ends with a configuration $(q, \varepsilon, \varepsilon)$, with $q \in F$. The language $L(A)$ recognized by the automaton is the set of all strings $w \in \Sigma^*$, on which there is at least one accepting computation.

An NIDPDA B becomes a DIDPDA, if $|Q_0| = 1$ and its transition functions give at most one possible action for each input symbol, state and top stack symbol. If there are only neutral symbols in the alphabet ($\Sigma_{+1} = \Sigma_{-1} = \varnothing$), then an NIDPDA is a nondeterministic finite automaton (NFA).

2.3   Limited Nondeterminism

This paper considers nondeterministic input-driven automata in which the number of computations on every input string is bounded by a constant. The simplest way of how this can happen is if the nondeterminism occurs only in the beginning of the computation (that is, in the choice of the initial states) whereas the transition function is deterministic. This leads to the notion of multiple-entry DIDPDAs.

Definition 3. A *multiple-entry DIDPDA (MDIDPDA) is a 6-tuple* $(\Sigma, Q, \Gamma, Q_0, [\delta_a]_{a \in \Sigma}, F)$, *which contains a set of initial states* $Q_0 \subseteq Q$ *instead of a single initial state, with the rest of the components as in Definition 1. If* $|Q_0| = k$, *then such an automaton is called a* k*-entry DIDPDA.*

An MDIDPDA with no stack operations (with $\Sigma_{+1} = \Sigma_{-1} = \varnothing$) is a multiple-entry DFA (MDFA), as studied by Holzer, Salomaa and Yu [13].

On each input string, a k-entry DIDPDA has exactly k computations. For general NIDPDAs that may make nondeterministic choices anywhere in a computation, one can consider limitations on the amount of nondeterminism, measured as the maximum number of computations on an input string.

Definition 4. *An NIDPDA $A = (\Sigma, Q, \Gamma, Q_0, [\delta_a]_{a \in \Sigma}, F)$ is called k-path if, for every input string $w \in \Sigma^*$, A has at most k computations on w. An NIDPDA that is k-path for some k is called a finite-path NIDPDA.*

2.4   Lower Bound Methods

The study of descriptional complexity typically involves proving that every automaton of a certain kind recognizing a certain language must have at least a certain number of states. Such statements are proved by specific proof techniques known as lower bound methods. The lower bound method presented in this section applies to DIDPDAs and employs a well-known idea commonly used for DFAs.

For every string w over an alphabet $\Sigma_{+1} \cup \Sigma_{-1} \cup \Sigma_0$, the difference between the number of left brackets and the number of right brackets in w is called the *depth of w*. Let d be the depth of w; then, after processing w, any input-driven automaton will have exactly d symbols in the stack.

Definition 5. *Let $\Sigma = \Sigma_{+1} \cup \Sigma_{-1} \cup \Sigma_0$ be an alphabet and let $L \subseteq \Sigma^*$. Let $S = \{x_1, \ldots, x_m\}$ be a set of strings over Σ, and let $d \geqslant 1$ be an integer.*

The set S is said to be a d-separator set for L if

(i) *each string x_i, with $i \in \{1, \ldots, m\}$, is of depth d and is a prefix of some string in L;*
(ii) *for every two strings x_i and x_j, with $1 \leqslant i < j \leqslant m$, there exists a string $w_{i,j}$, such that exactly one of the strings $x_i w_{i,j}$ and $x_j w_{i,j}$ is in L.*

The notion of a d-separator set was used by Okhotin, Piao and Salomaa [20] to obtain certain lower bounds for IDPDAs. The following new lower bound condition involving this notion is adapted for the particular type of separator sets used in this paper.

Lemma 1. *Let L be a language over an alphabet $\Sigma = \Sigma_{+1} \cup \Sigma_{-1} \cup \Sigma_0$, and let S be a 1-separator set for L of the form $S = \{<x_1, \ldots, <x_m\}$, for a fixed left bracket $< \in \Sigma_{+1}$ and for $x_1, \ldots, x_m \in \Sigma_0^*$. Then, every DIDPDA for L must have at least m states.*

Proof. Let A be an arbitrary DIDPDA for L. Each string $<x_i$ in S is a prefix of a string in L, and hence the automaton A must reach the end of $<x_i$ in some state q_i. Since all these strings begin with the same left bracket $<$ and continue with only neutral symbols, the stack contents of A after reading each $<x_i$ is independent of i.

It is claimed that all states $q_1, \ldots, q_m$ are pairwise distinct. Indeed, if $q_i = q_j$, for some $i \neq j$, then the configurations of A after reading $<x_i$ and $<x_j$ are identical, and therefore, for any continuation $w \in \Sigma^*$, the string $<x_i w$ is accepted if and only if $<x_j w$ is accepted. This means that for any string w, $<x_i w \in L$ if and only if $<x_j w \in L$, which contradicts the assumption that S is a separator set for L. $\square$

3   From Multiple-entry DIDPDA to Ordinary DIDPDA

The precise succinctness trade-off between k-entry DIDPDAs and ordinary DIDPDAs is determined in two steps. First, there is a transformation of an arbitrary k-entry DIDPDA to an ordinary DIDPDA that recognizes the same language. Later it will be proved that this transformation is optimal with respect to the number of states.

3.1   Construction

On a given input, a k-entry DIDPDA A carries out k different computations corresponding to the choice of the initial state. An ordinary DIDPDA B simulating A carries out a single computation, in which, at every point, it keeps track of all k ongoing computations of A, remembering their states. On a left bracket, each of the k instances of A has its own stack symbol to push, and thus B pushes a k-tuple of stack symbols. Later, when a matching right bracket is read, B pops this k-tuple, and each of the simulated instances of A receives its own stack symbol. In order to match the stack symbols to the current states, both the states and the stack symbols of B are *ordered k-tuples*, rather than just sets of k elements.

A further detail of the construction is that any of A's computations may prematurely end by reaching an undefined transition. In this case, B puts a special symbol for "failed" $(-)$ in the corresponding component of its state.

Lemma 2. *For every k-entry DIDPDA A, with a set of states Q and a stack alphabet Γ, there exists and can be effectively constructed a standard DIDPDA B with the set of states $(Q \cup \{-\})^k \setminus \{-\}^k$ and the stack alphabet Γ^k, which recognizes the same language.*

Proof. Let $A = (\Sigma, Q, \Gamma, \{q_0^1, \ldots, q_0^k\}, [\delta_a]_{a \in \Sigma}, F)$ be the given k-entry DIDPDA, where $q_0^1, \ldots, q_0^k$ is any enumeration of its initial states.

Construct a standard DIDPDA $B = (\Sigma, (Q \cup \{-\})^k \setminus \{-\}^k, \Gamma^k, (q_0^1, \ldots, q_0^k), [\delta_a']_{a \in \Sigma}, F')$, where the transitions are defined as follows. A transition on each neutral symbol $c \in \Sigma_0$ simply applies the transition function of A to each component of the state.

$$\delta_c'(q_1, \ldots, q_k) = \big(\delta_c(q_1), \ldots, \delta_c(q_k)\big)$$

If $q_i = -$, or if $q_i \in Q$ and $\delta_c(q_i)$ is undefined, then the corresponding component of the result is set to "failed" $(-)$.

For each left bracket $< \in \Sigma_{+1}$, if the state of the simulating automaton B is a k-tuple $(q_1, \ldots, q_k) \in (Q \cup \{-\})^k$, let $\delta_<(q_i) = (q_i', \gamma_i)$ be A's transition for each component. Then define

$$\delta_<'(q_1, \ldots, q_k) = ((q_1', \ldots, q_k'), (\gamma_1, \ldots, \gamma_k))$$

As in the previous case, if $q_i = -$, or if $\delta_c(q_i)$ is undefined, then the corresponding component of the result is set to "failed" $(-)$, whereas the stack symbol γ_i is defined arbitrarily.

The transition for each right-bracket $> \in \Sigma_{-1}$ is again defined componentwise.

$$\delta_>'((q_1, \ldots, q_k), (\gamma_1, \ldots, \gamma_k)) = (\delta_>(q_1, \gamma_1), \ldots, \delta_>(q_k, \gamma_k))$$

Finally, a k-tuple of states is an accepting state of B if one of its components is an accepting state of A.

$$F' = \{ (q_1, \ldots, q_k) \mid \exists i : q_i \in F \}$$

Note that the k-tuple $(-, \ldots, -)$ is excluded from B's set of states, because if all simulated computations have failed, then B may reject straight away by an undefined transition. $\qquad\square$

3.2 Lower Bound

The construction in Lemma 2 transforms an n-state k-entry DIDPDA to an ordinary DIDPDA with $(n+1)^k - 1$ states.

The first lower bound argument proves that the construction in Lemma 2 is optimal with respect to the number of states, that is, that $(n+1)^k - 1$ states are necessary in the worst case, using an alphabet growing exponentially with n.

Lemma 3. *For every $k \geqslant 1$ and $n \geqslant k$, there exists an alphabet $\Sigma^{k,n}$ and a language $L_{k,n}$ over $\Sigma^{k,n}$ recognized by a k-entry DIDPDA with n states and k stack symbols, such that any DIDPDA for $L_{k,n}$ needs $(n+1)^k - 1$ states.*

Proof. The alphabet consists of one left bracket $(\Sigma_{+1}^{k,n} = \{<\})$, one right bracket $(\Sigma_{-1}^{k,n} = \{>\})$, and a large number of neutral symbols indexed by partial functions mapping numbers to numbers: $\Sigma_0^{k,n} = X_{\text{func}} \cup Y_{\text{func}}$, where

$$X_{\text{func}} = \{ a_f \mid f : \{1, \ldots, k\} \to \{1, \ldots, n, \text{undefined}\} \},$$
$$Y_{\text{func}} = \{ b_g \mid g : \{1, \ldots, n\} \to \{1, \ldots, k, \text{undefined}\} \}.$$

Consider the following language, in which every string consists of two such functions, that their composition has a fixed point.

$$\widehat{L}_{k,n} = \{ <a_f b_g> \mid a_f, b_g \in \Sigma_0^{k,n}, \exists s \in \{1, \ldots, k\} : g(f(s)) = s \}.$$

A k-entry DIDPDA recognizing precisely the language $\widehat{L}_{k,n}$ would need to use more than n states, because, besides verifying the fixed point of the composition, it would have to ensure that the given string is exactly of the form $<a_fb_g>$. This proof uses a slightly different language, which is defined by a k-entry DIDPDA A with n states and k stack symbols recognizing it. This automaton shall accept all well-formed strings from $\widehat{L}_{k,n}$, and possibly some strings not of the form $<a_fb_g>$:

$$L(A) \cap \{ <a_fb_g> \mid a_f \in X_{\text{func}}, \, b_g \in Y_{\text{func}} \} = \widehat{L}_{k,n}.$$

Define the desired k-entry automaton $A = (\Sigma, \Gamma, Q, Q_0, [\delta_a]_{a \in \Sigma}, F)$, with the set of states $Q = \{q_1, \ldots, q_n\}$ and the stack alphabet $\Gamma = \{t_1, \ldots, t_k\}$. The first k of its states are initial ($Q_0 = \{q_1, \ldots, q_k\}$), and q_1 is the only accepting state ($F = \{q_1\}$). The transitions of A are defined as follows.

$$\begin{aligned}
\delta_<(q_i) &= (q_i, t_i), & &\text{for all } i \in \{1, \ldots, k\} \\
\delta_{a_f}(q_i) &= q_{f(i)}, & &\text{for } a_f \in X_{\text{func}}, \, i \in \{1, \ldots, k\} \\
\delta_{b_g}(q_j) &= q_{g(j)}, & &\text{for } b_g \in Y_{\text{func}}, \, j \in \{1, \ldots, n\} \\
\delta_>(q_j, t_i) &= \begin{cases} q_1, & \text{if } i = j, \\ \text{undefined}, & \text{otherwise}, \end{cases} & &\text{for all } i, j \in \{1, \ldots, k\}
\end{aligned}$$

All transitions not listed above are undefined. In particular, whenever f is undefined on i or g is undefined on j, the above transitions by a_f and b_g are not defined.

When A begins its computation in any initial state q_i, with $1 \leqslant i \leqslant k$, it guesses a fixed point of $g \circ f$. At the first left bracket $<$, the automaton remembers the guess in the stack, as a symbol t_i, and stays in state q_i. After reading the two neutral symbols a_f and b_g, the state q_i is transformed into $q_{g(f(i))}$, and the transition at the last right bracket $>$ pops t_i from the stack and checks that $g(f(i)) = i$. Thus, A accepts an input of the form $w = <a_fb_g>$ if and only if $w \in L'_{k,n}$.

The lower bound argument uses the following 1-separator set.

$$S = \{ <a_f \mid a_f \in X_{\text{func}}, \, \exists i \in \{1, \ldots, k\} : f(i) \text{ is defined} \}$$

To see that S is a 1-separator set for $L(A)$, consider any two distinct strings $<a_{f_1}, <a_{f_2} \in S$, with $f_1 \neq f_2$. Then there exists an argument $i \in \{1, \ldots, k\}$, on which $f_1(i)$ is defined, and $f_2(i)$ is either undefined or defined differently from $f_1(i)$. Let g be a partial function from $\{1, \ldots, n\}$ to $\{1, \ldots, k\}$, defined, for $1 \leqslant j \leqslant n$, as

$$g(j) = \begin{cases} i, & \text{if } j = f_1(i), \\ \text{undefined}, & \text{otherwise}. \end{cases}$$

Then $g(f_1(i)) = i$ and hence $<a_{f_1}b_g> \in L(A)$. On the other hand, for any ℓ, if $g(f_2(\ell))$ is defined, then its value can only be i, whereas its argument ℓ cannot be i because g is not defined on $f_2(i)$. Therefore, $g(f_2(\ell)) \neq \ell$ for all arguments $\ell \in \{1, \ldots, k\}$, which means that the string $<a_{f_2}b_g>$ is not in $L(A)$.

By Lemma 1, it follows that any DIDPDA for $L(A)$ needs at least $|S| = (n+1)^k - 1$ states. □

The above lemma shows that the construction in Lemma 2 cannot be improved, at least not for unbounded alphabets. The next lower bound construction uses multiple-entry DIDPDAs over a small fixed alphabet. The construction, generally speaking, encodes the language in Lemma 3 over a fixed alphabet, so that the functions f and g are implemented by listing their values in unary notation. Unfortunately, a multiple-entry DIDPDA for this language needs to use more states than in Lemma 3, leading to a weaker lower bound.

Lemma 4. *For every $k \geqslant 1$ and $n \geqslant k$, there exists a language $L_{k,n}$ over an alphabet Σ of size 6, which is recognized by a k-entry DIDPDA A with $2n + 2k$ states and k stack symbols, whereas every DIDPDA recognizing $L_{k,n}$ needs at least n^k states.*

Proof. The alphabet Σ consists of a single pair of brackets and four neutral symbols: $\Sigma_{+1} = \{<\}$, $\Sigma_{-1} = \{>\}$ and $\Sigma_0 = \{a, b, c, d\}$. Before defining the promised witness language $L_{k,n}$, consider the following base language.

$$\widehat{L}_{k,n} = \{ <a^{i_1}b \cdots a^{i_r}bc^{j_1}d \cdots c^{j_s}d> \mid r, s \geqslant 0, (\exists z \leqslant k)\, j_{i_z} = z \text{ and } i_z \leqslant n \}$$

All strings in $\widehat{L}_{k,n}$ belong to the set $M_0 = <(a^*b)^*(c^*d)^*>$ of well-formed strings. The k-entry DIDPDA $A_{n,k}$ constructed in this proof shall accept all strings from $\widehat{L}_{k,n}$ and possibly some strings from outside of M_0, so that

$$L(A_{n,k}) \cap M_0 = \widehat{L}_{k,n}. \tag{1}$$

(Note that a k-entry DIDPDA recognizing exactly $\widehat{L}_{k,n}$ would need more than $2n + 2k$ states.)

For an input string $<a^{i_1}b \cdots a^{i_r}bc^{j_1}d \cdots c^{j_s}d>$, its substring a^{i_z} is referred to as *the zth a-component*. Similarly, the substring c^{j_z} is called *the zth c-component*.

The set of states of $A_{n,k}$ is chosen to be

$$Q = \{p_1, \ldots, p_k\} \cup \{p'_1, \ldots, p'_k\} \cup \{q_1, \ldots, q_n\} \cup \{q'_1, \ldots, q'_n\},$$

of which the initial states are $\{p_1, \ldots, p_k\}$ and q_1 is the only accepting state. The set of stack symbols is $\{1, \ldots, k\}$. At the beginning of the computation, $A_{k,n}$ nondeterministically choses an initial state p_z, and this is the only nondeterminism in the computation. When reading the first left bracket $<$, the automaton pushes z to the stack. Then, using the states $p_z, p_{z-1}, \ldots p_1$ as a counter, $A_{k,n}$ finds the zth a-component a^{i_z}, and then uses the states $q_1, \ldots, q_n$ to count its length i_z.

On the next symbol b, the automaton enters the state q'_{i_z}. If $i_z > n$, the computation fails. Otherwise, the computation proceeds through the remaining symbols a and b without changing its state. Note that at this stage of the computation, the length of the a-component is stored in a "primed" state q'_{i_z}, because the computation needs to remember that it has finished counting the length of the a-component.

Eventually, the automaton reaches the first c-component in the state q'_{i_z}. This second part of the input is formed of the symbols c, d rather than a, b, and hence the automaton can now reuse its states for a different purpose. First, it uses the states q'_{i_z}, q'_{i_z-1}, ..., q'_1 as a counter to locate the i_zth c-component $c^{j_{iz}}$. Next, using the states p_1, ..., p_k, it measures the length of $c^{j_{iz}}$ and stores it in a state $p'_{j_{iz}}$. If $j_{i_z} > k$, the computation fails. Finally, the computation by-passes the remaining c-components in the state $p'_{j_{iz}}$, and the pop-transition at the right bracket $>$ matches the state to the stack symbol to verify that $j_{i_z} = z$. If the verification is successful, the automaton enters the accepting state q_1.

The lower bound on the size of a DIDPDA equivalent to $A_{k,n}$ is established using the following 1-separator set for $L(A_{k,n})$.

$$S = \{ <a^{i_1}b \cdots a^{i_k}b \mid i_1, \ldots, i_k \in \{1, \ldots, n\} \}$$

Consider any two distinct strings $u = <a^{i_1}b \cdots a^{i_k}b$ and $u' = <a^{i'_1}b \cdots a^{i'_k}b$ in S, and let them be different in some ℓ-th component, that is, $i_\ell \neq i'_\ell$ for $\ell \in \{1, \ldots, k\}$. Define another string $w = (c^{n+1}d)^{i_\ell-1}c^\ell d>$. Then both concatenations uw and $u'w$ are well-formed strings in M_0, and uw is in $\widehat{L}_{k,n}$, whereas $u'w$ is not. This, by (1), means that $uw \in L(A_{k,n})$ and $u'w \notin L(A_{k,n})$. Since such a separator string w exists for any two distinct elements of S, it follows that S is a 1-separator set for $L(A_{k,n})$. Then, by Lemma 1, every DIDPDA for $L(A_{k,n})$ needs at least $|S| = n^k$ states.   □

The lower bound construction in Lemma 4 uses an alphabet with 4 neutral symbols, as well as one left bracket and one right bracket. By doubling the number of states of the original k-entry DIDPDA, the number of neutral symbols could be reduced to two.

The results of Lemmata 2, 3 and 4 are summarized in the following theorem.

Theorem 1. *A k-entry DIDPDA with n states and m stack symbols can be simulated by a DIDPDA with $(n+1)^k - 1$ states and m^k stack symbols.*

For every $n \geqslant k$, there exists a k-entry DIDPDA with n states and k stack symbols, defined over an alphabet depending on n and k, such that any equivalent DIDPDA needs at least $(n+1)^k - 1$ states.

For every $n \geqslant k$, there exists a k-entry DIDPDA over a 6-symbol alphabet, with n states and k stack symbols, such that any equivalent DIDPDA needs at least $(\frac{n}{4})^k$ states.

When k is viewed as a constant, the descriptional complexity of converting k-entry DIDPDAs to DIDPDAs coincides asymtotically with the corresponding bound for finite automata [13,23,24].

Corollary 1. *Let $k \geqslant 1$ be a constant. The worst case size of a DIDPDA equivalent to a k-entry DIDPDA of size n is $\Theta(n^k)$.*

4   Converting a k-path NIDPDA to a DIDPDA

The transformation of a k-entry DIDPDA A to an ordinary DIDPDA B involves simulating k instances of A. The states of B are k-tuples of states of A. On a left bracket, each of the k instances of A has its own stack symbol to push, and thus B pushes a k-tuple of stack symbols. Later, when a matching right bracket is read, B pops this k-tuple, and each of the simulated instances of A receives its own stack symbol.

When simulating a k-path NIDPDA A by a DIDPDA B, in the beginning, B has a k_0-tuple of instances of A to simulate, where k_0 is the number of its initial states, and later on, as A has to make nondeterministic decisions, B extends this tuple by spawning additional computations off the existing ones. This entails the following complication with simulating the stack operations in each of these instances. Consider B encountering a left bracket with k_1 ongoing computations. Then, each of the k_1 instances of A pushes its own symbol, and B simulates that by pushing a k_1-tuple. Later on, inside the brackets, some of the simulated instances of A may have to make nondeterministic decisions, so that B arrives to the matching right bracket with a k_2-tuple of states, where $k_2 \geqslant k_1$. At this point, B pops a k_1-tuple of stack symbols from the stack, and if $k_2 > k_1$, then the question is, how those k_1 symbols are to be matched to the k_2 states?

Consider that each of these k_2 states occurs in one of the computations of A, and the stack symbol pushed in that computation is among the k_1 symbols now popped from the stack. In order to match states to stack symbols, it is sufficient to maintain an additional data structure that records, for each currently traced computation, from which computation it has spawned off. This data structure is implemented in the set of states of the automaton constructed in the following lemma.

Lemma 5. *For every k-path NIDPDA A, with a set of states Q and a stack alphabet Γ, there exists and can be effectively constructed a DIDPDA B with the set of states $Q' \subseteq \bigcup_{\ell=1}^{k} \left((Q \cup \{-\}) \times \{1, \ldots, \ell\} \right)^{\ell}$ and with the stack alphabet $\Gamma' = \Gamma^{\leqslant k}$, which recognizes the same language.*

Proof. Let $A = (\Sigma, Q, Q_0, \Gamma, [\delta_a]_{a \in \Sigma}, F)$ be the given k-path NIDPDA, and let $Q_0 = \{q_0^1, \ldots, q_0^{\ell_0}\}$ be its set of initial states. Each state of the simulating DIDPDA $B = (\Sigma, Q', q_0', \Gamma', [\delta_a']_{a \in \Sigma}, F')$ contains between ℓ_0 and k numbered components, each corresponding to one of the currently simulated computations of A. Every component is comprised of a state and of a pointer to another computation, from which this computation has stemmed off. The state may be replaced by a failure marker $(-)$ indicating that this branch of computation has already rejected. A computation is referenced by the number of the component. Accordingly, the set of states of B is defined as follows.

$$Q' = \left\{ \begin{pmatrix} q_1 \cdots q_\ell \\ p_1 \cdots p_\ell \end{pmatrix} \;\middle|\; \ell_0 \leqslant \ell \leqslant k,\; q_i \in Q \cup \{-\},\; p_i \in \{1, \ldots, i\} \right\}$$

To be more precise, for every i-th component, if $i \leqslant \ell_0$, then p_i always points to itself, being fixed as $p_i = i$, and if $i > \ell_0$, then p_i points to some earlier component, that is, $p_i \in \{1, \ldots, i-1\}$.

The pushdown alphabet $\Gamma' = \Gamma^{\leqslant k}$ of the simulating automaton allows communicating any ℓ-tuples of stack symbols, for $\ell \leqslant k$.

The initial state of B initializes ℓ_0 simulated computations, and marks each of them as stemming off itself.

$$Q_0 = \begin{pmatrix} q_0^1 & q_0^2 & \cdots & q_0^{\ell_0} \\ 1 & 2 & \ldots & \ell_0 \end{pmatrix}$$

For every neutral symbol $c \in \Sigma_0$, the transition from a state $\begin{pmatrix} q_1 & \cdots & q_\ell \\ p_1 & \cdots & p_\ell \end{pmatrix}$ by c is defined as follows. For each i-th component, let the nondeterministic transitions of A be $\delta_c(q_i) = \{q'_{i,1}, \ldots, q'_{i,m_i}\}$. The automaton B keeps the first target state $q'_{i,1}$ in the same i-th component; the remaining target states $q'_{i,2}, \ldots, q'_{i,m_i}$ are put into newly created components, provided with pointers to the i-th component from which they are stemmed off. Thus, the transition of B leads to a state with $\sum_{i=1}^{\ell} m_i$ components.

$$\delta'_c \begin{pmatrix} q_1 & \cdots & q_\ell \\ p_1 & \cdots & p_\ell \end{pmatrix} = \begin{pmatrix} q'_{1,1} & \cdots & q'_{\ell,1} & q'_{1,2} & \cdots & q'_{1,m_1} & \cdots & q'_{\ell,2} & \cdots & q'_{\ell,m_\ell} \\ p_1 & \cdots & p_\ell & 1 & \ldots & 1 & \ldots & \ell & \ldots & \ell \end{pmatrix}$$

(vertical lines separate the new components added in this transition)

The transition of B from a state $\begin{pmatrix} q_1 & \cdots & q_\ell \\ p_1 & \cdots & p_\ell \end{pmatrix}$ by a left bracket $< \, \in \Sigma_{+1}$ is defined very similarly. For each i-th component, let $\delta_<(q_i) = \{(q'_{i,1}, \gamma_{i,1}), \ldots, (q'_{i,m_i}, \gamma_{i,m_i})\}$. Then the transition of B leads to a state with $\sum_{i=1}^{\ell} m_i$ components and pushes a stack symbol with $\sum_{i=1}^{\ell} m_i$ components, where the first ℓ components contain all first nondeterministic choices, and the new computations occupy the rest of the components.

$$\delta'_< \begin{pmatrix} q_1 & \cdots & q_\ell \\ p_1 & \cdots & p_\ell \end{pmatrix} = \left(\begin{pmatrix} q'_{1,1} & \cdots & q'_{\ell,1} & q'_{1,2} & \cdots & q'_{1,m_1} & \cdots & q'_{\ell,2} & \cdots & q'_{\ell,m_\ell} \\ p_1 & \cdots & p_\ell & 1 & \ldots & 1 & \ldots & \ell & \ldots & \ell \end{pmatrix}, \right.$$

$$\left. (\gamma_{1,1}, \ldots, \gamma_{\ell,1}, \gamma_{1,2}, \ldots, \gamma_{1,m_1}, \ldots, \gamma_{\ell,2}, \ldots, \gamma_{\ell,m_\ell}) \right)$$

For every right bracket $> \, \in \Sigma_{-1}$, let $(\gamma_1, \ldots, \gamma_n)$ be the stack symbol pushed at the matching left bracket, and let $\begin{pmatrix} q_1 & \cdots & q_\ell \\ p_1 & \cdots & p_\ell \end{pmatrix}$ be the current state, which must satisfy $\ell_0 \leqslant n \leqslant \ell \leqslant k$. Define a function $f \colon \{1, \ldots, \ell\} \to \{1, \ldots, n\}$ that matches the components of the current state to the components of the stack symbol by first setting $f(i) = i$ for all $i \leqslant n$ (these are the components that already existed at the time the stack symbol was pushed), and then inductively defining $f(p_i) = f(i)$ whenever $f(i)$ is already defined. Then, for every i-th component of the current state, A's transition should use the matching stack symbol $\gamma_{f(i)}$; let $\delta_>(q_i, \gamma_{f(i)}) = \{q'_{i,1}, \ldots, q'_{i,m_i}\}$. Then B's transition from this state upon popping this stack symbol leads to the following state, in which all nondeterministic choices are listed in the same order as in the previous two cases.

$$\delta'_< \left(\begin{pmatrix} q_1 & \cdots & q_\ell \\ p_1 & \cdots & p_\ell \end{pmatrix}, (\gamma_1, \ldots, \gamma_n) \right) = \begin{pmatrix} q'_{1,1} & \cdots & q'_{\ell,1} & q'_{1,2} & \cdots & q'_{1,m_1} & \cdots & q'_{\ell,2} & \cdots & q'_{\ell,m_\ell} \\ p_1 & \cdots & p_\ell & 1 & \ldots & 1 & \ldots & \ell & \ldots & \ell \end{pmatrix}$$

Finally, a state $((q_1, p_1), \ldots, (q_\ell, p_\ell))$ is marked as accepting, if $q_i \in F$ for some i.   □

At the moment, there is no lower bound on the complexity of determinizing k-path NIDPDAs that would precisely match the upper bound in Lemma 5. However, the lower bound for multiple-entry automata, given earlier in Lemma 3, also applies in this case. The resulting bounds, presented in the next theorem, are fairly close to each other.

Theorem 2. *A k-path NIDPDA with n states and m stack symbols can be simulated by a DIDPDA with*

$$\sum_{i=1}^{k}(n+1)^i \cdot i^i \text{ states and } \sum_{i=1}^{k} m^i \text{ stack symbols.}$$

For every $n \geqslant k$, there exists a k-path NIDPDA A with n states and k stack symbols, such that any equivalent DIDPDA needs at least $(n+1)^k - 1$ states.

It remains open whether k-path NIDPDAs can be used to establish a better lower bound than MDIDPDAs from Lemma 3. It is also possible that the construction in Lemma 5 may be unoptimal.

5   From NIDPDA to k-path NIDPDA

The next result is that converting a general NIDPDA to a k-path automaton entails, for any $k \geqslant 1$, the same $2^{\Omega(n^2)}$ size blow-up as determinizing an arbitrary NIDPDA. This result is established by a modified argument of Goldstine et al. [12], who used it to obtain an exponential size blow-up of converting a general NFA to an NFA with limited nondeterminism. The main idea of the argument is that for languages of a certain form presented below, k-path automata are no more succinct than deterministic automata.

Lemma 6. *Let A be an NIDPDA over an alphabet Σ, such that L(A) consists only of well nested strings, and let \$ be a new neutral symbol not in Σ.*

Assume that the language $(\$L(A)\$)^*$ is recognized by a k-path NIDPDA with n states and m stack symbols. Then $(\$L(A)\$)^*$ can be recognized by a DIDPDA with n states and m stack symbols.

Proof. The proof by Goldstine et al. [12, Lemma 4.1] demonstrates, for every regular language L, that if $(\$L\$)^*$ has a k-path NFA C with n states, then $(\$L\$)^*$ has a DFA D with n states.[1] The proof is constructive and D is obtained simply by taking the "deterministic restriction" of the transition function of C (i.e., by eliminating from C all transitions with more than one outcome) and choosing the initial state appropriately. The same argument, word for word, works for input-driven automata, because A accepts only well nested strings, and hence the stack of A is always empty at the end of an accepted string.   □

---

[1] More precisely, the proof by Goldstine et al. [12] considers a so-called *NFA with finite branching:* a finite-path NFA is guaranteed to have finite branching, however, the reverse implication does not hold.

The proof of the following result is inspired by another argument by Goldstine et al. [12, Thm 4.2].

Theorem 3. *For any* $k \geqslant 1$, *the worst-case number of states in a* k-*path NIDPDA equivalent to an NIDPDA with* n *states is* $2^{\Theta(n^2)}$.

Proof. The upper bound follows from the upper bound for determinizing an NIDPDA [2]. So, it is sufficient to show that there exists an NIDPDA with $O(n)$ states, such that an equivalent k-path NIDPDA needs at least 2^{n^2} states. This is done, roughly speaking, by applying the result of Lemma 6 to a language exhibiting the worst-case size blow-up of determinization.

Choose $\Sigma_{+1} = \{<\}$, $\Sigma_{-1} = \{>\}$ and $\Sigma_0 = \{a, b, \#, \$\}$. Consider the following language, due to Alur and Madhusudan [2, Thm. 3.4].

$$L'_n = \{<a^{i_1}b^{j_1} \cdots a^{i_m}b^{j_m}\#b^s>a^r \mid i_x, j_x \geqslant 1, \, r, s \in \{1, \ldots, n\}, \text{ and,}$$
$$\exists z \in \{1, \ldots, m\}: \, b_{j_z} = s, \, a_{i_z} = r\}$$

The language L'_n is recognized by an NIDPDA A with $O(n)$ states and n stack symbols [2]. Let $L_n = (\$L'_n\$)^*$. Since L'_n consists of only well-nested strings, an NIDPDA for L_n can be obtained from A by adding one new state.

For any binary relation R on the set $\{1, \ldots, n\}$, denote

$$w_R = a^{i_1}b^{j_1}a^{i_2}b^{j_2} \cdots a^{i_{|R|}}b^{j_{|R|}},$$

where $(i_1, j_1), \ldots, (i_{|R|}, j_{|R|})$ is a listing of the elements of R in an arbitrary order. Choose

$$S = \{ \$<w_R \mid R \subseteq \{1, \ldots, n\} \times \{1, \ldots, n\} \}.$$

Every string in S is a prefix of a string in L_n. Note that $w_\varnothing = \varepsilon$ and $\$<$ is a prefix of L_n. Consider any two distinct strings $\$<w_{R_1}$, $\$<w_{R_2} \in S$, and let $(i, j) \in R_1 \setminus R_2$ be an element on which the relations differ. These two strings are distinguished by continuing them with $\#b^j>a^i\$$: indeed, $\$<w_{R_1}\#b^j>a^i\$ \in L_n$, but $\$<w_{R_2}\#b^j>a^i\$ \notin L_n$. Hence, S is a 1-separator set for L_n and, by Lemma 1, any DIDPDA for L_n needs at least $|S| = 2^{n^2}$ states. By Lemma 6, any k-path NIDPDA for L_n also needs the same number of states. $\square$

6   Decision Problems

The k-path property of an NIDPDA refers to its computations on all possible inputs, and hence, based on the syntactic specification of an NIDPDA, it is not at all clear whether or not it has this property. The following lemma describes an algorithm for testing a given NIDPDA for being k-path.

Lemma 7. *Given an NIDPDA* A *with* n *states and a number* $k \geqslant 1$, *one can decide in time* $poly(k^k \cdot n^k)$ *whether or not* A *has the* k-*path property.*

Proof. Let $A = (\Sigma, \Gamma, Q, Q_0, [\delta_a]_{a \in \Sigma}, F)$ be an NIDPDA, and let $Q_0 = \{q_1^I, \ldots, q_m^I\}$ be its set of initial states. If $|Q_0| > k$, then A does not have the k-path property. Assume that $m \leqslant k$.

The proposed algorithm works by constructing a new deterministic IDPDA B, which will accept an input string if and only if A has more than k computations on this string. Then A has the k-path property if and only if B generates the empty language, and the known polynomial-time algorithm for the DIDPDA emptiness problem [2] computes the desired answer.

The DIDPDA B is constructed generally according to Lemma 5, so that on an input w, it simulates up to k computation paths of A on w. The important difference is that in Lemma 5, the simulated automaton A is assumed to have the k-path property, whereas in this case it is not guaranteed to have it. If there are at most k computations on the given input w, the simulation proceeds exactly as defined in Lemma 5, but then, in the end of the computation, B always rejects. If A has more than k computations on w, then the simulation eventually requests to spawn more computations than B can handle; at this point, B enters a special state, in which it ignores the rest of the input and accepts in the end.

The number of states of B is less than $k^k \cdot (n+2)^k + 1$, and hence is polynomial in $k^k \cdot n^k$. As the running time of the emptiness test for B is polynomial in the size of B, this proves the time bound. $\qquad\square$

Theorem 4. *For a fixed $k \geqslant 1$, checking whether or not a given NIDPDA has the k-path property is P-complete.*

Proof. If k is treated as a constant, then the algorithm in Lemma 7 runs in time polynomial in n. The P-hardness is proved by reduction (in logspace) from the DIDPDA emptiness problem, which is known to be P-complete [15]. The reduction begins with modifying a given DIDPDA, so that from any accepting state it makes a nondeterministic transition on a new neutral symbol, with $k+1$ choices. The resulting NIDPDA is k-path if and only if the original DIDPDA does not accept any strings. $\qquad\square$

The algorithm in Lemma 7 allows testing an automaton for the k-path property only for a given value of k. Another decision problem is whether a given NIDPDA is *finite-path*, that is, k-path for some unknown value of k. The decidability status of this problem remains unknown.

Problem 1. Is it decidable whether or not a given NIDPDA has the finite path property?

To compare with the case of finite automata, the finite path property for a given NFA can be checked by analyzing its state transition graph [23,24]. However, this approach does not seem to work in the presence of stack operations.

Another point of comparison is the case of nondeterministic pushdown automata of the general form (NPDA), to which the definition of the k-path property is directly extended. Here the k-path property becomes undecidable already for $k = 3$.

Theorem 5. *It is undecidable to determine whether a given NPDA is 3-path.*

Proof. The proof is by reduction from the Post Correspondence Problem (PCP) [27, Sect. 6.5]. Consider a PCP instance (u_1, v_1), ..., (u_m, v_m), where $m \geqslant 1$, all u_i, v_i are strings over an alphabet $\{a, b\}$, and the question is, whether there is such a sequence $i_1, \ldots, i_\ell \in \{1, \ldots, m\}$, that $u_{i_1} u_{i_2} \cdots u_{i_\ell} = v_{i_1} v_{i_2} \cdots v_{i_\ell}$.

Construct an NPDA A with an input alphabet $\{\#, a, b, 1, \ldots, m\}$, that processes inputs of the following general form.

$$i_1 u_{i_1} i_2 u_{i_2} \cdots i_\ell u_{i_\ell} \# j_j v_{j_h}^R j_{h-1} v_{j_{h-1}}^R \cdots j_1 v_1^R \#, \quad i_1, \ldots, i_\ell, j_1, \ldots, j_h \in \{1, \ldots, m\}$$

Given such an input, in the beginning of the computation, A nondeterministically chooses to check either whether $u_{i_1} u_{i_2} \cdots u_{i_\ell} = v_{i_1} v_{i_2} \cdots v_{i_h}$, or whether $\ell = h$ and $i_1 i_2 \cdots i_\ell = j_1 j_2 \cdots j_h$. In each of the two nondeterministic branches, assuming that the verification was successful, A makes a nondeterministic step with two choices on the last symbol $\#$. If the check fails, then A rejects without making any further nondeterministic choices. If the PCP instance has a solution, then on the encoding of that solution, A has 4 paths. Therefore, A has the 3-path property if and only if the given PCP instance does not have a solution.   □

Besides testing a given NIDPDA for the k-path property, one can also consider the standard decision problems for k-path NIDPDAs, such as emptiness, equivalence and inclusion. All these problems can be solved is polynomial time by first determinizing the given automata with a polynomial blow-up (Theorem 2), and then applying an algorithm by Alur and Madhusudan [2]. Furthermore, these problems are P-complete, which can be proved by a reduction from the emptiness problem for DIDPDAs [15].

7   Conclusion

Though the trade-off between k-entry input-driven automata and deterministic input-driven automata in terms of the number of states has been determined exactly (Theorem 1), many other contributions of this paper call for more precise arguments. Already for this trade-off, one could consider how this transformation affects the number of stack symbols; for the NIDPDA determinization, results of this kind were obtained by Okhotin, Piao and Salomaa [20]. For another transformation studied in the paper—the determininization of k-path NIDPDAs—the given lower and upper bounds on the number of states do not precisely match (Theorem 2). In order to improve the estimation, it is suggested to give a new lower bound construction using k-path NIDPDAs, rather than simply multiple-entry automata. Another open question is whether the finite path property for NIDPDAs (that is, with the number of paths not specified) can be effectively decided.

A natural extension of this work is to limit the amount of nondeterminism as a function of input length, analogously to the existing work on finite automata employing limited nondeterminism [16,19,25].

References

1. Alur, R., Madhusudan, P.: Visibly pushdown languages. In: ACM Symposium on Theory of Computing, STOC 2004, Chicago, USA, June 13-16, pp. 202–211 (2004)
2. Alur, R., Madhusudan, P.: Adding nesting structure to words. Journal of the ACM 56(3) (2009)
3. Björklund, H., Martens, W.: The tractability frontier of NFA minimization. J. Comput. System Sci. 78, 198–210 (2012)
4. von Braunmühl, B., Verbeek, R.: Input driven languages are recognized in $\log n$ space. North-Holland Mathematics Studies 102, 1–19 (1985)
5. Chervet, P., Walukiewicz, I.: Minimizing variants of visibly pushdown automata. In: Kučera, L., Kučera, A. (eds.) MFCS 2007. LNCS, vol. 4708, pp. 135–146. Springer, Heidelberg (2007)
6. Chistikov, D., Majumdar, R.: A uniformization theorem for nested word to word transductions. In: Konstantinidis, S. (ed.) CIAA 2013. LNCS, vol. 7982, pp. 97–108. Springer, Heidelberg (2013)
7. Crespi-Reghizzi, S., Mandrioli, D.: Operator precedence and the visibly pushdown property. In: Dediu, A.-H., Fernau, H., Martín-Vide, C. (eds.) LATA 2010. LNCS, vol. 6031, pp. 214–226. Springer, Heidelberg (2010)
8. Debarbieux, D., Gauwin, O., Niehren, J., Sebastian, T., Zergaoui, M.: Early nested word automata for XPath query answering on XML streams. In: Konstantinidis, S. (ed.) CIAA 2013. LNCS, vol. 7982, pp. 292–305. Springer, Heidelberg (2013)
9. Dymond, P.W.: Input-driven languages are in $\log n$ depth. Information Processing Letters 26, 247–250 (1988)
10. Gauwin, O., Niehren, J., Roos, Y.: Streaming tree automata. Information Processing Letters 109, 13–17 (2008)
11. Goldstine, J., Kappes, M., Kintala, C.M.R., Leung, H., Malcher, A., Wotschke, D.: Descriptional complexity of machines with limited resources. Journal of Universal Computer Science 8, 193–234 (2002)
12. Goldstine, J., Kintala, C.M.R., Wotschke, D.: On measuring nondeterminism in regular languages. Information and Computation 86(2), 179–194 (1990)
13. Holzer, M., Salomaa, K., Yu, S.: On the state complexity of k-entry deterministic finite automata. J. Automata, Languages, and Combinatorics 6, 453–466 (2001)
14. Hromkovič, J., Seibert, S., Karhumäki, J., Klauck, H., Schnitger, G.: Communication complexity method for measuring nondeterminism in finite automata. Information and Computation 172, 202–217 (2002)
15. Lange, M.: P-hardness of the emptiness problem for visibly pushdown automata. Inf. Proc. Lett. 111(7), 338–341 (2011)
16. Leung, H.: Separating exponentially ambiguous finite automata from polynomially ambiguous finite automata. SIAM J. Comput. 27, 1073–1082 (1998)
17. Leung, H.: Descriptional complexity of NFA of different ambiguity. Internat. J. Foundations Comput. Sci. 16, 975–984 (2005)
18. Mehlhorn, K.: Pebbling mountain ranges and its application to DCFL-recognition. In: de Bakker, J., van Leeuwen, J. (eds.) ICALP 1980. LNCS, vol. 85, pp. 422–435. Springer, Heidelberg (1980)
19. Okhotin, A.: Unambiguous finite automata over a unary alphabet. Inform. Comput. 212, 15–36 (2012)
20. Okhotin, A., Piao, X., Salomaa, K.: Descriptional complexity of input-driven pushdown automata. In: Bordihn, H., Kutrib, M., Truthe, B. (eds.) Languages Alive 2012. LNCS, vol. 7300, pp. 186–206. Springer, Heidelberg (2012)

21. Okhotin, A., Salomaa, K.: Descriptional complexity of unambiguous nested word automata. In: Dediu, A.-H., Inenaga, S., Martín-Vide, C. (eds.) LATA 2011. LNCS, vol. 6638, pp. 414–426. Springer, Heidelberg (2011)
22. Okhotin, A., Salomaa, K.: Complexity of input-driven pushdown automata. In: Hemaspaandra, L.A. (ed.) SIGACT News Complexity Theory Column 82. SIGACT News (to appear, 2014)
23. Palioudakis, A., Salomaa, K., Akl, S.G.: State complexity and limited nondeterminism. In: Kutrib, M., Moreira, N., Reis, R. (eds.) DCFS 2012. LNCS, vol. 7386, pp. 252–265. Springer, Heidelberg (2012)
24. Palioudakis, A., Salomaa, K., Akl, S.G.: State complexity of finite tree width NFAs. J. Automata, Languages and Combinatorics 17, 245–264 (2012)
25. Palioudakis, A., Salomaa, K., Akl, S.G.: Comparisons between measures of nondeterminism on finite automata. In: Jurgensen, H., Reis, R. (eds.) DCFS 2013. LNCS, vol. 8031, pp. 217–228. Springer, Heidelberg (2013)
26. Salomaa, K.: Limitations of lower bound methods for deterministic nested word automata. Information and Computation 209, 580–589 (2011)
27. Shallit, J.: A Second Course in Formal Languages and Automata Theory. Cambridge University Press (2009)
28. Yu, S.: Regular languages. In: Rozenberg, G., Salomaa, A. (eds.) Handbook of Formal Languages, vol. I, pp. 41–110 (1997)

How to Remove the Look-Ahead
of Top-Down Tree Transducers

Joost Engelfriet[1], Sebastian Maneth[2], and Helmut Seidl[3]

[1] LIACS, Leiden University, The Netherlands
engelfri@liacs.nl
[2] School of Informatics, University of Edinburgh, United Kingdom
smaneth@inf.ed.ac.uk
[3] Institut für Informatik, Technische Universität München, Germany
seidl@in.tum.de

Abstract. For a top-down tree transducer with regular look-ahead we introduce the notion of difference bound, which is a number bounding the difference in output height for any two look-ahead states of the transducer. We present an algorithm that, for a given transducer with a known difference bound, decides whether it is equivalent to a transducer without regular look-ahead, and constructs such a transducer if the answer is positive. All transducers are total and deterministic.

1 Introduction

Many simple tree transformations can be modeled by top-down tree transducers, as recently used in XML database theory (e.g., [6,11,13,15]), in computational linguistics (e.g., [12,14]) and in picture generation [4]. A *top-down tree transducer* is a finite-state device that scans the input tree in a (parallel) top-down fashion, simultaneously producing the output tree. The more expressive (but also more complex) top-down tree transducer *with regular look-ahead* [5] consists of a top-down tree transducer and a finite-state bottom-up tree automaton, called the look-ahead automaton. At each input node, the transducer can inspect the look-ahead state (i.e., the state of the automaton) of each child of that node. Consider, e.g., a transducer M_{ex} of which the look-ahead automaton checks whether the input tree has a leaf labeled a; if so, M_{ex} outputs a, otherwise it outputs a copy of the input tree. Clearly, there is no transducer without look-ahead that realizes the same translation as M_{ex}. In general, is there a method to determine for a given top-down tree transducer *with look-ahead* (dtla), whether or not there is an equivalent top-down tree transducer *without look-ahead* (dtop)?

In this paper we provide a general method as discussed above, for total deterministic transducers. However, part of the method is not automatic: it depends on additional knowledge about the given transducer with look-ahead. For transducers with some restrictions concerning the power to copy and erase, that knowledge can also be obtained automatically.

The main notion on which our method is based, is that of a *difference tree* of a dtla M. Consider two trees obtained from one input tree by replacing one leaf

A.M. Shur and M.V. Volkov (Eds.): DLT 2014, LNCS 8633, pp. 103–115, 2014.

by two different look-ahead states of M. Compare now the two output trees of M on these input trees, where M treats the look-ahead state as representing an input subtree for which the look-ahead automaton arrives in that state at the root of the subtree.[1] Removing the largest common prefix of these two output trees (i.e., every node of which every ancestor has the same label in each of the two trees), we obtain a number of output subtrees that we call difference trees of M. Intuitively, the largest common prefix is the part of the output that does not depend on the two possible look-ahead states of the subtree, whereas a difference tree is a part of the output that can be produced because M knows that look-ahead state. Thus, the set diff(M) of all difference trees of M can be viewed as a measure of the impact of the look-ahead on the behaviour of M. E.g., diff(M_{ex}) is infinite: it consists of the one-node tree a and all trees of which no leaf is labeled a (with one leaf representing a subtree without a-labeled leaves).

The idea of our method is as follows. For any dtop an equivalent canonical dtop can be constructed [6]. Canonical means that each output node is produced as early as possible, and that different states of the transducer are inequivalent. We can generalize that result to dtlas. Thus, if there is a (canonical) dtop N equivalent to the (canonical) dtla M, then M is at least as early as N: at each moment of the translation, the output of N is a prefix of that of M. The output of N is the part of M's output that does not depend on the look-ahead state. Thus, when removing the output of N from that of M, the remaining trees are difference trees of M. Since N is able to simulate M, it has to store these difference trees in its states. Hence, diff(M) is finite. In fact, it turns out that the above description of N's behaviour completely determines N, and so N can be constructed from M and diff(M), and then tested for equivalence with M [9].

A natural number h is a *difference bound* for a dtla M if the following holds: if M has finitely many difference trees, then h is an upper bound on their height. Our first main result is that it is decidable for a given dtla M for which a difference bound is also given, whether M is equivalent to a dtop N, and if so, such a dtop N can be constructed. We do not know whether a difference bound can be computed for every dtla M, but the designer of M will usually be able to determine diff(M) and hence a difference bound for M. Our second main result is that a difference bound can be computed for dtlas that are linear and nonerasing (or even ultralinear and bounded erasing); the proof is too involved to be presented here. The full version of this paper can be found in [7].

Related Work. For deterministic string transducers it is decidable whether a given transducer with look-ahead is equivalent to a transducer without look-ahead, and if so, such a transducer can be constructed. This was proved in [3] (see also [2, Theorem IV.6.1]), for so-called subsequential functions. For macro tree transducers [8] and streaming tree transducers [1], regular look-ahead can always be removed. The same is true for nondeterministic visibly pushdown transducers [10]; for deterministic visibly pushdown transducers the addition of regular look-ahead increases their power, but the decidability of look-ahead removal for these transducers is not studied in [10].

---

[1] Since M is total and deterministic, the output trees exist and are unique, respectively.

2   Top-Down Tree Transducers and Difference Trees

We assume the reader to be familiar with top-down tree transducers working on ranked trees: the number of children of a tree node is determined by its label.

A *deterministic top-down tree transducer with regular look-ahead* (*dtla* for short) is a tuple $M = (Q, \Sigma, \Delta, R, A, P, \delta)$ where Q is a finite set of states of rank 1, Σ and Δ are the ranked input and output alphabets, and P is a finite nonempty set of look-ahead states. For every $p \in P$, $A(p)$ is a tree in $\mathcal{T}_\Delta(Q(\{x_0\}))$ called the p-axiom of M.[2] For every $q \in Q$, $a \in \Sigma$ of rank $k \geq 0$, and $p_1, \ldots, p_k \in P$, the set R contains at most one rule $q(a(x_1 : p_1, \ldots, x_k : p_k)) \to \zeta$ where ζ is a tree in $\mathcal{T}_\Delta(Q(X_k))$ denoted by $\mathsf{rhs}(q, a, p_1, \ldots, p_k)$. Finally, δ is the transition function of the (total deterministic bottom-up) look-ahead automaton (P, δ), i.e., $\delta(a, p_1, \ldots, p_k) \in P$ for every $a \in \Sigma$ of rank $k \geq 0$ and $p_1, \ldots, p_k \in P$. The extension of δ to a mapping from $\mathcal{T}_\Sigma$ to P, also denoted by δ, is defined by $\delta(a(s_1, \ldots, s_k)) = \delta(a, \delta(s_1), \ldots, \delta(s_k))$ for $a \in \Sigma$ of rank $k \geq 0$ and $s_1, \ldots, s_k \in \mathcal{T}_\Sigma$. For $p \in P$ we define $[\![p]\!]_M = \{s \in \mathcal{T}_\Sigma \mid \delta(s) = p\}$. The dtla M *realizes* the partial function $[\![M]\!] : \mathcal{T}_\Sigma \to \mathcal{T}_\Delta$, called its *translation*, defined for $s \in \mathcal{T}_\Sigma$ by $[\![M]\!](s) = A(\delta(s))[q(x_0) \leftarrow [\![q]\!]_M(s) \mid q \in Q]$.[3] For $q \in Q$ the partial function $[\![q]\!]_M : \mathcal{T}_\Sigma \to \mathcal{T}_\Delta$ is defined for $s \in \mathcal{T}_\Sigma$ of the form $a(s_1, \ldots, s_k)$ by $[\![q]\!]_M(s) = \mathsf{rhs}(q, a, \delta(s_1), \ldots, \delta(s_k))[q'(x_i) \leftarrow [\![q']\!]_M(s_i) \mid q' \in Q, 1 \leq i \leq k]$. We write $M(s)$ for $[\![M]\!](s)$, and $q_M(s)$ for $[\![q]\!]_M(s)$. Two dtlas M_1 and M_2 are *equivalent* if they realize the same translation, i.e., $\Sigma_{M_1} = \Sigma_{M_2}$, $\Delta_{M_1} = \Delta_{M_2}$ and $[\![M_1]\!] = [\![M_2]\!]$.

Convention. We (can) assume that all states and look-ahead states of M are *reachable*: $p \in P$ is reachable if $[\![p]\!]_M \neq \emptyset$; $q \in Q$ is reachable if q occurs in an axiom, or in the right-hand side of a rule of which the left-hand side starts with a reachable state.

A dtla M is *total* if its translation $[\![M]\!]$ is a total function, i.e., its domain is $\mathcal{T}_\Sigma$. **From now on we only consider total dtlas.**

A *deterministic top-down tree transducer* (*dtop* for short) is a dtla M such that P is a singleton, i.e., $P = \{p\}$. For convenience, we drop (P, δ) from the tuple defining M, write a rule as $q(a(x_1, \ldots, x_k)) \to \zeta$ rather than $q(a(x_1 : p, \ldots, x_k : p)) \to \zeta$, identify A with the unique axiom $A(p)$, and denote ζ by $\mathsf{rhs}(q, a)$.

A dtla M is *proper* (a *dtpla* for short) if it is not a dtop, i.e., if $|P| \geq 2$.

---

[2] We use variables x_i with $i \in \mathbb{N}$, of rank 0. The set $\{x_0, x_1, x_2, \ldots\}$ is denoted X; for $k \in \mathbb{N}$, $X_k := \{x_1, \ldots, x_k\}$. For a set of trees $\mathcal{T}$, $Q(\mathcal{T})$ is the set of trees $q(t)$ with $q \in Q$ and $t \in \mathcal{T}$, and $\mathcal{T}_\Delta(\mathcal{T})$ is the smallest set of trees $\mathcal{T}'$ containing $\mathcal{T}$ such that $d(t_1, \ldots, t_k) \in \mathcal{T}'$ if $d \in \Delta$ and $t_1, \ldots, t_k \in \mathcal{T}'$ (where d has rank k). We denote $\mathcal{T}_\Delta(\emptyset)$ by $\mathcal{T}_\Delta$. For a tree $t \in \mathcal{T}_\Delta$, we denote by $V(t)$ the set of nodes of t, which are strings of positive natural numbers, i.e., $V(t) \subseteq \mathbb{N}_+^*$ with $\mathbb{N}_+ = \mathbb{N} - \{0\}$. The empty string ε is the root node and, for $i \in \mathbb{N}_+$, vi is the ith child of the node v. Every node $v \in V(t)$ has a label in Δ, denoted $\mathsf{lab}(t, v)$; the subtree of t rooted at v is denoted by t/v. For $d \in \Delta$, we define $V_d(t) = \{v \in V(t) \mid \mathsf{lab}(t, v) = d\}$.

[3] For sets of trees $\mathcal{S}, \mathcal{T}$, a tree $t \in \mathcal{T}$ and a partial function $\psi : \mathcal{S} \to \mathcal{T}$, we define $t[s \leftarrow \psi(s) \mid s \in \mathcal{S}]$ to be the result of replacing every subtree s of t by $\psi(s)$, for every $s \in \mathcal{S}$ (assuming that no tree in $\mathcal{S}$ has a proper subtree in $\mathcal{S}$).

Look-Ahead States in Input Trees. Let $M = (Q, \Sigma, \Delta, R, A, P, \delta)$ be a (total) dtla. To analyze the behaviour of M for different look-ahead states, we consider input trees $\bar{s}$ with occurrences of $p \in P$, viewed as input symbol of rank zero, representing an absent subtree s with $\delta(s) = p$. Intuitively, when M arrives in state q at a p-labeled leaf of $\bar{s}$, we let M output the new symbol $\langle q, p \rangle$ of rank zero, representing the absent output tree $q_M(s)$; thus, input trees $\bar{s} \in \mathcal{T}_\Sigma(P)$ are translated to output trees in $\mathcal{T}_\Delta(Q \times P)$. Formally, we extend M to a dtla $M^\circ = (Q, \Sigma^\circ, \Delta^\circ, R^\circ, A, P, \delta^\circ)$ where $\Sigma^\circ = \Sigma \cup P$, every element of P has rank zero, $\Delta^\circ = \Delta \cup (Q \times P)$, every element of $Q \times P$ has rank zero, $R^\circ = R \cup \{q(p) \to \langle q, p \rangle \mid q \in Q, p \in P, \exists s \in [\![p]\!]_M : q_M(s) \text{ is defined}\}$, and δ° is the extension of δ with $\delta^\circ(p) = p$ for every $p \in P$. For notational simplicity, we will denote $\delta^\circ(\bar{s})$, $M^\circ(\bar{s})$ and $q_{M^\circ}(\bar{s})$ by $\delta(\bar{s})$, $M(\bar{s})$ and $q_M(\bar{s})$, respectively, for every $\bar{s} \in \mathcal{T}_\Sigma(P)$. But note that $[\![p]\!]_M$, $[\![M]\!]$ and $[\![q]\!]_M$ keep their meaning.

A Σ-*context* is a tree in $\mathcal{T}_\Sigma(\{\bot\})$ that contains exactly one occurrence of $\bot$ (which is a new symbol of rank 0). The set of all Σ-contexts is denoted $\mathcal{C}_\Sigma$. For $C \in \mathcal{C}_\Sigma$ and a tree s, the tree $C[s]$ is obtained from the context C by replacing the unique occurrence of $\bot$ in C by s. We consider in particular trees $C[p]$ where $C \in \mathcal{C}_\Sigma$ and $p \in P$. Note that the tree $M(C[p])$ is in $\mathcal{T}_\Delta(Q \times \{p\})$.

Lemma 1. *Let M be a dtla. Let $C \in \mathcal{C}_\Sigma$, $s \in \mathcal{T}_\Sigma(P)$ and $p \in P$ such that $\delta(s) = p$. Then $\delta(C[s]) = \delta(C[p])$ and $M(C[s]) = M(C[p]) \big[\langle q, p \rangle \leftarrow q_M(s) \mid q \in Q \big]$.*

Difference Trees and Difference Bounds. For a ranked alphabet Ω, an Ω-*pattern* (or just pattern) is a tree in $\mathcal{T}_\Omega(\{\bot\})$, where $\bot \notin \Omega$ has rank 0. Intuitively, an Ω-pattern is a prefix of a tree in $\mathcal{T}_\Omega$. If $t_0 \in \mathcal{T}_\Omega(\{\bot\})$ contains exactly k occurrences of $\bot$, and $t_1, \ldots, t_k \in \mathcal{T}_\Omega(\{\bot\})$, then the pattern $t = t_0[t_1, \ldots, t_k]$ is obtained from t_0 by replacing the ith occurrence of $\bot$ (in left-to-right order) by t_i. On the set $\mathcal{T}_\Omega(\{\bot\})$ we define a partial order: for patterns t and t', t' is a *prefix* of t, denoted $t' \sqsubseteq t$, if $t = t'[t_1, \ldots, t_k]$ for some patterns $t_1, \ldots, t_k$; equivalently, $V_b(t') \subseteq V_b(t)$ for every $b \in \Omega$. In [6] the inverse of $\sqsubseteq$ is used. Note that $\bot \sqsubseteq t$ for every pattern t. Every nonempty set Π of Ω-patterns has a greatest lower bound $\sqcap \Pi$ in $\mathcal{T}_\Omega(\{\bot\})$, called the *largest common prefix* of the patterns in Π; it is the unique pattern t' such that for every $v \in \mathbb{N}_+^*$ and $b \in \Omega$, $v \in V_b(t')$ if and only if (1) $v \in V_b(t)$ for every $t \in \Pi$ and (2) every proper ancestor of v is in $V(t')$. For instance, $\sqcap\{\sigma(\tau(a), b), \sigma(b, b)\} = \sigma(\tau(a), b) \sqcap \sigma(b, b) = \sigma(\bot, b)$.

We wish to decide whether the dtla M is equivalent to a dtop. Let C be a Σ-context and let $p, p' \in P$. As explained in the Introduction, we are interested in the difference between $M(C[p])$ and $M(C[p'])$, cf. Lemma 1. Intuitively, a dtop N that is equivalent to M does not know whether the subtree s of an input tree $C[s]$ has look-ahead state p or p', and hence, when reading the context C, it can output at most the largest common prefix $M(C[p]) \sqcap M(C[p'])$ of the output trees $M(C[p])$ and $M(C[p'])$. Let v be a node of $M(C[p]) \sqcap M(C[p'])$ with label $\bot$. Then we say that $M(C[p])/v$ is a *difference tree* of M (and hence, by symmetry, so is $M(C[p'])/v$). Thus, a difference tree is a part of the output that can be produced by M because it knows that s has look-ahead state p (or p'). Intuitively, to simulate M, the dtop N must store the difference trees in its state.

Hence, for N to exist, there should be finitely many difference trees (Corollary 1). We denote the set of all difference trees of M by $\mathsf{diff}(M)$, for varying C, p, p' and v. Thus we define $\mathsf{diff}(M) = \{M(C[p])/v \mid C \in \mathcal{C}_\Sigma, p \in P, \exists p' \in P : v \in V_\perp(M(C[p]) \sqcap M(C[p']))\}$ which is a subset of $\mathcal{T}_\Delta(Q \times P)$. We define the number $\mathsf{maxdiff}(M) \in \mathbb{N} \cup \{\infty\}$ to be the maximal height of all difference trees of M, i.e., $\mathsf{maxdiff}(M) = \sup\{\mathsf{ht}(t) \mid t \in \mathsf{diff}(M)\}$. Intuitively, $\mathsf{maxdiff}(M)$ gives a measure of how much M makes use of its look-ahead information. Obviously, $\mathsf{maxdiff}(M)$ is finite if and only if $\mathsf{diff}(M)$ is finite. A number $h(M) \in \mathbb{N}$ is a *difference bound* for M if either $\mathsf{diff}(M)$ is infinite or $\mathsf{maxdiff}(M) \leq h(M)$. Our first main result is that if a difference bound for M is known, then we can decide whether M is equivalent to a dtop, and if so, construct such a dtop from M (Theorem 2).

Example 1. Let $\Sigma = \Delta = \{\sigma^{(1)}, a^{(0)}, b^{(0)}\}$, the ranked alphabet $\{\sigma, a, b\}$ such that σ has rank 1 and a, b have rank 0. For $n \in \mathbb{N}$, the tree $\sigma(\sigma(\cdots \sigma(a) \cdots))$ with n occurrences of σ is denoted by $\sigma^n a$. Consider a dtla $M = (Q, \Sigma, \Delta, R, A, P, \delta)$ such that $M(\sigma^n a) = a$ and $M(\sigma^n b) = \sigma^n b$ for every $n \in \mathbb{N}$. It is, in fact, the dtla M_{ex} of the Introduction, for this particular input alphabet. Its set of look-ahead states is $P = \{p_a, p_b\}$ with transition function δ defined by $\delta(a) = p_a$, $\delta(b) = p_b$, $\delta(\sigma, p_a) = p_a$ and $\delta(\sigma, p_b) = p_b$. Its set of states is $Q = \{q\}$, its axioms are $A(p_a) = a$ and $A(p_b) = q(x_0)$, and R contains the rules $q(\sigma(x_1 : p_b)) \to \sigma(q(x_1))$ and $q(b) \to b$.

For $C = \sigma^n \perp$, $M(C[p_a]) = a$ and $M(C[p_b]) = \sigma^n \langle q, p_b \rangle$. Since $M(C[p_a]) \sqcap M(C[p_b]) = \perp$, the only node of $M(C[p]) \sqcap M(C[p'])$ with label $\perp$ is ε. Hence, $\mathsf{diff}(M) = \{a\} \cup \{\sigma^n \langle q, p_b \rangle \mid n \in \mathbb{N}\}$ and $\mathsf{maxdiff}(M) = \infty$. Since $\mathsf{diff}(M)$ is infinite, M is not equivalent to a dtop, as will be shown in Corollary 1. $\square$

Example 2. Let $\Sigma = \{\sigma^{(2)}, aa^{(0)}, ab^{(0)}, ba^{(0)}, bb^{(0)}\}$ where we view aa, ab, ba and bb as symbols, and let $\Delta = \Sigma \cup \{\#^{(2)}, a^{(0)}, b^{(0)}\}$ with $\sigma^{(3)}$ instead of $\sigma^{(2)}$. We consider a dtla M such that $M(yz) = yz$ for $y, z \in \{a, b\}$; moreover, $M(\sigma(s_1, s_2)) = \sigma(M(s_1), M(s_2), \#(y, z))$ where $y \in \{a, b\}$ is the first letter of the label of the left-most leaf of $\sigma(s_1, s_2)$ and $z \in \{a, b\}$ is the second letter of the label of its right-most leaf. It has four look-ahead states p_{yz} with $y, z \in \{a, b\}$, such that $\delta(yz) = p_{yz}$ and $\delta(\sigma, p_{wx}, p_{yz}) = p_{wz}$ for all $w, x, y, z \in \{a, b\}$. It has one state q, its axioms are $A(p_{yz}) = q(x_0)$, and its rules are $q(yz) \to yz$ and $q(\sigma(x_1 : p_{wx}, x_2 : p_{yz})) \to \sigma(q(x_1), q(x_2), \#(w, z))$ for all $w, x, y, z \in \{a, b\}$.

Consider a Σ-context C and the trees $M(C[p_{aa}])$ and $M(C[p_{ba}])$. Let u be the node of C with $\mathsf{lab}(C, u) = \perp$. It is easy to see that the nodes of $M(C[p]) \sqcap M(C[p'])$ with label $\perp$ are the node u and all nodes $v \cdot (3, 1)$ such that $v \neq u$ is a node of C and u is the left-most leaf of C/v. That gives the difference trees $M(C[p_{aa}])/u = \langle q, p_{aa} \rangle$, $M(C[p_{ba}])/u = \langle q, p_{ba} \rangle$, $M(C[p_{aa}])/v \cdot (3, 1) = a$ and $M(C[p_{ba}])/v \cdot (3, 1) = b$. Thus, $\mathsf{diff}(M) = \{a, b\} \cup \{\langle q, p_{yz} \rangle \mid y, z \in \{a, b\}\}$ and $\mathsf{maxdiff}(M) = 0$.

A dtop N equivalent to M has states q_0, q_1, q_2, axiom $q_0(x_0)$, and rules $q_0(yz) \to yz$, $q_1(yz) \to y$, $q_2(yz) \to z$ for $y, z \in \{a, b\}$, $q_2(\sigma(x_1, x_2)) \to q_2(x_2)$, $q_1(\sigma(x_1, x_2)) \to q_1(x_1)$, and $q_0(\sigma(x_1, x_2)) \to \sigma(q_0(x_1), q_0(x_2), \#(q_1(x_1), q_2(x_2)))$. $\square$

3   Normal Form

In this section we state a normal form for (total) dtlas M, together with its effect on $\mathsf{maxdiff}(M)$. We start by requiring a simple and technically convenient property so that every state of M only translates input trees that have the same look-ahead state; moreover, the rules satisfy a completeness condition.

A dtla M is *look-ahead uniform* (for short, *la-uniform*) if there is a mapping $\rho : Q \to P$ (called *la-map*) satisfying the following conditions, for $p \in P$ and $q, \bar{q} \in Q$:

(1) If $q(x_0)$ occurs in $A(p)$, then $\rho(q) = p$.

(2) For every rule $q(a(x_1 : p_1, \ldots, x_k : p_k)) \to \zeta$ in R: $\rho(q) = \delta(a, p_1, \ldots, p_k)$, and if $\bar{q}(x_i)$ occurs in ζ then $\rho(\bar{q}) = p_i$.

(3) For every $q \in Q$, $a \in \Sigma$ of rank $k \geq 0$, and $p_1, \ldots, p_k \in P$ such that $\delta(a, p_1, \ldots, p_k) = \rho(q)$, there is a rule $q(a(x_1 : p_1, \ldots, x_k : p_k)) \to \zeta$ in R.

If M is la-uniform, then the domain of $[\![q]\!]_M$ is $[\![\rho(q)]\!]_M$ for every $q \in Q$. This implies that M° is la-uniform with the same la-map ρ as M.

Clearly, the dtla M of Example 1 is la-uniform with $\rho(q) = p_b$, but the dtla M of Example 2 is not la-uniform.

Example 3. We change the dtla M of Example 2 into an la-uniform dtla (still calling it M), with the same look-ahead automaton as M, by adding look-ahead information to its states. Thus, it has set of states $Q = \{q_{yz} \mid y, z \in \{a, b\}\}$ with $\rho(q_{yz}) = p_{yz}$, axioms $A(p_{yz}) = q_{yz}(x_0)$, and rules $q_{yz}(yz) \to yz$ and $q_{wz}(\sigma(x_1 : p_{wx}, x_2 : p_{yz})) \to \sigma(q_{wx}(x_1), q_{yz}(x_2), \#(w, z))$ for all $w, x, y, z \in \{a, b\}$.   □

A dtla M is *earliest* if it is la-uniform and, for every state q of M, $\mathsf{rlabs}_M(q) := \{\mathsf{lab}(q_M(s), \varepsilon) \mid s \in [\![\rho(q)]\!]_M\} \subseteq \Delta$ is not a singleton. Thus, M is *not* earliest if it has a state q for which the roots of all output trees $q_M(s)$, $s \in \mathcal{T}_\Sigma$, have the same label; intuitively, the node with that label could be produced earlier by M. A dtla M is *canonical* if it is earliest and $[\![q]\!]_M \neq [\![q']\!]_M$ for all distinct states q, q' of M.

It is easy to see that the dtla M of Example 3 is canonical: for all $y, z \in \{a, b\}$, $\mathsf{rlabs}_M(q_{yz}) = \{yz, \sigma\}$ and $[\![q_{yz}]\!]_M$ is the restriction of $[\![M]\!]$ to $[\![p_{yz}]\!]_M$.

We now present (without proof) the fact that canonicalness is a normal form for dtlas, generalizing the normal form for dtops in [6] for the total case.

Theorem 1. *For every total dtla M, one can construct an equivalent canonical dtla $\mathsf{can}(M)$, with the same look-ahead automaton as M, such that $\mathsf{maxdiff}(M) - 8^{|M|^3} \leq \mathsf{maxdiff}(\mathsf{can}(M)) \leq \mathsf{maxdiff}(M) + 8^{|M|^3}$, where $|M|$ is the size of M.*

4   Difference Tuples

Let M be a dtpla and let $P = \{\hat{p}_1, \ldots, \hat{p}_n\}$, where the order of the look-ahead states is fixed as indicated. Recall that a dtpla is a dtla that is not a dtop, hence $n \geq 2$. For a given context C consider the trees $M(C[\hat{p}_1]), \ldots, M(C[\hat{p}_n])$. Intuitively, the largest common prefix of these trees does *not* depend on the look-ahead. In contrast, the subtrees that are not part of the largest common prefix, *do* depend on the look-ahead information.

For trees $t_1, \ldots, t_n \in \mathcal{T}_\Delta(Q \times P)$ we define a subset of $\mathcal{T}_\Delta(Q \times P)^n$ as follows: $\mathsf{diftup}(t_1, \ldots, t_n) := \{(t_1/v, \ldots, t_n/v) \mid v \in V_\bot(\sqcap\{t_1, \ldots, t_n\})\}$. We define the *set of difference tuples* of M as $\mathsf{diftup}(M) := \bigcup_{C \in \mathcal{C}_\Sigma} \mathsf{diftup}(M(C[\hat{p}_1]), \ldots, M(C[\hat{p}_n]))$. For a Σ-context C we define the Δ-*pattern* $\mathsf{pref}(M, C) := \sqcap\{M(C[p]) \mid p \in P\}$.

We wish to decide whether M is equivalent to a dtop. If there exists such a dtop N, then we may expect intuitively for any $s \in \mathcal{T}_\Sigma$, that $N(C[s]) = t[(q_1)_N(s), \ldots, (q_r)_N(s)]$ where $t = \mathsf{pref}(M, C) = \sqcap\{M(C[\hat{p}_1]), \ldots, M(C[\hat{p}_n])\}$ and $r = |V_\bot(t)|$; in other words, since N does not know the look-ahead state $\delta_M(s)$ of s, it translates C into the largest common prefix of the output trees $M(C[\hat{p}_1]), \ldots, M(C[\hat{p}_n])$. Moreover, if the ith occurrence of $\bot$ is at node v_i of t, $1 \leq i \leq r$, then we expect the difference tuple $(M(C[\hat{p}_1])/v_i, \ldots, M(C[\hat{p}_n])/v_i)$ to be stored in the state q_i of N; in this way N is prepared to continue its simulation of M on the subtree s. This is shown in Lemma 4, for canonical M and earliest N. If N is canonical, then its states are in one-to-one correspondence with the difference tuples of M, as shown in Lemma 5.

It is easy to show that every component of a difference tuple is a difference tree, and every difference tree is a subtree of a component of a difference tuple. Consequently, the maximal height of the components of the difference tuples of M is $\mathsf{maxdiff}(M)$, see [7, Lemma 17]. This implies that $\mathsf{diftup}(M)$ is finite if and only if $\mathsf{diff}(M)$ is finite.

Example 4. For the dtla M of Example 1, with the order $P = \{p_a, p_b\}$, we obtain that $\mathsf{diftup}(M) = \{(a, \sigma^n \langle q, p_b \rangle) \mid n \in \mathbb{N}\}$.

For the dtla M of Example 3 (which is the la-uniform version of the dtla of Example 2) it is not difficult to see that $\mathsf{diff}(M) = \{a, b\} \cup \{\langle q_{yz}, p_{yz} \rangle \mid y, z \in \{a, b\}\}$, and that the set $\mathsf{diftup}(M)$ consists of the three 4-tuples (a, a, b, b), (a, b, a, b) and $(\langle q_{aa}, p_{aa} \rangle, \langle q_{ab}, p_{ab} \rangle, \langle q_{ba}, p_{ba} \rangle, \langle q_{bb}, p_{bb} \rangle)$, with $P = \{p_{aa}, p_{ab}, p_{ba}, p_{bb}\}$. $\square$

In the next lemmas, M is a canonical dtpla (with la-map ρ_M) and N a canonical dtop equivalent to M, i.e., $[\![M]\!] = [\![N]\!]$. We assume that the unique look-ahead state of N is $\bot$; for a Σ-context C we of course write C instead of $C[\bot]$.

We first formalize the fact that the translation of an input tree by M is always ahead of its translation by N, in a uniform way. An *aheadness mapping* from N to M is a function $\varphi : Q_N \times P_M \to \mathcal{T}_\Delta(Q_M \times P_M)$ such that for every $C \in \mathcal{C}_\Sigma$ and $p \in P_M$,

$$M(C[p]) = N(C)[\langle q, \bot \rangle \leftarrow \varphi(q, p) \mid q \in Q_N]. \tag{1}$$

Note that $\varphi(q, p)$ is in $\mathcal{T}_\Delta(\{\langle \bar{q}, p \rangle \mid \bar{q} \in Q_M, \rho_M(\bar{q}) = p\})$. Intuitively, φ defines the exact amount in which M is ahead of N, which is independent of C.

For the next lemma it is essential that M is canonical.

Lemma 2. *There is a unique aheadness mapping φ from N to M.*

Proof. We first show that M is ahead of N, i.e., that all output symbols produced by N on a given input context are also produced by M. Let $p \in P_M$ and $C \in \mathcal{C}_\Sigma$.

Claim 1. $V_d(N(C)) \subseteq V_d(M(C[p]))$ for every $d \in \Delta$.
Equivalently, $N(C)[\langle q, \perp \rangle \leftarrow \perp \mid q \in Q_N] \sqsubseteq M(C[p])$.

Proof: By induction on the length of the nodes of $N(C)$. Let $v \in V_d(N(C))$ with $d \in \Delta$. Since the labels of v's proper ancestors are in Δ, $v \in V(M(C[p]))$ by induction. Consider an arbitrary $s \in [\![p]\!]_M$. By Lemma 1, $v \in V_d(N(C[s]))$. Since $[\![M]\!] = [\![N]\!]$, $M(C[s]) = N(C[s])$ and so $v \in V_d(M(C[s]))$. Suppose that $v \notin V_d(M(C[p]))$. Then, again by Lemma 1, v has some label $\langle q, p \rangle$ in $M(C[p])$ such that $q_M(s)$ has root label d. Since this holds for every $s \in [\![p]\!]_M$, we obtain that $\mathsf{rlabs}_M(q) = \{d\}$ contradicting the fact that M is earliest. Note that, since M is la-uniform, $\rho_M(q) = p$.

Next we show that the amount in which M is ahead of N, is independent of C. Let $p \in P_M$, $C_1, C_2 \in \mathcal{C}_\Sigma$, $v_1, v_2 \in \mathbb{N}_+^*$ and $q \in Q_N$.

Claim 2. If $N(C_1)/v_1 = N(C_2)/v_2 = \langle q, \perp \rangle$, then $M(C_1[p])/v_1 = M(C_2[p])/v_2$.

Proof: By Claim 1, v_i is a node of $M(C_i[p])$. Let $t_i \in \mathcal{T}_\Delta(Q_M \times \{p\})$ denote the tree $M(C_i[p])/v_i$. For every $s \in [\![p]\!]_M$, $N(C_1[s])/v_1 = N(C_2[s])/v_2 = q_N(s)$ by Lemma 1, and so $M(C_1[s])/v_1 = M(C_2[s])/v_2$. Hence, again by Lemma 1, $t_1 \Psi_s = t_2 \Psi_s$ for all $s \in [\![p]\!]_M$, where $\Psi_s = [\langle q, p \rangle \leftarrow q_M(s) \mid q \in Q_M]$. Suppose that $t_1 \neq t_2$. Then there is a leaf v of, e.g., t_1 with label $\langle q_1, p \rangle$ such that v is a node of t_2 with $t_2/v \neq \langle q_1, p \rangle$. If the root label of t_2/v is $d \in \Delta$, then $(q_1)_M(s)$ has root label d for all $s \in [\![p]\!]_M$, contradicting the fact that M is earliest. If t_2/v equals $\langle q_2, p \rangle$ with $q_1 \neq q_2$, then $(q_1)_M(s) = (q_2)_M(s)$ for all $s \in [\![p]\!]_M$. Since $[\![p]\!]_M$ is the domain of both $[\![q_1]\!]_M$ and $[\![q_2]\!]_M$, we obtain that $[\![q_1]\!]_M = [\![q_2]\!]_M$, contradicting the fact that M is canonical.

An aheadness mapping from N to M can now be defined as follows. Let $q \in Q_N$ and $p \in P_M$. Since, by convention, q is reachable, there is a Σ-context C such that $N(C)$ has a node v labeled $\langle q, \perp \rangle$. By Claim 1, v is a node of $M(C[p])$ and we define $\varphi(q, p) = M(C[p])/v$. By Claim 2, the definition of φ does not depend on C and v. It is easy to see that φ is an aheadness mapping, and that it is unique. $\qquad\square$

Lemma 3. *For every $s \in \mathcal{T}_\Sigma$ and $q \in Q_N$,*
if $\delta_M(s) = p$, then $q_N(s) = \varphi(q, p)[\langle \bar{q}, p \rangle \leftarrow \bar{q}_M(s) \mid \bar{q} \in Q_M]$.

Proof. Since q is reachable, there exist C, v such that $N(C)/v = \langle q, \perp \rangle$. By (1), $M(C[p])/v = \varphi(q, p)$. Since M and N are equivalent, $N(C[s]) = M(C[s])$. Applying Lemma 1 twice, we obtain that $q_N(s) = N(C[s])/v = M(C[s])/v = (M(C[p])/v)[\langle \bar{q}, p \rangle \leftarrow \bar{q}_M(s) \mid \bar{q} \in Q_M]$, which proves the equation. $\qquad\square$

The next lemma expresses our intuition that the output of N on input C is the largest common prefix of the outputs of M on all inputs $C[p]$, $p \in P$, such that the difference tuples of M are stored in the states of N. Its proof uses that N is earliest. For a tree $t \in \mathcal{T}_\Delta(Q_N \times \{\perp\})$ we define the Δ-pattern $t\Phi := t[\langle q, \perp \rangle \leftarrow \perp \mid q \in Q_N]$; similarly, for $t \in \mathcal{T}_\Delta(Q_N(X))$, we define $t\Phi := t[q(x_i) \leftarrow \perp \mid q \in Q_N, i \in \mathbb{N}]$.

Lemma 4. *For every* $C \in \mathcal{C}_\Sigma$, $N(C)\Phi = \mathsf{pref}(M, C)$; *moreover, for every* $v \in \mathbb{N}_+^*$, $q \in Q_N$ *and* $p \in P_M$, *if* $N(C)/v = \langle q, \bot \rangle$ *then* $\varphi(q, p) = M(C[p])/v$.

Proof. By Equation (1), $N(C)\Phi \sqsubseteq M(C[p])$ for every $p \in P_M$ (cf. Claim 1 in the proof of Lemma 2), and so $N(C)\Phi \sqsubseteq \mathsf{pref}(M, C)$. To show equality, we prove for every $v \in V_\bot(N(C)\Phi)$ that $v \in V_\bot(\mathsf{pref}(M, C))$. Let $N(C)/v = \langle q, \bot \rangle$ for $q \in Q_N$. Then, by Equation (1), $M(C[p])/v = \varphi(q, p)$ for every $p \in P_M$ (which proves the second part of this lemma). Suppose that $v \in V_d(\mathsf{pref}(M, C))$ with $d \in \Delta$. Then $v \in V_d(M(C[p]))$ and so $\mathsf{lab}(\varphi(q, p), \varepsilon) = d$ for every $p \in P_M$. Then, by Lemma 3, $\mathsf{lab}(q_N(s), \varepsilon) = d$ for every $s \in \mathcal{T}_\Sigma$, contradicting the fact that N is earliest. $\square$

If M is equivalent to a dtop, then it is equivalent to a canonical dtop by Theorem 1. By [6, Theorem 15], equivalent canonical dtops are the same (modulo a renaming of states). Thus, if M is equivalent to a dtop, then it is equivalent to a *unique* canonical dtop $\mathsf{td}(M)$. In the next three lemmas we give another proof of this, and we show that $\mathsf{td}(M)$ can be constructed from M and $\mathsf{diftup}(M)$. We start by showing that $Q_{\mathsf{td}(M)}$ can be identified with $\mathsf{diftup}(M)$. The proof uses that N is canonical.

Lemma 5. *For a state* $q \in Q_N$, *let* $\psi(q) = (\varphi(q, \hat{p}_1), \ldots, \varphi(q, \hat{p}_n))$, *where* $P_M = \{\hat{p}_1, \ldots, \hat{p}_n\}$. *Then* ψ *is a bijection between* Q_N *and* $\mathsf{diftup}(M)$.

Proof. (i) $\psi(q) \in \mathsf{diftup}(M)$. Proof: There are C, v such that $N(C)/v = \langle q, \bot \rangle$. By Lemma 4, $v \in V_\bot(\mathsf{pref}(M, C))$ and $M(C[\hat{p}_i])/v = \varphi(q, \hat{p}_i)$ for every i. Thus $\psi(q) \in \mathsf{diftup}(M)$. (ii) ψ is surjective. Proof: If $(t_1, \ldots, t_n) \in \mathsf{diftup}(M)$ then there are C, v such that $\mathsf{pref}(M, C)/v = \bot$ and $M(C[\hat{p}_i])/v = t_i$ for every i. By Lemma 4, $N(C)/v = \langle q, \bot \rangle$ for some $q \in Q_N$, and $M(C[\hat{p}_i])/v = \varphi(q, \hat{p}_i)$. Thus, $t_i = \varphi(q, \hat{p}_i)$ for every i. (iii) ψ is injective. Proof: Let $\psi(q_1) = \psi(q_2)$. By Lemma 3, $(q_1)_N(s) = (q_2)_N(s)$ for all $s \in \mathcal{T}_\Sigma$, i.e., $[\![q_1]\!]_N = [\![q_2]\!]_N$, so $q_1 = q_2$ because N is canonical. $\square$

Corollary 1. *Let* M *be a total dtla. If* M *is equivalent to a dtop, then* $\mathsf{diff}(M)$ *is finite.*

Proof. If M is a dtop, then $\mathsf{diff}(M) = \emptyset$. Now let M be a dtpla, equivalent to a dtop. By Theorem 1, the canonical dtla $\mathsf{can}(M)$ is equivalent to a canonical dtop. By Lemmas 2 and 5, $\mathsf{diftup}(\mathsf{can}(M))$ is finite, and so $\mathsf{diff}(\mathsf{can}(M))$ is finite. Hence $\mathsf{diff}(M)$ is finite because $\mathsf{maxdiff}(M) \leq \mathsf{maxdiff}(\mathsf{can}(M)) + 8^{|M|^3}$, cf. Theorem 1. $\square$

Next we show how to compute the axiom of $\mathsf{td}(M)$, representing the states of $\mathsf{td}(M)$ by difference tuples. For a tree $t \in \mathcal{T}_\Delta(Q_M(X))$ we define $t\Omega \in \mathcal{T}_\Delta(Q_M \times P_M)$ by $t\Omega := t[q(x_i) \leftarrow \langle q, \rho_M(q) \rangle \mid q \in Q_M, i \in \mathbb{N}]$; similarly, for $t \in \mathcal{T}_\Delta(Q_N(X))$, we define $t\Omega := t[q(x_i) \leftarrow \langle q, \bot \rangle \mid q \in Q_N, i \in \mathbb{N}]$.

Lemma 6. $A_N\Phi = \sqcap\{A_M(p)\Omega \mid p \in P_M\}$; *moreover, for every* $v \in \mathbb{N}_+^*$, $q \in Q_N$ *and* $p \in P_M$, *if* $A_N/v = q(x_0)$ *then* $\varphi(q, p) = A_M(p)\Omega/v$.

Proof. Clearly, $N(\bot) = A_N \Omega$ and $M(p) = A_M(p)\Omega$ for every $p \in P_M$. Hence by Lemma 4, with $C = \bot$, $A_N\Phi = A_N\Omega\Phi = N(\bot)\Phi = \text{pref}(M, \bot) = \sqcap\{M(p) \mid p \in P_M\} = \sqcap\{A_M(p)\Omega \mid p \in P_M\}$. If $A_N/v = q(x_0)$ then $N(\bot)/v = A_N\Omega/v = \langle q, \bot \rangle$; so by Lemma 4, with $C = \bot$, $\varphi(q, p) = M(p)/v = A_M(p)\Omega/v$ for every $p \in P_M$. $\square$

Finally we show, without proof, how to compute the rules of $\text{td}(M)$. Let M be an la-uniform dtla, Q_N a finite set and $\varphi : Q_N \times P_M \to \mathcal{T}_\Delta(Q_M \times P_M)$ a mapping such that $\varphi(q, p) \in \mathcal{T}_\Delta(\{\langle \bar{q}, p \rangle \mid \bar{q} \in Q_M, \rho_M(\bar{q}) = p\})$ for every $q \in Q_N$ and $p \in P_M$. Then we define for every $q \in Q_N$, $a \in \Sigma$ of rank $k \geq 0$, and $p_1, \ldots, p_k \in P_M$, the tree $\text{rhs}_{M,\varphi}(q, a, p_1, \ldots, p_k) := \varphi(q, p)[\langle \bar{q}, p \rangle \leftarrow \text{rhs}_M(\bar{q}, a, p_1, \ldots, p_k) \mid \bar{q} \in Q_M]$ where $p = \delta_M(a, p_1, \ldots, p_k)$. For $k \in \mathbb{N}$, let $[k] = \{1, \ldots, k\}$.

Lemma 7. (1) *For every $q \in Q_N$ and $a \in \Sigma$ of rank $k \geq 0$,*
    $\text{rhs}_N(q, a)\Phi = \sqcap\{\text{rhs}_{M,\varphi}(q, a, p_1, \ldots, p_k)\Omega \mid p_1, \ldots, p_k \in P_M\}$.
(2) *Let $q \in Q_N$, $a \in \Sigma$ of rank $k \geq 0$, and $i \in [k]$. For $j \in [k]$, $j \neq i$, let $s_j \in \mathcal{T}_\Sigma$ and $p_j = \delta_M(s_j)$. Let $\Psi_{iM} = [\bar{q}(x_j) \leftarrow \bar{q}_M(s_j) \mid \bar{q} \in Q_M, j \in [k], j \neq i]\Omega$. For every $v \in V_\bot(\text{rhs}_N(q, a)\Phi)$,*
    (a) *$\text{rhs}_N(q, a)/v \in Q_N(\{x_i\})$ if and only if*
        $v \in V_\bot(\sqcap\{\text{rhs}_{M,\varphi}(q, a, p_1, \ldots, p_{i-1}, p, p_{i+1}, \ldots, p_k)\Psi_{iM} \mid p \in P_M\})$, and
    (b) *for every $\bar{q} \in Q_N$ and $p \in P_M$, if $\text{rhs}_N(q, a)/v = \bar{q}(x_i)$ then*
        $\varphi(\bar{q}, p) = \text{rhs}_{M,\varphi}(q, a, p_1, \ldots, p_{i-1}, p, p_{i+1}, \ldots, p_k)\Psi_{iM}/v$.

By the last three lemmas, every dtpla M that is equivalent to a dtop, is equivalent to a unique canonical dtop $\text{td}(M)$, modulo a renaming of states. Based on these same lemmas, we can now construct $\text{td}(M)$ from any given canonical dtpla M for which $\text{diftup}(M)$ is a given finite set. The construction returns the answer 'no' if M is not equivalent to any dtop. We construct the dtop $N = \text{td}(M)$, if it exists, by taking $Q_N = \text{diftup}(M)$, defining $\varphi : Q_N \times P_M \to \mathcal{T}_\Delta(Q_M \times P_M)$ as $\varphi((t_1, \ldots, t_n), \hat{p}_i) = t_i$ for $i \in [n]$ (in accordance with Lemma 5), and constructing the axiom and rules of N according to Lemmas 6 and 7, respectively (i.e., by viewing the statements of these lemmas as definitions). In Lemma 7(2) we choose s_j arbitrarily but fixed. If the construction of an axiom or a rule fails because a possible state occurring in it (which is a tuple in $\mathcal{T}_\Delta(Q_M \times P_M)^n$) is not a difference tuple of M, then the answer is 'no'. The construction of a rule can also fail (and produce the answer 'no') when a node $v \in V_\bot(\text{rhs}_N(q, a)\Phi)$ is not an element of $V_\bot(\sqcap\{\text{rhs}_{M,\varphi}(q, a, p_1, \ldots, p_{i-1}, p, p_{i+1}, \ldots, p_k)\Psi_{iM} \mid p \in P_M\})$ for any i, see Lemma 7(2)(a). If the construction of the dtop N succeeds, then it remains to test whether M and N are equivalent (because, by Lemmas 5, 6 and 7, if M is equivalent to a dtop then it is equivalent to N). If they are equivalent then the construction returns the dtop $N = \text{td}(M)$, otherwise the answer is 'no'. Equivalence of dtlas is decidable by [9] (see also [6, Corollary 19]). It is shown in [7, Section 6.1] that there is a simple direct test for equivalence of M and N.

Unfortunately, we do not know whether it is decidable if $\text{diftup}(M)$ is finite, and whether it can be computed if it is finite. We now show that, to determine whether a dtla M is equivalent to a dtop, it suffices to have an upper bound for $\text{maxdiff}(M)$.

Theorem 2. *It is decidable for a given total dtla M and a given difference bound for M whether there exists a dtop N such that $[\![M]\!] = [\![N]\!]$, and if so, such a dtop N can be constructed.*

Proof. Let M be a (total) dtla and let $h(M)$ be a difference bound for M. We may, of course, assume that M is a dtpla. By Theorem 1 we may assume that M is canonical, because $h(M) + 8^{|M|^3}$ is a (computable) difference bound for can(M).

So, let M be a canonical dtpla and let $h(M)$ be a difference bound for M. This means that if diftup(M) is finite, then the height of the components of the difference tuples of M is at most $h(M)$. We now decide whether M is equivalent to a dtop by constructing td(M) as described before this theorem. However, since diftup(M) is not given, we construct $N = \text{td}(M)$ incrementally, using a variable Q_N to accumulate its states (which are all assumed to be reachable). In accordance with Lemma 5 we take $Q_N \subseteq \mathcal{T}_\Delta(Q_M \times P_M)^n$ and $\varphi((t_1, \ldots, t_n), \hat{p}_i) = t_i$ for $i \in [n]$. We first construct the axiom A_N according to Lemma 6 and initialize the set Q_N with the states, i.e., the tuples in $\mathcal{T}_\Delta(Q_M \times P_M)^n$, that occur in that axiom. If the height of one of the components of one of those tuples is larger than $h(M)$, then either diftup(M) is infinite or that tuple is not a difference tuple of M, and we stop the construction with answer 'no', indicating that M is not equivalent to any dtop. Then, repeatedly, for every $q \in Q_N$ and $a \in \Sigma$ we construct $\text{rhs}_N(q, a)$ according to Lemma 7, and we add to Q_N the states that occur in that right-hand side. If the construction of $\text{rhs}_N(q, a)$ fails or if the height of one of the components of its states is larger than $h(M)$, then the answer is 'no'. If the construction of the dtop N succeeds, then it remains to test whether M and N are equivalent. $\qquad\square$

In the next example we show that without the tests on height, the construction may not halt; in such a case it can be viewed as computing an *infinite* dtop equivalent to M. In Example 6 the construction of N succeeds and N is equivalent to M.

Example 5. Consider the dtla M of Example 1. It is easy to see that M is canonical. In Example 4 we have seen that diftup(M) $= \{(a, \sigma^n\langle q, p_b\rangle) \mid n \in \mathbb{N}\}$.

We now apply to M the construction of N in the proof of Theorem 2, without the tests on height. By Lemma 6, $A_N\Phi = a \sqcap \langle q, p_b\rangle = \bot$ and so $A_N = q_0(x_0)$ with $\varphi(q_0, p_a) = a$ and $\varphi(q_0, p_b) = \langle q, p_b\rangle$, i.e., $q_0 = (a, \langle q, p_b\rangle)$. Assume now that the algorithm has constructed the state q_n with $\varphi(q_n, p_a) = a$ and $\varphi(q_n, p_b) = \sigma^n\langle q, p_b\rangle$, i.e., q_n is the difference tuple $(a, \sigma^n\langle q, p_b\rangle)$ of M. By Lemma 7(1), $\text{rhs}_N(q_n, b) = \text{rhs}_{M,\varphi}(q_n, b) = \varphi(q_n, p_b)[\langle \bar{q}, p_b\rangle \leftarrow \text{rhs}_M(\bar{q}, b) \mid \bar{q} \in Q_M] = \varphi(q_n, p_b)[\langle q, p_b\rangle \leftarrow b] = \sigma^n b$. Thus, N has the rule $q_n(b) \to \sigma^n b$. Similarly, $\text{rhs}_N(q_n, a) = \text{rhs}_{M,\varphi}(q_n, a) = \varphi(q_n, p_a) = a$ and so N has the rule $q_n(a) \to a$. Next, we compute $\text{rhs}_N(q_n, \sigma)$. To do so we need $\text{rhs}_{M,\varphi}(q_n, \sigma, p)$ for every $p \in P_M$. For $p = p_b$ we have $\text{rhs}_{M,\varphi}(q_n, \sigma, p_b) = \varphi(q_n, p_b)[\langle q, p_b\rangle \leftarrow \text{rhs}_M(q, \sigma, p_b)] = \sigma^n \sigma q(x_1) = \sigma^{n+1}q(x_1)$, and for $p = p_a$ we have $\text{rhs}_{M,\varphi}(q_n, \sigma, p_a) = \varphi(q_n, p_a) = a$. Thus, by Lemma 7(1), $\text{rhs}_N(q_n, \sigma)\Phi = a \sqcap \sigma^{n+1}\langle q, p_b\rangle = \bot$. Hence, $\text{rhs}_N(q_n, \sigma) = q(x_1)$ for some $q \in Q_N$. By Lemma 7(2)(b), $\varphi(q, p_y) = \text{rhs}_{M,\varphi}(q_n, \sigma, p_y)\Omega$ for

$y \in \{a, b\}$ and so $\varphi(q, p_a) = a$ and $\varphi(q, p_b) = \sigma^{n+1}\langle q, p_b \rangle$. Thus, $q = q_{n+1}$ and N has the rule $q_n(\sigma(x_1)) \to q_{n+1}(x_1)$. This shows that the construction does not halt. It can be viewed as constructing the *infinite* dtop N with $Q_N = \{q_n \mid n \in \mathbb{N}\} = \mathsf{diftup}(M)$, $A_N = q_0(x_0)$ and rules $q_n(a) \to a$, $q_n(b) \to \sigma^n b$ and $q_n(\sigma(x_1)) \to q_{n+1}(x_1)$ for every $n \in \mathbb{N}$. Clearly, N is equivalent to M. *With a given difference bound h, the construction halts when constructing q_{h+1}.*   □

Example 6. Consider the dtla M of Example 3. As observed after Example 3, M is canonical. We have seen in Example 4 that $\mathsf{diftup}(M)$ consists of the three 4-tuples (a, a, b, b), (a, b, a, b) and $(\langle q_{aa}, p_{aa} \rangle, \langle q_{ab}, p_{ab} \rangle, \langle q_{ba}, p_{ba} \rangle, \langle q_{bb}, p_{bb} \rangle)$.

We construct N as in the proof of Theorem 2; since $\mathsf{maxdiff}(M) = 0$, the construction is the same for every difference bound $h(M)$. By Lemma 6, $A_N = q_0(x_0)$ with $\varphi(q_0, p_{yz}) = \langle q_{yz}, p_{yz} \rangle$ for $y, z \in \{a, b\}$. Hence, $q_0 = (\langle q_{aa}, p_{aa} \rangle, \langle q_{ab}, p_{ab} \rangle, \langle q_{ba}, p_{ba} \rangle, \langle q_{bb}, p_{bb} \rangle)$. Then Lemma 7(1) implies the equalities $\mathsf{rhs}_N(q_0, yz)\Phi = \mathsf{rhs}_{M,\varphi}(q_0, yz) = \varphi(q_0, p_{yz})[\langle q_{yz}, p_{yz} \rangle \leftarrow \mathsf{rhs}_M(q_{yz}, yz)] = \mathsf{rhs}_M(q_{yz}, yz) = yz$, so N has the rules $q_0(yz) \to yz$ for all $y, z \in \{a, b\}$. To compute $\mathsf{rhs}_N(q_0, \sigma)$, observe that for every $w, x, y, z \in \{a, b\}$, $\mathsf{rhs}_{M,\varphi}(q_0, \sigma, p_{wx}, p_{yz}) = \varphi(q_0, p_{wz})[\langle q_{wz}, p_{wz} \rangle \leftarrow \mathsf{rhs}_M(q_{wz}, \sigma, p_{wx}, p_{yz})] = \sigma(q_{wx}(x_1), q_{yz}(x_2), \#(w, z))$. By Lemma 7(1), $\mathsf{rhs}_N(q_0, \sigma)\Phi = \sigma(\bot, \bot, \#(\bot, \bot))$. Thus, N may have a rule of the form

$$q_0(\sigma(x_1, x_2)) \to \sigma(q_3(x_{i_3}), q_4(x_{i_4}), \#(q_1(x_{i_1}), q_2(x_{i_2}))).$$

Let $s_1 = s_2 = aa$. From Lemma 7(2)(a) we obtain for $v = (3, 1)$ that

$$i_1 = 1 \iff v \in V_\bot(\sqcap\{\mathsf{rhs}_{M,\varphi}(q_0, \sigma, p_{wx}, p_{aa})\Psi_{1M} \mid w, x \in \{a, b\}\})$$
$$\iff v \in V_\bot(\sqcap\{\sigma(\langle q_{wx}, p_{wx} \rangle, aa, \#(w, a)) \mid w, x \in \{a, b\}\})$$

if and only if $v \in V_\bot(\sigma(\bot, aa, \#(\bot, a)))$, which is true. So $i_1 = 1$ and $\varphi(q_1, p_{wx}) = w$ for all $w, x \in \{a, b\}$ by Lemma 7(2)(b). Thus, $q_1 = (a, a, b, b)$. Similarly we obtain for $v = (3, 2)$ that $i_2 = 2$ and $\varphi(q_2, p_{yz}) = z$, for $v = 1$ that $i_3 = 1$ and $\varphi(q_3, p_{wx}) = \langle q_{wx}, p_{wx} \rangle$, and for $v = 2$ that $i_4 = 2$ and $\varphi(q_4, p_{yz}) = \langle q_{yz}, p_{yz} \rangle$. Hence $q_2 = (a, b, a, b)$, $q_3 = q_4 = q_0$ and N has the rule

$$q_0(\sigma(x_1, x_2)) \to \sigma(q_0(x_1), q_0(x_2), \#(q_1(x_1), q_2(x_2))).$$

Next we consider q_2. Clearly, both $\mathsf{rhs}_{M,\varphi}(q_2, yz)$ and $\mathsf{rhs}_{M,\varphi}(q_2, \sigma, p_{wx}, p_{yz})$ equal z. Thus, N has the rules $q_2(yz) \to z$ and it may have a rule of the form $q_2(\sigma(x_1, x_2)) \to q(x_i)$. Taking again $s_1 = s_2 = aa$, we get that $i = 2$ if and only if ε has label $\bot$ in $\sqcap\{\mathsf{rhs}_{M,\varphi}(q_2, \sigma, p_{aa}, p_{yz}) \mid y, z \in \{a, b\}\}$ if and only if $\varepsilon \in V_\bot(a \sqcap b)$, which is true. So $i = 2$ and $\varphi(q, p_{yz}) = z$, which means that $q = q_2$. Hence, N has the rule $q_2(\sigma(x_1, x_2)) \to q_2(x_2)$. Similarly it has the rules $q_1(yz) \to y$ and $q_1(\sigma(x_1, x_2)) \to q_1(x_1)$. So, the construction ends with the dtop N given at the end of Example 2.   □

5   Conclusion

A dtla M is *linear* if no right-hand side of a rule contains the same variable twice, and *nonerasing* if no right-hand side of a rule is in $Q(X)$. Our two example dtlas are both.

Theorem 3. *It is decidable for a total linear nonerasing dtla M whether there exists a dtop N such that $[\![M]\!] = [\![N]\!]$, and if so, such a dtop N can be constructed.*

The proof uses (involved) pumping arguments to show that $37 \cdot |M|^5$ is a difference bound for such a dtla M. The same proof holds for dtlas with less stringent restrictions on copying and erasing: total dtlas that are ultralinear and bounded erasing, see [7].

We would like to extend the above result to the nontotal case where a dtla realizes a partial function, to the case where the dtla and the dtop are restricted to a given regular tree language, and to more general dtlas (preferably to all dtlas, of course). Even more generally, we would like to have an algorithm that for a given dtla constructs an equivalent dtla with a minimal number of look-ahead states.

References

1. Alur, R., D'Antoni, L.: Streaming tree transducers. In: Czumaj, A., Mehlhorn, K., Pitts, A., Wattenhofer, R. (eds.) ICALP 2012, Part II. LNCS, vol. 7392, pp. 42–53. Springer, Heidelberg (2012)
2. Berstel, J.: Transductions and Context-Free Languages. Teubner-Verlag (1979)
3. Choffrut, C.: Une caractérisation des fonctions séquentielles et des fonctions sous-séquentielles en tant que relations rationnelles. Theor. Comput. Sci. 5(3), 325–337 (1977)
4. Drewes, F.: Grammatical Picture Generation – A Tree-Based Approach. Springer (2006)
5. Engelfriet, J.: Top-down tree transducers with regular look-ahead. Mathematical Systems Theory 10, 289–303 (1977)
6. Engelfriet, J., Maneth, S., Seidl, H.: Deciding equivalence of top-down XML transformations in polynomial time. J. Comput. Syst. Sci. 75(5), 271–286 (2009)
7. Engelfriet, J., Maneth, S., Seidl, H.: Look-ahead removal for top-down tree transducers. CoRR abs/1311.2400 (2013)
8. Engelfriet, J., Vogler, H.: Macro tree transducers. J. Comput. Syst. Sci. 31(1), 71–146 (1985)
9. Ésik, Z.: Decidability results concerning tree transducers I. Acta Cybern. 5, 1–20 (1980)
10. Filiot, E., Servais, F.: Visibly pushdown transducers with look-ahead. In: Bieliková, M., Friedrich, G., Gottlob, G., Katzenbeisser, S., Turán, G. (eds.) SOFSEM 2012. LNCS, vol. 7147, pp. 251–263. Springer, Heidelberg (2012)
11. Hosoya, H.: Foundations of XML Processing – The Tree-Automata Approach. Cambridge University Press (2010)
12. Knight, K., Graehl, J.: An overview of probabilistic tree transducers for natural language processing. In: Gelbukh, A. (ed.) CICLing 2005. LNCS, vol. 3406, pp. 1–24. Springer, Heidelberg (2005)
13. Lemay, A., Maneth, S., Niehren, J.: A learning algorithm for top-down XML transformations. In: PODS, pp. 285–296 (2010)
14. Maletti, A., Graehl, J., Hopkins, M., Knight, K.: The power of extended top-down tree transducers. SIAM J. Comput. 39(2), 410–430 (2009)
15. Martens, W., Neven, F., Gyssens, M.: Typechecking top-down XML transformations: Fixed input or output schemas. Inf. Comput. 206(7), 806–827 (2008)

Scope-Bounded Pushdown Languages

Salvatore La Torre[1], Margherita Napoli[1], and Gennaro Parlato[2]

[1] Dipartimento di Informatica, Università degli Studi di Salerno, Italy
[2] School of Electronics and Computer Science, University of Southampton, UK

Abstract. We study the formal language theory of multistack pushdown automata (MPA) restricted to computations where a symbol can be popped from a stack S only if it was pushed within a bounded number of contexts of S (*scoped* MPA). We contribute to show that scoped MPA are indeed a robust model of computation, by focusing on the corresponding theory of *visibly* MPA (MVPA). We prove the equivalence of the deterministic and nondeterministic versions and show that scope-bounded computations of an n-stack MVPA can be simulated, rearranging the input word, by using only one stack. These results have several interesting consequences, such as, the closure under complement, the decidability of universality, inclusion and equality, and a Parikh theorem. We also give a logical characterization and compare the expressiveness of the scope-bounded restriction with MVPA classes from the literature.

1 Introduction

Pushdown automata working with multiple stacks (*multistack pushdown automata*, MPA for short) are the automata-theoretic model of concurrent programs with recursion and shared memory. Within the domain of formal verification of programs, program executions are analyzed against correctness properties, that may refer to the stack operations in the model such as for stack inspection properties and Hoare-like pre/post conditions. Such visibility of stack operations is captured in the formal languages by the notion of *visibly pushdown language* [1].

The class of *multistack visibly pushdown languages* (MVPL) is defined via the model of *multistack visibly pushdown automaton* (MVPA), that is a MPA where the push and pop operations on each stack are made visible in the input symbols, by a partition of the input alphabet into calls, returns and internals. Though visibility allows to synchronize the stack usage in the constructions, thus gaining interesting properties such as the closure under intersection, in general, it does not limit the expressiveness up to gaining decidability: the language of the executions (i.e., the sequence of transitions) of a MPA is a MVPL, and MPAs are equivalent to Turing machines already with two stacks.

In this paper, we study the formal language theory of MVPA restricted to *scoped computations* [13]: for a positive integer, a computation is k-scoped if for each stack i, each popped symbol was pushed within the last k contexts of i (a context is a continuous portion of the computation where only one stack is used). The notion of scope-bounded computations was introduced in [12] to

A.M. Shur and M.V. Volkov (Eds.): DLT 2014, LNCS 8633, pp. 116–128, 2014.

extend the analysis of MPA to unboundedly many context switches. The original notion of scope-bounded is significantly less expressive than the one used in this paper. The notion of scoped computations naturally extends to infinite words and temporal logic model checking [13,3]. Also, global reachability was solved for concurrent collapsible pushdown automata restricted to scoped computations [8].

Our first main contribution is to prove that deterministic and nondeterministic scoped MVPA are language equivalent. The main notion used in our construction is the *switching mask*. A switching mask summarizes the states of a MVPA at context-switches. We show that for scope-bounded computations also the switching masks are bounded. The resulting deterministic MVPA has size doubly exponential in both the number of stacks and the bound k. By this construction we gain the closure under complement, and by the effectiveness of closure under intersection and the decidability of emptiness, we also get the decidability of universality, inclusion, and equality. In general, MVPA and most of the already studied classes of MVPA are not determinizable [9].

As a second main contribution, we show a sequentialization construction for scoped MVPA. Namely, we give a mapping π that rearranges the contexts in a scoped word w s.t. it can be read by using only one stack (all the calls and returns of the starting alphabet are interpreted as calls and returns of the only available stack). We show a construction that starting from a MVPA A builds a visibly pushdown automaton A_{seq} that accepts all the scoped words in $\pi(L(A))$. Sequentialization of concurrent programs is nowadays one of the emerging techniques for building model-checkers for concurrent programs. As a corollary of this result, we can show a Parikh theorem for scoped MVPL.

Closure under union and intersection can be shown via standard constructions, and since the reachability problem is PSPACE-COMPLETE [13], we also get that emptiness is PSPACE-COMPLETE. Decidability of membership is straightforward: guess and check a run over the input word. We also give an MSO characterization of scoped MVPL. To the best of our knowledge this class is the largest subclass of MVPL with all the above properties.

As a further result we compare scoped MVPL with the main MVPL classes from the literature and show that it is incomparable with the most expressive ones, and strictly subsumes the others.

Related Work: In the literature several classes of MVPL have been studied: *phase* [10,11], *ordered* [6,7], and *path-tree* [14] are not determinizable and incomparable with scoped MVPL. The class of *round* MVPL [9], which is based on the notion of bounded-context switching [19], has the same properties as scoped MVPL (checking emptiness is NP-COMPLETE) but it is strictly included in it.

Parikh theorem was originally given for context-free languages in [18]. Visibility of stack operations was first introduced for input-driven pushdown automata [22] (see also [17] and references therein). A fixed-point algorithm for the reachability problem and a sequentialization are given in [15] for MPA under the restriction from [12]. The bounded context-switching restriction was proposed in [19] for under-approximate analysis of multi-threaded programs. More work on decision problems is done in [20,5] for phase MPA and ordered MPA [4].

2   Preliminaries

For $i, j \in \mathbb{N}$, we denote with $[i, j] = \{d \in \mathbb{N} \mid i \leq d \leq j\}$, and with $[j] = [1, j]$.

Words over Call-Return Alphabets. Given an integer $n > 0$, an n-stack call-return alphabet $\widetilde{\Sigma}_n$ is $(\Sigma^{int}, \langle \Sigma_h^c, \Sigma_h^r \rangle_{h \in [n]})$, where $\Sigma^{int}, \Sigma_1^c, \Sigma_1^r, \ldots, \Sigma_n^c, \Sigma_n^r$ are pairwise disjoint finite alphabets; Σ^{int} is the set of *internals*, and for $h \in [n]$, Σ_h^r is the set of *stack-h returns* and Σ_h^c is the set of *stack-h calls*.

In the following, for an n-stack call-return alphabet $\widetilde{\Sigma}_n$, we let $\Sigma_h = \Sigma_h^c \cup \Sigma_h^r \cup \Sigma^{int}$, $\Sigma^c = \bigcup_{h \in [n]} \Sigma_h^c$, $\Sigma^r = \bigcup_{h \in [n]} \Sigma_h^r$ and with $\Sigma = \Sigma^{int} \cup \Sigma^r \cup \Sigma^c$.

For a word $w = a_1 \ldots a_m$ over $\widetilde{\Sigma}_n$, denoting $C_h = \{i \in [m] \mid a_i \in \Sigma_h^c\}$ and $R_h = \{i \in [m] \mid a_i \in \Sigma_h^r\}$, the *matching relation* $\sim_h$ defined by w is such that (1) $\sim_h \subseteq C_h \times R_h$, (2) if $i \sim_h j$ then $i < j$, (3) for each $i \in C_h$ and $j \in R_h$ s.t. $i < j$, there is an $i' \in [i, j]$ s.t. either $i' \sim_h j$ or $i \sim_h i'$, and (4) for each $i \in C_h$ (resp. $i \in R_h$) there is at most one $j \in [m]$ s.t. $i \sim_h j$ (resp. $j \sim_h i$). When $i \sim_h j$, we say that positions i and j *match* in w (they are *matching call and return* in w). If $i \in C_h$ and $i \not\sim_h j$ for any $j \in R_h$, then i is an *unmatched call*. Analogously, if $i \in R_h$ and $j \not\sim_h i$ for any $j \in C_h$, then i is an *unmatched return*.

Multi-stack Visibly Pushdown Languages. A multi-stack visibly pushdown automaton pushes a symbol on stack h when it reads a stack-h call, and pops a symbol from stack h when it reads a stack-h return. Moreover, it just changes its state, without reading or modifying any stack, when reading an internal symbol. A special bottom-of-stack symbol $\bot$ is used: it is never pushed or popped, and is in the stack when computation starts. Fix a call-return alphabet $\widetilde{\Sigma}_n$.

Definition 1. (MULTI-STACK VISIBLY PUSHDOWN AUTOMATON) *A multi-stack visibly pushdown automaton (MVPA) over $\widetilde{\Sigma}_n$, is a tuple $A = (Q, Q_I, \Gamma, \delta, Q_F)$ where Q is a finite set of states, $Q_I \subseteq Q$ is the set of initial states, Γ is a finite stack alphabet containing the symbol $\bot$, $\delta \subseteq (Q \times \Sigma_c \times Q \times (\Gamma \setminus \{\bot\})) \cup (Q \times \Sigma_r \times \Gamma \times Q) \cup (Q \times \Sigma_{int} \times Q)$ is the transition function, and $Q_F \subseteq Q$ is the set of final states. Moreover, A is deterministic if $|Q_I| = 1$, and $|\{(q, a, q') \in \delta\} \cup \{(q, a, q', \gamma') \in \delta\} \cup \{(q, a, \gamma, q') \in \delta\}| \leq 1$, for each $q \in Q$, $a \in \Sigma$ and $\gamma \in \Gamma$.*

A *configuration* of an MVPA A over $\widetilde{\Sigma}_n$ is a tuple $\alpha = \langle q, \sigma_1, \ldots, \sigma_n \rangle$, where $q \in Q$ and each $\sigma_h \in (\Gamma \setminus \{\bot\})^* . \{\bot\}$ is a *stack content*. Moreover, α is *initial* if $q \in Q_I$ and $\sigma_h = \bot$ for every $h \in [n]$, and *accepting* if $q \in Q_F$. A transition $\langle q, \sigma_1, \ldots, \sigma_n \rangle \xrightarrow{a}_A \langle q', \sigma_1', \ldots, \sigma_n' \rangle$ is such that one of the following holds:

[Push] $a \in \Sigma_h^c$, $\exists \gamma \in \Gamma \setminus \{\bot\}$ such that $(q, a, q', \gamma) \in \delta$, $\sigma_h' = \gamma \cdot \sigma_h$, and $\sigma_i' = \sigma_i$ for every $i \in ([n] \setminus \{h\})$.

[Pop] $a \in \Sigma_h^r$, $\exists \gamma \in \Gamma$ such that $(q, a, \gamma, q') \in \delta$, $\sigma_i' = \sigma_i$ for every $i \in ([n] \setminus \{h\})$, and either $\gamma \neq \bot$ and $\sigma_h = \gamma \cdot \sigma_h'$, or $\gamma = \sigma_h = \sigma_h' = \bot$.

[Internal] $a \in \Sigma^{int}$, $(q, a, q') \in \delta$, and $\sigma_h' = \sigma_h$ for every $h \in [n]$.

For a word $w = a_1 \ldots a_m$ in Σ^*, a *run* of A on w from α_0 to α_m, denoted $\alpha_0 \xrightarrow{w}_A \alpha_m$, is a sequence of transitions $\alpha_{i-1} \xrightarrow{a_i}_A \alpha_i$ for $i \in [m]$. Word w is accepted by A if there is an initial configuration α and an accepting configuration α' such that $\alpha \xrightarrow{w}_A \alpha'$. The language accepted by A is denoted with $L(A)$.

A language L is a *multi-stack visibly pushdown language* (MVPL) if it is accepted by an MVPA over a call-return alphabet $\widetilde{\Sigma}_n$.

A visibly pushdown automaton (VPA) [1] is an MVPA with just one stack, and a *visibly pushdown language* (VPL) is an MVPL accepted by a VPA.

Scope-Bounded Matching Relations [12,13]. A *stack-h context* is a word in Σ_h^+. We say that w has at most k *maximal* contexts of stack h if $w \in \Sigma_h^* (\Sigma_{\neq h}^* \Sigma_h^*)^{k-1}$ where $\Sigma_{\neq h} = \bigcup_{h' \neq h} \Sigma_{h'}$.

For a word $w = a_1 \ldots a_m \in \Sigma^*$ we denote with $w[i,j]$ the subword $a_i \ldots a_j$. A word w is *k-scoped* (according to $\widetilde{\Sigma}_n$) if for each $h \in [n]$ and $i,j \in [m]$ s.t. $i \sim_h j$, $w[i,j]$ has at most k maximal contexts of stack h, i.e., each matching call and return of stack h occur within at most k stack-h maximal contexts.

In all the examples, we assume $\Sigma_1^c = \{a\}$, $\Sigma_2^c = \{b\}$, $\Sigma_1^r = \{c\}$, and $\Sigma_2^r = \{d\}$. Consider a sample word $\nu_1 = a^3 bd^2 c^2 a b^3 c^2$. Fig. 1 illustrates its splitting into contexts and the matching relations with edges. Note that the only pair of matching b's and d's is in the same stack-2 context. Moreover, the first a occurs in the first stack-1 context and is matched

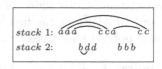

by the last c which occurs in the third stack-1 context. Any other matching pair of a's and c's occur within two stack-1 contexts. Therefore, ν_1 is k-scoped for any $k \geq 3$ but it is not 2-scoped.

With $Scoped(\widetilde{\Sigma}_n, k)$, we denote the set of all the k-scoped words over $\widetilde{\Sigma}_n$. A language $L \subseteq \Sigma^*$ is a *scoped* MVPL (SMVPL) if $L = Scoped(\widetilde{\Sigma}_n, k) \cap L(A)$ for some MVPA A over the call-return alphabet $\widetilde{\Sigma}_n$.

3   Properties of MVPA Runs over Scoped Words

Fix an integer $k > 0$ and an MVPA $A = (Q, Q_0, \Gamma, \delta, F)$ over $\widetilde{\Sigma}_n$.

k-scoped Splitting. For a word w over $\widetilde{\Sigma}_n$ and $h \in [n]$, a *cut* of w is $w_1 : w_2$ s.t. $w = w_1 w_2$. Such a cut is *consistent* with the matching relation $\sim_h$ ($\sim_h$-*consistent*, for short) if in w no call of stack h occurring in the prefix w_1 is matched with a return occurring in the suffix w_2.

A ($\sim_h$-*consistent*) *splitting* of w is defined by a set of ($\sim_h$-*consistent*) cuts of w, that is, it is an ordered tuple $\langle w_i \rangle_{i \in [d]}$ s.t. $w = w_1 \ldots w_d$, w_i is non-empty for $i \in [d]$ and $w_1 \ldots w_i : w_{i+1} \ldots w_d$ is a ($\sim_h$-*consistent*) cut for $i \in [d-1]$.

A *context-splitting* of w is a splitting $\langle w_i \rangle_{i \in [d]}$ where w_i is a stack-h_i context for $i \in [d]$. The *canonical* context-splitting of w is the only context-splitting $\langle w_i \rangle_{i \in [d]}$ s.t., for each $i \in [2,d]$, stack-h_i context w_i starts with a call or a return, and $h_{i-1} \neq h_i$. For example, Fig. 1 gives the canonical context-splitting η of ν_1 that splits ν_1 into: aaa, bdd, cca, bbb, and cc.

The *h-projection* of a context-splitting $\chi = \langle w_i \rangle_{i \in [d]}$ is obtained from χ by deleting all the w_i that are not stack-h contexts. For example, the 2-projection of η is: bdd, bbb. Note that a h-projection is trivially a context-splitting.

An ordered tuple $\chi = \langle w_i \rangle_{i \in [d]}$ of stack-h contexts is *k-bounded* if there is a $\sim_h$-consistent splitting $\xi = \langle v_i \rangle_{i \in [m]}$ of $w_1 \ldots w_d$ s.t. each v_i is the concatenation of at most k consecutive contexts of χ. In the following, we refer to such a ξ as a k-*bounding* splitting for χ and will denote with χ_{v_i} the ordered tuple of the contexts from χ that form v_i, for $i \in [m]$.

A k-*scoped splitting* χ of w is the canonical context-splitting of w refined with additional cuts s.t. for $h \in [n]$, the h-projection of χ is k-bounded.

Consider a sample word $\nu_2 = a^2bd^2c^2a^2b^3c^2ad^2bc^2$.
Fig. 2 illustrates a 2-scoped splitting χ that refines the canonical context-splitting of ν_2 by further cutting it at the dashed vertical lines. Thus, χ splits ν_2 into: $aa, bdd, cc, aa, bbb, cc, a, ddb, cc$. We observe that the dashed lines define a $\sim_1$-consistent splitting of word $a^2 c^2 a^2 c^2 a c^2$ where each portion is the con-

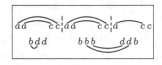

Fig. 2. k-scoped splitting

catenation of two contexts of the 1-projection of χ. Moreover, by cutting the word $bd^2 b^3 d^2b$ at the first dashed line, we get a $\sim_2$-consistent splitting where each portion has at most two contexts of the 2-projection of χ.

Lemma 1. *A word w is k-scoped iff there is a k-scoped splitting of w.*

Scope-Bounded Switching-Vector VPA. Fix $h \in [n]$. We start by recalling the definition of switching vector [9]. Intuitively, a switching vector summarizes the computations of an MVPA across several consecutive stack-h contexts.

Let A^h be the VPA over Σ_h obtained by restricting A to use only stack h. For $d > 0$, a tuple $I = \langle in_i, out_i \rangle_{i \in [d]}$ is a stack-h d-*switching vector* (d-SV, for short, d is omitted when we do not need to refer to its size) if there is an ordered tuple $\langle w_i \rangle_{i \in [d]}$ of stack-h contexts such that, for $i \in [d]$, $\langle in_i, \sigma_{i-1} \rangle \xrightarrow{w_i}_{A^h} \langle out_i, \sigma_i \rangle$ where $\sigma_0 = \perp$. We also define $st(I) = in_1$ and $cur(I) = out_d$, and say that $\langle w_i \rangle_{i \in [d]}$ *witnesses* I.

A stack-h k-*scoped* switching vector is a SV I that can be witnessed by a k-bounded ordered tuple of stack-h contexts.

Let χ be a k-bounded ordered tuple of stack-h contexts and $\xi = \langle v_i \rangle_{i \in [m]}$ be a k-bounding splitting for χ. Denote with I a stack-h k-scoped SV witnessed by χ. From the definition, I is given by the concatenation $I_1 \ldots I_m$ where each I_i is a stack-h d_i-SV witnessed by χ_{v_i} and $d_i \in [k]$ is the number of contexts of χ_{v_i}. Note that not all the concatenations of SV's with at most k pairs form a k-scoped SV. In fact, by concatenating two witnesses a call from one could match a return from the other, thus the resulting tuple could not be k-bounded.

We now define a VPA A_k^h that if the input is an encoding of a k-bounded tuple χ of stack-h contexts then it *computes* all the stack-h k-scoped SV's of A witnessed by χ. Essentially, A_k^h nondeterministically guesses any k-bounding splitting for χ and for each resulting portion, say formed by $d \le k$ contexts, it computes a corresponding d-SV while mimicking the behavior of A^h.

We encode a tuple of stack-h contexts by marking the first symbol of each context. Namely, for each $a \in \Sigma$, we add a fresh symbol $\bar{a}$ that is a call (resp. return, internal) if a is a call (resp. return, internal). Let $\bar{\Sigma}_h$ denote the set of all such new symbols. For a word $u = a_1a_2 \ldots a_d$, we denote with $\bar{u}$ the word $\bar{a}_1a_2 \ldots a_d$. We encode a tuple of stack-h contexts $u_1, \ldots, u_m$ as $\bar{u}_1\bar{u}_2 \ldots \bar{u}_m$. The new symbols $\bar{a}$ are interpreted as a when mimicking the moves of A^h.

By assuming the input $\bar{u}_1\bar{u}_2\ldots\bar{u}_m$, in a typical run, A_k^h starts from any $(p,p) \in Q^2$ and on reading the first symbol of $\bar{u}_1$, it updates the second component in this pair according to an A^h move. Now, assume a stored pair (p,p'). On any other symbol of $\bar{u}_1$, for any move of A^h from p' to p'' there are two nondeterministic moves of A_k^h: one updating p' to p'' in the stored pair (as before), the other starting a new SV by storing (p',p'') and thus guessing a cut. On the first symbol of $\bar{u}_2$, for any $q \in Q$ and for any move of A^h from q to q', again there are two nondeterministic moves as before: one updating the stored pair to $(p,p')(q,q')$, the other starting a new SV by replacing the stored pair with (q,q'). Then, the run continues similarly on the rest of the input.

There are two more aspects that A_k^h needs to take care of.

First, we only store d-SV's for $d \leq k$: when context-switching (i.e., reading a symbol $\bar{a} \in \bar{\Sigma}_h$), appending a new pair to the stored SV I must not be allowed if I already contains k pairs. By Lemma 1, this is sufficient for our purposes.

Second, we need to ensure that A_k^h uses only the portion of the stack that has been pushed since the computation of the current SV started; moreover, if it attempts to pop a symbol that was pushed when computing the previous SV, then the guessed splitting is clearly wrong (a guessed cut is not consistent with $\sim_h$) and the computation should halt. To ensure this, we store a bit e^s in the states of A_k^h and maintain the invariant: $e^s = 1$ iff the stack does not contain symbols pushed after the last guessed cut. Also, since pop transitions on an empty stack are allowed in VPAs, even if the portion of the stack currently in use is empty, we should allow them only if the whole stack is also empty. Thus, we store another bit e^g and maintain the invariant: $e^g = 1$ iff the stack is empty.

A state of A_k^h is thus (e^g, e^s, I) where $e^g, e^s \in \{0,1\}$ and $I \in (Q \times Q)^m$, $m \in [k]$. All the states are final and all the states of the form $(1,1,(q,q))$ for $q \in Q$ are initial. We leave to the reader the formal definition of the transitions.

Let w be a word over the alphabet $\Sigma_h \cup \bar{\Sigma}_h$. With $\mathcal{I}_k^h(w)$, we denote the set of the SV's $I \in \bigcup_{d>0}(Q \times Q)^d$ s.t. there exists a run ρ of A_k^h on w and I is the concatenation of $I_1,\ldots,I_j,I_{j+1}$ where: I_{j+1} is the SV stored in the state of the last configuration of ρ and $I_1,\ldots,I_j$ is the sequence of the SV's of all the states occurring at the configurations of ρ from which a transition that starts a new SV is taken (in the order they appear in ρ).

Lemma 2. *I is a stack-h k-scoped switching vector of A iff $I \in \mathcal{I}_k^h(w)$ for some $w \in (\Sigma_h \cup \bar{\Sigma}_h)^*$.*

Let ρ be a run of an MVPA A over a 3-stack call-return alphabet given as $\langle q_{i-1}, \bar{\sigma}_{i-1}\rangle \xrightarrow{u_i} \langle q_i, \bar{\sigma}_i\rangle$ with $i \in [11]$ and contexts u_i. Let $v_1 = \bar{u}_1\bar{u}_7\bar{u}_9$ be accepted by A_3^1, $v_2 = \bar{u}_2\bar{u}_4\bar{u}_{10}$ by A_3^2 and $v_3 = \bar{u}_3\bar{u}_5\bar{u}_6\bar{u}_8\bar{u}_{11}$ by A_3^3. Ac-

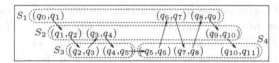

Fig. 3. Sample switching vectors and 3-scoped switching mask

cording to ρ, A_3^3 computes on v_3 the concatenation of the 2-SV S_3 over $\bar{u}_3\bar{u}_5$ and the 3-SV S_4 over $\bar{u}_6\bar{u}_8\bar{u}_{11}$. The 3-SV's computed on v_1 and v_2 are respectively S_1 and S_2 (Fig. 3).

Switching Masks. We use the SV's to summarize the runs of an MVPA over scoped words. For a k-scoped splitting χ of a word w over $\widetilde{\Sigma}_n$ and $h \in [n]$, denote with d_h the number of contexts in the h-projection χ^h of χ. Moreover, for $h, h' \in [n]$, $j \in [d_h]$ and $j' \in [d_{h'}]$, we define $next_\chi(h, j) = (h', j')$ s.t. the j'-th context of $\chi^{h'}$ is the context following in w the j-th context of χ^h.

For a word w over $\widetilde{\Sigma}_n$, a tuple $M = (I_1, \dots, I_n)$ is a k-scoped switching mask for w if there is a k-scoped splitting χ of w s.t. for $h \in [n]$: (1) $I_h = \langle in_y^h, out_y^h \rangle_{y \in [x_h]}$ is a stack-h k-scoped SV of A and (2) $out_y^h = in_{y'}^{h'}$ for each h, y, h', y' for which $next_\chi(h, y) = (h', y')$. Moreover, we let $st(M) = st(I_{h_1})$ and $cur(M) = cur(I_{h_d})$, where each w_i in χ is a stack-h_i context.

In Fig. 3, we give the 3-scoped switching mask according to the sample run ρ given above. The edges denote the mapping $next_\chi$.

Thus, by the given definitions and Lemmas 1 and 2, the following holds:

Lemma 3. *Suppose that $A = (Q, Q_0, \Gamma, \delta, F)$ is an MVPA over $\widetilde{\Sigma}_n$ and $w \in Scoped(\widetilde{\Sigma}_n, k)$. Then $w \in L(A)$ if and only if there exists a k-scoped switching mask M for w such that $st(M) \in Q_0$ and $cur(M) \in F$.*

4   Determinization, Sequentialization and Parikh Theorem

Determinization. We show that, when restricting to k-scoped words, deterministic and nondeterministic MVPAs are equivalent.

For an MVPA A, we define a deterministic MVPA A^D that, for a k-scoped input word w, constructs the set of all switching masks according to any k-scoped splitting of w. Thus, A^D accepts w iff it constructs a switching mask as in Lemma 3, and by supposing $w \in Scoped(\widetilde{\Sigma}_n, k)$, iff $w \in L(A)$.

For $h \in [n]$, let $D_k^h = (S_D, S_{D,0}, \Gamma_D^h, \delta_D^h, F_D)$ be the deterministic VPA equivalent to $A_k^h = (S, S_0, \Gamma, \delta^h, S)$ and obtained through the construction given in [1]. We recall that, according to that construction, the set of states S_D is $2^{S \times S} \times 2^S$, and the second component of a state is updated in a run as in the standard subset construction for finite automata. For $\hat{q} \in S_D$, denote with $R(\hat{q})$ the set of SV's contained in the A_k^h states stored as the second component of $\hat{q}$.

We construct $A^D = (Q^D, Q_0^D, \Gamma^D, \delta^D, F^D)$ building on the cross product of $D_k^1, \dots, D_k^n$; a state of A^D is $(h, \hat{q}_1, \dots, \hat{q}_n, \mathcal{M})$, where $h > 0$ denotes the stack that is active in the current context, $h = 0$ denotes the initial state, $\hat{q}_h$ is a state of D_k^h, and $\mathcal{M}$ is a set of tuples $(I_1, \dots I_n)$ where for $h \in [n]$, I_h is from $R(\hat{q}_h)$. The idea is to accumulate in the $\mathcal{M}$ component the tuples corresponding to the current SV's that are tracked in the states of $A_k^1, \dots, A_k^n$ while mimicking a run of A on the input word. Therefore, in each tuple $(I_1, \dots I_n)$ in the $\mathcal{M}$ component, on reading input a, we update I_h according to any transition of A_k^h on a if this is not the first symbol of the context, and on $\bar{a}$, otherwise (when context-switching into a stack-h context). The components $\hat{q}_1, \dots, \hat{q}_n$ are updated essentially by mimicking each deterministic automaton D_k^h on the stack-h contexts of the input word by dealing with the first symbol of each context as before. The accepting states are of the form $(h, \hat{q}_1, \ldots, \hat{q}_n, \mathcal{M})$ s.t. there is $(I_1, \ldots I_n) \in \mathcal{M}$ with $cur(I_h) \in F$.

The tuples in the component $\mathcal{M}$ of A^D states of a run can be composed by concatenating the component switching vectors I_h as done for the single A_k^h to define $\mathcal{I}_k^h(z)$. Thus, for each run ρ of A^D, we define a set $\mathcal{L}_\rho$ of tuples obtained in this way. We can show that $\mathcal{L}_\rho$ is exactly the set of all the k-scoped switching masks for the input word. Also, from the above description, we get that for each switching mask $M \in \mathcal{L}_\rho$, $st(M) \in Q_0$ holds, and if ρ is accepting, then there is at least a switching mask $M \in \mathcal{L}_\rho$ such that $cur(M) \in F$. Therefore, by Lemma 3:

Theorem 1. *For any n-stack call-return alphabet $\widetilde{\Sigma}_n$ and any* MVPA *A over $\widetilde{\Sigma}_n$, there exists a deterministic* MVPA *A^D over $\widetilde{\Sigma}_n$ such that* $Scoped(\widetilde{\Sigma}_n, k) \cap L(A^D) = Scoped(\widetilde{\Sigma}_n, k) \cap L(A)$. *Moreover, the size of A^D is exponential in the number of the states of A and doubly exponential in k and n.*

Sequentialization. We show that when restricting to k-scoped words, we can mimic the computations of an n-stack MVPA A using only one stack (*sequentialization*). We start by describing how the input word is rearranged.

Fix a k-scoped word w over $\widetilde{\Sigma}_n$, and let $\chi = \langle w_i \rangle_{i \in [d]}$ be a k-scoped splitting of w. For $h \in [n]$, denote with $\chi^h = \langle w_i^h \rangle_{i \in [x_h]}$ the h-projection of χ. Since χ is k-scoped, χ^h is k-bounded and let $\xi^h = \langle v_i^h \rangle_{y_h}$ be a k-bounding splitting for χ^h.

We define a total order $\preceq_w$ over all the v_j^h according to the position of their first symbol in w, that is, $v_j^h \preceq_w v_{j'}^{h'}$ iff $r \leq s$ where r is the position in w of the first symbol of v_j^h and s is that of the first symbol of $v_{j'}^{h'}$.

We denote with $\pi_\chi(w)$ the concatenation of all the v_j^h in the ordering given by $\preceq_w$. For example, consider the word u and the k-scoped splitting ξ resulting from the example of Fig. 3. The word $\pi_\xi(u)$ is $u_1 u_7 u_9.u_2 u_4 u_{10}.u_3 u_5.u_6 u_8 u_{11}$.

We define $\pi(w)$ as the set of all words $\pi_\chi(w)$ for any possible k-scoped splitting χ of w. We extend π to languages in the usual way.

We show that L is a k-scoped MVPL iff $\pi(L)$ is VPL (all calls and returns are interpreted as calls and returns of the unique stack). In fact, since ξ^h is k-bounding for χ^h, we get to process consecutively each set of (at most k) contexts that share the same stack content. Thus, when entering the next portion, we can start as the stack were empty (all that is left in the stack is not needed any more). Moreover, all the stack-h contexts, for a given h, occur in the same order as in w. Thus, we can process them by using A_k^h, and construct the VPA A_{seq} starting from the cross product of A_k^h for $h \in [n]$.

A second main feature of π is that when reading an input word $v \in \pi(w)$, we can reconstruct w by using only bounded memory: at any time, we keep a summary of each already processed portion of w (i.e., starting and ending states of corresponding portions of an A run) and a partial order of all such portions.

Observe that while parsing v, we know neither w nor a run on it. We reconstruct them on-the-fly by making nondeterministic guesses and ruling out the wrong guesses as soon as we realize it. For simplicity, we illustrate our idea on our running example by assuming that we know instead the run and the word

u. We refer to in Fig. 4 and for $i \in [4]$, S_i is as in Fig. 3. The input word to A_{seq} is $u_1 u_7 u_9 . u_2 u_4 u_{10} . u_3 u_5 . u_6 u_8 u_{11} \in \pi(u)$. After parsing $u_1 u_7 u_9$, we compute S_1 according to the considered run, and store the partial order shown on the edge from S_1 to S_2. Now after parsing $v_2 = u_2 u_4 u_{10}$, we compute S_2. Since u_2 follows u_1 and u_{10} follows u_9, by the ordering in v_2 and the fact that u_7 and u_4 are not consecutive, we get the partial order labeling the edge from S_2 to S_3, and so on. We succeed in reconstructing w iff in the end the maintained partial order collapses to just one summary (i.e., all the portions get connected). To keep the size of the stored partial order small, when the computation of a stack-h d-SV I starts,

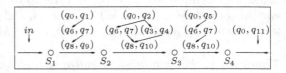

Fig. 4. A_{seq} run for the running example

we ensure that all the previously computed stack-h SV's are entirely *hidden* in the summaries (i.e., each pair of such SV's has been glued on both sides to other pairs) except for at most the second component of the last pair. In this case, we impose that the first pair of I starts with such a second component (as for S_3 and S_4 in the running example).

This is indeed sufficient to accept all the words in $\pi(w)$ for a k-scoped word w. In fact, assume as input $\pi_\chi(w)$ for a k-scoped splitting χ, and also the notation given in the beginning of this subsection. By definition of π_χ, the v_i^h's forming the k-bounding splittings ξ^h, for $h \in [n]$ and $i \in [y_h]$, are ordered according to their first contexts. Thus, when processing a v_i^h all the $v_{i'}^h \preceq_w v_i^h$ have been already read by A_{seq} and hence, the first contexts of all such $v_{i'}^h$ belong to a prefix of w that has been already processed. Therefore, the computed partial orders can be restricted to those that have a unique pair that precedes all the others. Moreover, for each $v_{i'}^h \preceq_w v_i^h$, since ξ^h is a splitting of the concatenation of the stack-h contexts of w (in the order they appear in w), also all the contexts of $v_{i'}^h$ must be in the already processed prefix of w. Hence, the number of pairs in the considered class of partial orders is bounded by $(n-1)(k-1)+1$.

Intuitively, A_{seq} mimics the cross product of $A_k^1, \dots, A_k^n$ and maintains the partial orders of the summaries (pairs of control states) of the starting MVPA A as observed above. The partial orders are updated at any context switch by using nondeterminism to guess how the next context is related to the summaries in the partial order. The nondeterminism of each A_k^h is reduced by ruling out all the moves that are not consistent with the stored partial order. The accepting states of A_{seq} are those with a partial order that is a single pair.

We omit the formal definition of A_{seq}. We only observe further that since the input of each A_k^h is over $\Sigma_h \cup \bar{\Sigma}_h$, we first need to transform them into corresponding VPAs B_k^h over Σ_h. This is done by modifying A_k^h such that the starting symbol of each context is now guessed nondeterministically (which is quite standard). Thus, denoting as B_k^h the resulting VPAs, we get that $\bar{w}_1 \dots \bar{w}_d \in L(A_k^h)$ iff $w_1 \dots w_d \in L(B_k^h)$. Also, the call-return alphabet of A_{seq} is $\tilde{\Sigma}_{seq}$ where

$\Sigma^c_{seq} = \bigcup_{h \in [n]} \Sigma^c_h$, $\Sigma^r_{seq} = \bigcup_{h \in [n]} \Sigma^r_h$ and $\Sigma^{int}_{seq} = \Sigma^{int}$ (recall that the alphabets from $\widetilde{\Sigma}_n$ are pairwise disjoint). The following lemma holds:

Lemma 4. *For an* MVPA *A and a k-scope word w over $\widetilde{\Sigma}_n$, $\pi(L(A)) = L(A_{seq})$. The size of A_{seq} is exponential in k and n, and polynomial in the size of A.*

Parikh's Theorem. The Parikh mapping associates a word with the vector of the numbers of the occurrences of each symbol in the word. Formally, the Parikh image of a word w, over the alphabet $\{a_1, \ldots, a_\ell\}$, is $\Phi(w) = (\#a_1, \ldots, \#a_\ell)$ where $\#a_i$ is the number of occurrences of a_i in w. This mapping extends to languages in the natural way: $\Phi(L) = \{\Phi(w) | w \in L\}$.

Parikh's theorem [18] states that for each context-free language L a regular language L' can be effectively found such that $\Phi(L) = \Phi(L')$. Lemma 4 gives an effective way to translate a k-scoped MVPL to a VPL, and thus we get:

Theorem 2. *For every k-scoped* MVPL *L over $\widetilde{\Sigma}_n$, there is a regular language L' over Σ such that $\Phi(L') = \Phi(L)$. Moreover, L' can be effectively computed.*

5 Closure Properties, Decision Problems and Expressiveness

Closure Properties and Decision Problems. Language union and intersection are defined for languages over a same call-return alphabet. The closure under these set operations can be shown with standard constructions and by exploiting that the stacks are synchronized over the input symbols. Complementation is defined w.r.t. the set $Scoped(\widetilde{\Sigma}_n, k)$ for a call-return alphabet $\widetilde{\Sigma}_n$, that is the complement of L is $Scoped(\widetilde{\Sigma}_n, k) \setminus L$. The closure under complementation follows from determinizability (Theorem 1).

The *membership problem* can be solved in nondeterministic polynomial time by simply guessing the transitions on each symbol and then checking that they form an accepting run. A matching lower bound can be given by a reduction from the satisfiability of 3-CNF Boolean formulas: for a formula with k variables, we construct a k-stack MVPA that nondeterministically guesses a valuation by storing the value of each variable in a separate stack, then starts evaluating the clauses (when evaluating a literal the guessed value is popped and then pushed into the stack to be used for next evaluations); partial evaluations are kept in the finite control (each clause has just three literals and we evaluate one at each time; for the whole formula we only need to store if we have already witnessed that it is false or that all the clauses evaluated so far are all true); thus each stack is only used to store the variable evaluation, and since for each stack h, each pushed symbol is either popped in the next stack-h context or is not popped at all, the input word is 2-scoped.

Checking *emptiness* is known to be PSPACE-COMPLETE for SMVPL [12,13]. Note that Lemma 4 reduces this problem to checking the emptiness for VPAs, and thus provides an alternative decision algorithm. Decidability of *universality*, *inclusion* and *equivalence* follows from the effectiveness of the closure under

Table 1. Summary of the main results on MVPLs (new results are in bold). In the table, NP-C stands for NP-COMPLETE, and so on.

	Closure properties				Decision Problems		
	∪	∩	Compl.	Determin.	Membership	Emptiness	Univ./ Equiv./Incl.
VPL [1]	Yes	Yes	Yes	Yes	PTIME-C	PTIME-C	EXPTIME-C
CFL	Yes	No	No	No	PTIME-C	PTIME-C	Undecidable
RMVPL [9]	Yes	Yes	Yes	Yes	NP	NP-C	2EXPTIME
SMVPL	**Yes**	**Yes**	**Yes**	**Yes**	**NP-c**	PSPACE-C	**2Exptime**
TMVPL [14]	Yes	Yes	Yes	No	NP-C	ETIME-C	2EXPTIME
PMVPL [10,14]	Yes	Yes	Yes	No	NP-C	2ETIME-C	3EXPTIME
OMVPL [2,14]	Yes	Yes	Yes	No	NP-C	2ETIME-C	3EXPTIME
CSL	Yes	Yes	Yes	Unknown	NLINSPACE	Undecidable	Undecidable

complementation and intersection, and the decidability of emptiness. This yields a double exponential upper bound. The best known lower bound is single exponential and comes from VPLs.

Comparisons with Known MVPL Classes. The class of SMVPL is incomparable with the main classes of MVPLs from the literature. In particular, we have compared it to RMVPL [19,9] (restricted to words with a bounded number of contexts), PMVPL [10] (restricted to words with a bounded number of phases where in each phase only the returns from one stack can occur), OMVPL [6,7,14] (restricted to words where a matched return of a stack i cannot occur between matching calls and returns of any stack j, for $j < i$), and PMVPL [14] (restricted to words with a bounded tree decomposition of the form of a stack tree).

Theorem 3. *1)* RMVPL $\subset$ SMVPL $\cap$ PMVPL. *2)* TMVPL $\supset$ PMVPL $\cup$ OMVPL. *3)* RMVPL *and* OMVPL *are incomparable.* *4)* SMVPL *and* TMVPL *are incomparable.* *5)* SMVPL, OMVPL, PMVPL *are pairwise incomparable.*

A Logical Characterization. We show that monadic second order logic (MSO_μ) on scoped words has the same expressiveness of scoped MVPAs. Here a word $w \in \Sigma^*$ is a structure over the universe $\{1, \ldots, |w|\}$. The logic has in its signature a predicate P_a for each $a \in \Sigma$ where $P_a(i)$ is true if the i-th symbol of w is a, and n predicates μ_h with $h \in [n]$, such that $\mu_h(i,j)$ holds true iff $i \sim_h j$.

We convert MSO sentences to automata using standard techniques that rely on the closure under Boolean operations and projection (see [21]). We get:

Theorem 4. *Let k, n be two positive integers, $\widetilde{\Sigma}_n$ be a call-return alphabet, and $L \subseteq Scoped(\widetilde{\Sigma}_n, k)$. L is k-scoped MVPL iff there is an MSO_μ sentence φ over $\widetilde{\Sigma}_n$ with $L_k(\varphi) = L$.*

6   Conclusions

We have shown that the class of SMVPL is closed under all the Boolean operations, it has a logical characterization, the Parikh theorem holds and the main decision problems are decidable (see Table 1 for a summary of the results on closure properties and decision problems). Moreover, the class of scoped MVPAS is determinizable and sequentializable (sequentialization is an effective technique for model-checking concurrent programs, see tools as CSeq, Microsoft Corral). We extend the results from [15,16] to a larger class of languages: there, only computations under a bounded number of round-robin scheduling are allowed and thus only scoped words with a bounded number of contexts between any two consecutive contexts of a same stack can be captured. Our sequentialization construction also suggests a tree-decomposition of the multiply nested words corresponding to scoped words with bags of $O(nk)$ size. Thus, also for the more expressive definition considered here, we get that the class of graphs defined by scoped words (and thus computations of scoped MPA) has bounded tree-width.

References

1. Alur, R., Madhusudan, P.: Visibly pushdown languages. In: STOC, pp. 202–211. ACM (2004)
2. Atig, M.F., Bollig, B., Habermehl, P.: Emptiness of multi-pushdown automata is 2ETIME-complete. In: Ito, M., Toyama, M. (eds.) DLT 2008. LNCS, vol. 5257, pp. 121–133. Springer, Heidelberg (2008)
3. Atig, M.F., Bouajjani, A., Kumar, K.N., Saivasan, P.: Linear-time model-checking for multithreaded programs under scope-bounding. In: Chakraborty, S., Mukund, M. (eds.) ATVA 2012. LNCS, vol. 7561, pp. 152–166. Springer, Heidelberg (2012)
4. Atig, M.F., Kumar, K.N., Saivasan, P.: Adjacent ordered multi-pushdown systems. In: Béal, M.-P., Carton, O. (eds.) DLT 2013. LNCS, vol. 7907, pp. 58–69. Springer, Heidelberg (2013)
5. Bollig, B., Kuske, D., Mennicke, R.: The complexity of model checking multi-stack systems. In: LICS, pp. 163–172. IEEE Computer Society (2013)
6. Breveglieri, L., Cherubini, A., Citrini, C., Crespi-Reghizzi, S.: Multi-push-down languages and grammars. Int. J. Found. Comput. Sci. 7(3), 253–292 (1996)
7. Carotenuto, D., Murano, A., Peron, A.: 2-visibly pushdown automata. In: Harju, T., Karhumäki, J., Lepistö, A. (eds.) DLT 2007. LNCS, vol. 4588, pp. 132–144. Springer, Heidelberg (2007)
8. Hague, M.: Saturation of Concurrent Collapsible Pushdown Systems. In: FSTTCS. LIPIcs, vol. 24, pp. 313–325 (2013)
9. La Torre, S., Madhusudan, P., Parlato, G.: The language theory of bounded context-switching. In: López-Ortiz, A. (ed.) LATIN 2010. LNCS, vol. 6034, pp. 96–107. Springer, Heidelberg (2010)
10. La Torre, S., Madhusudan, P., Parlato, G.: A robust class of context-sensitive languages. In: LICS, pp. 161–170. IEEE Computer Society (2007)
11. La Torre, S., Madhusudan, P., Parlato, G.: An infinite automaton characterization of double exponential time. In: Kaminski, M., Martini, S. (eds.) CSL 2008. LNCS, vol. 5213, pp. 33–48. Springer, Heidelberg (2008)
12. La Torre, S., Napoli, M.: Reachability of multistack pushdown systems with scope-bounded matching relations. In: Katoen, J.-P., König, B. (eds.) CONCUR 2011. LNCS, vol. 6901, pp. 203–218. Springer, Heidelberg (2011)
13. La Torre, S., Napoli, M.: A temporal logic for multi-threaded programs. In: Baeten, J.C.M., Ball, T., de Boer, F.S. (eds.) TCS 2012. LNCS, vol. 7604, pp. 225–239. Springer, Heidelberg (2012)

14. La Torre, S., Napoli, M., Parlato, G.: A Unifying Approach for Multistack Push-down Automata. In: Csuhaj-Varjú, E., Dietzfelbinger, M., Ésik, Z. (eds.) MFCS 2014, Part I. LNCS, vol. 8634, pp. 373–384. Springer, Heidelberg (2014), http://users.ecs.soton.ac.uk/gp4/papers/MVPL14.pdf
15. La Torre, S., Parlato, G.: Scope-bounded multistack pushdown systems: Fixed-point, sequentialization, and tree-width. In: FSTTCS. LIPIcs, vol. 18, pp. 173–184 (2012)
16. Madhusudan, P., Parlato, G.: The tree width of auxiliary storage. In: POPL, pp. 283–294. ACM (2011)
17. Okhotin, A., Piao, X., Salomaa, K.: Descriptive Complexity of Input-Driven Pushdown Automata. In: Bordihn, H., Kutrib, M., Truthe, B. (eds.) Languages Alive 2012. LNCS, vol. 7300, pp. 186–206. Springer, Heidelberg (2012)
18. Parikh, R.: On context-free languages. J. ACM 13(4), 570–581 (1966)
19. Qadeer, S., Rehof, J.: Context-bounded model checking of concurrent software. In: Halbwachs, N., Zuck, L. (eds.) TACAS 2005. LNCS, vol. 3440, pp. 93–107. Springer, Heidelberg (2005)
20. Seth, A.: Global reachability in bounded phase multi-stack pushdown systems. In: Touili, T., Cook, B., Jackson, P. (eds.) CAV 2010. LNCS, vol. 6174, pp. 615–628. Springer, Heidelberg (2010)
21. Thomas, W.: Languages, automata, and logic. In: Handbook of Formal Languages, vol. 3, pp. 389–455. Springer (1997)
22. Melhorn, K.: Pebbling mountain ranges and its application to DCFL-recognition. In: de Bakker, J., van Leeuwen, J. (eds.) ICALP 1980. LNCS, vol. 85, pp. 422–435. Springer, Heidelberg (1980)

Visibly Pushdown Transducers
with Well-Nested Outputs

Pierre-Alain Reynier* and Jean-Marc Talbot*

Aix Marseille Université, CNRS, LIF UMR 7279, 13288, Marseille, France

Abstract. Visibly pushdown transducers (VPTs) are visibly pushdown
automata extended with outputs. They have been introduced to model
transformations of nested words, i.e. words with a call/return structure.
When outputs are also structured and well nested words, VPTs are a
natural formalism to express tree transformations evaluated in streaming. We prove the class of VPTs with well-nested outputs to be decidable
in PTIME. Moreover, we show that this class is closed under composition and that its type-checking against visibly pushdown languages is
decidable.

1 Introduction

Visibly pushdown automata (VPA) [1], first introduced as input-driven pushdown
automata [3], are pushdown machines whose stack behavior is synchronized with
the structure of the input word. More precisely, the input alphabet is partitioned
into call and return symbols; when reading a call symbol the machine must push
a symbol onto the stack, when reading a return symbol it must pop a symbol from
the stack and when reading an internal symbol the stack remains unchanged.
Such words over a structure alphabet are called nested words.

Visibly pushdown transducers (VPTs) [6,7,9,10] extend visibly pushdown automata with outputs. Each transition is equipped with an output word; a VPT
thus transforms an input word into an output word obtained as the concatenation of all the output words produced along a successful run (*i.e.* a sequence of
transitions) on this input. VPTs are a strict subclass of pushdown transducers
(PTs) and strictly extend finite state transducers. Several problems that are undecidable for PTs are decidable for VPTs similarly to finite state transducers:
functionality (in PTIME), k-valuedness (in co-NPTIME) and functional equivalence (EXPTIME-complete) [6]. However, some decidability results or valuable
properties of finite-state transducers do not hold for VPTs [7]: VPTs are not
closed under composition and type-checking against VPA is undecidable (deciding whether the range of a transducer is included into the language of a given
VPA).

Unranked trees and more generally hedges can be linearized into nested words
over a structured alphabet (such as XML documents). These words for which the

---

* Partly supported by the PEPS project "Synthesis of Stream Processors" funded by
  CNRS.

A.M. Shur and M.V. Volkov (Eds.): DLT 2014, LNCS 8633, pp. 129–141, 2014.

matching between call and return symbols is perfect are called well-nested words. So, VPTs are a suitable formalism to express hedge transformations. Moreover, as they process the linearization from left to right, they are also an adequate formalism to model and analyze transformations in streaming, as shown in [5]. VPTs output strings; operating on well-nested inputs, they define hedge-to-string transformations. If the output strings are well-nested too, they define hedge-to-hedge transformations [4].

In [6], by means of a syntactical restriction on transition rules, a class of VPTs whose range contains only well-nested words is presented. This class enjoys good properties: it is closed under composition and type-checking against visibly pushdown languages is decidable. One may then wonder whether these properties come from this particular subclass or from the fact that the range of these VPTs contains only well-nested words.

In this paper, we consider two classes of transductions (that is, of relations) over nested words definable by VPTs. First, the class of *globally well-nested* transductions, denoted $\mathcal{G}_{\mathsf{wn}}$, is the class of VPT transductions whose range contains only well-nested words. The second class, named *almost well-nested* and denoted $\mathcal{A}_{\mathsf{wn}}$, slightly generalizes the first one as follows: there must exist $k \in \mathbb{N}$ such that every output word contains at most k unmatched returns and at most k unmatched calls. These two classes of transductions naturally define two classes of transducers gwnVPT and awnVPT: a VPT is in gwnVPT (resp. in awnVPT) if the transduction it represents is in $\mathcal{G}_{\mathsf{wn}}$ (resp. in $\mathcal{A}_{\mathsf{wn}}$). While defined in a semantical way, we provide criteria on successful computations of VPTs characterizing precisely the classes gwnVPT and awnVPT. Then, based on these criteria, we prove the class awnVPT to be decidable in PSPACE. Regarding the class gwnVPT, using a recent result of [2], we prove it is decidable in PTIME. Finally, we prove that the two classes gwnVPT and awnVPT enjoy good properties: they are closed under composition and type-checking is decidable against visibly pushdown languages.

The paper is organized as follows: definitions and recalls of some basic properties on VPTs are presented in Section 2. We introduce in Section 3 the two classes of transductions we define in this paper as well as the corresponding classes of transducers. Considering additionally the (restricted) class introduced in [6], we prove also that they form a strict hierarchy. Then, we give in Section 4 a precise characterization of the classes gwnVPT and awnVPT by means of some criteria on VPTs. Section 5 describes decision procedure of the considered classes of transducers. Finally, the closure of the considered classes under composition and the decidability of type-checking are addressed in Section 6. Omitted details can be found in a technical report [8].

2   Preliminaries

(Well) Nested Words. The set of all finite words (resp. of all words of length at most n) over a finite alphabet Σ is denoted by Σ^* (resp. $\Sigma^{\leq n}$); the empty word is denoted by ϵ. A *structured alphabet* is a triple $\Sigma = (\Sigma_c, \Sigma_i, \Sigma_r)$ of disjoint alphabets, of call, internal and return symbols respectively. Given a structured

alphabet Σ, we always denote by Σ_c, Σ_i and Σ_r its implicit structure, and identify Σ with $\Sigma_c \cup \Sigma_i \cup \Sigma_r$. A *nested word* is a finite word over a structured alphabet.

The set of *well-nested words* over a structured alphabet Σ is the least set, denoted by Σ_{wn}^*, that satisfies (i) $\epsilon \in \Sigma_{wn}^*$, (ii) for all $i \in \Sigma_i$, $w \in \Sigma_{wn}^*$, $iw \in \Sigma_{wn}^*$, and (iii) for all $w, w' \in \Sigma_{wn}^*$, $c \in \Sigma_c$, $r \in \Sigma_r$, $cwrw' \in \Sigma_{wn}^*$. E.g. on $\Sigma = (\{c_1, c_2\}, \emptyset, \{r\})$, the nested word $c_1 r c_2 r$ is well-nested while $r c_1$ is not.

For a word w from Σ^*, we define its balance B as the difference between the number of symbols from Σ_c and of symbols from Σ_r occurring in w. Note that if $w \in \Sigma_{wn}^*$, then $B(w) = 0$; but the converse is false as exemplified by $r c_1$.

Lemma 1. *Let $u, v \in \Sigma^*$. We have $B(uv) = B(u) + B(v) = B(vu)$.*

For any word w from Σ^*, we denote by $Oc(w)$ (resp. $Or(w)$) the number of open calls (resp. open returns) in w. Formally,

$$Or(w) = -\min\{B(w') \mid w'w'' = w\} \qquad Oc(w) = B(w) + Or(w)$$

We define, for any word w, $O(w)$ as the pair $(Or(w), Oc(w)) \in \mathbb{N}^2$. Given $(n_1, n_2) \in \mathbb{N}^2$, we define $||(n_1, n_2)|| = \max(n_1, n_2)$. Note that, for a word w, we obtain $||O(w)|| = \max\{Or(w), Oc(w)\}$ and $w \in \Sigma_{wn}^*$ iff $||O(w)|| = 0$, that is $O(w) = (0, 0)$.

Given a word $w \in \Sigma^*$, we let $\mathsf{height}(w) = \max\{||O(w_1)|| \mid w = w_1 w_2\}$ be the height of w. We denote by $|w|$ the length of w, defined as usual.

Definition 1. *For any two pairs (n_1, n_2) and (n'_1, n'_2) of naturals from $\mathbb{N}^2$, we define $(n_1, n_2) \oplus (n'_1, n'_2)$ as the pair*

$$\begin{cases} (n_1, n_2 - n'_1 + n'_2) & \text{if } n_2 \geq n'_1 \\ (n_1 + n'_1 - n_2, n'_2) & \text{if } n'_1 > n_2 \end{cases}$$

Proposition 1. *$(\mathbb{N}^2, \oplus, (0, 0))$ is a monoid, and the mapping O is a morphism from $(\Sigma^*, ., \epsilon)$ to $(\mathbb{N}^2, \oplus, (0, 0))$; in particular, for any two words u_1, u_2 from Σ^*, $O(u_1 u_2) = O(u_1) \oplus O(u_2)$.*

Transductions – Transducers. Let Σ be a structured (input) alphabet, and Δ be a structured (output) alphabet. A relation over $\Sigma^* \times \Delta^*$ is a *transduction*. We denote by $\mathcal{T}(\Sigma, \Delta)$ the set of these transductions. For a transduction T, the set of words u (resp. v) such that $(u, v) \in T$ is called the *domain* (resp. the *range*) of T.

A *visibly pushdown transducer* from Σ to Δ (the class is denoted $\mathsf{VPT}(\Sigma, \Delta)$) is a tuple $A = (Q, I, F, \Gamma, \delta)$ where Q is a finite set of states, $I \subseteq Q$ the set of initial states, $F \subseteq Q$ the set of final states, Γ a (finite) stack alphabet, and $\delta = \delta_c \uplus \delta_r \uplus \delta_i$ is the transition relation where:

- $\delta_c \subseteq Q \times \Sigma_c \times \Gamma \times \Delta^* \times Q$ are the *call transitions*,
- $\delta_r \subseteq Q \times \Sigma_r \times \Gamma \times \Delta^* \times Q$ are the *return transitions*.
- $\delta_i \subseteq Q \times \Sigma_i \times \Delta^* \times Q$ are the *internal transitions*.

A stack (content) is a word over Γ. Hence, Γ^* is a monoid for the concatenation with $\bot$ (the empty stack) as neutral element. A configuration of A is a pair (q, σ) where $q \in Q$ and $\sigma \in \Gamma^*$ is a stack content. Let $u = a_1 \ldots a_l$ be a (nested) word on Σ, and $(q, \sigma), (q', \sigma')$ be two configurations of A. A *run* of the VPT A over u from (q, σ) to (q', σ') is a (possibly empty) sequence of transitions $\rho = t_1 t_2 \ldots t_l \in \delta^*$ such that there exist $q_0, q_1, \ldots q_l \in Q$ and $\sigma_0, \ldots \sigma_l \in \Gamma^*$ with $(q_0, \sigma_0) = (q, \sigma)$, $(q_l, \sigma_l) = (q', \sigma')$, and for each $0 < k \leq l$, we have either (i) $t_k = (q_{k-1}, a_k, \gamma, w_k, q_k) \in \delta_c$ and $\sigma_k = \sigma_{k-1}\gamma$, or (ii) $t_k = (q_{k-1}, a_k, \gamma, w_k, q_k) \in \delta_r$, and $\sigma_{k-1} = \sigma_k \gamma$, or (iii) $t_k = (q_{k-1}, a_k, w_k, q_k) \in \delta_i$, and $\sigma_{k-1} = \sigma_k$. When the sequence of transitions is empty, $(q, \sigma) = (q', \sigma')$.

The length (resp. height) of a run ρ over some word $u \in \Sigma^*$, denoted $|\rho|$ (resp. $\mathsf{height}(\rho)$) is defined as the length of u (resp. as the height of u).

The *output* of ρ (denoted $\mathsf{output}(\rho)$) is the word $v \in \Delta^*$ defined as the concatenation $w = w_1 \ldots w_l$ when the sequence of transitions is not empty and ϵ otherwise. We write $(q, \sigma) \xrightarrow{u|w} (q', \sigma')$ when there exists a run on u from (q, σ) to (q', σ') producing w as output. Initial (resp. final) configurations are pairs $(q, \bot)$ with $q \in I$ (resp. with $q \in F$). A configuration (q, σ) is *reachable* (resp. *co-reachable*) if there exists some initial configuration $(i, \bot)$ (resp. some final configuration $(f, \bot)$) and a run from $(i, \bot)$ to (q, σ) (resp. from (q, σ) to $(f, \bot)$). A run is *accepting* if it starts in an initial configuration and ends in a final configuration.

A transducer A defines relation/transduction from nested words to nested words, denoted by $[\![A]\!]$, and defined as the set of pairs $(u, v) \in \Sigma^* \times \Delta^*$ such that there exists an accepting run on u producing v as output. Note that since both initial and final configurations have empty stack, A accepts only well-nested words, i.e. $[\![A]\!] \subseteq \Sigma_{\mathsf{wn}}^* \times \Delta^*$.

We denote $\mathcal{VP}(\Sigma, \Delta)$ the class of transductions defined by VPTs over the structured alphabets Σ (as input alphabet) and Δ (as output alphabet).

Given a VPT $A = (Q, I, F, \Gamma, \delta)$, we let $\mathsf{O}_{\max}^A$ be the maximal number of open calls and of open returns in a word produced as output of a call or of a return transition in A. Formally, we have:

$$\mathsf{O}_{\max}^A = \max\{\|\mathsf{O}(w)\| \mid (p, \alpha, w, \gamma, q) \in \delta_c \cup \delta_r\}$$

Visibly Pushdown Automata. We define visibly pushdown automata (VPA) simply as a particular case of VPT; we may think of them as transducers with no output. Hence, only the domain of the transduction matters and is called the language defined by the visibly pushdown automaton. For an automaton A, this language will be denoted $L(A)$.

Properties of Computations in VPA/VPT. We recall two standard results on runs of visibly pushdown machines.

Lemma 2. *Let A be a VPA with set of states Q and $\rho : (p, \bot) \xrightarrow{u} (q, \bot)$ be a run of A over some word $u \in \Sigma_{\mathsf{wn}}^*$. Let $h \in \mathbb{N}_{>0}$. We have:*

(i) *if* $\text{height}(u) < h$ *and* $|u| \geq |Q|^h$, *then* ρ *can be decomposed as follows:*

$$\rho : (p, \perp) \xrightarrow{u_1} (p_1, \sigma) \xrightarrow{u_2} (p_1, \sigma) \xrightarrow{u_3} (q, \perp)$$

with $u_1 u_3$ *and* u_2 *well-nested words and* $u_2 \neq \epsilon$.

(ii) *if* $\text{height}(u) \geq |Q|^2$, *then* ρ *can be decomposed as follows:*

$$\rho : (p, \perp) \xrightarrow{u_1} (p_1, \sigma) \xrightarrow{u_2} (p_1, \sigma\sigma') \xrightarrow{u_3} (p_2, \sigma\sigma') \xrightarrow{u_4} (p_2, \sigma) \xrightarrow{u_5} (q, \perp)$$

with $u_1 u_5$, $u_2 u_4$ *and* u_3 *well-nested words, and* $\sigma' \neq \perp$.

3 Classes of VPT Producing (almost) Well-Nested Outputs

In this section, after recalling the definition of (locally) well-nested VPT, we introduce the new classes of globally and almost well-nested VPT. Then, we prove relationships between these classes.

3.1 Definitions

Locally Well-nested VPTs (lwnVPT). In [6], the class of (locally) well-nested VPT has been introduced. For this class, the enforcement of the well-nestedness of the output is done locally and syntactically at the level of transition rules.

Definition 2 (Locally Well-nested). *Let* $A = (Q, I, F, \Gamma, \delta)$ *be a* VPT. *A is a locally well-nested* VPT (lwnVPT) *if:*

- *for any pair of transitions* $(q, a, v, \gamma, q') \in \delta_c$, $(p, b, w, \gamma, p') \in \delta_r$, *the word* vw *is well nested, and*
- *for any transition* $(q, a, v, q') \in \delta_i$, *the word* v *is well-nested.*

A VPT *transduction* T *is* locally well-nested *if there exists a* lwnVPT *A that realizes* T *(* $[\![A]\!] = T$ *). The class of locally well-nested* VPT *transductions is denoted* $\mathcal{L}_{\text{wn}}$.

It is straightforward to prove that

Proposition 2. *Let* A *be a locally well-nested* VPT *and* (p, σ), (q, σ) *two configurations of* A. *For all well-nested word* u, *if* $(p, \sigma) \xrightarrow{u/v} (q, \sigma)$ *then* $v \in \Sigma_{\text{wn}}^*$.

Thus any locally well-nested VPT transduction T is included into $\Sigma_{\text{wn}}^* \times \Delta_{\text{wn}}^*$.

Globally Well-nested VPT *Transduction – Almost Well-nested* VPT *Transduction.* In this section, we introduce the class of globally well-nested transductions and its weaker variant of "almost" well-nested transductions. Unlike the definition of $\mathcal{L}_{\text{wn}}$ which is done at the level of transducers, these definitions are done at the level of transductions and thus, as a semantical property.

Definition 3 (Globally Well-nested). *A* VPT *transduction T is globally well-nested if* $T(\Sigma^*_{\text{wn}}) \subseteq \Delta^*_{\text{wn}}$. *The class of globally well-nested* VPT *transductions is denoted* $\mathcal{G}_{\text{wn}}$.

A VPT *A is globally well-nested if its transduction* $[\![A]\!]$ *is. The class of globally well-nested* VPT *is denoted* gwnVPT.

Definition 4 (Almost Well-nested). *A* VPT *transduction T is almost well-nested if there is* $k \in \mathbb{N}$ *such that every pair of words* $(u, v) \in T$ *satisfies* $||O(v)|| \leq k$. *The class of almost well-nested* VPT *transductions is denoted* $\mathcal{A}_{\text{wn}}$.

A VPT *A is almost well-nested if its transduction* $[\![A]\!]$ *is. The class of almost well-nested* VPT *is denoted* awnVPT.

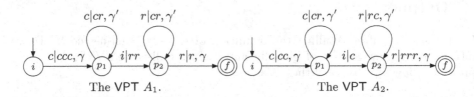

The VPT A_1.   The VPT A_2.

Fig. 1. Two VPTs in $\mathcal{VP}(\Sigma, \Sigma)$ with $\Sigma_c = \{c\}$, $\Sigma_r = \{r\}$ and $\Sigma_i = \{i\}$

3.2   Comparison of the Different Classes

Classes of transductions $\mathcal{G}_{\text{wn}}$ and $\mathcal{A}_{\text{wn}}$ are defined by semantical conditions on the defined relations. This yields a clear correspondence between the classes $\mathcal{G}_{\text{wn}}$ and gwnVPT on one side and $\mathcal{A}_{\text{wn}}$ and awnVPT on the other side. This is not the case for $\mathcal{L}_{\text{wn}}$: two examples of VPTs are given in Figure 1. It is easy to verify that $A_1, A_2 \in$ gwnVPT. Moreover, none of these transducers belongs to lwnVPT. However, one can easily build a transducer A'_2 such that $[\![A_2]\!] = [\![A'_2]\!]$ and $A'_2 \in$ lwnVPT. Indeed one can perform the following modifications:

- the transition (p_1, i, c, p_2) becomes (p_1, i, ϵ, p_2)
- the transition $(p_2, r, rc, \gamma', p_2)$ becomes $(p_2, r, cr, \gamma', p_2)$
- the transition (p_2, r, rrr, γ, f) becomes $(p_2, r, crrr, \gamma, f)$

On the contrary, as we prove below, the transduction $[\![A_1]\!]$ does not belong to $\mathcal{L}_{\text{wn}}$: there exists no transducer $A'_1 \in$ lwnVPT such that $[\![A'_1]\!] = [\![A_1]\!]$.

To summarize, we prove the following proposition.

Proposition 3. *The following inclusion results hold:*

- *For transducers:* lwnVPT $\subsetneq$ gwnVPT $\subsetneq$ awnVPT
- *For transductions:* $\mathcal{L}_{\text{wn}} \subsetneq \mathcal{G}_{\text{wn}} \subsetneq \mathcal{A}_{\text{wn}}$

Proof (Sketch). The non-strict inclusions are straightforward. The two strict inclusions gwnVPT $\subsetneq$ awnVPT and $\mathcal{G}_{\text{wn}} \subsetneq \mathcal{A}_{\text{wn}}$ follow from the constraint on the range. The strict inclusion lwnVPT $\subsetneq$ gwnVPT is witnessed by A_2 from Figure 1, as explained above.

To sketch the proof of the strict inclusion $\mathcal{L}_{\mathsf{wn}} \subsetneq \mathcal{G}_{\mathsf{wn}}$, we consider the transducer A_1 on Figure 1. Observe that $[\![A_1]\!] \in \mathcal{G}_{\mathsf{wn}}$, we show that $[\![A_1]\!] \notin \mathcal{L}_{\mathsf{wn}}$. First note that $[\![A_1]\!] = \{(cc^k ir^k r, ccc(cr)^k rr(cr)^k r) \mid k \in \mathbb{N}\}$ and that

- (Fact 1) The transduction defined by A_1 is injective
- (Fact 2) Any word of the output can be decomposed as $w_1 rr w_2$ where $w_1 = ccc(cr)^k$ and $w_2 = (cr)^k r$ for some natural k and for each w_1 with fixed k there exists a unique w_2 such that $w_1 rr w_2$ is in the range of A_1 (and conversely).

By contradiction, suppose that there exists $A_1' \in \mathsf{lwnVPT}$ such that $[\![A_1']\!] = [\![A_1]\!]$. Now, for k sufficiently large and depending only on the fixed size of A_1', A_1' has an accepting run for the input $cc^k ir^k r$ of the form given in Lemma 2(ii). Let us denote by u_i (resp. v_i), $i \in \{1, \ldots, 5\}$ the corresponding decomposition of the input (resp. output) word. Due to Proposition 2, words $v_1 v_5$, $v_2 v_4$ and v_3 are well-nested.

Now assume $v_2 = v_4 = \epsilon$. Then, using a simple pumping argument over the pair (u_2, u_4), one would obtain a different input producing the same output, contradicting the injectivity of A_1' (due to (Fact 1)). So, $v_2 \neq \epsilon$ or $v_4 \neq \epsilon$.

Using a case analysis on the presence of the previously mentioned pattern rr in the outputs of A_1', using the fact that $v_2 v_4 \neq \epsilon$, (Fact 2) and a pumping argument over the pair of words (u_2, u_4), one obtains a contradiction. $\quad\square$

4 Characterizations

In this section we give criteria on VPTs that aim to characterize the classes gwnVPT and awnVPT.

Definition 5. *Let A be a VPT. Let us consider the following criteria:*

(C1) For all states p, i, f such that i is initial and f is final, for any stack σ, then any accepting run

$$(i, \bot) \xrightarrow{u_1/v_1} (p, \sigma) \xrightarrow{u_2/v_2} (p, \sigma) \xrightarrow{u_3/v_3} (f, \bot)$$

with $u_1 u_3, u_2 \in \Sigma_{\mathsf{wn}}^*$ satisfies $\mathsf{B}(v_2) = 0$.

(C2) For all states p, q, i, f such that i is initial and f is final, for any stack σ, σ', then any accepting run

$$(i, \bot) \xrightarrow{u_1/v_1} (p, \sigma) \xrightarrow{u_2/v_2} (p, \sigma\sigma') \xrightarrow{u_3/v_3} (q, \sigma\sigma') \xrightarrow{u_4/v_4} (q, \sigma) \xrightarrow{u_5/v_5} (f, \bot)$$

with $u_2 u_4, u_3 \in \Sigma_{\mathsf{wn}}^*$ and $\sigma' \neq \bot$ satisfies $\mathsf{B}(v_2) + \mathsf{B}(v_4) = 0$ and $\mathsf{B}(v_2) \geq 0$.

The following result follows from Propositions 4 and 5 that we prove below.

Theorem 1. *A VPT A is almost well-nested iff it verifies (C1) and (C2).*

Lemma 3. *Let $X \subseteq \Sigma^*$ such that the set $\mathsf{B}(X) = \{\mathsf{B}(u) \mid u \in X\}$ is infinite. Then the set $\{\mathsf{O}(u) \mid u \in X\}$ is infinite as well.*

Lemma 4. *Let $u \in \Sigma^*$ and k be a strictly positive integer. Then $O(u^k)$ is equal to $(Or(u), (Oc(u) - Or(u)) * (k - 1) + Oc(u))$ if $Oc(u) \geq Or(u)$ and to $(Or(u) + (Or(u) - Oc(u)) * (k - 1), Oc(u))$ otherwise.*

Proof. By definition of $\oplus$ and by induction on k.   □

Proposition 4. *Let A be a VPT. If A does not satisfy $(C1)$ or $(C2)$, then $A \notin$ awnVPT.*

Proof. Let us assume that A does not satisfy $(C1)$. Hence there exists an accepting run as described in criterion $(C1)$ such that $B(v_2) \neq 0$. We then build by iterating the loop on word u_2 accepting runs for words of the form $u_1(u_2)^k u_3$ for any natural k, producing output words $v_1(v_2)^k v_3$. Let us denote this set by X. As $B(v_2) \neq 0$ and by Lemma 1, the set $B(X)$ is infinite. Lemma 3 entails that A is not almost well-nested.

Assume now that A does not satisfy $(C2)$. Hence, there exists an accepting run as described in the statement of the proposition such that either (i) $B(v_2) + B(v_4) = b \neq 0$ or (ii) $B(v_2) < 0$. In the case of (i), from this run, one can build by pumping accepting runs for words of the form $u_1(u_2)^k u_3(u_4)^k u_5$ for any natural k, producing output words $v_1(v_2)^k v_3(v_4)^k v_5$. As before, Lemmas 1 and 3 imply that A is not almost well-nested.

Now, for (ii) assuming that $B(v_2) + B(v_4) = 0$. As $B(v_2) < 0$, it holds that $B(v_4) > 0$ and thus, $Or(v_2) > Oc(v_2)$, $Or(v_4) < Oc(v_4)$. From the run of the statement, one can build by pumping accepting runs for words of the form $u_1(u_2)^k u_3(u_4)^k u_5$ for any natural k, producing output words $v_1(v_2)^k v_3(v_4)^k v_5$. Now, we consider $O(v_1(v_2)^k v_3(v_4)^k v_5)$ which, by associativity of $\oplus$, is equal to $O(v_1) \oplus O((v_2)^k) \oplus O(v_3) \oplus O((v_4)^k) \oplus O(v_5))$. Now, by Lemma 4, it is equal to

$$O(v_1) \oplus (Or(v_2) + (Or(v_2) - Oc(v_2)) * (k - 1), Oc(v_2)) \oplus O(v_3) \oplus$$
$$(Or(v_4), (Oc(v_4) - Or(v_4)) * (k - 1) + Oc(v_4)) \oplus O(v_5)$$

It is easy to see that for k varying, the described pairs are unbounded.   □

Given a VPT $A = (Q, I, F, \Gamma, \delta)$, we define the integer $N_A = 2|Q|^{2|Q|^2}$.

Lemma 5. *Let A be a VPT. If A satisfies the criteria $(C1)$ and $(C2)$, then for any accepting run ρ such that $|\rho| \geq N_A$, there exists an accepting run ρ' such that $|\rho'| < |\rho|$ and $||O(\text{output}(\rho'))|| \geq ||O(\text{output}(\rho))||$.*

Proof (Sketch). Let $A = (Q, I, F, \Gamma, \delta)$ and ρ be an accepting run such that $|\rho| \geq N_A$. We distinguish two cases, depending on $\text{height}(\rho)$:

- when $\text{height}(\rho) < 2|Q|^2$: by definition of N_A, we can apply Lemma 2.(i) twice and prove that ρ is of the following form:

$$(i, \bot) \xrightarrow{u_1/v_1} (p, \sigma) \xrightarrow{u_2/v_2} (p, \sigma) \xrightarrow{u_3/v_3} (q, \sigma') \xrightarrow{u_4/v_4} (q, \sigma') \xrightarrow{u_5/v_5} (f, \bot)$$

with $u_2, u_4 \in \Sigma^*_{wn} \setminus \{\epsilon\}$. Then, by criterion $(C1)$, we have $B(v_2) = B(v_4) = 0$.

One can prove that at least one of u_2 and u_4 can be removed from u while preserving the value $\mathsf{Or}(u)$. Let us denote by v' the resulting output word. Observe also that removing this part of the run does not modify the balance $\mathsf{B}(.)$ of the run, as $\mathsf{B}(v_2) = \mathsf{B}(v_4) = 0$. As $\mathsf{Oc}(v) = \mathsf{B}(v) + \mathsf{Or}(v)$, we obtain $\mathsf{O}(v) = \mathsf{O}(v')$, yielding the result.

- when $\mathsf{height}(\rho) \geq 2|Q|^2$: in this case, we can apply Lemma 2.(ii) twice and prove that ρ is of the following form:

$$(i, \perp) \xrightarrow{u_1/v_1} (p_1, \sigma) \xrightarrow{u_2/v_2} (p_1, \sigma\sigma_1) \xrightarrow{u_3/v_3} (q_1, \sigma\sigma_1\sigma_2) \xrightarrow{u_4/v_4} (q_1, \sigma\sigma_1\sigma_2\sigma_3)$$
$$\xrightarrow{u_5/v_5} (q_2, \sigma\sigma_1\sigma_2\sigma_3) \xrightarrow{u_6/v_6} (q_2, \sigma\sigma_1\sigma_2) \xrightarrow{u_7/v_8} (p_2, \sigma\sigma_1) \xrightarrow{u_8/v_8} (p_2, \sigma) \xrightarrow{u_9/v_9}$$

$(f, \perp)$, with $u_1u_9, u_2u_8, u_3u_7, u_4u_6, u_5 \in \Sigma^*_{\mathsf{wn}}$ and $\sigma_1, \sigma_3 \neq \perp$.

Then the two following runs can be built: the one obtained by removing the parts of ρ on u_2 and u_8, and the one obtained by removing the parts of ρ on u_4 and u_6, yielding runs whose length is strictly smaller than $|\rho|$. Let us denote these two runs by ρ' and ρ'' respectively, and their outputs by v' and v''. As A verifies the criterion $(C2)$, we have that $\mathsf{B}(v) = \mathsf{B}(v') = \mathsf{B}(v'')$, as $\mathsf{B}(v_2) + \mathsf{B}(v_8) = \mathsf{B}(v_4) + \mathsf{B}(v_6) = 0$ and B is commutative. In order to obtain the result, we study $\mathsf{Or}(v)$. Considering different cases, we manage to prove that either $\mathsf{Or}(v') \geq \mathsf{Or}(v)$ or $\mathsf{Or}(v'') \geq \mathsf{Or}(v)$. The result follows as for any word w we have $\mathsf{Oc}(w) = \mathsf{B}(w) + \mathsf{Or}(w)$.   □

Proposition 5. *Let A be a VPT. If A satisfies $(C1)$ and $(C2)$, then every accepting run $\rho : (i, \perp) \xrightarrow{u|v} (f, \perp)$ of A verifies $\|\mathsf{O}(v)\| \leq N_A.\mathsf{O}^A_{\max}$.*

Proof. If $|\rho| \leq N_A$ the result is trivial; otherwise, assuming the existence of a minimal counterexample of this statement, a contradiction follows from Lemma 5.   □

Now we can show a precise characterization of transducers from gwnVPT amongst those in awnVPT.

Definition 6. *Let A be a VPT. We consider the following criterion:*

(D) For all $(u, v) \in [\![T]\!]$, if $|u| \leq N_A$ then $v \in \Sigma^*_{\mathsf{wn}}$.

Theorem 2. *A VPT A is globally well-nested iff it verifies criteria $(C1)$, $(C2)$ and (D).*

Proof. The direct implication is trivial, the other one follows from Lemma 5.   □

5   Deciding the Classes of Almost and Globally Well-Nested VPT

In this section, we prove that given a VPT A, it is decidable to know whether $[\![A]\!] \in \mathcal{A}_{\mathsf{wn}}$ and whether $[\![A]\!] \in \mathcal{G}_{\mathsf{wn}}$. It is known that

Proposition 6. *Given a VPT $A = (Q, I, F, \Gamma, \delta)$ and states p, q of A, deciding whether there exists some stack σ such that (p, σ) is reachable and (q, σ) is co-reachable can be done in* PTIME.

Theorem 3. *Let A be a* VPT. *Whether* $[\![A]\!] \in \mathcal{A}_{wn}$ *can be decided in* PSPACE.

Proof (Sketch). By Theorem 1, deciding the class awnVPT amounts to decide criteria $(C1)$ and $(C2)$. Therefore we propose a non-deterministic algorithm running in polynomial space, yielding the result thanks to Savitch theorem.

We claim that A verifies $(C1)$ and $(C2)$ if and only if it verifies these criteria on "small instances", defined as follows:

- Criterion $(C1)$: consider only words u_2 such that $\mathsf{height}(u_2) \leq |Q|^2$ and $|u_2| \leq 2.|Q|^{|Q|^2}$.
- Criterion $(C2)$: consider only stacks σ' such that $|\sigma'| \leq |Q|^2$ and words u_2, u_4 of height at most $2.|Q|^2$ and length at most $|Q|^2.|Q|^{|Q|^2}$.

The non-deterministic algorithm follows from the claim: in order to exhibit a witness of the fact that $A \notin$ awnVPT, the algorithm guesses whether $(C1)$ or $(C2)$ is violated; then, the claim implies the existence of a witness of at most exponential size. This witness can be guessed on-the-fly in polynomial space. Proposition 6 is then used to check that the witness can be completed into an accepted run.

To prove this claim, we show, by induction on $u \in \Sigma^*_{wn}$, that for every run $(p, \perp) \xrightarrow{u|v} (q, \perp)$ that can be completed into an accepting run, and for every decomposition of this run according to criterion (C_1) or $(C2)$, the property stated by the corresponding criterion is fulfilled. $\qquad\square$

The previous algorithm could be extended to handle in addition criterion (D), yielding a PSPACE algorithm to decide whether a VPT A is globally well-nested. However, we can use a recent result to prove that this problem can be solved in PTIME.

Theorem 4. *Let A be a* VPT. *Whether* $[\![A]\!] \in \mathcal{G}_{wn}$ *can be decided in* PTIME.

Proof. This proof relies on results from [2] showing that deciding whether a context-free language is included into a Dyck language can be solved in PTIME.

We first erase the precise symbols of the produced outputs keeping track only of the type of the symbols: we build from A a VPT A' defined on the structured output alphabet Σ' with $\Sigma'_c = \{(\}, \Sigma'_r = \{)\}$ and $\Sigma'_i = \emptyset$. A transition of A' is obtained from a transition of A by replacing in output words of the transition of A call symbols by (and return symbols by) and removing internal symbols. It is then easy to see that A is in gwnVPT iff A' is in gwnVPT (actually, for each run in A producing v, its corresponding run in A' produces some v' such that $O(v) = O(v')$). Then, as shown in [9], one can build in polynomial time a context-free grammar $G_{A'}$ generating the range of A'. Finally, we appeal to [2] to conclude. $\qquad\square$

6 Closure under Composition and Type-Checking

6.1 Definitions and Existing Results

We consider two natural problems for transducers : the first one is related to composition of transductions. The second problem is the type-checking problem that aims to verify that any output of a transformation belongs to some given type/language. For VPT, the obvious class of "types" to consider is the class of languages defined by VPA.

Definition 7 (Closure under composition). *A class $\mathcal{T}$ of transductions included in $\Sigma^* \times \Sigma^*$ is closed under composition if for all T, T' in $\mathcal{T}$, the transduction $T \circ T'$ is also in $\mathcal{T}$. It is effectively closed under composition if for any transducers A, A' such that $[\![A]\!], [\![A']\!] \in \mathcal{T}$, $A \circ A'$ is computable and $[\![A \circ A']\!]$ is in $\mathcal{T}$.*

A class of transducers T is effectively closed under composition if for any two transducers A, A' in T, $A \circ A'$ is computable and $A \circ A'$ is in T.

Definition 8 (Type-checking (against VPA)). *Given a VPT A and two VPA B, C, decide whether $[\![A]\!](L(B)) \subseteq L(C)$.*

The following results give the status of these properties for arbitrary VPTs and for lwnVPT:

Theorem 5 ([6, 7]). *Regarding closure under composition, we have:*

- *The class $\mathcal{VP}(\Sigma, \Sigma)$ is not closed under composition.*
- *The class lwnVPT is effectively closed under composition.*

In addition, the problem of type checking against VPA is undecidable for (arbitrary) VPT and decidable for lwnVPT.

6.2   New Results

Actually, regarding the closure under composition of the class lwnVPT, though not explicitly stated, the result proved in [6] is slightly stronger. It is indeed shown that for any VPT A, B such that $A \in$ lwnVPT, there exists an (effectively computable) VPT C satisfying $[\![C]\!] = [\![A]\!] \circ [\![B]\!]$. In addition, if $B \in$ lwnVPT, then $C \in$ lwnVPT.

We extend this positive result to any almost well-nested transducer.

One of the main ingredients of the proof of this result is the set $\mathcal{UPS}_A$ defined for any VPT transducer $A = (Q_A, I_A, F_A, \Gamma_A, \delta^A)$ as

$$\left\{ (p, p', n_1, n_2) \,\middle|\, \begin{array}{l} \exists \sigma \in \Gamma^*, \ (p, \sigma) \text{ is reachable and } (p', \sigma) \text{ is co-reachable and} \\ \exists u \in \Sigma_{\mathsf{wn}}^*, (p, \bot) \xrightarrow{u|v} (p', \bot) \text{ and } \mathsf{O}(v) = (n_1, n_2) \end{array} \right\}$$

Proposition 7. *Let A in awnVPT. Then the set $\mathcal{UPS}_A$ is finite and computable in exponential time in the size of A.*

Theorem 6. *Let A, B be two VPTs. If A is almost-well nested, then one can compute in exponential time in the size of A and B a VPT C such that $[\![C]\!] = [\![A]\!] \circ [\![B]\!]$. Moreover, if B is also almost well-nested, then so is C, and if A and B are globally well-nested, then so is C.*

Proof (Sketch). We present the construction of C. By Proposition 7, $\mathcal{UPS}_A$ is finite and we denote by K the computable integer value $\max\{\||(n_1, n_2)\|| \mid (p, p', n_1, n_2) \in \mathcal{UPS}_A\}$.

Given $B = (Q_B, I_B, F_B, \Gamma_B, \delta^B)$, we define $C = (Q_C, I_C, F_C, \Gamma_C, \delta^C)$ as

$$Q_C = Q_A \times Q_B \times \Gamma_B^{\leq K} \qquad I_C = I_A \times I_B \times \{\bot\}$$
$$\Gamma_C = \Gamma_A \times \Gamma_B^{\leq O_{\max}^A + K} \qquad F_C = F_A \times F_B \times \{\bot\}$$

Now for the transition rules δ^C:

- $((p, q, \sigma), i, w, (p', q', \sigma')) \in \delta_i^C$ if there exist a word $v \in \Delta^*$ and a stack $\sigma_0 \in \Gamma_B^*$ such that $\sigma = \sigma_0 \sigma_1$, $\sigma' = \sigma_0 \sigma_1'$, $O(v) = (|\sigma_1|, |\sigma_1'|)$, and $(p, i, v, p') \in \delta_i^A$ and there exists a run $(q, \sigma_1) \xrightarrow{v|w} (q', \sigma_1')$ in B,
- $((p, q, \sigma), c, w, (\gamma, \sigma_3), (p', q', \sigma_4)) \in \delta_c^C$ if there exist a word $v \in \Delta^*$, two stacks $\sigma_0, \sigma_2 \in \Gamma_B^*$ and a stack symbol $\gamma \in \Gamma_A$ such that $\sigma = \sigma_0 \sigma_1$, $O(v) = (|\sigma_1|, |\sigma_2|)$, $\sigma_0 \sigma_2 = \sigma_3 \sigma_4$, $(p, c, v, \gamma, p') \in \delta_c^A$ and there exists a run $(q, \sigma_1) \xrightarrow{v|w} (q', \sigma_2)$ in B such a transition exists provided the bounds on the sizes of the different stacks are fulfilled, *i.e.* $|\sigma| \leq K$, $|\sigma_4| \leq K$, and $|\sigma_3| \leq O_{\max}^A + K$,
- $((p, q, \sigma), r, w, (\gamma, \sigma_3), (p', q', \sigma')) \in \delta_r^C$ if there exist a word $v \in \Delta^*$, a stack $\sigma_0 \in \Gamma_B^*$ such that $\sigma_0 \sigma_1 = \sigma_3 \sigma$, $\sigma_0 \sigma_2 = \sigma'$, $O(v) = (|\sigma_1|, |\sigma_2|)$, $(p, r, v, \gamma, p') \in \delta_r^A$ and there exists a run $(q, \sigma_1) \xrightarrow{v|w} (q', \sigma_2)$ in B such a transition exists provided the bounds on the sizes of the different stacks are fulfilled, *i.e.* $|\sigma| \leq K$, $|\sigma'| \leq K$, and $|\sigma_3| \leq O_{\max}^A + K$.

In a state of C, we store the current states of A and B. In addition, a part of the top of the stack of B is also stored in the state of C to allow the simulation of B. The (finite) amount that needs to be stored in the state is identified using the set $\mathcal{UPS}_A$. □

Corollary 1. *The classes $\mathcal{G}_{\mathsf{wn}}$ and $\mathcal{A}_{\mathsf{wn}}$ are (effectively) closed under composition.*

Theorem 7 (Type-checking against VPA). *Given an almost well-nested VPT A and two visibly pushdown automata B, C, whether $[\![A]\!](L(B)) \subseteq L(C)$ is decidable in $2 - \mathrm{EXPTIME}$.*

Proof. Restricting the domain of A to $L(B)$ is easy: it suffices to compute the product VPA of A and B. Then, VPA being closed under complementation, we compute $\overline{C}$, the complement of C. Note that the size of $\overline{C}$ is at most exponential in the size of C. We then turn $\overline{C}$ into a transducer C' defining the identity relation over $L(\overline{C})$ (this is obvious by simply transforming rules of $\overline{C}$ into rules of transducers outputting their input). Now, by Theorem 6, one can build a transducer defining the composition of $[\![A]\!] \circ [\![C']\!]$. This can be done in doubly exponential time in the size of A and C. Now, it is sufficient to test whether the VPA underlying this transducer is empty or not. □

7   Conclusion

In this paper, we have considered and precisely characterized the class of VPT with well-nested outputs. We have shown that this class is closed under composition and that its type-checking against VPA is decidable. We have restricted ourselves in this paper to transducers with well-nested domains. We conjecture that this restriction can be easily relaxed and thus, one could consider transducers based on nested word automata [1]. We left open the problem of deciding the class $\mathcal{L}_{wn}$. As we have described on some examples, this problem is far from being trivial. In [4], a clear relationship between the class lwnVPT and hedge-to-hedge transducers is described; investigating such a relationship for gwnVPT is also an interesting problem.

References

1. Alur, R., Madhusudan, P.: Adding Nesting Structure to Words. Journal of the ACM 56(3), 1–43 (2009)
2. Bertoni, A., Choffrut, C., Radicioni, R.: The inclusion problem of context-free languages: Some tractable cases. International Journal of Foundations of Computer Science 22(2), 289–299 (2011)
3. von Braunmühl, B., Verbeek, R.: Input-driven Languages are Recognized in log n Space. In: Karpinski, M. (ed.) FCT 1983. LNCS, vol. 158, pp. 40–51. Springer, Heidelberg (1983)
4. Caralp, M., Filiot, E., Reynier, P.A., Servais, F., Talbot, J.M.: Expressiveness of visibly pushdown transducers. In: Second International Workshop on Trends in Tree Automata and Tree Transducers. EPTCS, vol. 134, pp. 17–26 (2013)
5. Filiot, E., Gauwin, O., Reynier, P.A., Servais, F.: Streamability of Nested Word Transductions. In: IARCS Annual Conference on Foundations of Software Technology and Theoretical Computer Science. LIPIcs, vol. 13, pp. 312–324 (2011)
6. Filiot, E., Raskin, J.-F., Reynier, P.-A., Servais, F., Talbot, J.-M.: Properties of Visibly Pushdown Transducers. In: Hliněný, P., Kučera, A. (eds.) MFCS 2010. LNCS, vol. 6281, pp. 355–367. Springer, Heidelberg (2010)
7. Raskin, J.-F., Servais, F.: Visibly pushdown transducers. In: Aceto, L., Damgård, I., Goldberg, L.A., Halldórsson, M.M., Ingólfsdóttir, A., Walukiewicz, I. (eds.) ICALP 2008, Part II. LNCS, vol. 5126, pp. 386–397. Springer, Heidelberg (2008)
8. Reynier, P.A., Talbot, J.M.: Visibly Pushdown Transducers with Well-nested Outputs. Tech. rep. (2014), http://hal.archives-ouvertes.fr/hal-00988129/PDF/wnVPT.pdf
9. Servais, F.: Visibly Pushdown Transducers. Ph.D. thesis. Université Libre de Bruxelles (2011), http://theses.ulb.ac.be/ETD-db/collection/available/ULBetd-09292011-142239/
10. Staworko, S., Laurence, G., Lemay, A., Niehren, J.: Equivalence of deterministic nested word to word transducers. In: Kutyłowski, M., Charatonik, W., Gębala, M. (eds.) FCT 2009. LNCS, vol. 5699, pp. 310–322. Springer, Heidelberg (2009)

Characterising REGEX Languages by Regular Languages Equipped with Factor-Referencing

Markus L. Schmid

Universität Trier, FB IV–Abteilung Informatikwissenschaften,
D-54286 Trier, Germany
MSchmid@uni-trier.de

Abstract. A (factor-)reference in a word is a special symbol that refers to another factor in the same word; a reference is dereferenced by substituting it with the referenced factor. We introduce and investigate the class ref-REG of all languages that can be obtained by taking a regular language R and then dereferencing all possible references in the words of R. We show that ref-REG coincides with the class of languages defined by regular expressions as they exist in modern programming languages like Perl, Python, Java, etc. (often called REGEX languages).

Keywords: REGEX languages, regular languages, memory automata.

1 Introduction

It is well known that most natural languages contain at least some structure that cannot be described by context-free grammars and also with respect to artificial languages, e. g., programming languages, it is often necessary to deal with structural properties that are inherently non-context-free (Floyd's proof (see [10]) that *Algol 60* is not context-free is an early example). Hence, as Dassow and Păun [8] put it, "the world seems to be non-context-free." On the other hand, the full class of context-sensitive languages, while powerful enough to model the structures appearing in natural languages and most formal languages, is often, in many regards, simply *too* much. Therefore, investigating those properties of languages that are inherently non-context-free is a classical research topic, which, in formal language theory is usually pursuit in terms of *restricted* or *regulated rewriting* (see Dassow and Păun [8]), and in computational linguistics *mildly context-sensitive* languages are investigated (see, e. g., Kallmeyer [13]).

In [9], Dassow et al. summarise the three most commonly encountered non-context-free features in formal languages as *reduplication*, leading to languages of the form $\{ww \mid w \in \Sigma^*\}$, *multiple agreements*, modelled by languages of the form $\{a^n b^n c^n \mid n \geq 1\}$ and *crossed agreements*, as modeled by $\{a^n b^m c^n d^m \mid n, m \geq 1\}$. In this work, we solely focus on the first such feature: reduplication.

The concept of reduplication has been mainly investigated by designing language generators that are tailored to reduplications (e. g., L systems (see Kari et al. [14] for a survey), Angluin's pattern languages [2] or H-systems by Albert and

A.M. Shur and M.V. Volkov (Eds.): DLT 2014, LNCS 8633, pp. 142–153, 2014.
© Springer International Publishing Switzerland 2014

Wegner [1]) or by extending known generators accordingly (e. g., Wijngaarden grammars, macro grammars, Indian parallel grammars or deterministic iteration grammars (cf. Albert and Wegner [1] and Bordihn et al. [3] and the references therein)). A more recent approach is to extend regular expressions with some kind of copy operation (e. g., pattern expressions by Câmpeanu and Yu [6], synchronized regular expressions by Della Penna et al. [15], EH-expressions by Bordihn et al. [3]). An interesting such variant are regular expressions with backreferences (REGEX for short), which play a central role in this work. REGEX are regular expressions that contain special symbols that refer to the word that has been matched to a specific subexpression. Unlike the other mentioned language descriptors, REGEX seem to have been invented entirely on the level of software implementation, without prior theoretical formalisation (see Friedl [12] for their practical relevance). An attempt to formalise and investigate REGEX and the class of languages they describe from a theoretical point of view has been started recently (see [4, 6, 16, 11]). This origin of REGEX from application render their theoretical investigation difficult. As pointed out by Câmpeanu and Santean in [5], "we observe implementation inconsistencies, ambiguities and a lack of standard semantics." Unfortunately, to at least some extend, these conceptional problems inevitably creep into the theoretical literature as well.

Regular expressions often serve as an user interface for specifying regular languages, since finite automata are not easily defined by human users. On the other hand, due to their capability of representing regular languages in a concise way, regular expressions are deemed inappropriate for implementations and for proving theoretical results about regular languages (e. g., closure properties or decision problems). We encounter a similar situation with respect to REGEX (which, basically, are a variant of regular expressions), i. e., their widespread implementations suggest that they are considered practically useful for specifying languages, but the theoretical investigation of the language class they describe proves to be complicated. Hence, we consider it worthwhile to develop a characterisation of this language class, which is independent from actual REGEX.

To this end, we introduce the concept of *unresolved* reduplications on the word level. In a fixed word, such a reduplication is represented by a pointer or reference to a factor of the word and resolving or dereferencing such a reference is done by replacing the pointer by the value it refers to, e. g.,

$$w = \mathtt{aba}\overbrace{\underbrace{\mathtt{c}\,\mathtt{b}\,\mathtt{c}}_{x}\,\overset{y}{\overbrace{x}}\,\mathtt{c}\,\overset{z}{\overbrace{\mathtt{b}}}\,z\,y\,\mathtt{a}},$$

where the symbols x, y and z are pointers to the factors marked by the brackets labelled with x, y and z, respectively. Resolving the references x and y yields $\mathtt{abacbcbaccbzcba}$ and resolving reference z leads to $\mathtt{abacbcbaccbbaccbcba}$. Such words are called ref-words and sets of ref-words are ref-languages. For a ref-word w, $\mathcal{D}(w)$ denotes the word w with all references resolved and for a ref-language L, $\mathcal{D}(L) := \{\mathcal{D}(w) \mid w \in L\}$. We shall investigate the class of ref-regular languages, i. e., the class of languages $\mathcal{D}(L)$, where L is both regular and a ref-language, and, as our main result, we show that it coincides with the class of REGEX languages. Furthermore, by a natural extension of classical finite automata, we obtain a very simple automaton model, which precisely describes the class of ref-regular languages (= REGEX languages). This automaton model is used in order to introduce a subclass of REGEX languages, that, in contrast to other recently investigated such subclasses, has a polynomial time membership problem and we investigate the closure properties of this subclass. As a side product, we obtain a very simple alternative proof for the closure of REGEX languages under intersection with regular languages; a known result, which has first been shown by Câmpeanu and Santean [5] by much more elaborate techniques.

Due to space restrictions, all formal proofs are omitted, but we give brief proof sketches for some of our results.

2   Definitions

Let $\mathbb{N} := \{1, 2, 3, \ldots\}$ and $\mathbb{N}_0 := \mathbb{N} \cup \{0\}$. For an alphabet B, the symbol B^+ denotes the set of all non-empty words over B and $B^* := B^+ \cup \{\varepsilon\}$, where ε is the empty word. For the *concatenation* of two words w_1, w_2 we write $w_1 \cdot w_2$ or simply $w_1 w_2$. We say that a word $v \in B^*$ is a *factor* of a word $w \in B^*$ if there are $u_1, u_2 \in B^*$ such that $w = u_1 v u_2$. For any word w over B, $|w|$ denotes the length of w, for any $b \in B$, by $|w|_b$ we denote the number of occurrences of b in w and for any $A \subseteq B$, we define $|w|_A := \sum_{b \in A} |w|_b$.

We use regular expressions as they are commonly defined (see, e. g., Yu [17]). By DFA and NFA, we refer to the set of deterministic and nondeterministic finite automata. Depending on the context, by DFA and NFA we also refer to an individual deterministic or nondeterministic automaton, respectively.

For any language descriptor D, by $L(D)$ we denote the language described by D and for any class $\mathfrak{D}$ of language descriptors, let $\mathcal{L}(\mathfrak{D}) := \{L(D) \mid D \in \mathfrak{D}\}$. In the whole paper, let Σ be an arbitrary finite alphabet with $\{\mathsf{a}, \mathsf{b}, \mathsf{c}, \mathsf{d}\} \subseteq \Sigma$.

2.1   References in Words, Languages and Expressions

References in Words. Let $\Gamma := \{[_{x_i},]_{x_i}, x_i \mid i \in \mathbb{N}\}$, where, for every $i \in \mathbb{N}$, the pairs of symbols $[_{x_i}$ and $]_{x_i}$ are *parentheses* and the symbols x_i are *variables*. For the sake of convenience, we shall also use the symbols x, y and z to denote arbitrary variables. A *reference-word over* Σ (or *ref-word*, for short) is a word over the alphabet $(\Sigma \cup \Gamma)$. For every $i \in \mathbb{N}$, let $h_i : (\Sigma \cup \Gamma)^* \to (\Sigma \cup \Gamma)^*$ be the morphism with $h_i(z) := z$, $z \in \{[_{x_i},]_{x_i}, x_i\}$, and $h_i(z) := \varepsilon$, $z \notin \{[_{x_i},]_{x_i}, x_i\}$. A reference word is *valid* if, for every $i \in \mathbb{N}$,

$$h_i(w) = x_i^{\ell_1} [_{x_i}]_{x_i} x_i^{\ell_2} [_{x_i}]_{x_i} x_i^{\ell_3} \ldots x_i^{\ell_{k_i}-1} [_{x_i}]_{x_i} x_i^{\ell_{k_i}}, \tag{1}$$

for some $k_i \in \mathbb{N}$ and $\ell_j \in \mathbb{N}_0$, $1 \leq j \leq k_i$. Intuitively, a reference-word w is valid if, for every $i \in \mathbb{N}$, there is a number of matching pairs of parentheses $[_{x_i}$ and $]_{x_i}$ that are not nested and, furthermore, no occurrence of x_i is enclosed by such a matching pair of parentheses. However, it is not required that w is a well-formed parenthesised expression with respect to *all* occurring parentheses.

The set of valid reference-words is denoted by $\Sigma^{[*]}$. A factor $[_x u]_x$ of a $w \in \Sigma^{[*]}$ where the occurrences of $[_x$ and $]_x$ are matching parentheses is called a *reference for variable* x, and u is the *value* of this reference. A reference is a *first order reference*, if its value does not contain another reference and it is called *pure*, if it is a first order reference and its value does not contain variables. Two references of some ref-word w are *overlapping* if one reference contains exactly one of the delimiting parentheses of the other reference, e. g., in $w_1[_xw_2[_yw_3]_xw_4]_yw_5$ the references $[_xw_2[_yw_3]_x$ and $[_yw_3]_xw_4]_y$ are overlapping. Let $w \in \Sigma^{[*]}$ and let x be a variable that occurs in w. An occurrence of a variable x in w that is not preceded by a reference for x is called *undefined*. Every occurrence of a variable x in w that is not undefined *refers* to the reference for x, which precedes this occurrence. This definition is illustrated by Equation 1, where all $k_i - 1$ references for variable x_i are shown and, for every j, $1 \le j \le k_i - 1$, the ℓ_{j+1} occurrences of x_i between the jth and $(j + 1)^{\text{th}}$ reference for x_i are exactly the occurrences of x_i that refer to the jth reference for variable x_i.

We consider the following examples:

$$w_1 := [_x\text{ab}]_x[_y\text{c}x\text{b}]_y[_x\text{bby}]_x\text{c}y\text{by}[_yx]_y\text{cc}, \quad w_2 := [_x\text{ba}x]_x\text{a}x[_x\text{bc}]_x[_x\text{ba}[_x\text{a}]_x\text{a}]_xx,$$
$$w_3 := [_x[_y\text{b}]_x\text{c}x[_x\text{b}]_y\text{xzyb}[_y\text{cz}]_yz[_z\text{cc}]_x]_z, \quad w_4 := [_x\text{a}[_y\text{b}[_z\text{bba}]_z\text{c}]_y\text{byb}]_x\text{xy}.$$

The words w_1, w_3 and w_4 are valid ref-words, whereas w_2 is not valid. Moreover, all references of w_1 are first order references, the reference for variable x in w_4 is not a first order reference and the first reference in w_3 is pure. The word w_3 contains an undefined occurrence of a variable and overlapping references. For the sake of convenience, from now on, we call valid ref-words simply ref-words. If a word over $(\Sigma \cup \Gamma)$ is not a ref-word, then we always explicitly state this.

Next, we define how a ref-word over Σ can be dereferenced, i. e., how it can be transformed into a (normal) word over Σ. To this end, let $w \in \Sigma^{[*]}$. The *dereference* of w, denoted by $\mathcal{D}(w)$, is constructed by first deleting all undefined occurrences of variables in w and then substituting all pure references and their variables by its value (ignoring possible parentheses in the value), until there is no pure reference left. A formal proof that $\mathcal{D}(w)$ is well-defined and in fact a word over Σ is straightforward and left to the reader. Next, we illustrate this definition with an example:

$$\mathcal{D}(z\text{a}[_zx[_x\text{y}\text{b}[_y\text{c}]_x\text{b}x[_x\text{c}]_y\text{b}]_x\text{yc}]_zx\text{c}z) = \mathcal{D}(\text{a}[_z[_x\text{b}[_y\text{c}]_x\text{b}x[_x\text{c}]_y\text{b}]_x\text{yc}]_zx\text{c}z) =$$
$$\mathcal{D}(\text{a}[_z\underset{x}{\underline{\text{b}[_y\text{c b}}}\ \underset{x}{\underline{\text{bc}}}[_x\text{c}]_y\text{b}]_x\text{yc}]_zx\text{c}z) = \mathcal{D}(\text{a}[_z\text{b }\underset{y}{\underline{\text{cbbc}}}[_x\text{c b}]_x\ \underset{y}{\underline{\text{cbbcc}}}\ \text{c}]_zx\text{c}z) =$$
$$\mathcal{D}(\text{a}[_z\text{bcbbc }\underset{x}{\underline{\text{cb}}}\text{cbbccc}]_z\ \underset{x}{\underline{\text{cb}}}\text{c}z) = \text{a }\underset{z}{\underline{\text{bcbbccbcbbccc}}}\text{ cbc }\underset{z}{\underline{\text{bcbbccbcbbccc}}}.$$

We point out that ref-words are similar to Lempel-Ziv compression. However, here we are exclusively concerned with language theoretic aspects of ref-words.

References in Languages. For every $i \in \mathbb{N}$, let $\Gamma_i := \{[_{x_j},]_{x_j}, x_j \mid j \le i\}$. A set of ref-words L is a *ref-language* if $L \subseteq (\Sigma \cup \Gamma_i)^*$, for some $i \in \mathbb{N}$. For the sake of convenience, we simply write $L \subseteq \Sigma^{[*]}$ to denote that L is a ref-language. For every ref-language L, we define the *dereference* of L by $\mathcal{D}(L) := \{\mathcal{D}(w) \mid w \in L\}$ and, for any class $\mathfrak{L}$ of ref-languages, $\mathcal{D}(\mathfrak{L}) := \{\mathcal{D}(L) \mid L \in \mathfrak{L}\}$.

An $L \subseteq \Sigma^{[*]}$ is a *regular ref-language* if L is regular. A language L is called *ref-regular* if it is the dereference of a regular ref-language, i.e., $L = \mathcal{D}(L')$ for some regular ref-language L'. For example, the copy language $L_c := \{ww \mid w \in \Sigma^*\}$ is ref-regular, since $L_c = \mathcal{D}(L_c')$, where L_c' is the regular ref-language $\{[_x w]_x x \mid w \in \Sigma^*\}$. The class of ref-regular languages is denoted by ref-REG.

By definition, REG $\subseteq$ ref-REG and it can be easily shown that ref-REG is contained in the class of context-sensitive language. On the other hand, the class of context-free languages is not included in ref-REG, e.g., $\{a^n b^n \mid n \in \mathbb{N}\} \notin$ ref-REG (see Câmpeanu et al. [4]). Other interesting examples of ref-regular languages are the set of imprimitive words: $\mathcal{D}(\{[_x w]_x x^n \mid w \in \Sigma^*, n \geq 1\})$, the set of words a^n, where n is not prime: $\mathcal{D}(\{[_x a^m]_x x^n \mid m, n \geq 2\})$ and the set of bordered words: $\mathcal{D}(\{[_x u]_x v x \mid u, v \in \Sigma^*, |u| \geq 1\})$.

References in Expressions. If we use the concept of references directly in regular expressions, i.e., we use variables x in the expression and enclose subexpressions by parentheses $[_x$ and $]_x$, then we obtain *extended regular expressions with backreferences* (or REGEX for short). For more detailed definitions and further information on REGEX, we refer to [4, 16, 11, 5].

A convenient definition of the semantics of a REGEX can also be given in terms of classical regular expressions and ref-words. We can interpret a REGEX r as a classical regular expression r' over the alphabet $(\Sigma \cup \Gamma_k)$, where k is the number of backreferences in r. Now $L(r')$ is a ref-language and $\mathcal{D}(L(r'))$ is the REGEX language described by the REGEX r. This observation yields the following result.

Proposition 1. $\mathcal{L}(\text{REGEX}) \subseteq$ ref-REG.

On the other hand, a regular expression s with $L(s) \subseteq \Sigma^{[*]}$ does not translate into a REGEX in an obvious way, which is due to the fact that in s it is not necessarily the case that every occurrence of $[_x$ matches with an occurrence of $]_x$ and, furthermore, even matching pairs of parentheses do not necessarily enclose subexpressions. For example, the regular expression

$$s := [_{x_1}(([_{x_2}b^*]_{x_1}a^*x_1[_{x_1}) + ([_{x_2}a^*c^*)) ba]_{x_2}(x_2 + d)^*]_{x_1}ax_1$$

describes a ref-language, but it cannot be interpreted as a REGEX.

In the following, we say that a regular expression r over the alphabet $(\Sigma \cup \Gamma_k)$ has the REGEX *property* if it is also a valid REGEX.

3 Memories in Automata

By a natural extension of classical finite automata, we now define memory automata, which are the main technical tool for proving the results of this paper.

A memory automaton is a classical NFA that is equipped with a finite number of k memory cells, each capable of storing a word. Each memory is either *closed*, which means that it is not affected in a transition, or *open*, which means that it records the currently scanned input symbol. In a transition it is possible to *consult* a closed memory, which means that its content, if it is a prefix of the remaining input, is consumed in one step from the input and, furthermore, also stored in all the open memories. A closed memory can be opened again, but then it completely loses its previous content; thus, memories always store factors of the input. We shall now formally define the model of memory automata.

Definition 1. *For every $k \in \mathbb{N}$, a k-memory automaton, denoted by MFA(k), is a tuple $M := (Q, \Sigma, \delta, q_0, F)$, where Q is a finite set of* states, Σ *is a finite* alphabet, q_0 *is the* initial state, F *is the set of* final states *and*

$$\delta : Q \times (\Sigma \cup \{\varepsilon\} \cup \{1, 2, \ldots, k\}) \to \mathcal{P}(Q \times \{\mathsf{o}, \mathsf{c}, \diamond\}^k)$$

is the transition function *(where $\mathcal{P}(A)$ denotes the power set of a set A). The elements o, c and $\diamond$ are called* memory instructions.

A *configuration of M is a tuple $(q, w, (u_1, r_1), \ldots, (u_k, r_k))$, where $q \in Q$ is the current state, w is the remaining input, for every i, $1 \leq i \leq k$, $u_i \in \Sigma^*$ is the content of memory i and $r_i \in \{\mathsf{0}, \mathsf{C}\}$ is the status of memory i. For a memory status $r \in \{\mathsf{0}, \mathsf{C}\}$ and a memory instruction $s \in \{\mathsf{o}, \mathsf{c}, \diamond\}$, we define $r \oplus s = \mathsf{0}$ if $s = \mathsf{o}$, $r \oplus s = \mathsf{C}$ if $s = \mathsf{c}$ and $r \oplus s = r$ if $s = \diamond$. Furthermore, for a tuple $(r_1, \ldots, r_k) \in \{\mathsf{0}, \mathsf{C}\}^k$ of memory statuses and a tuple $(s_1, \ldots, s_k) \in \{\mathsf{o}, \mathsf{c}, \diamond\}^k$ of memory instructions, we define $(s_1, \ldots, s_k) \oplus (r_1, \ldots, r_k) = (s_1 \oplus r_1, \ldots, s_k \oplus r_k)$.*

M *can change from a configuration $c := (q, w, (u_1, r_1), \ldots, (u_k, r_k))$ to a configuration $c' := (p, w', (u_1', r_1'), \ldots, (u_k', r_k'))$, denoted by $c \vdash_M c'$, if there exists a transition $\delta(q, b) \ni (p, s_1, \ldots, s_k)$ such that $w = vw'$, where $v = b$ if $b \in (\Sigma \cup \{\varepsilon\})$ and $v = u_b$ and $r_b = \mathsf{C}$ if $b \in \{1, 2, \ldots, k\}$. Furthermore, $(r_1', \ldots, r_k') = (r_1, \ldots, r_k) \oplus (s_1, \ldots, s_k)$ and, for every i, $1 \leq i \leq k$, $u_i' = u_i v$ if $r_i' = r_i = \mathsf{0}$, $u_i' = v$ if $r_i' = \mathsf{0}$ and $r_i = \mathsf{C}$, $u_i' = u_i$ if $r_i' = \mathsf{C}$. The symbol $\vdash_M^*$ denotes the reflexive and transitive closure of $\vdash_M$. For any $w \in \Sigma^*$, a configuration $(p, v, (u_1, r_1), \ldots, (u_k, r_k))$ is* reachable *(on input w), if*

$$(q_0, w, (\varepsilon, \mathsf{C}), \ldots, (\varepsilon, \mathsf{C})) \vdash_M^* (p, v, (u_1, r_1), \ldots, (u_k, r_k)).$$

A $w \in \Sigma^*$ is accepted *by M if a configuration $(q_f, \varepsilon, (u_1, r_1), \ldots, (u_k, r_k))$ with $q_f \in F$ is reachable on input w and $L(M)$ is the set of words accepted by M.*

For any $k \in \mathbb{N}$, MFA(k) is the class of k-memory automata and MFA $:= \bigcup_{k \geq 0}$ MFA(k). We also use MFA(k) and MFA in order to denote an instance of a k-memory automaton or a memory automaton with some number of memories. The set $\mathcal{L}(\text{MFA}) := \{L(M) \mid M \in \text{MFA}\}$ is the *class of* MFA *languages*.

As an example, we observe that an MFA(3) can accept $L := \{v_1 v_2 v_3 v_1 v_2 v_2 v_3 \mid v_1, v_2, v_3 \in \{\mathsf{a}, \mathsf{b}\}^*\}$ by reading a prefix $u = v_1 v_2 v_3$ of the input and storing v_1, v_2 and v_3 in the memories 1, 2 and 3. The length of u as well as the factorisation $u = v_1 v_2 v_3$ is nondeterministically guessed. Now we can check whether the remaining input equals $v_1 v_2 v_2 v_3$ by consulting the memories. Alternatively, while reading the prefix $u = v_1 v_2 v_3$, we can also store $v_1 v_2$ and $v_2 v_3$ in only two memories, which is sufficient to check whether the remaining input equals $v_1 v_2 v_2 v_3$. We point out that this alternative, which needs one memory less, requires the factor v_2 of the input to be simultaneously recorded by both memories or, in other words, the factors recorded by the memories overlap in the input word.

Determinism in Memory Automata. Let $M := (Q, \Sigma, \delta, q_0, F)$ be a k-memory automaton. M is *pseudo deterministic* if, for every $q \in Q$ and $b \in (\Sigma \cup \{\varepsilon\} \cup \{1, 2, \ldots, k\})$, $|\delta(q, b)| \leq 1$. We call a memory automaton with this property pseudo deterministic, since it is still possible that for a $q \in Q$ and $b \in (\Sigma \cup \{\varepsilon\})$, $|\bigcup_{i=1}^k \delta(q, i)| + |\delta(q, b)| \geq 1$.

Any MFA can be transformed into an equivalent pseudo-deterministic one by applying a variant of the subset construction that also takes the memories into account, i. e., instead of $\mathcal{P}(Q)$ as the set of new states, we use $\mathcal{P}(Q) \times \{\mathtt{O}, \mathtt{C}\}^k$.

Lemma 1. *Let $k \in \mathbb{N}$. For every $M \in \mathrm{MFA}(k)$, there exists a pseudo deterministic $M' \in \mathrm{MFA}(k)$ with $L(M) = L(M')$.*

Analogously to the definition of determinism for classical finite automata, we can define determinism for memory automata as the situation that in every state there is at most one applicable transition. More precisely, a k-memory automaton $M := (Q, \Sigma, \delta, q_0, F)$ is *deterministic* if it is ε-free and, for every $q \in Q$ and $b \in \Sigma$, $|\bigcup_{i=1}^k \delta(q, i)| + |\delta(q, b)| \leq 1$. We denote the class of deterministic k-memory automata by $\mathrm{DMFA}(k)$ and $\mathrm{DMFA} := \bigcup_{k \in \mathbb{N}} \mathrm{MFA}(k)$. The class $\mathcal{L}(\mathrm{DMFA})$ shall be investigated in more detail in Section 5.

Normal Forms of Memory Automata. Intuitively speaking, a memory automaton is in normal form if every transition can either read a part of the input without changing the status of any memory or it changes the status of exactly one memory, but then it does not touch the input. Furthermore, in every accepting configuration, the automaton does not try to open or close memories that are already opened or closed, respectively.

Definition 2. *Let $M = (Q, \Sigma, \delta, q_0, F)$ be an $\mathrm{MFA}(k)$, $k \in \mathbb{N}$. We say that M is in* normal form *if the following conditions are satisfied. For every transition $\delta(q, b) \ni (p, s_1, \ldots, s_k)$, if $s_i \neq \diamond$ for some i, $1 \leq i \leq k$, then $b = \varepsilon$ and $s_j = \diamond$, for all j with $1 \leq j \leq k$ and $i \neq j$. If M reaches configuration $(q, w, (u_1, r_1), \ldots, (u_k, r_k))$ in a computation and transition $\delta(q, b) \ni (p, s_1, \ldots, s_k)$ is applied next, then, for every i, $1 \leq i \leq k$, if $r_i = \mathtt{O}$, then $s_i \neq \mathtt{o}$ and if $r_i = \mathtt{C}$, then $s_i \neq \mathtt{c}$.*

MFA can be transformed into normal form, by replacing transitions that change the status of more than one memory by several transitions that satisfy the conditions of the normal form. Furthermore, in the states we can store which memories are currently open and use this information to remove transitions that open or close memories that are already open or closed, respectively.

Lemma 2. *Let $k \in \mathbb{N}$. For every $M \in \mathrm{MFA}(k)$ there exists an $M' \in \mathrm{MFA}(k)$ with $L(M) = L(M')$ and M' is in normal form.*

Above, after Definition 1, we consider an example of an MFA that uses its memories in an overlapping way. We shall now formally define this situation. Let M be a k-memory automaton, let C be a computation of M with n steps and let $1 \leq i, j \leq k$ with $i \neq j$. We say that there is an i-j-overlap in C if there are p, q, r, s, $1 \leq p < q < r < s < n$, such that memory i is opened in step p and closed again in step r of C and memory j is opened in step q and closed again in step s of C. A computation C is said to be *nested* if, for every i, j, $1 \leq i \leq j \leq k$, $i \neq j$, there is no i-j-overlap in C and an MFA M is *nested* if every possible computation of M is nested.

For transforming MFA into REGEX, it is crucial to get rid of these overlaps, which is generally possible. A formal definition of the construction is omitted; here, we only give the general idea. We first modify M in such a way that in the finite state control, for every currently open memory i, we store the set A_i of currently open memories that have been opened after memory i. If memory i is closed, then, for every $j \in A_i$, an i-j-overlap occurs. Thus, we close all memories $j \in A_i$ together with memory i and then, for every $j \in A_i$, we open a new auxiliary memory instead. By doing this every time a memory is closed, we make sure that no more overlaps occur. Whenever the original MFA consults a memory, then we have to consult the right auxiliary memories in the right order instead. This strategy is only applicable because, for every original memory i, at most $k - 1$ auxiliary memories are needed.

Lemma 3. *Let $k \in \mathbb{N}$. For every $M \in \mathrm{MFA}(k)$ there exists a nested $M' \in \mathrm{MFA}(k^2)$ with $L(M) = L(M')$.*

We conclude this section by pointing out that every MFA can be transformed into an equivalent one that is pseudo-deterministic, in normal form and nested.

4   Equivalence of ref-REG, $\mathcal{L}(\mathrm{MFA})$ and $\mathcal{L}(\mathrm{REGEX})$

In this section, we show that the ref-regular languages, the MFA languages and the REGEX languages are identical. To this end, we shall first prove the equality $\mathcal{L}(\mathrm{MFA}) = $ ref-REG and then the inclusion $\mathcal{L}(\mathrm{MFA}) \subseteq \mathcal{L}(\mathrm{REGEX})$, which, together with Proposition 1, implies our main result:

Theorem 1. ref-REG $= \mathcal{L}(\mathrm{MFA}) = \mathcal{L}(\mathrm{REGEX})$.

Intuitively speaking, $\mathcal{L}(\mathrm{MFA}) = $ ref-REG follows from the fact that an NFA that accepts a regular ref-language can be translated into an MFA that accepts its dereference and any MFA in normal form can be translated into an NFA that accepts a regular ref-language, the dereference of which equals the language accepted by the MFA.

We recall that every transition $\delta(q, b) \ni (p, s_1, s_2, \ldots, s_k)$ of an $\mathrm{MFA}(k)$ in normal form is of one of the following four types:

Σ**-transition:** $b \in \Sigma$ and $s_i = \diamond$, $1 \leq i \leq k$.
o_i**-transition:** $b = \varepsilon$, $s_i = o$ and, for every j, $1 \leq j \leq k$, $i \neq j$, $s_j = \diamond$.

c_i-**transition:** $b = \varepsilon$, $s_i = \mathsf{c}$ and, for every j, $1 \leq j \leq k$, $i \neq j$, $s_j = \diamond$.
m_i-**transition:** $b \in \{1, 2, \ldots, k\}$ and $s_i = \diamond$, $1 \leq i \leq k$.

Let $\mathrm{NFA_{ref}} := \{M \mid M \in \mathrm{NFA}, L(M) \subseteq \Sigma^{[*]}\}$ and $\mathrm{MFA_{nf}} := \{M \mid M \in \mathrm{MFA}, M \text{ is in normal form}\}$. We define a mapping $\psi_{\mathcal{D}} : \mathrm{NFA_{ref}} \to \mathrm{MFA_{nf}}$. To this end, let $M := (Q, \Sigma \cup \Gamma_k, \delta, q_0, F) \in \mathrm{NFA_{ref}}$. We define a k-memory automaton $\psi_{\mathcal{D}}(M) := (Q, \Sigma, \delta', q_0, F)$, where δ' is defined as follows. For every transition $\delta(q, b) \ni p$ of M, we add a transition $\delta'(q, b) \ni (p, s_1, s_2, \ldots, s_k)$ to $\psi_{\mathcal{D}}(M)$, where this transition is an Σ-transition if $b \in \Sigma$, an o_i-transition if $b = [_{x_i}$, a c_i-transition if $b =]_{x_i}$ and an m_i-transition if $b = x_i$. This concludes the definition of $\psi_{\mathcal{D}}(M)$ and it can be easily seen that $\psi_{\mathcal{D}}(M) \in \mathrm{MFA_{nf}}$. Without loss of generality, we can assume that all elements of $\mathrm{NFA_{ref}}$ are such that the input alphabet does not contain symbols that do not occur in the accepted language. This implies that, for any two $M_1, M_2 \in \mathrm{NFA_{ref}}$ with $M_1 \neq M_2$, $\psi_{\mathcal{D}}(M_1) \neq \psi_{\mathcal{D}}(M_2)$ is implied, which means that $\psi_{\mathcal{D}}$ is injective. Furthermore, since every transition of some $\mathrm{MFA}(k)$ in normal form is of one of the four types described above, we can conclude that the reverse of $\psi_{\mathcal{D}}$ is an injective mapping $\psi_{\mathcal{D}}^{-1} : \mathrm{MFA_{nf}} \to \mathrm{NFA_{ref}}$, which implies that $\psi_{\mathcal{D}}$ is a bijection. The following lemma directly implies ref-REG $= \mathcal{L}(\mathrm{MFA})$.

Lemma 4. *Let $M \in \mathrm{NFA}$ with $L(M) \subseteq \Sigma^{[*]}$ and let $N \in \mathrm{MFA}$ be in normal form. Then $\mathcal{D}(L(M)) = L(\psi_{\mathcal{D}}(M))$ and $L(N) = \mathcal{D}(L(\psi_{\mathcal{D}}^{-1}(N)))$.*

In order to conclude the proof of Theorem 1, it only remains to show that every MFA language can be expressed as a REGEX language. To do this, we use the fact that every MFA language can be described by a nested MFA (see Lemma 3). Next, we observe an obvious, but important property of nested MFA:

Proposition 2. *Let M be a nested $\mathrm{MFA}(k)$ in normal form. There is no word $w \in L(\psi_{\mathcal{D}}^{-1}(M))$ with overlapping references.*

Let M be a fixed nested $\mathrm{MFA}(k)$ in normal form and let $N := \psi_{\mathcal{D}}^{-1}(M)$ with transition function δ. Without loss of generality, we can assume that every transition of the form $\delta(p, b) \ni q$ with $b \in \{[_{x_i},]_{x_i} \mid 1 \leq i \leq k\}$ is such that at least one accepting state is reachable from q and, furthermore, that every state of N is reachable from the initial state. For every i, $1 \leq i \leq k$, let $n_i, m_i \in \mathbb{N}$ be such that $\delta(p_{i,j}, [_{x_i}) \ni q_{i,j}$, $1 \leq j \leq n_i$, are exactly the transitions of N labeled with $[_{x_i}$ and $\delta(r_{i,\ell},]_{x_i}) \ni s_{i,\ell}$, $1 \leq \ell \leq m_i$, are exactly the transitions of N labeled with $]_{x_i}$. For every i, j, ℓ, $1 \leq i \leq k$, $1 \leq j \leq n_i$, $1 \leq \ell \leq m_i$, let $R_{i,j,\ell}$ be the set of words that can take N from $q_{i,j}$ to $r_{i,\ell}$ without reading any occurrence of $]_{x_i}$. If some $R_{i,j,\ell}$ contains a word that is not a ref-word, then, since an accepting state can be reached from $s_{i,\ell}$, N accepts a word that is not a ref-word or a ref-word with overlapping references, which is a contradiction:

Lemma 5. *For every i, j, ℓ, $1 \leq i \leq k$, $1 \leq j \leq n_i$, $1 \leq \ell \leq m_i$, $R_{i,j,\ell} \subseteq \Sigma^{[*]}$.*

In the following, we transform N into a regular expression r and, since we want r to have the REGEX property, this has to be done in such a way that in r

every $[_{x_i}$ matches a $]_{x_i}$ and such matching parentheses enclose a subexpression. The idea to achieve this is that, for each pair of transitions $\delta(p_{i,j}, [_{x_i}) \ni q_{i,j}$ and $\delta(r_{i,\ell},]_{x_i}) \ni s_{i,\ell}$, we transform the set of words that take N from $q_{i,j}$ to $r_{i,\ell}$, i.e., the set $R_{i,j,\ell}$, into a regular expression individually. For the correctness of this construction, it is crucial that M is nested and that we transform the $R_{i,j,\ell}$ into regular expressions in a specific order, which is defined next.

Let the binary relation $\prec$ over the set $\Phi := \{(i,j,\ell) \mid 1 \le i \le k, 1 \le j \le n_i, 1 \le \ell \le m_i\}$ be defined as follows. For every $(i_1,j_1,\ell_1), (i_2,j_2,\ell_2) \in \Phi$, we define $(i_1,j_1,\ell_1) \prec (i_2,j_2,\ell_2)$ if and only if there is a computation of N that starts in p_{i_2,j_2} and reaches s_{i_2,ℓ_2} and takes the transitions (1) $\delta(p_{i_2,j_2}, [_{x_{i_2}}) \ni q_{i_2,j_2}$, (2) $\delta(p_{i_1,j_1}, [_{x_{i_1}}) \ni q_{i_1,j_1}$, (3) $\delta(r_{i_1,\ell_1},]_{x_{i_1}}) \ni s_{i_1,\ell_1}$ and (4) $\delta(r_{i_2,\ell_2},]_{x_{i_2}}) \ni s_{i_2,\ell_2}$ in exactly this order and no $]_{x_{i_2}}$ is read between performing transitions 1 and 4 and no $]_{x_{i_1}}$ is read between performing transitions 2 and 3.

Lemma 6. *If $(i_1,j_1,\ell_1) \prec (i_2,j_2,\ell_2)$, then $i_1 \ne i_2$. The relation $\prec$ is irreflexive, transitive and antisymmetric.*

Next, we define a procedure that turns N into a regular expression that is based on the well-known state elimination technique (see Yu [17] for details). To this end, we need the concept of an *extended finite automaton*, which is an NFA whose transitions can be labeled by regular expressions over the input alphabet.

For every $(i,j,\ell) \in \Phi$, we define $\Delta(i,j,l) := \{(i',j',\ell') \mid (i',j',\ell') \prec (i,j,\ell)\}$ and $\Phi' := \Phi$. We iterate the following steps as long as Φ' is non-empty.

Step 1 For some $(i,j,\ell) \in \Phi'$ with $|\Delta(i,j,l)| = 0$, we obtain an automaton $K_{i,j,\ell}$ from (the current version of) N by deleting all transitions $\delta(r_{i,\ell'},]_{x_i}) \ni s_{i,\ell'}$, $1 \le \ell' \le m_i$, and by defining $q_{i,j}$ to be the initial state and $r_{i,\ell}$ the only accepting state. Then, we transform $K_{i,j,\ell}$ into a regular expression $t_{i,j,\ell}$ by applying the state elimination technique.

Step 2 For every $(i',j',\ell') \in \Phi$, we delete (i,j,ℓ) from $\Delta(i',j',\ell')$.

Step 3 We add the transition $\delta(p_{i,j}, [_{x_i} t_{i,j,\ell}]_{x_i}) \ni s_{i,\ell}$ to N.

Step 4 We delete (i,j,l) from Φ'.

Step 5 If, for every ℓ', $1 \le \ell' \le m_i$, $(i,j,\ell') \notin \Phi'$, then we delete the transition $\delta(p_{i,j}, [_{x_i}) \ni q_{i,j}$ from N.

Step 6 If, for every j', $1 \le j' \le n_i$, $(i,j',\ell) \notin \Phi'$, then we delete the transition $\delta(r_{i,j},]_{x_i}) \ni s_{i,\ell}$ from N.

In order to see that this procedure is well-defined, we observe that as long as $\Phi' \ne \emptyset$, there is at least one element $(i,j,\ell) \in \Phi'$ with $|\Delta(i,j,l)| = 0$, which follows directly from the transitivity and antisymmetry of $\prec$ (see Lemma 6). Furthermore, from the definition of the automata $K_{i,j,\ell}$ constructed in Step 1, it can be easily verified that $L(t_{i,j,\ell}) = R_{i,j,\ell}$ holds.

The automaton obtained by this procedure, denoted by N', does not contain any transitions labeled with symbols from $\{[_{x_i},]_{x_i} \mid 1 \le i \le k\}$. We can now transform N' into a regular expression r by the state elimination technique. The next lemma concludes the proof of Theorem 1.

Lemma 7. *The regular expression r has the REGEX-property and $L(r) = L(N)$.*

5   DMFA Languages

We now take a closer look at the class of languages accepted by DMFA. As an example, we consider $\{wcw \mid w \in \{a, b\}^*\}$, which can be accepted by a DMFA(1). However, $L_{copy} := \{ww \mid w \in \{a, b\}^*\} \notin \mathcal{L}(DMFA)$, for which we give a proof sketch. Let $M \in DMFA$ with $L(M) = L_{copy}$. Since $L_{copy} \notin REG$ there is a $w \in L_{copy}$, such that M first reads a prefix v of w and then consults a memory i that stores a non-empty word u. However, since M is deterministic, this means that it cannot accept any word with prefix vb, where b does not equal the first symbol of u, which is a contradiction.

Theorem 2. $\mathcal{L}(DMFA) \subset \mathcal{L}(MFA)$.

Next, we note that the membership problem of this language class can be solved in linear time by simply running the DMFA.

Theorem 3. *For a given DMFA M and a word $w \in \Sigma^*$, we can decide in time* $O(|w|)$ *whether or not $w \in L(M)$.*

This contrasts the situation that for other prominent subclasses of REGEX languages (e. g., the ones investigated in [16, 1, 3, 6]) the membership problem is usually NP-complete.[1]

REGEX languages are closed under union, but not under intersection or complementation [4, 7]. For the subclass $\mathcal{L}(DMFA)$, we observe a different situation:

Theorem 4. $\mathcal{L}(DMFA)$ *is closed under complementation, but it is not closed under union or intersection.*

For the non-closure under intersection, we can apply the example used by Carle and Narendran [7] to prove the non-closure of REGEX languages under intersection. The closure under complementation follows from the fact that interchanging the accepting and non-accepting states of a DMFA yields a DMFA that accepts its complement. Finally, the non-closure under union follows from $\{a^n ba^n \mid n \in \mathbb{N}\} \cup \{a^m ba^n ba^k \mid m, n, k \in \mathbb{N}_0\} \notin \mathcal{L}(DMFA)$, which can be shown by a similar argument used to prove that L_{copy} is not a DMFA language.[2]

In [4], which marks the beginning of the formal investigation of REGEX, Câmpeanu et al. ask whether REGEX languages are closed under intersection with regular languages, which has been answered in the positive by Câmpeanu and Santean in [5]. We can give an analogue with respect to DMFA languages:

Theorem 5. $\mathcal{L}(DMFA)$ *is closed under intersection with regular languages.*

Theorem 5 follows from the fact that we can simulate a DMFA(k) M and a DFA N in parallel by a DMFA M'. The difficulty that we encounter is that if N is currently in a state q and M consumes the content u_i of a memory i from the input, then we do not know in which state N needs to change. However, earlier

---

[1] Other exceptions are REGEX with a bounded number of backreferences (see [16]).

[2] This also proves non-closure of DMFA languages under union with regular languages.

in this computation, memory i is filled with content u_i and at the same time we can determine and store the state in which N needs to change if u_i is consumed from the input. Since at this time we do not yet know that q will be the current state of N when memory i is consulted, for *every* state of N, we have to store which state is reached by reading u_i. In order to update these informations we use the transition function of N (when single symbols are read) and these stored informations (when memories are consulted).

This also constitutes a much simpler alternative proof for the closure of REGEX languages under intersection with regular languages, which demonstrates that MFA are a convenient tool to handle REGEX languages.

References

1. Albert, J., Wegner, L.: Languages with homomorphic replacements. Theoretical Computer Science 16, 291–305 (1981)
2. Angluin, D.: Finding patterns common to a set of strings. In: Proc. 11th Annual ACM Symposium on Theory of Computing, pp. 130–141 (1979)
3. Bordihn, H., Dassow, J., Holzer, M.: Extending regular expressions with homomorphic replacement. RAIRO Theoretical Informatics and Applications 44, 229–255 (2010)
4. Câmpeanu, C., Salomaa, K., Yu, S.: A formal study of practical regular expressions. Int. Journal of Foundations of Computer Science 14, 1007–1018 (2003)
5. Câmpeanu, C., Santean, N.: On the intersection of regex languages with regular languages. Theoretical Computer Science 410, 2336–2344 (2009)
6. Câmpeanu, C., Yu, S.: Pattern expressions and pattern automata. Information Processing Letters 92, 267–274 (2004)
7. Carle, B., Narendran, P.: On extended regular expressions. In: Dediu, A.H., Ionescu, A.M., Martín-Vide, C. (eds.) LATA 2009. LNCS, vol. 5457, pp. 279–289. Springer, Heidelberg (2009)
8. Dassow, J., Păun, G.: Regulated Rewriting in Formal Language Theory. Springer, Berlin (1989)
9. Dassow, J., Păun, G., Salomaa, A.: Grammars with controlled derivations. In: Rozenberg, G., Salomaa, A. (eds.) Handbook of Formal Languages, vol. 2, pp. 101–154. Springer (1997)
10. Floyd, R.W.: On the nonexistence of a phrase structure grammar for algol 60. Communications of the ACM 5, 483–484 (1962)
11. Freydenberger, D.D.: Extended regular expressions: Succinctness and decidability. Theory of Computing Systems 53, 159–193 (2013)
12. Friedl, J.E.F.: Mastering Regular Expressions, 3rd edn. O'Reilly, Sebastopol (2006)
13. Kallmeyer, L.: Parsing Beyond Context-Free Grammars. Springer (2010)
14. Kari, L., Rozenberg, G., Salomaa, A.: L systems. In: Rozenberg, G., Salomaa, A. (eds.) Handbook of Formal Languages, vol. 1, ch. 5, pp. 253–328. Springer (1997)
15. Penna, G.D., Intrigila, B., Tronci, E., Zilli, M.V.: Synchronized regular expressions. Acta Informatica 39, 31–70 (2003)
16. Schmid, M.L.: Inside the class of regex languages. International Journal of Foundations of Computer Science 24, 1117–1134 (2013)
17. Yu, S.: Regular languages. In: Rozenberg, G., Salomaa, A. (eds.) Handbook of Formal Languages, vol. 1, ch. 2, pp. 41–110. Springer (1997)

Pumping Lemma and Ogden Lemma for Displacement Context-Free Grammars

Alexey Sorokin[1,2,*]

[1] Moscow State University, Faculty of Mathematics and Mechanics
[2] Moscow Institute of Physics and Technology,
Faculty of Innovations and High Technologies, Russia

Abstract. The pumping lemma and Ogden lemma offer a powerful method to prove that a particular language is not context-free. In 2008 Kanazawa proved an analogue of pumping lemma for well-nested multiple context-free languages. However, the statement of lemma is too weak for practical usage. We prove a stronger variant of pumping lemma and an analogue of Ogden lemma for this language family. We also use these statements to prove that some natural context-sensitive languages cannot be generated by tree-adjoining grammars.

1 Introduction

Since 1980s, context-free grammars are known to be too restrictive for syntactic description of natural languages ([10]). The class of mildly context-sensitive languages ([1]) was an informal attempt to capture the degree of context-sensitivity required for most common language phenomena preserving the advantages of context-free grammars. The principal properties to inherit were feasible polynomial parsing complexity, independence of derivation from the context (the notion of context had to be extended to handle long-distance dependencies) and existence of convenient normal forms. The class of well-nested multiple context-free languages (wMCFLs) is one of the candidates to satisfy these requirements[1]. The corresponding grammar formalism called well-nested multiple context-free grammars (well-nested MCFGs or wMCFGs), is defined as a subclass of multiple context-free grammars (MCFGs, [9]) with rules of special form providing the correct embedding of constituents. In particular, 2-wMCFGs are equivalent to tree-adjoining grammars (TAGs, [14], [2]) and then to head grammars ([8]).

We find it reasonable to think of wMCFGs as the generalization of head grammars, not the restriction of MCFGs. Our approach is based on two principal ideas. The first is to derive not words but terms whose values are the words of the language. Then the generative power of a grammar formalism essentially depends on the set of term connectives and their interpretation as language operations. If the only operation in use is concatenation, the terms are just strings of terminals and nonterminals and we get nothing but context-free grammars. Our approach

---

* Partially supported by RFFI Leading Scientific Schools grant NSh-1423.2014.1.
[1] See [4] for discussion.

A.M. Shur and M.V. Volkov (Eds.): DLT 2014, LNCS 8633, pp. 154–165, 2014.

seems redundant there, but is vital in more complex cases. The second idea is to extend the alphabet by a distinguished separator 1^2. Using the separators, well-nested MCFGs may be simulated with the help of intercalation connectives. The binary operation $\odot_j$ of j-intercalation replaces the jth separator in its first argument by its second argument (for example, $a1b1c \odot_2 a1b = a1ba1bc$). It is straightforward to prove that all "well-nested" combinations of constituents can be presented using only intercalation and concatenation operations.

The exact generative power of wMCFGs is not known. Moreover, some languages are supposed not to be wMCFLs, although they are not proved to be outside this family. The most known example is the MIX language $\{w \in \{a, b, c\}^* \mid |w|_a = |w|_b = |w|_c\}$. It was shown in [4] to be not a 2-MCFL, but the proof used combinatorial and geometric arguments which are troublesome to be generalized for the class of all wMCFGs. The pumping lemma for wMCFLs was presented in [3] is also too weak since it does not impose any conditions on the length and position of the pumped segment. We prove a stronger version of pumping lemma and weak Ogden lemma[3] for wMCFGs basing on the ideas already used in [3]. Our variant of Ogden lemma allows us to give a simple proof of the fact that MIX cannot be generated by a TAG.

Due to the space constraints, heavy technical proofs are omitted in the paper[4]. We suppose the reader to be familiar with the basics of formal languages theory though all necessary definitions are explicitly formulated.

2 Preliminaries

2.1 Terms and Their Equivalence

In this section we define displacement context-free grammars (DCFGs) which are a "purely logical" reformulation of well-nested MCFGs. The first subsection is devoted to the notions of term, context and generalized context that play the key role in the architecture of DCFGs, and contains some results on term equivalence which are extensively used in the paper. We follow the definitions of [12], but the purposes of this work require some technical complications.

Let Σ be a finite alphabet and $1 \notin \Sigma$ be a distinguished separator, let $\Sigma_1 = \Sigma \cup \{1\}$. For every word $w \in \Sigma_1^*$ we define its rank $rk(w) = |w|_1$. We define the jth intercalation operation $\odot_j$ which consists in replacing the jth separator in its first argument by its second argument. For example, $a1b11d \odot_2 c1c = a1bc1c1d$.

Let k be a natural number and N be the set of nonterminals. The function $rk \colon N \to \overline{0,k}$ assigns ranks to elements of N. Let $Op_k = \{\cdot, \odot_1, \ldots, \odot_k\}$ be the set of binary operation symbols, then the ranked set of k-correct terms $Tm_k(N, \Sigma)$ is defined in the following way (we write simply Tm_k in the further):

---

[2] This idea is inspired by the works of Morrill and Valentín on discontinuous Lambek calculus ([5], [13]).

[3] A stronger version of Ogden lemma for tree-adjoining languages, which form the first level of well-nested MCFLs hierarchy, was proved in [7], but the proof hardly can be generalized for entire class of wMCFLs.

[4] The version with full proofs is available on http://arxiv.org/abs/1403.6230

1. $N \subset Tm_k(N, \Sigma)$,
2. $\Sigma^* \subset Tm_k(N, \Sigma)$, $\forall w \in \Sigma^*\ rk(w) = 0$,
3. $1 \in Tm_k$, $rk(1) = 1$,
4. If $\alpha, \beta \in Tm_k$ and $rk(\alpha) + rk(\beta) \leq k$, then $(\alpha \cdot \beta) \in Tm_k$, $rk(\alpha \cdot \beta) = rk(\alpha) + rk(\beta)$.
5. If $j \leq k$, $\alpha, \beta \in Tm_k$, $rk(\alpha) + rk(\beta) \leq k + 1$, $rk(\alpha) \geq j$, then $(\alpha \odot_j \beta) \in Tm_k$, $rk(\alpha \cdot \beta) = rk(\alpha) + rk(\beta) - 1$.

We refer to the elements of the set $N \cup \Sigma^* \cup \{1\}$ as basic subterms. We will often omit the symbol of concatenation and assume that concatenation has greater priority than intercalation, so $Ab \odot_2 cD$ means $(A \cdot b) \odot_2 (c \cdot D)$. This simplification allows us to consider words in the alphabet Σ_1^* as terms. The set of k-correct terms includes all the terms of sort k or less that also do not contain subterms of rank greater than k.

Let $\mathrm{Var} = \{x_1, x_2, \ldots\}$ be a countable set of variables. We assume that every variable has a fixed rank and there are infinitely many variables of each rank. A context $C[x]$ is a term in which a variable x occurs in a leaf position, and the rank of x respects the constraints of term construction. Provided $\beta \in Tm_k$ and $rk(x) = rk(\beta)$, $C[\beta]$ denotes the result of substituting β for x in C. For example, $C[x] = b1 \odot_1 (a \cdot x)$ is a context and $C[A \cdot c] = b1 \odot_1 aAc$. The notion of multicontext is defined in the same way, except it may contain several distinct variables $x_1, \ldots, x_t$. In the case $t = 0$ a multicontext is just a term. If for any i it holds that $rk(\alpha_i) = rk(x_i)$, then $C[\alpha_1, \ldots, \alpha_t]$ denotes the result of substituting $\alpha_1, \ldots, \alpha_t$ for $x_1, \ldots, x_t$ in C.

We call a term (respectively, a context, a multicontext) ground if it contains no occurrences of nonterminals. Let μ be a valuation function, mapping each variable of rank l to some language of words of rank l. Then every ground multicontext α is assigned a value, interpreting the elements of Σ_1^* as themselves and the connectives from Op_k as corresponding language operations. Note that ground terms have the same value under all valuations.

Two ground multicontexts $C_1[x_1, \ldots, x_t]$ and $C_2[x_1, \ldots, x_t]$ with the same variables are equivalent if the expressions $C_1[\mu(x_1), \ldots, \mu(x_t)]$ and $C_2[\mu(x_1), \ldots, \mu(x_t)]$ have the same value under any valuation μ. The equivalence relation is denoted by $\sim$. Note that $\alpha \sim \mu(\alpha)$ for any ground term α. Further, $\sim$ is a congruence relation, which means that the equivalences $C' \sim C''$ and $\alpha_i \sim \beta_i$ for any $i \leq t$ imply $C'[\alpha_1, \ldots, \alpha_t] \sim C''[\beta_1, \ldots, \beta_t]$. Two terms (not necessarily ground) α_1 and α_2 are called equivalent if they can be represented in the form $\alpha_1 = C_1[A_1, \ldots, A_t]$ and $\alpha_2 = C_2[A_1, \ldots, A_t]$ for some equivalent ground multicontexts C_1 and C_2.

With every multicontext α we associate its syntactic tree $tree(\alpha)$ in a natural way. Then submulticontexts of α correspond to the nodes of this tree and vice versa. A submulticontext is internal if it corresponds to an internal node (it means the submulticontext contains a binary connective). A multicontext α is k-essential if its rank is at most k, as well as the rank of all its variables and nonterminals. The next lemma is proved in the full version of the paper.

Lemma 1. *Any k-essential multicontext C is equivalent to a k-correct multicontext C'.*

Corollary 1. *Any k-essential term α is equivalent to a k-correct term α'.*

2.2 Displacement Context-Free Grammars

This subsection introduces the notion of a displacement context-free grammar. In the definitions below, $GrTm_k$ denotes the set of all ground terms in Tm_k.

Definition 1. *A k-displacement context-free grammar (k-DCFG) is a quadruple $G = \langle N, \Sigma, P, S \rangle$, where Σ is a finite alphabet, N is a finite ranked set of nonterminals, $\Sigma \cap N = \emptyset$, $S \in N$ is a start symbol such that $rk(S) = 0$ and P is a set of rules of the form $A \to \alpha$. Here A is a nonterminal and α is a term from $Tm_k(N, \Sigma)$ such that $rk(A) = rk(\alpha)$.*

Definition 2. *The derivability relation $\vdash_G \in N \times Tm_k$ associated with the grammar G is the smallest reflexive transitive relation such that for any context C, the conditions $(B \to \beta) \in P$ and $A \vdash C[B]$ imply $A \vdash C[\beta]$. If $L_G(A) = \{\nu(\alpha) \mid A \vdash_G \alpha, \ \alpha \in GrTm_k\}$ is the set of words derivable from $A \in N$, then $L(G) = L_G(S)$.*

Example 1. Let the i-DCFG G_i be the grammar $G_i = \langle \{S, T\}, \{a, b\}, P_i, S \rangle$. Here P_i is the following set of rules (notation $A \to \alpha|\beta$ means $A \to \alpha, A \to \beta$):

$$S \to \underbrace{(\dots(}_{i-1 \text{ times}} aT \odot_1 a) + \dots) \odot_1 a \mid \underbrace{(\dots(}_{i-1 \text{ times}} bT \odot_1 b) + \dots) \odot_1 b$$

$$T \to \underbrace{(\dots(}_{i-1 \text{ times}} aT \odot_1 1a) + \dots) \odot_i 1a \mid \underbrace{(\dots(}_{i-1 \text{ times}} bT \odot_1 1b) + \dots) \odot_i 1b \mid 1^i$$

The grammar G_i generates the language $\{w^{i+1} \mid w \in \{a, b\}^+\}$. Below is the derivation of the word $(aba)^3$ in G_2:

$$S \to (aT \odot_1 a) \odot_1 a \to (a((bT \odot_1 1b) \odot_2 1b) \odot_1 a) \odot_1 a \to$$
$$(a((b((aT \odot_1 1a) \odot_2 1a) \odot_1 1b) \odot_2 1b) \odot_1 a) \odot_1 a \to$$
$$(a((b((a11 \odot_1 1a) \odot_2 1a) \odot_1 1b) \odot_2 1b) \odot_1 a) \odot_1 a =$$
$$(a(b(a1a1a \odot_1 1b) +_2 1b) +_1 a) \odot_1 a = (aba1ba1ba \odot_1 a) \odot_1 a = abaabaaba.$$

Two k-DCFGs are equivalent if they generate the same language. Since internal nodes of terms in a k-DCFG rules are also of rank k or less, the k-DCFGs can be binarized just like the context-free grammars to obtain a variant of the Chomsky normal form. Precisely, the following theorem holds (see [12]):

Theorem 1. *Every k-DCFG is equivalent to some k-DCFG $G = \langle N, \Sigma, P, S \rangle$ which has the rules only of the following form:*

1. $A \to B \cdot C$, where $A \in N$, $B, C \in N - \{S\}$,
2. $A \to B \odot_j C$, where $j \leq k$, $A \in N$, $B, C \in N - \{S\}$,
3. $A \to a$, where $A \in N$, $a \in \Sigma_1$,
4. $S \to \epsilon$.

We have already mentioned that k-DCFGs are equivalent to $(k+1)$-wMCFGs. In the case of $k = 1$ this statement is straightforward since both 1-DCFGs and 2-wMCFGs are just reformulations of Pollard wrap grammars [8]. We will not recall the definitions of a wMCFG, the interested reader may consult [9,3].

3   Terms and Derivations in DCFGs

In this section we investigate more thoroughly the properties of terms and derivation in DCFGs. First we give some fundamental notions. We assume that all the grammars are in the Chomsky normal form.

Definition 3. *A node v' in the syntactic tree is a direct descendant of a node v if $rk(v') = rk(v)$, v' is a descendant of v and all the nodes on the path from v to v' has the same rank as v and v'. A subterm β is a direct subterm of a term α, if its root node is the direct descendant of the root of α.*

Let α be a term of rank l. We denote[5] by $\alpha \otimes (u_1, \ldots, u_l)$ the result of simultaneous replacement of all the separators in α by $u_1, \ldots, u_l$.

Lemma 2. *Let $\alpha = C[\beta]$ for some ground context C and term β of rank l. There exist words $s_1, s_2, u_1, \ldots, u_l \in \Sigma_1^*$ depending only from the context C such that $\alpha \sim s_1(\beta \otimes (u_1, \ldots, u_l))s_2$ and $\mathrm{rk}(\alpha) = \mathrm{rk}(s_1) + \mathrm{rk}(s_2) + \sum_{i=1}^{l} \mathrm{rk}(u_i)$.*

Proof. By induction on the structure of the context C.

Lemma 3. *Let β be a direct subterm of a term α and C be the ground context such that $\alpha = C[\beta]$. Then the equivalence $\alpha \sim s_1(\beta \otimes (y_1 1 z_1, \ldots, y_l 1 z_l))s_2$ holds for some words $s_1, s_2, y_1, z_1 \ldots, y_l, z_l \in \Sigma^*$, depending only on the context C.*

Proof. The proof is by induction on the structure of the context C. The inductive step uses the fact that if v is the direct descendant of its ancestor v, then it is also a direct descendant of all the nodes between them.

Let D be the derivation of α from some nonterminal A of the grammar G (we denote it by $D \colon A \vdash \alpha$). We associate with D its derivation tree T_D obtained by attaching nonterminals to the nodes of $tree(\alpha)$. The labeling procedure is the following: if in the last step of D the rule $B \to \beta$ in the context C was applied, then we label by B the root node of the inserted subtree and keep other labels unchanged. Since G is in the Chomsky normal form, only the nonterminal leaves of $tree(\alpha)$ are unlabeled. Then we label every such node by the nonterminal it contains. The lemma below is proved by induction on derivation length.

---

[5] This notation is brought from discontinuous Lambek calculus.

Lemma 4. *Let $D: A \vdash \alpha$ and T_D be the corresponding derivation tree. For every representation $\alpha = C[\beta_1, \ldots, \beta_t]$ there are derivations $D_0: A \vdash C[B_1, \ldots B_t]$ and $D_i: B_i \vdash \beta_i$ for any $i \leq t$ such that T_D is obtained by replacing B_i with T_{D_i} in the multicontext C.*

Let G be a k-DCFG containing N_l nonterminals of rank l and T be a derivation tree in this grammar. We call an l-matryoshka[6] a subbranch of length $N_l + 1$ or more, containing only nodes of rank l. Note that all the elements of l-matryoshka are direct descendants of each other. By the pigeon-hole principle it contains two nodes with the same nonterminal label.

A rule $A \rightarrow \alpha$ is derivable in a grammar G if $A \vdash_G \alpha$. Adding derivable rules to a grammar does not change the language it generates. Rules $A \rightarrow \alpha$ and $A \rightarrow \alpha'$ are called equivalent if the terms α and α' are equivalent. If one of such rules is already in G, adding the other does not affect the generated language. Note that if every rule of G' is equivalent to some rule of G and vice versa, then the grammars themselves are also equivalent.

We denote the depth of a term β by $d(\beta)$. A term is called l-internal if all its internal nodes, possibly except the root, are of rank l. An l-internal term β which is l-essential and satisfies $d(\beta) \leq N_l + 1$, is called l-redundant. The grammar G is called l-duplicated, if for every derivable rule $A \rightarrow \alpha$ with α being l-redundant, there is an equivalent derivable rule $A \rightarrow \alpha'$ with an $(l-1)$-correct term α'.

Lemma 5. *For every k-DCFG G in the Chomsky normal form and every $l \leq k$ there is an equivalent l-duplicated grammar G' in the Chomsky normal form with the same set of nonterminals of rank $\geq l$.*

Proof (sketched). We use downward induction on l; for any derivable rule $A \rightarrow \alpha$ containing an l-redundant term α an equivalent rule $A \rightarrow \alpha'$ with an $(l-1)$-correct term α' is added. Such a term exists by Corollary 1. The process converges since there is a finite number of l-redundant terms and such terms do not appear in new derivable rules.

Definition 4. *A k-DCFL G' is m-compact if for every word w there is a derivation tree T_w such that for every node v of positive rank l in T_w there is an element v' of l'-matryoshka for some $l' \geq l$, such that the length of the path between v and v' is not greater than m and all the nodes in this path has rank l or greater.*

Theorem 2. *For every k-DCFL G there is an equivalent k-DCFL G', which is m-compact.*

Proof. See the full paper. Lemma 5 implies that every word possesses a derivation tree without l-redundant subtrees and such trees satisfy the definition of m-compactness.

---

[6] Matryoshka is a Russian souvenir consisting of several dolls nested one into another. We use this term since the yields of the subtrees, whose roots are the elements of an l-matryoshka, demonstrate the same nesting property.

4   Main Results

In this section we use Theorem 2 to prove a strong version of Pumping lemma.

Definition 5. *A pair of internal nodes v and v' of a derivation tree such that v and v' has the same label of rank $l-1$ and v' is the direct descendant of v, is called an l-pump. Here v is the top node of the pump and v' is its bottom node*[7].

Theorem 3. *For any k-DCFL L there exists a number n such that any word $w \in L$ with $|w| > n$ can be represented, for some $l \leq k+1$, in the form $w = s_0 y_1 u_1 z_1 s_1 y_2 u_2 z_2 s_2 \ldots y_l u_l z_l s_l$ satisfying the following requirements:*

1. $|y_1 z_1 \ldots y_l z_l| > 0$,
2. $|y_1 u_1 z_1 \ldots y_l u_l z_l| \leq n$,
3. *For any $p \in \mathbb{N}$ the word $s_0 y_1^p u_1 z_1^p s_1 \ldots y_l^p u_l z_l^p s_l$ belongs to $L(G)$.*

Proof. By Theorem 2 we assume that L is generated by a m-compact grammar G for some natural m. Let N_l be the number of nonterminals of rank l in this grammar, $N_+ = \max\{N_l \mid l > 0\}$ and $N = N_0 + N_+ + m$. We set $n = 2^N$.

Let $w \in L(G)$ be a word with $|w| \geq n$ and T_w be its derivation tree, deriving the term α and satisfying the requirement of Theorem 2. Then $d(T_w) \geq n+1$. Consider $N_0 + 1$ deepest nodes of the longest branch of T_w. If all of them are of rank 0, then some pair of nodes have the same label and hence form a 1-pump. If conversely, some node v_t is of rank $t > 0$, then there is an element of some l'-matryoshka on the distance not greater than m from v_t. Then the distance from v_t to the upper node of this matryoshka is at most $m + N_+$. This l'-matryoshka contains an $(l'+1)$-pump, and the depth of the top node of this pump differs from the depth of T_w by at most $N_0 + N_+ + m = n$. So we have proved an existence of such an l-pump for some $l \leq k+1$, that the depth of the subtree below its top node is at most n (in this case $l = l'+1$).

Let v and v' be the top and bottom nodes of this pump, B be their nonterminal label, and C_1 and C_2 be their outer contexts. Then $\alpha = C_1[C_2[\beta]]$ for some contexts C_1 and C_2 and term β such that $S \vdash C_1[B]$, $B \vdash C_2[B]$ and $B \vdash \beta$.

Let $\nu(\beta) = u_1 1 \ldots 1 u_l$. By Lemma 3, for any term γ of rank $l-1$ the context $C_2[\gamma]$ is equivalent to $y_1(\gamma \otimes (z_1 1 y_2, \ldots, z_{l-1} 1 y_l)) z_l$ for some $y_1, z_1, \ldots, y_l, z_l \in \Sigma^*$. Also $C_1[\eta] \sim s_0(\eta \otimes (s_1, \ldots, s_{l-1})) s_l$ for some words $s_0, \ldots, s_l \in \Sigma^*$. Then w is equivalent and hence equal to the word

$$s_0((y_1((u_1 1 \ldots 1 u_l) \otimes (z_1 1 y_2, \ldots, z_{l-1} 1 y_l)) z_l) \otimes (s_1, \ldots, s_{l-1})) s_l =$$
$$s_0((y_1 u_1 z_1 1 \ldots 1 y_l u_l z_l) \otimes (s_1, \ldots, s_{l-1})) s_l = s_0 y_1 u_1 z_1 s_1 \ldots y_l u_l z_l s_l.$$

The depth of $C_2[\beta]$ is not greater than N, so its value $y_1 u_1 z_1 1 \ldots 1 y_l u_l z_l$ cannot be longer than n. It remains to prove the third statement.

We write $C_2^1 = C_2$ and $C_2^p = C_2[C_2^{p-1}]$ for any $p > 1$. Repeating the derivation $B \vdash C_2[B]$ p times, we obtain the derivation $B \vdash C_2^p[B]$. Applying Lemma 3 to the context C_2 several times and using basic term equivalences, we get that $C_2^p[\gamma] \sim y_1^p(\gamma \otimes (z_1^p 1 y_2^p, \ldots, z_{l-1}^p 1 y_l^p)) z_l^p$. Setting $\gamma = \beta$

---

[7] Our definition of l-pump reformulates the definition of an even k-pump from [3].

yields $y_1^p u_1 z_1^p 1 \ldots 1 z_l^p u_l y_l^p \in L_{G'}(B)$ and consequently $s_0 y_1^p u_1 z_1^p s_1 \ldots z_l^p u_l y_l^p s_l \in L_{G'}(S)$. The theorem is proved.

Let the pair of nodes v and v' be an l-pump. The replacement of the subtree rooted at v by the subtree rooted at v' is called collapsing of the pump. The scope of an l-pump consists of the nodes that are descendants of v but not of v'; these are exactly the nodes removed by collapsing this pump.

Lemma 6. *Let T' be a tree obtained from T by collapsing some pump. If the nodes v_1 and v_2 form a pump in T', then they form a pump in T.*

Proof. Let v and v' be, respectively, the top and bottom nodes of the collapsed pump. If v' is not on the path from v_1 to v_2 in T' then v_2 has already been a direct descendant of v_1 in T. Otherwise (v_1, v') and (v', v_2) are the pairs of direct descendants in T', which means that (v_1, v) and (v', v_2) were the pairs of direct descendants in T. Using the fact that v' was a direct descendant of v in T and the transitivity of direct descendance, we obtain that v_2 is a direct descendant of v_1 in T. Hence they form a pump in T. The lemma is proved.

Lemma 6 implies that a terminal vertex in the scope of a pump in a collapsed derivation tree is also in the scope of this pump in the original tree. This fact allows us to prove a weaker analogue of the Ogden lemma ([6]).

Theorem 4 (Ogden lemma for 1-DCFGs). *For any k-DCFL L there is a number t such that for any word $w \in L$ with at least t selected positions there is a representation $w = s_0 y_1 u_1 z_1 s_1 \ldots y_l u_l z_l s_l$ for some $k \le l + 1$, satisfying the following conditions:*

1. *For any $p \in \mathbb{N}$ the word $s_0 y_1^p u_1 z_1^p \ldots y_l^p u_l z_l^p s_l$ belongs to $L(G)$.*
2. *At least one of the words $y_1, z_1, \ldots, y_l, z_l$ contains a selected position.*

Proof. We set t equal to n from pumping lemma. It suffices to show that one of the selected positions is in the scope of some pump. We use an induction on $|w|$; note that $|w| \ge n$. There is a representation $w = x_0' y_1' u_1' z_1' x_1' \ldots y_l' u_l' z_l' x_l'$ such that the word $x_0' u_1' x_1' \ldots u_l' x_l'$ is also in L. If the removed words contained a labeled position, the lemma is proved. Otherwise the word $w' = x_0' u_1' x_1' \ldots u_l' x_l'$ contains the same number of labeled positions and we can apply the induction hypothesis to its derivation tree T', which is obtained from T by collapsing. Then one of the selected positions is in the scope of some pump in T', which implies by Lemma 6 that it is in the scope of a pump in T. The theorem is proved.

5   Examples of Non 1-DCFLs

In this section we use the established theoretical results to give some examples of non-1-DCFLs. To address this question we need to investigate more thoroughly the properties of constituents of displacement context-free grammars. A constituent is the fragment of the word derived from a node of its derivation tree. In the context-free case every constituent is a continuous subword, hence it can be described by two numbers: the position of its first symbol and the position of its last symbol plus one (we add one to deal with empty constituents). Recall that context-free constituents must be correctly embedded which means they either do not intersect or one constituent is the part of another.

The situation is a bit more complex in the case of DCFGs. However, the results of [11] provide analogous geometrical intuition. The constituents of rank 1 are the words of the form $w_1 1 w_2$, where w_1, w_2 are continuous subwords of the derived word w. Then a constituent of rank 1 is characterized by four indexes $i_1 \leq j_1 \leq i_2 \leq j_2$ of the border of its subwords. We identify a constituent with a tuple of its characterizing indexes in the ascending order. The proofs of the statements below, as well as illustrative figures, can be found in the full version of the paper.

Lemma 7. *One of the possibilities below hold without loss of generality for any pair of constituents (i_1, j_1, i_2, j_2) and (i_1', j_1', i_2', j_2'):*

1. $j_2 \leq i_1'$,
2. $j_1 \leq i_1' \leq j_2' \leq i_2$,
3. $i_1 \leq i_1' \leq j_2' \leq j_1$ or $i_2 \leq i_1' \leq j_2' \leq j_2$,
4. $i_1 \leq i_1' \leq j_1' \leq j_1 \leq i_2 \leq i_2' \leq j_2' \leq j_2$.

Since every pump is just a pair of properly embedded constituents labeled by the same nonterminal, Lemma 7 helps to specify the mutual positions of different 2-pumps. The scope of the pump contains exactly the positions which are in the top constituent but not in the bottom, so every 2-pump is described by eight indexes $i_1 \leq j_1 \leq k_1 \leq l_1 \leq i_2 \leq j_2 \leq k_2 \leq l_2$, such that (i_1, l_1, i_2, l_2) is the tuple of indexes of its top constituent and (j_1, k_1, j_2, k_2) — of the bottom.

Lemma 8. *One of the possibilities below hold without loss of generality for any pair of 2-pumps $(i_1, j_1, k_1, l_1, i_2, j_2, k_2, l_2)$ and $(i_1', j_1', k_1', l_1', i_2', j_2', k_2', l_2')$:*

1. $l_2 \leq i_1'$,
2. $i_1 \leq i_1' \leq l_2' \leq j_1$ or $k_2 \leq i_1' \leq l_2' \leq l_2$,
3. $i_1 \leq i_1' \leq j_1' \leq j_1 \leq k_1 \leq k_1' \leq l_1' \leq l_1 \leq i_2 \leq i_2' \leq j_2' \leq j_2 \leq k_2 \leq k_2' \leq l_2' \leq l_2$,
4. $i_1 \leq i_1' \leq j_1' \leq k_1' \leq j_1 \leq k_1 \leq l_1' \leq l_1 \leq i_2 \leq i_2' \leq j_2 \leq k_2 \leq j_2' \leq k_2' \leq l_2' \leq l_2$,
5. $i_1 \leq i_1' \leq j_1 \leq k_1 \leq j_1' \leq k_1' \leq l_1' \leq l_1 \leq i_2 \leq i_2' \leq j_2' \leq k_2' \leq j_2 \leq k_2 \leq l_2' \leq l_2$,
6. $i_1 \leq i_1' \leq j_1 \leq j_1' \leq k_1' \leq k_1 \leq l_1' \leq l_1 \leq i_2 \leq i_2' \leq j_2 \leq j_2' \leq k_2' \leq k_2 \leq l_2' \leq l_2$,
7. $k_1 \leq i_1' \leq l_1' \leq l_1 \leq i_2 \leq i_2' \leq l_2' \leq j_2$,
8. $i_1 \leq i_1' \leq l_1' \leq j_1 \leq k_2 \leq i_2' \leq l_2' \leq l_2$,
9. $k_1 \leq i_1' \leq l_2' \leq l_1$ or $i_2 \leq i_1' \leq l_2' \leq j_2$,
10. $j_1 \leq i_1' \leq l_1' \leq k_1 \leq j_2 \leq i_2' \leq l_2' \leq k_2$,
11. $j_1 \leq i_1' \leq l_2' \leq k_1$ or $j_2 \leq i_1' \leq l_2' \leq k_2$,
12. $l_1 \leq i_2' \leq l_2' \leq i_2$.

Let $\pi_1 = (i_1, j_1, k_1, l_1, i_2, j_2, k_2, l_2)$ and $\pi_2 = (i_1', j_1', k_1', l_1', i_2', j_2', k_2', l_2')$ be two 2-pumps. We call a pair of π_1 and π_2 linear if $l_2 \leq i_1'$ or $l_2' \leq i_1$. We call π_1 outer for the pump π_2 if $i_1 \leq i_1' \leq l_2' \leq l_2$. Note that if a pair of 2-pumps is not linear, then one of its elements is the outer pump for another. We call π_1 embracing for π_2 if $l_1 \leq i_1' \leq l_2' \leq i_2$.

Corollary 2. Let $(i_1, j_1, k_1, l_1, i_2, j_2, k_2, l_2)$ and $(i'_1, j'_1, k'_1, l'_1, i'_2, j'_2, k'_2, l'_2)$ be 2-pumps such that one of the segments of the second pump is a proper subset of the segment $[l_1; i_2]$. Then either the second pump is outer for the first (which means $i'_1 \leq i_1 \leq l_2 \leq l'_2$) or the first pump is embracing for the second.

Lemma 8 allows us to give some examples of non 1-DCFLs. The first example is the language $4\text{MIX} = \{w \in \{a, b, c, d\}^* \mid |w|_a = |w|_b = |w|_c = |w|_d\}$.

Theorem 5. *The language 4MIX cannot be generated by any 1-DCFG.*

Proof. Since wMCFLs are closed under intersection with regular languages, it suffices to prove that the language $4\text{MIX} \cap (a^+ b^+ c^+ d^+)^2$ is not a 1-DCFL. Assume the contrary, let t be the number from Ogden's lemma applied to this language. Let the word $w = a^{m_1} b^{m_2} c^{m_3} d^{m_4} a^{n_1} b^{n_2} c^{n_3} d^{n_4}$ satisfy the following conditions:

1. $\min(m_j, n_j) \geq t$,
2. $m_1 \geq (3M + 1)(M + t)$, where $M = \max(m_2, m_4, n_3)$,
3. $m_4 \geq (n_1 + 1)(n_1 + t)$.

Note that every 2-pump contains an equal number of a's, b's, c's and d's, and every continuous segment of it consists of identical symbols (we call such segments homogeneous). We enumerate the maximal continuous homogeneous subwords of w from 1 to 8. Then every 2-pump intersects exactly four such segments. We call a pump intersecting the segments with numbers $d_1, \dots, d_l$ (and possibly some others) a $[d_1, \dots, d_l]$-pump.

We select $3M+1$ segments of length t in the first segment of the word w such that any two segments are separated by at least M symbols. By Theorem 4 each such segment intersects with some 2-pump. We want to prove that some of them intersects with a $[1, 3, 6, 8]$-pump. Any two points from different segments cannot belong to the same $[1, 7]$-pump, since in this case there is a continuous segment of at least $M + 1$ a's in the pump. Then the pump contains at least $M + 1$ c's, which exceeds the length of the 7th segment. It follows that there are at most M $[1, 7]$-pumps. By the same argument there are at most M $[1, 2]$-pumps and at most M $[1, 4]$-pumps, therefore the number of $[1]$-pumps which are not $[1, 3, 6, 8]$-pumps is less than $3M + 1$ which proves the existence of a $[1, 3, 6, 8]$-pump.

By the same argument there is at least one $[4]$-pump, which is not a $[4, 5]$-pump. By corollary 2 applied to the $[1, 3, 6, 8]$-pump it is either a $[1, 4, 8]$-pump or it is embraced by the $[1, 3, 6, 8]$-pump. In the first case there are two d-segments in the pump, in the second case it should be a $[3, 4, 5, 6]$-pump which contradicts our assumption. So we have reached a contradiction and the theorem is proved.

Our technique of embedding different 2-pumps also works in a more complex case. Consider the language $MIX = \{w \in \{a, b, c\}^* \mid |w|_a = |w|_b = |w|_c\}$. It is expected to be not a $DCFL$ since it demonstrates an extreme degree of non-projectivity. It is proved in [4] that MIX is not a 2-wMCFL (and hence not a 1-DCFL). The proof extensively uses geometrical arguments and is therefore very difficult to be generalized for similar languages or wMCFGs of higher order. Our proof uses only the Ogden's lemma for DCFGs and is much shorter.

Theorem 6. *The MIX language is not a 1-DCFL.*

Proof. We use the same method and notation as in the case of 4MIX language. Again, it suffices to prove that the language $L = MIX \cap a^+ b^+ c^+ b^+ c^+ a^+$ is not a 1-DCFL. Let t be the number from Ogden's lemma for L. Consider the word $w = a^{m_1} b^{m_2} c^{m_3} b^{n_2} c^{n_3} a^{n_1}$ satisfying the following properties:

1. $\min(m_j, n_j) \geq t$,
2. $m_1 \geq (2M + 1)(M + t)$, where $M = \max(m_3, n_2)$,
3. $n_1 \geq (2M + 1)(M + t)$, where $M = \max(m_3, n_2)$,
4. $m_3 \geq (n_2 + 1)(n_2 + t)$.

By the same arguments as in Theorem 5 we establish the existence of $[1, 2, 5]$-and $[2, 5, 6]$-pumps. Since they cannot form a linear pair of 2-pumps, one of them is an outer pump for another, which implies one of them is a $[1, 2, 5, 6]$-pump. The condition $m_3 \geq (n_2 + 1)(n_2 + t)$ implies the existence of a $[2, 3]$-pump. By Corollary 2 applied to the $[1, 2, 5, 6]$-pump and the $[2, 3]$-pump, the $[2, 3]$-pump is actually a $[1, 2, 3, 6]$-pump since it contains some a's.

The condition $n_2 \geq t$ implies there is a $[4]$-pump, which is a $[1, 4, 6]$-pump again by Corollary 2. To be correctly embedded with the $[1, 2, 3, 6]$-pump it should be a $[1, 2, 4, 6]$-pump but there is no c's in such pump. Hence we reached the contradiction and the MIX language cannot be generated by a 1-DCFG. The theorem is proved.

6   Conclusions and Future Work

We have proved a strong version of the pumping lemma and a weak Ogden lemma for the class of DCFLs which is also the class of well-nested multiple context-free languages. These statements allow us to prove that some languages, like the well-known MIX language, do not belong to the family of 1-DCFLs or, in other terms, the family of tree adjoining languages. We hope to adopt the proof for the case of half-blind three-counter language $\{w \in \{a, b, c\}^* \mid |w|_a = |w|_b = |w|_c, \forall u \sqsubseteq w \, |u|_a \geq |u|_b \geq |u|_c\}$ to prove that a shuffle iteration of a one-word language may lie outside the family of 1-DCFLs. The author supposes that the technique used in the article will work not only in the case of 2-pumps, but also in a more complex cases. We hope that our results will help to understand better the structure of well-nested MCFLs and, in particular, to prove the Kanazawa-Salvati conjecture stating that MIX is not a well-nested MCFL.

Acknowledgments. The author thanks Makoto Kanazawa for his helpful suggestions and the anonymous referees of DLT 2014 conference, whose thoughtful comments essentially improved the paper.

References

1. Joshi, A.K.: Tree adjoining grammars: How much context-sensitivity is required to provide reasonable structural descriptions? University of Pennsylvania, Moore School of Electrical Engineering, Department of Computer and Information Science (1985)
2. Joshi, A.K., Schabes, Y.: Tree-adjoining grammars. In: Rozenberg, G., Salomaa, A. (eds.) Handbook of Formal Languages, vol. 3, pp. 69–123. Springer (1997)
3. Kanazawa, M.: The pumping lemma for well-nested multiple context-free languages. In: Diekert, V., Nowotka, D. (eds.) DLT 2009. LNCS, vol. 5583, pp. 312–325. Springer, Heidelberg (2009)
4. Kanazawa, M., Salvati, S.: MIX is not a tree-adjoining language. In: Proceedings of the 50th Annual Meeting of the Association for Computational Linguistics: Long Papers, vol. 1, pp. 666–674. Association for Computational Linguistics (2012)
5. Morrill, G., Valentín, O., Fadda, M.: The displacement calculus. Journal of Logic, Language and Information 20(1), 1–48 (2011)
6. Ogden, W.: A helpful result for proving inherent ambiguity. Theory of Computing Systems 2(3), 191–194 (1968)
7. Palis, M.A., Shende, S.M.: Pumping lemmas for the control language hierarchy. Mathematical Systems Theory 28(3), 199–213 (1995)
8. Pollard, C.: Generalized phrase structure grammars, head grammars, and natural languages. PhD thesis. Stanford University, Stanford (1984)
9. Seki, H., Matsumura, T., Fujii, M., Kasami, T.: On multiple context-free grammars. Theoretical Computer Science 88(2), 191–229 (1991)
10. Shieber, S.M.: Evidence against the context-freeness of natural language. In: The Formal Complexity of Natural Language, pp. 320–334. Springer (1987)
11. Sorokin, A.: Monoid automata for displacement context-free languages. In: ESSLLI Student Session 2013 Preproceedings, pp. 158–167 (2013), http://loriweb.org/uploads/ESSLLIStuS2013.pdf#page=162, Extended version to appear in ESSLLI Student Session 12-13 Selected Papers, http://arxiv.org/abs/1403.6060
12. Sorokin, A.: Normal forms for multiple context-free languages and displacement Lambek grammars. In: Artemov, S., Nerode, A. (eds.) LFCS 2013. LNCS, vol. 7734, pp. 319–334. Springer, Heidelberg (2013)
13. Valentın, O., Morrill, G.: Theory of discontinuous Lambek calculus. PhD thesis. Universitat Autonoma de Barcelona (2012)
14. Vijay-Shanker, K., Weir, D.J., Joshi, A.K.: Tree adjoining and head wrapping. In: Proceedings of the 11th Conference on Computational Linguistics, pp. 202–207. Association for Computational Linguistics (1986)

Aperiodic Tilings and Entropy[*]

Bruno Durand[1], Guilhem Gamard[2], and Anaël Grandjean[3]

[1] Université Montpellier 2, Lirmm, Montpellier, France
www.lirmm.fr/~bdurand
[2] Lirmm and ENS Paris, Montpellier and Paris, France
[3] Lirmm and ENS Lyon, Montpellier and Lyon, France

Abstract. In this paper we present a construction of Kari-Culik aperiodic tile set, the smallest known until now. Our construction is self-contained and organized to allow reasoning on properties of the resulting sets of tilings. With the help of this construction, we prove that this tileset has positive entropy. We also explain why this result was not expected.

1 Introduction

In this paper we focus on aperiodic tilesets. These tilesets can be used to tile the plane but none of the obtained tilings has a period. The role of aperiodic tilesets is crucial in different fields such as logics (see for instance [1]) or for the study of quasi-periodic structures such as quasi-crystals. Furthermore these aperiodic tilesets are a classical tool to prove undecidability problems for planar structures or dynamical systems. We work with the formalization that Wang proposed in [2].

A classical question about a tileset is to measure its entropy. Roughly speaking the entropy of a tileset is positive if "points of freedom are dense". One can easily make it positive for any aperiodic tileset by a cartesian product with a free bit. The number of tiles is then multiplicated by a factor two but the resulting tiling has positive entropy. It is easy to observe that for classical self similar tilesets such as Berger [3], Robinson [4] the entropy is zero. The main question we address is the entropy of the smallest known aperiodic tileset: it was conjectured that its entropy is zero but we prove it is positive. This entropy zero conjecture comes from other works on this tileset and some algorithmic remarks developed in Sect. 4.

Our paper is organized as follows: first we explain exactly the same tileset as Kari and Culik in [5,6]. Our explanation makes it easier to analyze (repeating the proof of aperiodicity). Then in Sect. 3 we formulate a substitutive property that guarantee positive entropy of Kari-Culik tileset. The rest of the section is devoted to the proof. The last section is focused on more refined approaches to the entropy of a tileset.

---

[*] Supported by ANR project EMC NT09 555297.

2   Presentation of the Tileset

2.1   Source of Aperiodicity

Let us start with an observation. Consider a bi-infinite sequence x_n of positive real numbers, such that either $x_{n+1} = 2x_n$ or $x_{n+1} = x_n/3$ for every n.

Every such sequences are aperiodic. Indeed, for all n and all $k > 0$, we have $x_{n+k} = x_n \times 2^i/3^j$ for some $i, j > 0$. If we had $x_{n+k} = x_n$, then we would have $1 = 2^i/3^j$ for $i, j > 0$, which is a contradiction.

Moreover, there exist some such sequences x_n which lie in the interval $[1/3; 2]$. Starting from some x_0 in this interval, we can always take $x_{n+1} = 2x_n$ if $x_n < 1$, and $x_{n+1} = x_n/3$ otherwise. The same argument works in the opposite direction.

2.2   Aperiodic Sequences and Tilings

A tile is an unit square with colored sides. Consider the (geometric) plane with a unit grid; a tiling is an assignation of a tile to each square of the grid, in such manner that matching borders have the same color. Thus, in a tiling, we have a bi-infinite sequence of colors along any horizontal or vertical line of the grid.

We are going to focus on the horizontal lines of our tilings. If we use three colors (say, 0, 1 and 2) for the top and bottom sides of the tiles, we will get bi-infinite sequences over the alphabet $\{0, 1, 2\}$. Such sequences might have an **average**, i.e. a limit of averages over finite parts as the length of the parts increases (In our tiling we prove that unique average always exists see Prop. 2).

Our goal is to construct a set of tiles with the following two properties:

1. for every tiling, if the averages of all horizontal lines exist, they form a sequence x_n with the property defined in Sect. 2.1.
2. for every such sequence x_n, we can find a tiling where averages exist and are equal to x_n.

This tile set will be aperiodic. If it had a periodic tiling, it would also have a bi-periodic tiling. In a bi-periodic tiling, all horizontal lines have an average (due to horizontal periodicity), and form a periodic sequence (due to vertical periodicity), which is impossible. The existence of tilings is a consequence of the second claim.

2.3   Tilings and Automata

Our tileset should guarantee that some relation holds between any two consecutive lines of a tiling (namely, "$x = 2y$ or $x = y/3$"). Thus, let us consider a stripe (a horizontal line of tiles), as displayed on Fig. 1. We call the sequence of bottom numbers a_n, top numbers b_n, and the matching left and right numbers q_n. Such a stripe can be viewed as a run of a non-deterministic automaton, where q_n are the traversed states, and (a_n, b_n) are the input.

More precisely, each tile (q', a, q, b) correspond to a transition $q \xrightarrow{(a,b)} q'$, where (a, b) is the input. This is illustrated by Fig. 2. Since a tileset only have a

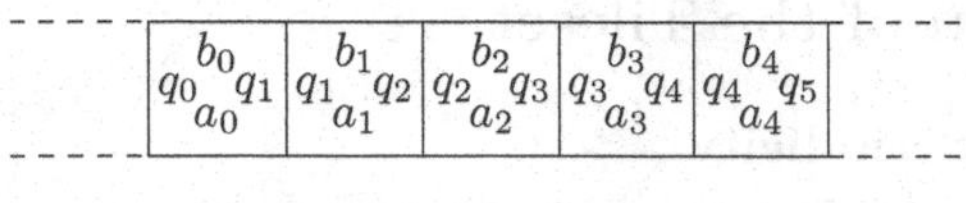

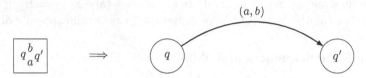

Fig. 1. Example of a horizontal tiled line

finite number of colors and tiles, the running automaton must be a finite-state automaton reading pairs of letters. Note that our automaton has no initial state; it runs infinitely in both directions.

$$q \, {}^{b}_{a} \, q' \quad \Longrightarrow \quad q \xrightarrow{(a,b)} q'$$

Fig. 2. Translation of a tile to a transition

We can see that there exists a bi-infinite run of the automaton on the sequence

$$\ldots (a_{-2}, b_{-2}), (a_{-1}, b_{-1}), (a_0, b_0), (a_1, b_1), (a_2, b_2) \ldots$$

if and only if there exists a tile horizontal stripe that carries the sequence $\ldots a_{-2}, a_{-1}, a_0, a_1, a_2 \ldots$ on the bottom and $\ldots b_{-2}, b_{-1}, b_0, b_1, b_2 \ldots$ on the top.

If we try to extract a set of tiles from several automata, and take the union of the results, we will get a tileset which performs a run of one of the automata on each line. We have to ensure that the set of states of the several automata are disjoint, which guarantees that automata will never be mixed within a single line.

2.4   Construction of Actual Automata

Let us construct a finite-state automaton which reads sequences of couples (a_n, b_n) and checks if $|\sum_i b_i - 2 \sum_i a_i|$ is bounded; and another one which checks if $|3 \sum_i b_i - \sum_i a_i|$ is bounded. The sequences a_n and b_n are on alphabets of two integers, for instance a_n is on $\{0, 1\}$ and b_n on $\{1, 2\}$.

These automata are constructed in the following way: fix a set of states Q, and have all transitions $q \xrightarrow{(a,b)} q'$ to satisfy the following relation:

$$q' = q + 2a - b \quad \text{(automaton for } b = 2a)$$
$$q' = q + a/3 - b \quad \text{(automaton for } b = a/3)$$

Thus, the automata will compute the cumulative sum of a_n and the cumulative sum of $2b_n$ (resp. $b_n/3$), and hold the difference into its current state. Since the number of states is finite, the difference must be bounded. As a consequence, if a couple of sequences (a_n, b_n) is accepted by the first (resp. second) automaton and a_n have an average, then b_n have an average which is twice (resp. one third of) a_n's one.

It only remains to set the alphabet for the a_n and b_n sequences. These alphabets are directly connected to the allowed range for averages of a_n and b_n. For instance, if a_n is on alphabet $\{1, 2\}$, its average can be any real number between 1 and 2. Likewise, we have to set an alphabet for b_n. As an additional restriction, we can make automata in such manner that they reject some finite patterns, like 000. Sequences on alphabet $\{0, 1\}$ without any pattern "000" cannot have an average lesser than $1/3$. Using this fact, we can restrict allowed ranges for averages in a more precise way.

We can describe our piecewise linear function with two automata on alphabet $\{0; 1; 2\}$ (one for each linear piece).

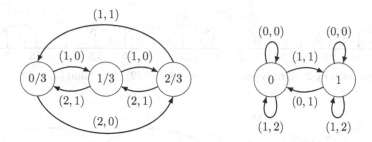

Fig. 3. Automata $M_{1/3}$ and M_2

These automata have 6 transitions each, yielding an almost aperiodic set of 12 tiles. Indeed one slight change must be made to avoid the tiling with only the "0" tile. This change will be presented further in this paper.

2.5 Existence of a Tiling

Remark that if we want to tile the whole plane, our automata cannot be fed with any sequences, even if those sequences have averages. Indeed, when the M_2 automaton reads with the sequence "$a_n = 0011$", the output sequence b_n contains both a 0 and a 2. However, automata $M_{1/3}$ only accepts sequences over $\{1, 2\}$ for a_n, and M_2 only accepts $\{0, 1\}$. Thus, if a stripe has 0011 on its bottom line, then it has both a 0 and a 2 on its top line. The next stripe cannot be a run of M_2 nor $M_{1/3}$, and the tiling does not go to infinity.

As a consequence, we need to show that there exists sequences of average x, for each positive real x, which are accepted by our automata. In order to achieve this, we will use Sturmian sequences. Define:

$$B_x(k) = \lfloor x(k+1) \rfloor - \lfloor xk \rfloor$$

B_x is the **Sturmian sequence** of slope x, and $B_x(k)$ is its kth letter. This sequence is bi-infinite over alphabet $\{\lfloor x \rfloor, \lceil x \rceil\}$. Since the sum over k of $B_x(k)$

is telescopic, it is easy to calculate the average of this sequence and check it is actually x.

Let us think a bit about Sturmian sequences. Fix a real number x. Imagine you are on an infinite, measured line, and you are making jumps of length x along the line. Whenever you make a jump, write down the number of integers you jumped over: this is the Sturmian sequence of slope x.

We can get the Sturmian sequence of $2x$ by making jumps of length x, and counting the number of multiples of 0.5 we jumped over. There are twice more multiples of 0.5 than integers; thus, in the long run, we actually get the Sturmian sequence of $2x$. This idea is illustrated on Fig. 4a.

If we want to get the Sturmian sequence of $x/3$ from the sequence of x, we have to consider multiples of 3. They actually occur three times less often than integers. This is illustrated on Fig. 4b.

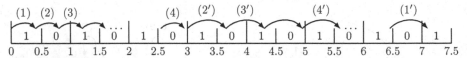

(a) Multiplication by 2: eight types of transitions, but $(2) = (2')$ and $(3) = (3')$, yielding six distinct types

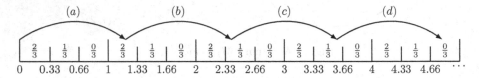

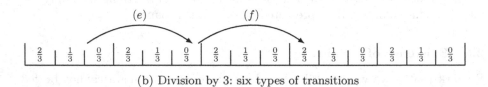

(b) Division by 3: six types of transitions

Fig. 4. Multiplications of Sturmian sequences

Note that, in Fig. 4a, jumping over a non-integer multiple of 0.5 (the "small obstacles") increments the difference between "twice number of integers jumped" and "number of multiples of 0.5 jumped" by 1. By contrast, jumping over an integer (big obstacles) decrements this difference by 1. Since we want it as close to 0 as possible, two states are enough (before and after the small obstacle).

This works the same for Fig. 4b. Jumping over "small obstacles" increments the difference between "one third of the obstacles jumped" and "number of big obstacles jumped". Jumping over "big obstacles" decrements this difference by 2. As a conclusion, only 3 states are needed.

One can finally check that all possible types of jumps are displayed on Fig. 4, and that each of them corresponds to a transition of our automata (Fig. 3). For instance, type (1) corresponds to $1 \xrightarrow{(0,0)} 1$, and (2) corresponds to $1 \xrightarrow{(0,1)} 0$. More generally, in M_2, a_n corresponds to the number of jumped obstacles (of any size) by an arrow, and b_n corresponds to the number of jumped big obstacles. In $M_{1/3}$, it is permuted: a_n is the number of jumped big obstacles, and M_2 the number of jumped obstacles.

As a conclusion, (B_x, B_{2x}) is always accepted by M_2 and (B_{3x}, B_x) is always accepted by $M_{1/3}$. Thus one can take any sequence x_n from Sect. 2.1, write the Sturmian sequence of x_n on line n of the tiling, and get valid runs for automata. Thus one get valid tilings.

2.6 Aperiodicity

This construction ensures that each tiling corresponds to a specific sequence which is identically null or aperiodic. Then we just have to avoid this null sequence. Culik presented one way to achieve that in [6]. The idea is to forbid three consecutive uses of the M_2 automaton. This can be done by adding only one tile. Consider a new color $0'$ which value is 0 in the average, such that above a 0 there can be either a 1 or a $0'$, and above a $0'$ there always is a 1. Thus there cannot be three consecutive $not - 1$ in a row, ensuring that the all zero configuration is forbidden. All tilings with this tileset are aperiodic. This tileset is displayed on Fig. 6.

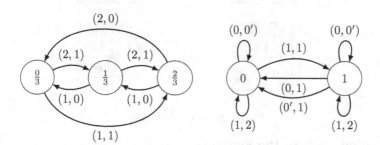

Fig. 5. Kari + Culik automata

3   Positive Entropy

3.1   Introduction

Let S be a palette and $C_S(n)$ be the number of different patterns of size $n \times n$ which appear in a tiling. Then the *entropy* of S is defined by $H(S) = \lim_n \dfrac{\log C_S(n)}{n^2}$ (the limit always exists).

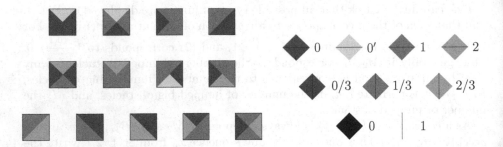

Fig. 6. Kari + Culik Tileset and colors meaning

The question we address is whether the Kari-Culik tileset has positive entropy. A usual method for proving such a fact is to exhibit a *substitutive pair*, i.e., a couple of different patterns with the same borders.

Our method is a small variation of the above: we prove that our tileset contains two substitutive pairs and for each sufficiently large square one of the pair items appears. Our substitutive pairs are shown in Fig. 7.

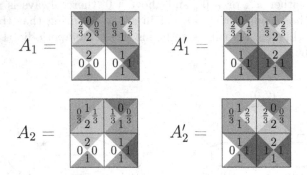

Fig. 7. Substitutive pairs

3.2   Coming Back to the Function

Recall that our function is $f : [\frac{1}{3}; 2] \mapsto [\frac{1}{3}; 2]$ such that

$$f(x) = \begin{cases} 2x & \text{if } x \in [\frac{1}{3}; 1] \\ \frac{1}{3}x & \text{if } x \in [1; 2] \end{cases}$$

Lemma 1. *The orbits of f are dense.*

Proof. It is well-known that irrational rotations on the circle have dense orbits. Thus, we map the interval $[\frac{1}{3}; 2]$ on the unit circle in such manner that the function f corresponds to a rotation of irrational angle.

We consider the following mapping:

$$\phi : [\frac{1}{3}; 2] \to [0; 1]$$

$$\phi(x) = \frac{\log(x) + \log 3}{\log(2) + \log 3} \mod 1$$

We view the interval $[0; 1]$ as a circle by identifying point 0 with point 1.

$$\phi(2x) = \frac{\log(2) + \log(x) + \log 3}{\log(2) + \log 3} \mod 1 = \phi(x) + \frac{\log(2)}{\log(2) + \log 3} \mod 1 \quad (1)$$

$$\phi(\frac{x}{3}) = \frac{\log(x)}{\log(2) + \log 3} \mod 1 = \phi(x) + \frac{\log(2)}{\log(2) + \log 3} \mod 1 \quad (2)$$

Both transitions map to the same irrational rotation of angle $\frac{\log 2}{\log 2 + \log 3}$.

Proposition 1. *Given any interval the maximal number of iterations of f between two occurrences in this interval is bounded.*

Proof. Consider any interval $I =]a; a + \alpha[$ in $[0; 1] \mod 1$. As the orbits of f are dense from starting point a, there exists N such that $f^N(a) \mod 1$ is in $]a; a + \alpha/2[$. Thus $f^N(a) = a + \beta$ with $\beta < \alpha/2$. From any point x in I, either $x + \beta$ or $x - \beta$ is in I. Thus either $f^{N\lfloor 1/\beta \rfloor}(x)$ or $f^{N\lceil 1/\beta \rceil}(x)$ is in I. Hence our required bound on the number of iterations of f is $N\lceil 1/\beta \rceil$.

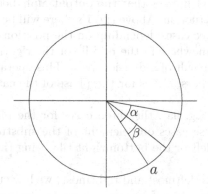

Fig. 8. Crossing intervals

Now let us examine the colors that appear on any horizontal line of the tiling. On this line colors represent 0 and 1 or 1 and 2 (0 and 0' being interpreted as the same zero).

Proposition 2. *Any horizontal line in any tiling has an average (in the sense of frequencies of numbers defined above).*

Proof. Our proof is based on the remark that on each line we have either 0 and 1 or 1 and 2 but never 0 and 2.

The average of any segment has a value in $[0; 2]$ which is a compact set. Consider two non overlapping growing sequences of subpatterns. Suppose their averages have different limits. Then we can take subpatterns of different averages as large as we want. With enough runs of the automata, one of this subpattern will have average less than 1 and the other, greater than 1, on the same line. But no automaton can read such a line, which makes a contradiction.

Remark that Kari's basic idea is that the averages of the lines obey the function f.

We prove below that a specific family of patterns appears dense in our tiling. The following lemma gives us the horizontal density, and the vertical density is obtained by combination of this lemma and Proposition 1.

Lemma 2. *The family of patterns* $\{01^\alpha 0 | \alpha > 3\}$ *appear with positive density in each line that has average in* $]\frac{4}{5}; \frac{9}{10}[$.

The choice of $\frac{9}{10}$ is arbitrary, the result remain true for any value in $]\frac{4}{5}; 1[$.

Proof. On each line with average greater than 4/5 the pattern 1111 must appear with positive density. On each line with average less than 9/10, 0 appear with a positive density, otherwise we would have a contradiction with Proposition 2.

We now have a family of linear patterns that appear in a dense way in our tiling. Let us prove that each time one of this patterns appear, one element of our substitutive pairs appear.

Let us consider the two lines above this pattern. Above the 0's will always be 1's because reading 11 implies that the output alphabet is $\{1; 2\}$, and a 2 is never above a 0 by construction. Above the 1's there will be 2's, expect for one 1. We then distinguish three cases depending on the position of the 1 in the block of 2's : the leftmost, somewhere in the middle or the rightmost. The line above this block of two is the result of a division by 3. This operation is deterministic, there are a priori three possibilities for the phase of the carry on the whole line : 0, 1/3, 2/3.

In the middle case (Fig. 9a), the three cases for the phase of the carry are possible. In each of those cases one element of the substitutive pairs appears, either with the bottomleft or the bottomright tile being the apparition of the 1 in the block of 2's.

In the two other cases (leftmost and rightmost) we have two consecutive ones. This prevents the appearance of one of the phases (one can check that block 0011 cannot be continued, drown in red in the pictures).

In the leftmost case (Fig. 9b), only two of the three possibilities for the phase of the carry may appear. In both cases, one element of the substitutive pairs appear above the two leftmost 1's of the first line.

In the rightmost case (Fig. 9c), only two of the three phases of the carry are possible. In both cases, one element of the substitutive pairs appear above the two rightmost 1's of the first line.

The colored vertical bar correspond to the code of bits presented in Fig. 6.

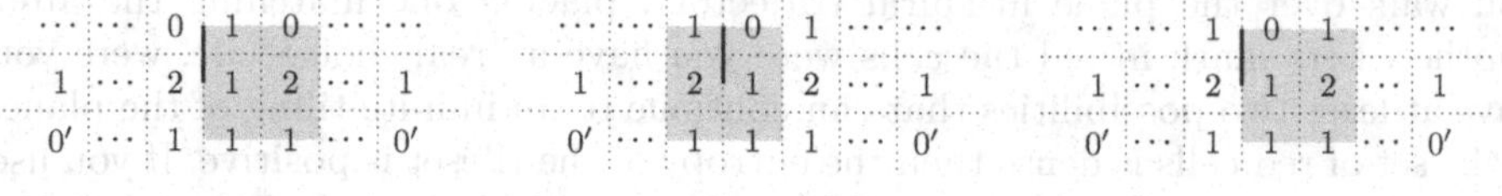

```
·····  0 │ 1   0 ·····         ·····  1 │ 0   1 ·····         ·····  1 │ 0   1 ·····
1 ··· 2 │1   2 ··· 1         1 ··· 2 │1   2 ··· 1         1 ··· 2 │1   2 ··· 1
0' ··· 1 │1   1 ··· 0'        0' ··· 1 │1   1 ··· 0'        0' ··· 1 │1   1 ··· 0'
```

(a) Middle case

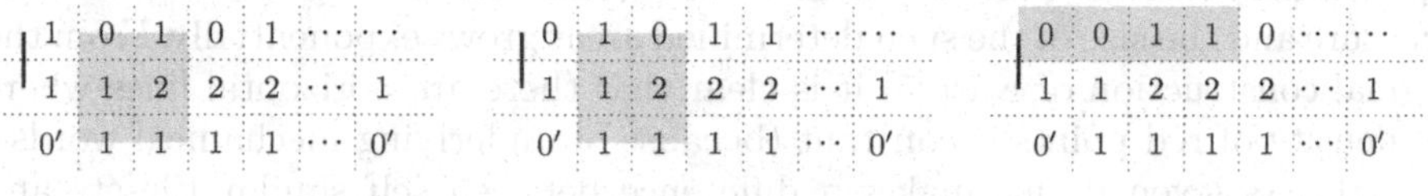

```
│ 1   0   1   0   1 ·····       │ 0   1   0   1   1 ·····       │ 0   0   1   1   0 ·····
│ 1   1   2   2   2 ··· 1       │ 1   1   2   2   2 ··· 1       │ 1   1   2   2   2 ··· 1
0' │ 1   1   1   1 ··· 0'        0' │ 1   1   1   1 ··· 0'        0' │ 1   1   1   1 ··· 0'
```

(b) Leftmost case

```
·····  0 │ 1   0   1         ·····  1 │ 1   0   0         ·····  1 │ 0   1   0
1 ··· 2   2 │2   1   1         1 ··· 2   2 │2   1   1         1 ··· 2   2 │2   1   1
0' ··· 1   1 │1   1   0'        0' ··· 1   1   1   1   0'        0' ··· 1   1 │1   1   0'
```

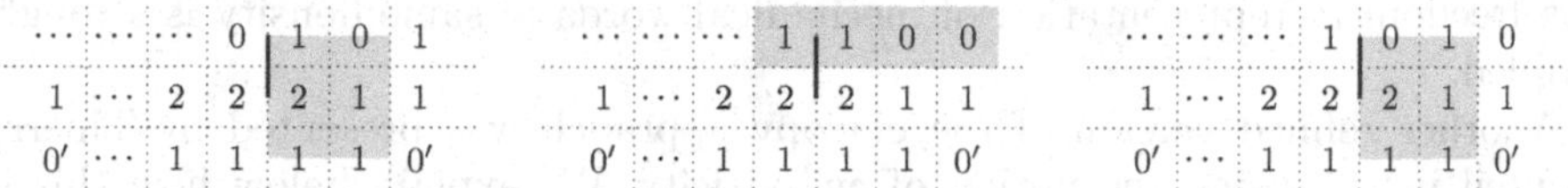

(c) Rightmost case

Fig. 9. Our case analysis

Theorem 1. *The Kari-Culik tileset have positive entropy.*

Proof. The theorem is a consequence of all the other results of this section : we have presented two substitutive pairs that together appear in a dense way in any tiling of the plane.

3.3   Open Problems

We proved that our two pairs are together dense in any tiling. Is one of those pairs dense alone in a given tiling?

Consider the extended tileset where we forbid one pattern in each of the presented pairs. The obtained tileset is still a palette. Has this tileset a positive entropy? If the answer is positive, is it possible to exclude a finite number of patterns so that the resulting tileset has zero entropy?

Using substitutive pairs we proved that there are tilings which horizontal lines do not represent mechanical words. Is it possible to better characterize the Kari-words : the language of the lines that can appear in a tiling ?

Major Open Problem: Does a sub-shift of finite type A exists such that

– A has positive entropy;
– any sub-shift of finite type B included in A, has positive entropy.

This question addresses spaces of dimension at least 2 and any finite alphabet.

4   Choices, "cylindricity" and Entropy

Positive entropy is a very rough method for understanding the quantity of choice that you meet when you effectively construct a tiling of the plane. Imagine that you walk over the plane in spiral trajectory, placing one matching tile after another. Lets mark in red the cells were you have a "real choice" *i.e.* were you have at least two possibilities that can continue to an infinite tiling of the plane. If the set of red cells is dense then the entropy of the tileset is positive. If you use this approach on selfsimilar tilesets as usually constructed, then the red cells are exponentially rare : as soon as you fix a tile on this cell you impose the next level structure and the size of the such determined areas grows exponentially. From the original construction of Kari [5] it is clear that there are horizontal lines where the density of red points is constant (because of underlying mechanical words). Nevertheless, even if this makes a difference between self similar tilesets and Kari's, this does not prove positive entropy, furthermore it was conjectured that this freedom in representation of mechanical words of same density is strongly coupled.

Another refined version of the entropy appraoch was presented by Thierry Monteil in [7] using the notion of cylindricity. We explain below how this is related with our work.

Consider a vertical cylinder of size n. If you can tile this cylinder with a tileset then two of the horizontal rings are identical, thus one can tile a torus which correspond to a periodic tiling of the plane. If the tileset is aperiodic, for each n there exists a maximal vertical size for a portion of the cylinder to be tilable. The smallest growing function greater than this vertical size is called the *cylindricity* function of the tileset.

Consider any self similar tileset (for instance use the generic approach of Nicolas Ollinger in [8] for generating a Wang tileset from a substitution). If one can tile a cylinder of given size, then we can rewrite all the tiles into blocks and thus obtain a larger cylinder with about the same proportions: depending on the proportion between horizontal and vertical factors. The cylindricity function is greater than x^α with α positive.

Remark that the cylindricity is always greater than a logarithm : consider a tiling of the plane and a vertical segment of size n in an horizontal stripe. The minimal distance for seeing twice the same vertical segment is bounded by an exponential in n (because the tileset is finite).

It would be interesting to study this function for more sophisticated tilesets, for instance the most complicated one [9] or the robust to errors version in [10]. In Kari's tiling if you have a periodic configuration then its image after a few steps will have a period three times larger because when we divide by three, consecutive periods assume different carry phases. Note that the ×2 operation cannot diminish the period. Thus if we have a cylinder of length n and height h, then the period of the first line is at most $(n/3)^{\alpha h}$ where α is a constant (some easy technical adjustments are needed to transform this argument into a complete proof).

From this result it was conjectured in [7] that the logarithm of number of patterns of size $n \times n$ was of order n. This would have produced an entropy zero tiling with strictly more choices than for self similar case. But it is not the case. We proved that the bound given by the cylindricity approach is not tight.

Acknowledgments and Related Results. The authors thank Alexander Shen for his help in stating the above results in a clear way.

Since the pre-publication of our paper in arXiv we received some comments from people working in this area :

- Emmanuel Jeandel used a brute-force program to prove that one of our pairs of entropic patterns always appears starting from a line of ones before a fixed number of steps;
- Nathalie Aubrun and Michael Rao worked on substitutive pairs and discovered the same pairs as ours. Their approach could be promising for solving our open problems but is not yet published;
- Uwe Grimm pointed to us an article by N. Nikola, D. Hexner and D. Levine [11]. Contrarily to what they write in their paper they do not provide a proof of positive entropy with only one of our entropic pairs, the proof is supposed to be in a paper of the same authors without title and labeled as "to be published". If their proof is correct then one of our open problem is solved.

References

1. Börger, E., Grädel, E., Gurevich, Y.: Classical Decision problem. Perspectives in Mathematical Logic. Springer (1997)
2. Wang, H.: Dominoes and the ∀∃∀ case of the decision problem. Mathematical Theory of Automata, 23–55 (1963)
3. Berger, R.: The Undecidability of the Domino Problem. PhD thesis. Harvard University (1964)
4. Robinson, R.: Undecidability and Nonperiodicity for Tilings of the Plane. Inventiones Mathematicae 12(3) (1971)
5. Kari, J.: A small aperiodic set of wang tiles. Discrete Mathematics 160, 259–264 (1996)
6. Culik II, K.: An aperiodic set of 13 wang tiles. Discrete Mathematics 160, 245–251 (1996)
7. Monteil, T.: Kari-Culik tile sets are too aperiodic to be substitutive. In: FRAC (2013)
8. Ollinger, N.: Two-by-two substitution systems and the undecidability of the domino problem. In: Beckmann, A., Dimitracopoulos, C., Löwe, B. (eds.) CiE 2008. LNCS, vol. 5028, pp. 476–485. Springer, Heidelberg (2008)
9. Durand, B., Levin, L.A., Shen, A.: Complex tilings. The Journal of Symbolic Logic 73(2), 593–673 (2008)
10. Durand, B., Romashchenko, A.E., Shen, A.: Fixed-point tile sets and their applications. J. Comput. Syst. Sci. 78(3), 731–764 (2012)
11. Nikola, N., Hexner, D., Levine, D.: Entropic commensurate-incommensurate transition. Phys. Rev. Lett. 110, 125701 (2013)

k-Abelian Pattern Matching*

Thorsten Ehlers, Florin Manea, Robert Mercaş, and Dirk Nowotka

Kiel University, Department of Computer Science, D-24098 Kiel, Germany
{the,flm,rgm,dn}@informatik.uni-kiel.de

Abstract. Two words are called k-abelian equivalent, if they share the same multiplicities for all factors of length at most k. We present an optimal linear time algorithm for identifying all occurrences of factors in a text that are k-abelian equivalent to some pattern P. Moreover, an optimal algorithm for finding the largest k for which two words are k-abelian equivalent is given. Solutions for various online versions of the k-abelian pattern matching problem are also proposed.

1 Introduction

The notion of k-abelian equivalence generalises the concepts of both the identity and the abelian equivalence of two words. Two words are called k-abelian equivalent, if they share the same multiplicities for all factors of length at most k. It is straightforward to see that two words of length n are identical if they are n-equivalent and they are abelian equivalent, if they are 1-abelian equivalent.

The notion of k-abelian equivalence was introduced in [1] and since then has captured more and more attention. The concept has been investigated with respect to repetitions [2–4], to periodicity properties [5], as well as to the complexity functions of the equivalence classes it determines [6]. In particular, it has been shown that in most situations, the concept oscillates between the two limit cases given above, identifying itself with one or the other.

Pattern matching is one of the most basic and well studied algorithmic problems: given a text T and a pattern P, we are interested in finding one or all occurrences of P in T. Besides the many obvious applications of pattern matching for finding a specific fragment in a larger sequential data structure, practical applications often emphasise the approximate variants of pattern matching like in the context of computational molecular biology [7]. Approximate pattern matching problems aim to find occurrences of factors of the text T that are equivalent to the pattern P by some given equivalence relation. In this paper, we investigate the approximate pattern matching problem with respect to k-abelian equivalence.

After fixing our notation, we show in Section 2 that the identification of all factors of a text T which are k-abelian equivalent to some pattern P can be done

---

* T. Ehlers is supported by the BMBF grant 01IS110355. F. Manea is supported by the *DFG* grant 596676. R. Mercaş is supported by the *DFG* grant 582014. D. Nowotka is supported by the *DFG Heisenberg* grant 590179.

A.M. Shur and M.V. Volkov (Eds.): DLT 2014, LNCS 8633, pp. 178–190, 2014.

in linear time with respect to the length of the text and the pattern, just as in the special cases of identity and abelian equivalence (Theorem 2). Moreover, we also show that identifying the largest k for which two given words are k-abelian equivalent takes time linear in the length of the words (Theorem 3). In Section 3 we investigate the pattern matching problem for k-abelian equivalences in the setting of online algorithms, and propose a series of real-time solutions of this problem (Theorem 4). Section 4 studies the same problem for an extended form of k-abelian equivalence. Finally, in Section 5 we give experimental results and discuss the problem of building index structures for k-abelian pattern matching.

Preliminaries. An *alphabet*, i.e., Σ, is a finite set of symbols. Let $\sigma = |\Sigma|$ denote its cardinality and take $\Sigma = \{1, \ldots, \sigma\}$; at times we will use the letter $0 \notin \Sigma$. By ε we denote the *empty symbol*. A *word* w is a finite sequence of letters from Σ. We denote by $|w|$ its *length* and by $|w|_u$ the number of occurrences of u in w.

The set of all words over Σ is denoted by Σ^*, while the set of all words of length n is denoted by Σ^n for any positive integer n. The *catenation* of two words u and v is the word uv obtained by adding to the right of u the letters of v. For a factorization $w = uxv$, we say that x is a *factor* of w. Whenever u is empty, x is a *prefix* of w, i.e., $x \leq_p w$, and when v is empty, x is a *suffix* of w. For w of length n, and the numbers $i \leq j \in \{1, \ldots, n\}$, we denote by $w[i]$ the ith symbol of w and by $w[i..j]$ the factor $w[i] \cdots w[j]$; clearly, $w = w[1..n] = w[1] \cdots w[n]$.

Considering the lexicographical order on Σ^*, for words u and v, we say that u is lexicographically smaller than v, i.e., $u \leq_{lex} v$, if either u is a prefix of v or there exist $a, b \in \Sigma$ such that $a < b$, $wa \leq_p u$, and $wb \leq_p v$ for some word w.

Let w be a word over Σ. The Parikh vector of the word w is an array $\pi_w[\cdot]$ with σ components, where $\pi_w[a] = |w|_a$ for all $a \in \Sigma$.

Words u and v are *abelian equivalent* if $|u|_a = |v|_a$ for all letters $a \in \Sigma$. That is, two words are *abelian equivalent* over Σ iff they have the same Parikh vector.

We say that u and v are *k-abelian equivalent*, i.e., $u \equiv_k v$, if either $u = v$ or $|u|, |v| \geq k$, $|u|_t = |v|_t$ for every $t \in \Sigma^k$, $u[1..k-1] = v[1..k-1]$ and $u[n-k+2..n] = v[n-k+2..n]$. According to [6], the suffix-equality requirement can be in fact dropped. An equivalent definition is that $u \equiv_k v$ if $|u|_t = |v|_t$ for every word t of length at most k. A *k-abelian n^{th} power* is a word $u_1 \ldots u_n$, where $u_1, \ldots, u_n$ are pairwise k-abelian equivalent. Obviously, 1-abelian equivalence is the same as abelian equivalence, while equality is equivalent to ∞-abelian equivalence.

Recall that a multi-set represents a set together with the multiplicity of each of the elements (i.e., both elements and multiplicities are present). We say that two words are *extended-k-abelian equivalent* if their multi-sets of factors of length k coincide (the condition of having the same prefixes is dropped).

In this paper, we solve a series of algorithmic problems related to k-abelian equivalence. The algorithms we propose use the RAM with logarithmic word size model. We also assume that whenever we are given as input of our problems a word w of length n, the alphabet of w is in fact included in $\{1, \ldots, n\}$ (i.e., $\sigma = |\Sigma| \leq n$). This is a common assumption in algorithmics on words and can be in fact replaced with a more general assumption, namely that Σ can be sorted in linear time by radix sort (see, e.g., the discussion in [8]). Clearly, for all results proved for integer alphabets our reasoning holds canonically for constant alphabets (i.e., with $\sigma \in \mathcal{O}(1)$), as well. Finally, note that most pattern matching problems that we deal with require searching for a word P inside another word T; generally, P is called pattern and T is called text.

The following result is well known.

Theorem 1. *Let a pattern $P \in \Sigma^m$ and a text $T \in \Sigma^n$, we can identify all factors of T that are abelian equivalent to P in $\mathcal{O}(n + m)$.*

Remark 1. The above result can be adapted to identify all the length $|P|$ factors P' of T that contain the same letters as P (not necessarily with same multiplicity as in P, so not abelian equivalent), and $\sum_{a \in \Sigma} |\pi_P[a] - \pi_{P'}[a]| \leq \Delta$, for some Δ.

We conclude the preliminaries section with a series of data structures.

For a string u of length n, over an alphabet $\Sigma \subseteq \{1, \ldots, n\}$, we define a suffix-array data structure that contains two arrays Suf_u, which is a permutation of $\{1, \ldots, n\}$, and lcp_u, with n elements from $\{0, \ldots, n-1\}$. Called the suffix array of u, Suf_u is defined such that $Suf_u[i] = j$ iff $u[j..n]$ is the ith suffix of u, in the lexicographical order. The following lemma is straightforward.

Lemma 1. *Let $w \in \Sigma^n$. If for $1 \leq i < j \leq n$ and $u \in \Sigma^*$ we have $u \leq_p w[Suf_w[i]..n]$ and $u \leq_p w[Suf_w[j]..n]$, then $u \leq_p w[Suf_w[\ell]..n]$ for any $i \leq \ell \leq j$.*

The array lcp_u is defined by $lcp_u[1] = 1$ and $lcp_u[r]$ is the length of the longest common prefix of the suffixes found on positions r and $r-1$ in the suffix array, i.e., $u[Suf_u[r-1]..n]$ and $u[Suf_u[r]..n]$. Both arrays Suf_u and lcp_u can be constructed in $\mathcal{O}(n)$ time (see [8], and the references therein). Moreover, lcp_u can be processed in $\mathcal{O}(n)$ time to produce a more general data structure that enables us to return in constant time the answer to longest common prefix (or, for short, LCP-) queries "$LCP(i, j)$: What is the length of the longest common prefix of $u[i..n]$ and $u[j..n]$?".

2 Offline k-Abelian Pattern Matching

The first step of our algorithms is to define the k-encoding of a word. For $w \in \Sigma^n$, we define the word $\#(w, k)$ of length $n - k + 1$ as follows:

- let $S = \{w[i+1..i+k] \mid 0 \leq i \leq n - k\}$ be the set of length k factors of w;
- sort S lexicographically and associate with each factor $w[i+1..i+k]$ its rank (position) in the sorted set, i.e., $rank(i + 1)$ for $0 \leq i \leq n - k$;
- let $\#(w, k)[i] = rank(i)$ for $1 \leq i \leq n - k + 1$.

Clearly, $\#(w, k)$ is defined over an alphabet included in $\{1, \ldots, n-k+1\}$. Moreover, w is uniquely defined by the set S and the word $\#(w, k)$.

It is important to note that $\#(w, k)$ can be computed in linear time.

Lemma 2. *Let $w \in \Sigma^n$ with $\sigma \leq n$. We can compute $\#(w, k)$ in $\mathcal{O}(n)$ time.*

Proof. We determine the ranks $rank(i)$ of the factors $w[i+1..i+k]$ in the set $S = \{w[i+1..i+k] \mid 0 \leq i \leq n-k\}$ by identifying in the suffix array of w the contiguous groups of suffixes that share a common prefix of length k, and then assigning to each of these groups (from left to right) consecutive numbers, starting with 1; the suffixes of length less than k are not taken into account. □

The following lemma, although straightforward, is essential to our algorithms.

Lemma 3. *Let $w_1, w_2 \in \Sigma^n$. If $w_1 \equiv_k w_2$ for some integer k, then $w_1[1..k-1] = w_2[1..k-1]$, $w_1[n-k+2..n] = w_2[n-k+2..n]$, and $\#(w_1, k) \equiv_1 \#(w_2, k)$.*

Proof. The equality $\#(w_1, k) \equiv_1 \#(w_2, k)$ follows from the fact that w_1 and w_2 have the same factors of length k, with the same multiplicities. □

If we take $w_1 = 1236$ and $w_2 = 1456$, both over the alphabet $\{1, \ldots, 6\}$, for $k = 1$ we have $w_1[1..k-1] = w_2[1..k-1] = \varepsilon$ and $\#(w_1, 1) = 1234 = \#(w_2, 1)$, but w_1 is not abelian equivalent to w_2. Hence, the converse implication of the lemma does not necessarily hold. In order for the converse to hold as well, we need to check that the two words have the same set of factors of length k.

We can test in linear time the k-abelian equivalence of two words:

Lemma 4. *Let $w_1, w_2 \in \Sigma^n$ and k be an integer with $1 \leq k \leq n$. We can decide whether $w_1 \equiv_k w_2$ in $\mathcal{O}(n)$ time.*

Proof. We construct $w = w_1 0 w_2$, and compute Suf_w. Next, we compute $\#(w, k)$ of length $2n - k + 2$, and set $w_1' = \#(w, k)[1..n - k + 1]$ and $w_2' = \#(w, k)[n + 2..2n - k + 2]$ (the two encodings are done using the ranking of w, and disconsider all letters of $\#(w, k)$ that contain a 0). Hence, w_1' and w_2' contain the same letters if w_1 and w_2 have the same multi-set of factors of length k. Note that w_1' and w_2' are computed in linear time, as they only require the computation of $\#(w, k)$. Finally, we remark that $w_1 \equiv_k w_2$ iff $w_1[1..k-1] = w_2[1..k-1]$ and $w_1' \equiv_1 w_2'$. This last equality can be tested in linear time, by Theorem 1. □

Lemma 4 together with its proof suggests a simple way to transform and solve in linear time the k-abelian pattern matching problem following the classical abelian pattern matching problem, solved in Theorem 1.

Problem 1. Given a text T and some pattern P, over an alphabet Σ, find all factors of T that are k-abelian equivalent to P.

Theorem 2. *Given a text $T \in \Sigma^n$ and some pattern $P \in \Sigma^m$, we can find all factors of T that are k-abelian equivalent to P in time $\mathcal{O}(n + m)$.*

Proof. As in the proof of Lemma 4, we construct the word $w = T0P$, and the encoding $\#(w, k) = w'$. Let $T' = w'[1..n - k + 1]$ and $P' = w'[n + 2..n + m - k + 2]$. Also, build LCP-data structures for $T0P$. Now, for any $i > 0$, a factor $T[i..i+m-1]$ is k-abelian equivalent to P iff $T[i..i+k-2] = P[1..k-1]$ (tested in $\mathcal{O}(1)$ time using LCP-queries), and $T'[i..i+m-1] \equiv_1 P'$.

Therefore, it is enough to find in linear time all the positions i of T' where factors that are abelian equivalent to P' occur, and then check, for each of them, whether $T[i..i+k-2] = P[1..k-1]$. All the positions fulfilling these conditions correspond to positions in T where a factor k-abelian equivalent to P occurs. □

Remark 2. Since Problem 1 is reducible to the classical abelian pattern matching problem, by Remark 1, our algorithm can be adapted to produce in linear time all the factors P' of length $|P|$ of T that contain the same factors of length k as P (not necessarily in the same numbers, so P' is not necessarily k-abelian equivalent to P) such that $\sum_{t \in \Sigma^k} | \, |P'|_t - |P|_t \, | \leq \Delta$, for some Δ.

Using the same reduction, we can extend a result from [9]:

Corollary 1. *For a word $w \in \Sigma^n$ and a positive integer k, we can identify all factors of w that are k-abelian powers in $\Theta((n-k+1)^2)$ time.*

The results shown so far help us answer a related, bit more difficult problem.

Problem 2. For words $u, v \in \Sigma^n$, find the largest integer k such that $u \equiv_k v$.

The immediate approach to solve this problem is to look through all possible k for the largest value such that $u \equiv_k v$. With the search for k implemented as a binary search, this approach takes $\mathcal{O}(n \log n)$ time, using the solution described in Lemma 4. However, this problem can also be solved in linear time.

Theorem 3. *Given two words $u, v \in \Sigma^n$, we can find the greatest positive integer k such that $u \equiv_k v$ in linear time $\mathcal{O}(n)$.*

Proof. As before, we construct $w = u0v$, the Suf_w and LCP_w data structures.

Due to Lemma 1, if there exists a positive integer k such that $u \equiv_k v$, then the suffixes of both u and v that share a common prefix of length at least k are grouped together in Suf_w. Also, if k is maximum such that $u \equiv_k v$, then the suffixes of length at most $k-1$ of u and v coincide, and if we truncate the suffixes of u and v to length k, then we should obtain the same multi-set for each word.

Following this remark, we split the suffix array of w into two separate new arrays: one contains suffixes that correspond to u and the other one the suffixes corresponding to v. Using the two arrays, we compute in linear time the maximum ℓ_1, resp. ℓ_2, such that the suffixes of length $\ell_1 - 1$, resp. prefixes of length $\ell_2 - 1$, of u and v coincide. We take $\ell = \min\{\ell_1, \ell_2\}$.

Further, going simultaneously through the sorted suffixes of u and v we compute the longest common prefix of the ith suffix of u (in lexicographic order) and of the ith suffix of v. Let ℓ' be the minimum value for which there exists i such that the ith suffix of u shares a common prefix of length exactly ℓ' with the ith suffix of v (the suffixes are again counted in lexicographical order), but both suffixes have length at least $\ell' + 1$. The value k we were looking for is $\min\{\ell', \ell\}$.

Therefore, to solve Problem 2, we first compute in linear time the values ℓ and ℓ', and then return the value k we were looking for as $\min\{\ell, \ell'\}$. The whole algorithm takes $\mathcal{O}(n)$ time, clearly, and produces the desired output.   □

3   Real-Time k-Abelian Pattern Matching

So far we discussed static problems, i.e., in Problem 1 both P and T are given at the beginning and we process them both in order to solve the problem. Now consider a different type of problem: we are given $P \in \Sigma^m$ and k, while T is read letter by letter (i.e., in an online manner). We want to preprocess P so we can tell, after each new letter, whether the prefix of T read so far ends with a factor k-abelian equivalent to P. We assume that $\Sigma = \{1, \ldots, \sigma\}$ with $\sigma \in \mathcal{O}(m)$.

Problem 3. Preprocess a pattern P and an integer k such that when given a text T, in letter by letter manner, to answer at each moment, efficiently, whether the prefix of T read so far ends with a factor k-abelian equivalent to P.

In general, an algorithm for such a problem is called online algorithm. For simplicity, when discussing this type of problem, we call the time needed to tell whether the prefix of T ends with a factor k-equivalent to P *query time*, while the time needed to preprocess P is called *preprocessing time*. If a solution has constant query time, then its algorithm is called *real-time*.

First, note that for $k = 1$, the result of Theorem 1 holds for the real-time version of Problem 3, as well.

The solution for $k > 1$ is based on the k-encoding strategy used already in the previous sections. We consider the sets of letters $\{1, \ldots, \ell_1\}$ of $\#(P, k - 1)$ and $\{1, \ldots, \ell_2\}$ of $\#(P, k)$, where $\ell_1 \leq \ell_2 \leq m - k + 2$. Recall that $i \in \{1, \ldots, \ell_1\}$ (resp., $i \in \{1, \ldots, \ell_2\}$) is the rank of a factor of length $k - 1$ (resp., k) of P, in the lexicographically ordered set of all factors of length $k - 1$ (resp., k) of P. Let $f_1[i]$ (resp., $f_2[i]$) be the length $k - 1$ (resp., k) factor of P, whose position in the lexicographically ordered set of factors of length $k - 1$ (resp., k) of P is i.

Further, we compute the list of triples $L = \{(i, a, j) \mid 1 \leq i \leq \ell_1, 1 \leq j \leq \ell_2, a \in \Sigma$, and $f_1[i]a = f_2[j]\}$. The suffixes of P that share the same prefix of length $k - 1$ form a contiguous subarray of the suffix array of P, according to Lemma 1. Moreover, each of these groups can be split into several subgroups, that come one after the other in the suffix array, based on the letter following the common prefix. Accordingly, these subgroups correspond to the groups of suffixes that share a common prefix of length k, ordered lexicographically. Computing the subgroups corresponding to a group of suffixes takes linear time, in the size of the group, thus $\mathcal{O}(m)$, altogether. Therefore, we can compute all elements of L in linear time, as well, and collect them in a linked list, for instance.

Alternatively, one can see L as the set of the (explicit or implicit) edges that connect (explicit or implicit) nodes of depth $k - 1$ to (explicit or implicit) nodes of depth k in the suffix tree constructed for P.

Now, we discuss several ways of implementing L so we can efficiently test whether there is a triple $(i, a, \cdot)$ in L, and, if so, quickly find its third component.

One way is to implement a $m \times \sigma$ table $M[\cdot][\cdot]$, where $M[i][a] = j$ iff $(i, a, j) \in L$. In this case, both operations mentioned above take constant time, while constructing this data structure takes $\mathcal{O}(m\sigma)$ time and space; however, the table M is sparse, so such an implementation of L is not practical.

As each component of the triples of L is a number between 1 and m, and L has at most m elements, we can also use perfect hashing to construct a hash table with satellite information and a dictionary search data structure, useful to do the above mentioned operations in $\mathcal{O}(1)$ time. The construction of these data structures can be done in $\mathcal{O}(m \log \log m)$ time deterministically (see [10, Theorem 1]) or in $\mathcal{O}(m)$ expected time, while the table itself takes $\mathcal{O}(m)$ space. Using these structures the two operations mentioned above take $\mathcal{O}(1)$ time.

Finally, allowing more than $\mathcal{O}(1)$ time for the operations on L, another implementation can be used. For each i we define a data structure in which the pairs (a, j), with $(i, a, j) \in L$, are stored. Assume that these pairs (a, j) are ordered by their first component, i.e., a. Now, each of the operations on L can be seen as a predecessor search among the pairs stored in the data structure associated to i, where the key on which the sorting/searching is done is the first component of the stored pairs. There are at most σ such components, so we can construct a van Emde Boas tree containing these pairs [11]. Since the time needed to construct such a tree is t_i (the number of triples having i on the first position) for each i, the time needed to construct the trees for all i's is $\mathcal{O}(m)$, and they can be stored in $\mathcal{O}(m)$ space. Further, doing predecessor search in each structure takes $\mathcal{O}(\log \log \sigma)$ time per query.

Once L constructed, we compute in $\mathcal{O}(m)$ time, using Suf_P and L, the values $suf[j]$ upper bounded by ℓ_1, such that $suf[j] = i$ iff $f_2[j] = af_1[i]$ for some $a \in \Sigma$.

As a first step for the real-time k-abelian pattern matching problem, using a real-time pattern matching algorithm, e.g., [12], each time we read a new letter of T we report whether $P[1..k-1]$ is a suffix of the prefix of T read so far.

Assume now that before reading the newest symbol of T, denoted a, the longest suffix of the text we already read, was P'. Furthermore, assume that $|P'| \leq |P|$ and its factors of length k are all factors of P. Then, $\#(P', k) = j_1 \ldots j_{m'-k}j_{m'-k+1}$ for some $m' \leq m$. Let $i = suf[j_{m'-k+1}]$, and check whether a triple $(i, a, \cdot)$ is in L.

If so, we return its third component, say j. The suffix of T becomes P'', with $\#(P'', k) = j_1 \ldots j_{m'-k}j_{m'-k+1}j$, if $m' < m$, or $\#(P'', k) = j_2 \ldots j_{m'-k}j_{m'-k+1}j$, otherwise. Using real-time abelian pattern matching we check whether $\#(P'', k)$ is abelian equivalent to $\#(P, k)$. If yes, we can decide in constant time whether the prefixes of length $k-1$ of P and P'' coincide, and, hence whether $P'' \equiv_k P$.

When no $(i, a, \cdot)$ is in L, we restart the procedure above when we find a new occurrence of $P[1..k-1]$, by reading the next letter and taking $i = \#(P, k-1)[1]$.

Therefore, the time we spend for each read letter is upper bounded by the time needed to find a triple in L. In conclusion, we obtain the following result.

Theorem 4. *Given a pattern $P \in \Sigma^m$ for $|\Sigma| = \sigma$, and a positive integer k, the online k-abelian pattern matching problem can be solved in:*

- $\mathcal{O}(m\sigma)$ *preprocessing time,* $\mathcal{O}(m\sigma)$ *space, and* $\mathcal{O}(1)$ *query time.*
- $\mathcal{O}(m \log \log m)$ *preprocessing time,* $\mathcal{O}(m)$ *space, and* $\mathcal{O}(1)$ *query time.*
- $\mathcal{O}(m)$ *expected preprocessing time,* $\mathcal{O}(m)$ *space, and* $\mathcal{O}(1)$ *query time.*
- $\mathcal{O}(m)$ *preprocessing time,* $\mathcal{O}(m)$ *space, and* $\mathcal{O}(\log \log \sigma)$ *query time.*

4   Online Extended-k-Abelian Pattern Matching

In this section, we consider the following more general problem.

Problem 4. Preprocess a pattern P and an integer k such that given a text T, in letter by letter manner, to answer at each moment, efficiently, whether the prefix of T read so far ends with a factor extended-k-abelian equivalent to P.

Although the idea used to solve Problem 3 cannot be directly applied in this setting, the strategy stays the same: for every letter of the text T we read we check if the suffix of length P can be encoded using the letters of $\#(P, k)$, and use a real-time abelian pattern matching algorithm to tell whether this suffix is extended-k-abelian equivalent to P. For this we maintain and update for each new letter read, a succinct representation of the longest factor of P, with length at most k, that is a suffix of the prefix of T read so far.

In addition to the previously mentioned data structures, we now also construct in $\mathcal{O}(m)$ time the suffix tree of P together with the suffix links (see, e.g., [13]). We say that a node of the suffix tree of P corresponds to the factor $P[i..j]$ iff the path from the root to that node is labelled with $P[i..j]$. For the succinct representation of the longest factor of P with length at most k, that is a suffix of the prefix of T read so far, we use the (explicit or implicit) node of the suffix-tree of P that corresponds to this factor, together with its length; when that node is implicit, we store the lowest explicit ancestor of this node.

For simplicity, we store the edges of the suffix tree of P using perfect hashing; we can check in $\mathcal{O}(1)$ time whether a node is the source of an edge labelled with a certain letter, and simultaneously get the respective target node (when it exists). Also, the longest factor of P with length at most k, that is a suffix of the prefix of T read so far is called *the P-suffix of that prefix of T*, or *the current P-suffix*.

Online Algorithm. We first remark that if $P[i..i+\ell-1]$ is the P-suffix of $T[1..j]$, then the P-suffix of $T[1..j+1]$ has length at most $\ell+1$. Therefore, when updating the representation of the current P-suffix, we just have to find the longest suffix of the previous P-suffix that can be extended by the letter $T[j+1]$.

We show first how to compute the P-suffix of $T[1..j+1]$ when the succinct representation defined above of the P-suffix of $T[1..j]$, namely $X = P[i..i+\ell-1]$, is known. If $\ell = k$ we use the approach in the previous section. If $\ell < k$ and $a = T[j+1]$, we first try to extend X with a, and see whether the new string Xa is a factor of P (an edge whose label starts with a leaves the node corresponding to X in the suffix tree of P). If yes, then Xa becomes the current P-suffix and we update the succinct representation of the factor of P according to the node where the aforementioned edge leads. If not, then we try extending the factor $X[2..\ell]$ with a. To compute its corresponding node in the suffix tree, we take the lowest explicit ancestor N_1 of the node N corresponding to X, and follow its suffix link. This takes us to an explicit node N_2 that corresponds to a prefix of $X[2..\ell]$, which is not necessarily the lowest explicit ancestor of the node corresponding to that factor. Thus, we use the letters that labelled the path from N_1 to N (i.e., the remaining suffix of $X[2..\ell]$) to advance in the suffix tree from N_2, until we reach the node N_3 corresponding to $X[2..\ell]$; we also compute in the same time the lowest explicit ancestor of N_3. Then we check again if this node is the source of an edge whose label begins with a. If yes, then we found the current P-suffix;

this is $P[i+1..i+\ell-1]a$, and we, therefore, also have its representation. If not, we repeat the procedure for the factor $X[3..\ell]$, now a suffix of $X[2..\ell]$, and so on.

It is simpler to compute the total time spent executing all the above algorithm, than upper bound each of the steps. Notice that each symbol of T is used only once to go through an edge of the tree. Thus, in total, we make at most $\mathcal{O}(n)$ steps for this action and only constant time for each of the other steps, which are executed at most $\mathcal{O}(n)$ times. Therefore, the overall complexity of maintaining the succinct representation of the longest suffix of T that is a factor of P with length at most k, is $\mathcal{O}(n)$. Using this, together with the approach of the previous section (for $\ell = k$), we get that Problem 4 can be solved in time $\mathcal{O}(n)$ and space $\mathcal{O}(m)$, to which the time and space needed to preprocess the pattern P should be added (either $\mathcal{O}(m \log \log m)$ in a deterministic implementation of the perfect hashing, or $\mathcal{O}(m)$ expected). The space needed remains $\mathcal{O}(m)$.

Other implementations of the set of edges of the suffix tree lead, as previously discussed, to other time complexities: with preprocessing time and space $\mathcal{O}(m\sigma)$ (resp. $\mathcal{O}(m)$) the algorithm runs in $\mathcal{O}(n)$ (resp. $\mathcal{O}(n \log \log \sigma)$) time.

Real-Time Algorithm. For a real-time algorithm, the idea is to report only the factors of T that are extended-k-abelian equivalent to P, thus not update the information after each new read letter, but have it ready whenever a length k factor of P is found. For this we also need the following data structures from [14]:

Lemma 5. *We can preprocess a word $P \in \Sigma^m$ in $\mathcal{O}(m \log k)$ time and linear space $\mathcal{O}(m)$ such that, for each i and j with $j - i \le k$, we can return in constant time the (explicit or implicit) node of the suffix tree of P corresponding to $P[i..j]$.*

Using this we skip the search of the tree for nodes corresponding to the suffixes of the current P-suffix. However, for a constant upper bound on the time needed to perform the update of the representation of the P-suffix that we try to maintain, we still have to deal with the unknown number of suffixes of the current P-suffix of $T[1..j]$, to determine the P-suffix of $T[1..j+1]$, when a new letter is read.

First, we maintain a queue of the letters read from T. In each step enqueue the new letter, and perform two more checks: 1) check whether the factor of P whose representation is stored can be extended by the first element in the queue; if yes, 2) update the P-suffix and its representation accordingly, delete the first letter from the queue, and repeat the previous check with the current P-suffix and its representation (also performing the eventual update of the P-suffix and of the queue); if no, 2') using the $\mathcal{O}(1)$ query in Lemma 5, get the node for the longest proper suffix of the factor of P whose representation we had, and check whether the first letter of the queue extends it.

Since when the current P-suffix has length ℓ and it cannot be extended the number of suffixes of factors of P we have to check until reaching again a P-suffix of length ℓ equals the number of letters read between these two moments, we use the lazy update algorithm described above to update the succinct representation of the current P-suffix, and output that we did not find a factor of T that is extended-k-abelian equivalent to P as long as its length is not k; when we reach a length k factor, the queue is empty and the succinct representation is that of the current P-suffix of the read prefix. Then we proceed just as in the case of the algorithm for Theorem 4, until the length decreases, and repeat the procedure.

Again, the previous discussion on the implementation of the suffix tree applies. Thus, for a real-time algorithm, the preprocessing uses $\mathcal{O}(m)$ space and the time needed is $\mathcal{O}(m(\log k + \log \log m))$ deterministically, or $\mathcal{O}(m \log k)$ expected.

5 Further Remarks

Experiments. We tested the algorithm for Theorem 2 on the E.coli and Vibrio cholera genomes with text sizes $4,638,690$ and $1,109,333$, respectively, for $\Sigma = \{A, C, G, T\}$. Having in mind the DNA sequencing process, one may see our pattern matching problem in the following way: given a template sequence P, we want to find out whether the long sequence T contains other sequences that may produce the same reads (of length k). For exemplification purposes, the pattern P was chosen so that $m = 100$ and $P = T[i..i+m-1]$ for random $i \in \{1, \ldots, n-m+1\}$. The values of k were chosen as multiples of 3, considering the normal length of a codon, with the assumption that removing an entire amino acid does not change the structure of a protein as much as the removal of one of the nucleotides from the translating RNA. We looked, as suggested in Remark 2, for factors P' of T such that the sum, over all factors of length k, of the absolute values of the differences between the number of occurrences of a factor in P' and those of the same factor in P is upper bounded by some Δ. For each values of k and Δ, we ran 100 tests. As P is a factor of T, each test finds at least one match. Fig. 1 reports the average number of matches except for this one occurrence.

Δ	E.coli			Vibrio cholera		
	k=3	k=6	k=9	k=3	k=6	k=9
0	0.04	0.04	0.04	0.01	0	0
2	0.04	0.04	0.04	0.01	0	0
4	0.04	0.04	0.04	0.02	0	0
8	0.06	0.04	0.04	0.03	0	0
16	0.1	0.05	0.04	0.11	0	0
32	0.16	0.05	0.05	0.27	0	0
64	353.2	0.05	0.05	98.26	0	0
128	18576.09	0.04	0.05	4742.9	0.02	0
256	18715.59	7.32	0.05	4747.54	1.45	0

k	E.coli	Vibrio cholera	$\sigma^k + k$
1	5	5	5
2	18	18	18
3	67	67	67
4	260	260	260
5	1,011	1,029	1,029
6	4,102	4,102	4,102
7	16,390	16,389	16,391
8	65,371	65,106	65,544
9	256,559	234,324	262,153

Fig. 1. Experiments with 100 runs each   **Fig. 2.** Alphabet size for $\#(w, k)$

For $k \geq 6$ the number of matches is quite low. Interestingly, the number of matches for $\Delta = 256$ is not too far from the expected number of matches, if these tests were run on random data. In that case, as $m \geq 2k - 2$ and Δ is sufficiently large, the problem is degraded to finding exact matches on two factors of length $k - 1$ each. Hence, the probability for a match is upper bounded by σ^{-2k+2}.

Additionally, we tested the extended version of k-abelian matching (Fig. 3). Here, we ignored the requirement of equal prefixes and suffixes of length $k - 1$.

Please note that for $\Delta = 256$ there is a match on every factor of T. Again, we found only few matches for small values of Δ. A reason for this may be the

Δ	E.coli			Vibrio cholera		
	k=3	k=6	k=9	k=3	k=6	k=9
0	0.07	0.04	0.04	0.02	0	0
2	2.34	2.12	2.12	2.19	2	2
4	4.74	4.2	4.2	4.41	4	4
8	9.38	8.36	8.36	9.09	8.01	8
16	20.44	16.76	16.68	19.38	16.05	16
32	48.06	33.74	33.47	46.05	32.21	32.01
64	78747.81	68.12	67.19	22283	64.98	64.02
128	4594162.28	139.14	135.21	1106909.91	139.68	131.16
256	4638591	4638591	4638591	1108151	1108151	1108151

Fig. 3. Disregarding matches on prefixes and suffixes, 100 runs each

increased number of different length k factors which appear, in close numbers, in the chosen texts. In Fig. 2 we give the number of different ranks that appeared in $\#(w, k)$, together with an upper bound $\sigma^k + k$. This bound is due to the fact that there are at most σ^k factors of length k, and additionally k factors containing 0. The number of occurrences are quite close to the upper bound.

A motivation for the very rare occurrences of some pattern in an arbitrary text comes also from an analytical analysis of the probability of a match in the latter setting; given the dependencies among the letters of $\#(T, k)$, the problem can be seen in terms of a Markov source of order $k - 1$. Another, more loose view of the problem, would be in the form of the string matching over reduced set of patterns problem, when the set of independent length k factors occurring at positions $kj + 1$ in the pattern, for $j \geq 0$, is the one that we look for within a factor of length $|P|$ in the text. However, this second model is not tight as it does not consider the fact that each factor of length k is influenced by the preceding $k - 1$ factors of length k. For more details, we recommend [15, Sect. 7.2–7.3].

Index Structures. A problem worth considering in this context, and that was recently considered in the context of abelian pattern matching [16, 17], is that of building index structures for k-abelian pattern matching. Basically, now we are given a positive integer k and a text T, and we want to preprocess the text such that we can answer quickly queries in which we are given a pattern P and have to report whether T has a factor that is k-abelian to P. Recall our assumption that the alphabet of T and of the query-patterns is integer.

Generally, we can approach the problem as follows. Following the solution of the online pattern matching problem, we consider the sets of letters $\{1, \ldots, \ell_1\}$ of $\#(T, k - 1)$ and $\{1, \ldots, \ell_2\}$ of $\#(T, k)$, where $\ell_1 \leq \ell_2 + 1 \leq n - k + 2$. Let $f_1[i]$ (resp., $f_2[i]$) be the factor of length $k - 1$ (resp., k) of T, whose rank in the lexicographically ordered set of all factors of length $k - 1$ (resp., k) of T is i. We construct Suf_T and the list $L = \{(i, a, j) \mid 1 \leq i \leq \ell_1, 1 \leq j \leq \ell_2, a \in \Sigma$, and $f_1[i]a = f_2[j]\}$, and implement L such that we can test efficiently whether there is a triple $(i, a, \cdot)$ in L, and, if so, to find efficiently its third component.

Assume we use perfect hashing to store the edges of the suffix tree of T and the list L. Then, given a pattern $P \in \Sigma^m$, we find in time $\mathcal{O}(k)$ an occurrence of $P[1..k - 1]$ in T (if none exists, then no factor of T is k-abelian equivalent to P). Next, reading $P[k..m]$ letter by letter, and checking the list L, we can produce $\#(P,k)$ in $\mathcal{O}(m-k)$ time. Hence the problem is reduced to producing an index of $\#(T,k)$, useful to check efficiently whether a factor abelian equivalent to $\#(P,k)$ occurs in $\#(T,k)$. Clearly, the classical abelian matching approach can be slightly adapted so we can check whether the length $k-1$ prefix of P matches the length $k-1$ prefix of the factor identified in T. Again, the preprocessing time depends on the implementation of the suffix tree of T and the list L (see the previous discussions); the query time is obtained by adding up the time needed to locate $P[1..k-1]$ in T, then to compute $\#(P,k)$, and, finally, to answer the abelian pattern matching query for the text $\#(T,k)$ and pattern $\#(P,k)$. Such an abelian pattern matching query is answered in $\mathcal{O}(n-k+1)$ time, for $\#(T,k)$ and $\#(P,k)$ over an integer alphabet. Note that even with Σ constant, $\#(T,k)$ and $\#(P,k)$ are words over large alphabets (the letters of the encodings are words in Σ^k, which is not of constant size when k is not a constant). Unfortunately, not much is known about building indexes for abelian pattern matching on integer alphabets. However, we hope that since the letters of $\#(T,k)$ or $\#(P,k)$ do not occur in an arbitrary order (the alphabet is not a random integer alphabet), we could solve the indexing problem faster than the naive approach.

References

1. Huova, M., Karhumäki, J., Saarela, A., Saari, K.: Local squares, periodicity and finite automata. In: Calude, C.S., Rozenberg, G., Salomaa, A. (eds.) Rainbow of Computer Science. LNCS, vol. 6570, pp. 90–101. Springer, Heidelberg (2011)
2. Huova, M., Karhumäki, J., Saarela, A.: Problems in between words and abelian words: *k*-abelian avoidability. Theor. Comput. Sci. 454, 172–177 (2012)
3. Mercaş, R., Saarela, A.: 3-abelian cubes are avoidable on binary alphabets. In: Béal, M.-P., Carton, O. (eds.) DLT 2013. LNCS, vol. 7907, pp. 374–383. Springer, Heidelberg (2013)
4. Rao, M.: On some generalizations of abelian power avoidability (2013) (preprint)
5. Karhumäki, J., Puzynina, S., Saarela, A.: Fine and Wilf's theorem for *k*-abelian periods. In: Yen, H.-C., Ibarra, O.H. (eds.) DLT 2012. LNCS, vol. 7410, pp. 296–307. Springer, Heidelberg (2012)
6. Karhumäki, J., Saarela, A., Zamboni, L.Q.: On a generalization of abelian equivalence and complexity of infinite words. J. Combin. Theory Ser. A 120(8), 2189–2206 (2013)
7. Gusfield, D.: Algorithms on strings, trees, and sequences: Computer science and computational biology. Cambridge University Press, New York (1997)
8. Kärkkäinen, J., Sanders, P., Burkhardt, S.: Linear work suffix array construction. Journal of the ACM 53, 918–936 (2006)
9. Cummings, L.J., Smyth, W.F.: Weak repetitions in strings. J. Combin. Math. Combin. Comput. 24, 33–48 (1997)
10. Ružić, M.: Constructing efficient dictionaries in close to sorting time. In: Aceto, L., Damgård, I., Goldberg, L.A., Halldórsson, M.M., Ingólfsdóttir, A., Walukiewicz, I. (eds.) ICALP 2008, Part I. LNCS, vol. 5125, pp. 84–95. Springer, Heidelberg (2008)
11. van Emde Boas, P.: Preserving order in a forest in less than logarithmic time. In: SFCS 16, pp. 75–84. IEEE Computer Society (1975)

12. Breslauer, D., Grossi, R., Mignosi, F.: Simple real-time constant-space string matching. In: Giancarlo, R., Manzini, G. (eds.) CPM 2011. LNCS, vol. 6661, pp. 173–183. Springer, Heidelberg (2011)
13. Maaß, M.G.: Computing suffix links for suffix trees and arrays. Inf. Process. Lett. 101(6), 250–254 (2007)
14. Gawrychowski, P., Lewenstein, M., Nicholson, P.K.: Weighted level ancestors in suffix trees (peprint, 2014)
15. Lothaire, M.: Applied Combinatorics on Words. Cambridge University Press, Cambridge (2005)
16. Kociumaka, T., Radoszewski, J., Rytter, W.: Efficient indexes for jumbled pattern matching with constant-sized alphabet. In: Bodlaender, H.L., Italiano, G.F. (eds.) ESA 2013. LNCS, vol. 8125, pp. 625–636. Springer, Heidelberg (2013)
17. Gagie, T., Hermelin, D., Landau, G.M., Weimann, O.: Binary jumbled pattern matching on trees and tree-like structures. In: Bodlaender, H.L., Italiano, G.F. (eds.) ESA 2013. LNCS, vol. 8125, pp. 517–528. Springer, Heidelberg (2013)

On k-Abelian Palindromic Rich and Poor Words

Juhani Karhumäki[1,*] and Svetlana Puzynina[1,2,**]

[1] Department of Mathematics, University of Turku, FI-20014 Turku, Finland
{karhumak,svepuz}@utu.fi
[2] Sobolev Institute of Mathematics, Russia

Abstract. A word is called a *palindrome* if it is equal to its reversal. In the paper we consider a k-abelian modification of this notion. Two words are called *k-abelian equivalent* if they contain the same number of occurrences of each factor of length at most k. We say that a word is a *k-abelian palindrome* if it is k-abelian equivalent to its reversal. A question we deal with is the following: how many distinct palindromes can a word contain? It is well known that a word of length n can contain at most $n + 1$ distinct palindromes as its factors; such words are called *rich*. On the other hand, there exist infinite words containing only finitely many distinct palindromes as their factors; such words are called *poor*. It is easy to see that there are no abelian poor words, and there exist words containing $\Theta(n^2)$ distinct abelian palindromes. We analyze these notions with respect to k-abelian equivalence. We show that in the k-abelian case there exist poor words containing finitely many distinct k-abelian palindromic factors, and there exist rich words containing $\Theta(n^2)$ distinct k-abelian palindromes as their factors. Therefore, for poor words the situation resembles normal words, while for rich words it is similar to the abelian case.

1 Introduction

The palindromicity of words is a widely studied area in formal languages. When a model of a computation is introduced, among the first questions is to ask whether the set of palindromes (or its infinite subset) can be recognized by the model. In other words, can the model identify whether it is irrelevant if words are read from left to right or from right to left? It is folklore that deterministic finite automata cannot do that. On the other hand it is among the simplest tasks for push-down automata, or on-line log-space Turing machines. A slightly different approach is to look at palindromic factors of words. They can be viewed as measuring how much the word is locally independent of the reading direction of a factor. The notion of palindromic complexity was formalized for infinite words in [6], and has been studied extensively ever since.

A problem related to our question of counting palindromes in a word is the problem of counting maximal repetitions in a word of length n, that is, runs in a

---

* Supported in part by the Academy of Finland under grant 257857.
** Supported in part by the Academy of Finland under grant 251371 and by Russian Foundation of Basic Research (grants 12-01-00448, 12-01-00089).

A.M. Shur and M.V. Volkov (Eds.): DLT 2014, LNCS 8633, pp. 191–202, 2014.

word. It was shown in [18] that the maximal number of runs in a word is linear in n. Subsequently, there was much research performed to find the bound [8], which led to a conjecture that the bound should be n. Very close lower and upper bounds have been proved; however, the conjecture still remains open. Not only runs, but also various other questions concerning counting squares in a word have been considered, see, e.g., [13,14,19].

We recall that a word is a *palindrome*, if it is equal to its reversal. It is well known that the maximal number of palindromes a word of length n can contain is equal to $n + 1$, and such words are called *rich* in palindromes [10]. In some papers the same class of words was called *full* words (see, e.g., [2,6]). Lately, there is an extensive number of papers devoted to the study of rich words and their generalizations (see, e.g., [7,12]). This notion can be extended to infinite words: an infinite word is rich if each of its factors is rich. For example, Sturmian words are known to be rich. Note also that Sturmian words can be characterized via palindromic closures [9].

Recently the notion of palindromic poorness has been considered in [5,11]. Namely, an infinite word is called *poor* in palindromes if it contains only finitely many distinct palindromes. In particular, it has been shown that there exist poor words with the set of factors closed under reversal. Besides that, in [11] the authors found the minimal number of palindromes an infinite word satisfying different conditions (uniform recurrence, closed under reversal, etc.) can contain. In a related paper [23] words avoiding reversed subwords were studied.

In this paper the k-abelian version of the notion of a palindrome is studied. Two words are called *abelian equivalent* if they contain the same number of occurrences of each letter, or, equivalently, if they are permutations of each other. In the recent years there is a growing interest in abelian properties of words, as well as modifications of the notion of abelian equivalence [1,4,17,21,23]. One such modification is the notion of k-abelian equivalence: two words are called *k-abelian equivalent* if they contain the same number of occurrences of each factor of length at most k. For $k = 1$, the notion of k-abelian equivalence coincides with the notion of abelian equivalence, and when k is greater than half of the length of the words, k-abelian equivalence means equality. Therefore, the notion of k-abelian equivalence is an intermediate notion between abelian equivalence and equality of words. For more on k-abelian equivalence we refer to [15,16].

In analogy with normal palindromes, we say that a word v is a *k-abelian palindrome* if its reversal is k-abelian equivalent to v. For example, the word $aabaaaabbaa$ is a 3-abelian palindrome. We are interested in the maximal and minimal numbers of k-abelian palindromes a word can contain.

For $k = 1$, clearly, each word is an abelian palindrome, since it is abelian equivalent to its reversal. Therefore, there are no infinite 1-abelian poor words. But for $k > 1$ this no longer holds. We build infinite k-abelian poor words for $k > 1$ and sufficiently large alphabets. In fact, we provide a complete characterization of pairs (k, Σ) for which k-abelian poor words over the alphabet Σ exist.

Since a word of length n contains $\Theta(n^2)$ factors in total, a k-abelian rich word cannot contain more than $\Theta(n^2)$ abelian palindromes. However, it can indeed contain $\Theta(n^2)$ inequivalent abelian palindromes. We show that this extends to k-abelian palindromes when k is small compared to n.

The maximal and minimal numbers of inequivalent palindromes in the case of the equality, the k-abelian equality and the abelian equality are summarized in Table 1 (here C is a constant). We remark that in the minimal case, that is for poor words, infinite words are considered, while in the maximal case, that is in rich words, only finite words are considered. The message of the table is that in the big picture k-abelian equivalence behaves like equality for poor words, while it behaves like abelian equivalence for rich words.

Table 1. Maximal and minimal numbers of palindromes in the case of equality, abelian and k-abelian equivalence

	equality	k-abelian	abelian
poor	C	C	∞
rich	$n + 1$	$\Theta(n^2)$	$\Theta(n^2)$

2   Definitions and Notation

Given a finite non-empty set Σ (called the alphabet), we let Σ^* and Σ^ω, respectively, denote the set of finite words and the set of (right) infinite words over the alphabet Σ. We will always assume $|\Sigma| \geq 2$. A word v is a *factor* (resp., a *prefix*, resp., a *suffix*) of a word w, if there exist words x, y such that $w = xvy$ (resp., $w = vy$, resp., $w = xv$). The set of factors of a finite or infinite word w is denoted by $F(w)$. The prefix and suffix of length k of w are denoted by $\mathrm{pref}_k(w)$ and $\mathrm{suff}_k(w)$, respectively. Given a finite word $u = u_1 u_2 \ldots u_n$ with $n \geq 1$ and $u_i \in \Sigma$, we let $|u| = n$ denote the length of u. The empty word is denoted by ε and we set $|\varepsilon| = 0$. An infinite word is called *recurrent* if each of its factors occurs infinitely often in it. An infinite word w is called *uniformly recurrent* if for each $v \in F(w)$ there exists N such that $v \in F(w_i \cdots w_{i+N})$ for every i. In other words, in a uniformly recurrent word each factor occurs with bounded gaps.

For each $v \in \Sigma^*$, we let $|u|_v$ denote the number of occurrences of the factor v in u. Two words u and v in Σ^* are said to be *abelian equivalent*, denoted $u \sim_{ab} v$, if and only if $|u|_a = |v|_a$ for all $a \in \Sigma$. For example, the words aba and aab are abelian equivalent. Clearly, abelian equivalence is an equivalence relation on Σ^*.

Let k be a positive integer. Two words u and v are *k-abelian equivalent*, denoted by $u \sim_k v$, if $|u|_t = |v|_t$ for every word t of length at most k. This is equivalent to the following conditions:

- $|u|_t = |v|_t$ for every word t of length k,
- $\mathrm{pref}_{k-1}(u) = \mathrm{pref}_{k-1}(v)$ and $\mathrm{suff}_{k-1}(u) = \mathrm{suff}_{k-1}(v)$ (or $u = v$, if $|u| < k - 1$ or $|v| < k - 1$).

For instance, $aabab \sim_2 abaab$, but $aabab \not\sim_2 aaabb$. It is easy to see that k-abelian equivalence implies k'-abelian equivalence for every $k' < k$. In particular, it implies abelian equivalence, that is, 1-abelian equivalence.

For a finite word $v = v_1 \cdots v_n$ we let $v^R = v_n \cdots v_1$ denote its reversal. A word v is a *palindrome* if $v = v^R$. A word is a *k-abelian palindrome* (or briefly k-palindrome) if $v \sim_k v^R$. The empty word ε is considered as a palindrome and k-palindrome.

An infinite word is *k-abelian palindromic poor* (briefly k-poor) if there exists a constant C such that the word contains at most C inequivalent (in the sense of k-abelian equivalence) k-abelian palindromes. Clearly, it makes sense to consider only words having the set of factors closed under reversal, otherwise the example can be built in the obvious way, e.g., one can take $(abc)^\omega$ containing only 4 k-palindromes.

A word of length n is called *k-abelian palindromic rich* (briefly k-rich), if it contains at least $n^2/4k$ inequivalent k-abelian palindromes. Notice that the total number of factors contained in a word of length n is equal to $1 + \frac{n(n+1)}{2}$. Therefore, for a fixed k a k-abelian rich word contains the number of k-palindromes of the same order as the total number of factors when k is small relatively to n.

We emphasize that for poor words we consider infinite words, and for rich words we consider finite ones, and this is caused by the nature of the problem. Indeed, for poor words, since there exist infinite words containing only finitely many palindromes, all their factors have a uniformly bounded number of palindromes. On the other hand, the closed under reversal condition is not applicable to finite poor words, since it would imply a growing number of palindromes. Concerning rich words, an infinite word could easily contain infinitely many palindromes, so we are interested in maximal number of palindromes in finite ones. However, we propose an open problem concerning a modification of k-palindromic richness for infinite words (see Problem 3 in Section 5). In the next two sections we consider k-abelian poor and rich words, respectively.

3 k-Abelian Poor Words

In this section we show that there exist k-abelian palindromic poor words. This holds for almost all values of k and $|\Sigma|$, and we characterize those.

Theorem 3.1. *Let $S = \{(1, l)|l \in \mathbb{N}\} \cup \{(2, 2), (2, 3), (4, 2), (3, 2)\}$.*
I. *For $(k, |\Sigma|) \notin S$ there exist k-abelian palindromic poor words over Σ having a set of factors that is closed under reversal.*
II. *For $(k, |\Sigma|) \in S$ there are no k-abelian palindromic poor words over Σ having a set of factors that is closed under reversal.*

The results can be summarized in Table 2. Here $+$ means that there exist k-abelian poor words having a set of factors that is closed under reversal over an alphabet Σ, and $-$ indicates that there are no such words. In what follows, we will write (k, l)-poor words for k-abelian poor words over an alphabet of cardinality l for brevity.

Table 2. The classification of $(k, |\Sigma|)$ for the existence of k-poor words

$k \backslash |\Sigma|$	2	3	4	...
1	−	−	−	−
2	−	−	+	
3	−	+	+	
4	−	+	+	...
5	+	+	+	
...		...	+	

Proof. First we prove Part I of the theorem by providing constructions of poor words, and then prove the non-existence for Part II of the theorem.

I. We remark that the existence of a (k, l)-poor word implies the existence of a (k', l')-poor word for each $k' \geq k$ and $l' \geq l$. Indeed, for $l' > l$ to build a (k, l')-poor word from a (k, l)-poor word one could split any letter into several letters in any way (i.e., for a chosen letter a, some of occurrences of a are substituted by one of the $l' - l$ new letters). The word remains k-poor, and closed under reversal condition can be preserved. For $k' > k$, the statement follows from the fact that every k'-abelian palindrome is also a k-abelian palindrome for any $k \leq k'$. Therefore, it is enough to build $(5, 2)$-, $(3, 3)$- and $(2, 4)$-poor words. We will provide the construction and a proof for the $(2, 4)$ case. The other cases are similar, so we just outline the constructions.

We construct an infinite recurrent $(2, 4)$-poor word as follows:

$$U_0 = abca\ abda\ acda,$$

$$U_n = U_{n-1}(abca)^{2^{2^n}} (abda)^{2^{2^n}} (acda)^{2^{2^n}} U_{n-1}^R. \tag{1}$$

The required word is obtained as the limit $u = \lim_{n \to \infty} U_n$:

$$u = abca\ abda\ acda(abca)^4(abda)^4(acda)^4adca\ adba\ acba(abca)^{16}(abda)^{16} \ldots$$

The set of factors of this word is closed under reversal since for each prefix the word contains its reversal as a factor by the construction.

To prove that it contains only finitely many 2-abelian palindromes, we will show that each factor of length greater than 12 contains either a unequal number of occurrences of factors bc and cb, or a unequal number of occurrences of factors bd and db, or a unequal number of occurrences of factors cd and dc.

A factor of the form $(abca)^t(abda)^t(acda)^t$ or $(adca)^t(adba)^t(acba)^t$ for $t = 2^{2^i}$ or $t = 1$ is called a *block*. In fact, the word u is a concatenation of blocks. A *subblock* is a factor of the form $(axya)^t$ for $t = 2^{2^i}$ or $t = 1$ and for $(x, y) \in \{(b, c), (c, b), (b, d), (d, b), (c, d), (d, c)\}$. Notice that each factor xy appears only in the corresponding subblock, and its reversal yx appears only only in the reversal of the corresponding subblock. Therefore, we have three pairs of subblocks, where a pair consists of a block and its reversal. We will say that subblocks $(axya)^{2^{2^i}}$ and $(ayxa)^{2^{2^i}}$ are of type (x, y).

Basically, the idea of the construction is based on the fact that we have three types of subblocks, each of them containing a specific factor, such that this factor and its reversal are contained only in subblocks of this type. Each long enough factor will contain a unequal number of occurrences of one of these specific factors and its reversals. Notice that it is important to have three pairs of such factors; two is not enough to guarantee absence of long k-abelian palindromes.

Take a factor v of length $|v| \geq 13$. The following cases are possible. The first case is the principal case when v intersects subblocks of all the three types. The second case is a special case when v intersects at most two different blocks.

Case 1: v intersects subblocks of all the three types. In this case the ends of v can cut at most two subblocks, and hence v contains only full subblocks of at least one type. We let (x, y) denote this type (or one of the types, if there are several of those). The sequence of numbers of occurrences of xy and its reversal yx in full subblocks of v is given by

$$z = 1, 4, -1, 16, 1, -4, -1, 256, 1, 4, -1, -16, 1, -4, -1 \ldots$$

Here positive numbers indicate occurrences of xy, and negative numbers mean the occurrences of yx. In fact, this sequence corresponds to the sequence of exponents of the corresponding block. For a subsequence $v = z(i), \ldots, z(j)$ of z we let $-v$ denote the sequence obtained by changing the sign of all numbers in it: $-v = -z(i), \ldots, -z(j)$. More formally, z is defined recursively as the infinite sequence starting with Z_n for all n:

$$Z_0 = 1,$$
$$Z_n = Z_{n-1}, 2^{2^n}, (-Z_{n-1})^R. \tag{2}$$

If a factor v of u containing only full subblocks of type (x, y) is a palindrome, then the sum of consecutive elements corresponding to the subsequence v in the sequence z is equal to 0. We will prove the following auxiliary claim:

Claim. The sum of consecutive elements in the sequence z is never equal to 0.

Consider any subsequence of consecutive elements $z(i), \ldots, z(i + k)$ of z, and take the element $z(j)$ in it with the largest absolute value, $i \leq j \leq i+k$. We will prove that

$$|z(j)| > \sum_{l=i}^{l=i+k} |z(l)| - |z(j)|.$$

In other words, $|z(j)|$ is greater than the sum of absolute values of all other elements of the subsequence, and hence the sum of all elements in $z(i), \ldots, z(i+k)$ cannot be 0.

Let $|z(j)| = 2^{2^m}$ for some integer m. Clearly, either $z(i), \ldots, z(i + k)$ or $(-z(i), \ldots, -z(i + k))^R$ is a factor of the prefix Z_m of z, including the middle element 2^{2^m} of Z_m. It is enough to prove that

$$22^{m} > 2 \sum_{z(l) \in Z_{m-1}} |z(l)|. \tag{3}$$

By the construction,

$$\sum_{z(l) \in Z_i} |z(l)| = 2^{2^i} + 2 \sum_{z(l) \in Z_{i-1}} |z(l)|. \tag{4}$$

We prove (3) by induction. Straightforward computation shows that it holds for $m = 1$ and 2. Assume it holds for $m = i$. Then, using the induction hypothesis, we obtain

$$2^{2^{i+1}} = 2^{2^i} \cdot 2^{2^i} > 2^{2^i} \cdot 2 \sum_{z(l) \in Z_{i-1}} |z(l)| = 2^{2^i} \cdot \sum_{z(l) \in Z_{i-1}} |z(l)| + 2^{2^i} \cdot \sum_{z(l) \in Z_{i-1}} |z(l)|.$$

For $i \geq 2$ one has $\sum_{z(l) \in Z_{i-1}} |z(l)| \geq 2$, and $2^{2^i} \geq 4$. Applying these two inequalities to the first and the second summands, correspondingly, and then applying (4), we get

$$2^{2^i} \cdot \sum_{z(l) \in Z_{i-1}} |z(l)| + 2^{2^i} \cdot \sum_{z(l) \in Z_{i-1}} |z(l)| \geq 2^{2^i} \cdot 2 + 4 \cdot \sum_{z(l) \in Z_{i-1}} |z(l)| = 2 \sum_{z(l) \in Z_i} |z(l)|.$$

Combining the above inequalities, we obtain

$$2^{2^{i+1}} > 2 \sum_{z(l) \in Z_i} |z(l)|,$$

i.e., we get that (3) holds for $m = i + 1$, and hence we have the induction step. The claim is proved.

Therefore, the factor v contains different numbers of occurrences of xy and yx, and hence is not a 2-palindrome.

Case 2: The factor v intersects subblocks of at most two types.

Case 2.1: The factor v is contained entirely in a block, i.e., we have $v \in F((abca)^{2^{2^i}} (abda)^{2^{2^i}} (acda)^{2^{2^i}})$ or $v \in F((adca)^{2^{2^i}} (adba)^{2^{2^i}} (acba)^{2^{2^i}})$. In this case since $|v| \geq 13$, the word contains at least one of the factors xy for $(x, y) \in \{(b, c), (c, b), (b, d), (d, b), (c, d), (d, c)\}$ and does not contain its reversal, so v is not a 2-palindrome.

Case 2.2: The factor v intersects two blocks and subblocks of at most two types. By construction, every other block in u is of exponent 1; therefore, one of the two blocks is of exponent 1. So, in this case v is a factor of $(axya)^{2^{2^m}} (ax'y'a)$ (or that of its reversal) or a factor of $(ax'y'a)^{2^{2^m}} (axya)^{2^{2^m}} (ayxa)(ay'x'a)$, $m \geq 1$, $x \neq x'$, $y \neq y'$ (or that of its reversal). In the first case since $|v| \geq 13$, the word contains at least one occurrence of the factor xy and does not contains its reversal, so v is not a 2-palindrome. In the second case the same argument works if v contains at least two copies of xy. Otherwise v is a factor of $yaaxyaayxaay'x'a$. Straightforward checking shows that this word does not contain 2-abelian palindromes of length greater or equal to 13.

So, the case $(k, |\Sigma|) = (2, 4)$ is proved.

The proofs in the cases $(k, |\Sigma|) = (3, 3)$ and $(5, 2)$ are similar, so we only provide the constructions.

An infinite recurrent $(3, 3)$-poor word can be constructed as follows:

$$V_0 = bbacc\ aabcc\ bbcaa,$$
$$V_n = V_{n-1}(bbacc)^{2^{2^n}}(aabcc)^{2^{2^n}}(bbcaa)^{2^{2^n}}V_{n-1}^R.$$

The word is given by the limit $v = \lim_{n\to\infty} V_n$:

$$v = bbacc\ aabcc\ bbcaa(bbacc)^4(aabcc)^4(bbcaa)^4aacbb\ ccbaa\ ccabb(bbacc)^{16}\ldots$$

The proof is based on the fact that each sufficiently long factor contains either unequal numbers of occurrences of factors bac and cab, or unequal numbers of occurrences of factors abc and cba, or unequal numbers of occurrences of factors bca and acb, and hence is not a 3-palindrome. In other words, two letter factors of the case $(2, 4)$ are now replaced by suitable three-letter factors over ternary alphabet.

An infinite recurrent $(5, 2)$-poor word can be constructed as follows:

$$W_0 = bbbabaaabbb\ bbbabbaabbb\ bbbabaabbbb,$$
$$W_n = W_{n-1}(bbbabaaabbb)^{2^{2^n}}(bbbabbaabbb)^{2^{2^n}}(bbbabaabbbb)^{2^{2^n}}W_{n-1}^R.$$

The word is given by the limit $w = \lim_{n\to\infty} W_n$. The proof is based on the fact that each sufficiently long factor contains either unequal numbers of occurrences of factors $abaaa$ and $aaaba$, or unequal numbers of occurrences of factors $abbaa$ and $aabba$, or unequal numbers of occurrences of factors $abaab$ and $baaba$, and hence is not a 5-palindrome. Now the specific factors of two previous cases are five-letter binary words.

II. For the proof we split the set S of pairs into two parts with different types of proofs.

Case 1: $(k, |\Sigma|) \in \{(1, l)|l \in \mathbb{N}\} \cup \{(2, 2), (3, 2)\}$. For $k = 1$ (i.e., the abelian equivalence) each word is an abelian palindrome, since every word is abelian equivalent to its reversal. Therefore, all factors of any infinite word are abelian palindromes, and hence there are no abelian palindromic poor words.

In the 2-abelian case, each word starting and ending in the same letter is a 2-abelian palindrome. Indeed, without loss of generality let a word v start and end with a, and let it contains m blocks of b's. Then v contains m occurrences of the factor ab and m occurrences of the factor ba. Factors aa and bb do not affect 2-abelian palindromicity; hence v is a 2-palindrome. Since any infinite binary word contains infinitely many factors starting and ending with the same letter, there are no 2-abelian poor binary infinite words.

In the 3-abelian binary case the proof is similar, just a bit more technical. We omit the details of the proof.

Case 2: $(k, |\Sigma|) \in \{(2,3), (4,2)\}$. We provide a detailed sketch of the proof for the $(2,3)$ case. The idea of the proof in the case $(4,2)$ is similar, although it requires more thorough analysis.

First we introduce rewriting rules which do not affect the 2-palindromicity:

(1) for $x \in \Sigma$, substitute $xx \to x$
(2) for $x, y \in \Sigma$, substitute $xyx \to x$

Claim (i). Let v be ternary word, and let v' be obtained from v by applying a rewriting rule (1) or (2). Then v is a 2-palindrome if and only if v' is a 2-palindrome.

Indeed, after applying the rewriting rule (1), the multiset (the set with multiplicities) of factors of length 2 of v' is obtained from the multiset of factors of length 2 of v by removing one factor xx. Clearly, the resulting set coincides with its reversal if and only if the original set does. After applying the rewriting rule (2), the multiset of factors of length 2 of v' is obtained from the multiset of factors of length 2 of v by removing two factors xy and yx. Again, the resulting set coincides with its reversal if and only if the original set does. The claim follows.

Now take a ternary word v and apply rewriting rules until the word does not contain factors of the form xx and xyx. We call the resulting word the *reduced form* of v. We note that the reduced form of v is unique.

The following claim is straightforward:

Claim (ii). 1. The reduced form of any ternary word v is a factor of $(abc)^\infty$ or $(cba)^\infty$. 2. A ternary word v is a 2-palindrome if and only if its reduced form is a letter.

Now assume that an infinite ternary word w with its set of factors closed under reversal does not contain 2-palindromes of length greater than N for some integer N. Take a factor $v = w_i \cdots w_{i+N}$ of length $N+1$. Since the set of factors of w is closed under reversal, there exists an occurrence of $v^R = w_j \cdots w_{j+N}$. Without loss of generality we can assume that $j > i + N$ and that the reduced form of v is a word u of the form $(abc)^m(\text{pref}(abc))$ for some $m \geq 0$. Then the reduced form of v^R equals u^R. Now consider the factor $w_{i+N+1} \cdots w_{j-1}$; again without loss of generality its reduced form is of the form $(\text{suff}(cba))(cba)^r(\text{pref}(cba))$ for some $r \geq 0$. Consider the factor $w_{i+N+1} \cdots w_{j+N+1}$, which was rewritten to $(\text{suff}(cba))(cba)^r(\text{pref}(cba))(\text{suff}(cba))(cba)^m$. So, there exists s, $i + N + 1 \leq s \leq j + N + 1$, such that the reduced form of $w_{i+N+1} \cdots w_s$ is equal to $u^R = (\text{suff}(cba))(cba)^m$. It is straightforward that $w_i \cdots w_s$ is reduced to a, and hence is a 2-palindrome. The length of this 2-palindrome is greater than N, a contradiction. □

Remark 1. Notice that the examples of k-abelian poor words we build are recurrent, but not uniformly recurrent.

Remark 2. In the construction (1) in fact the powers can be made smaller (although growing), it is convenient for us to use 2^{2^n} in the proofs.

Remark 3. Our constructions are modifications of the so-called sesquipowers, see, e.g., [20, Chapter 4].

4 k-Abelian Rich Words

In this section we show that there exist words of length n which have the number of inequivalent k-abelian palindromic factors of the same order as the total number of their factors $\Theta(n^2)$. In this sense these words contain "many" k-palindromes and hence are considered as rich.

Theorem 4.1. *Let k be a natural number. There exists a positive constant C such that for each $n \geq k$ there exists a word of length n containing at least Cn^2 k-abelian palindromes. Actually, we can choose $C = 1/4k$.*

Proof. The word is defined by

$$v = a^l(ba^{k-1})^m,$$

where l and m are chosen to give maximal number of k-palindromes among words of this type. We let $\lfloor r \rceil$ denote the closest integer to r, we can take $m = \lfloor \frac{n-k+1}{2k} \rceil$. Let us count the number of inequivalent k-palindromes in the word $v = v_1 \cdots v_n$, $n = km + l$. The k-palindromes are the following:

- Starting from position 1, we get the following k-palindromes
  - ε (empty word)
  - $v_1, v_1v_2, \ldots, v_1 \cdots v_l$ (l k-palindromes consisting of only a's)
  - $v_1 \cdots v_{l+k}, v_1 \cdots v_{l+2k}, \ldots, v_1 \cdots v_{l+mk}$ (m k-palindromes starting with a^{k-1}, of length $l + ik$ and containing i letters b, $i = 1, \ldots, m$)
- Starting from each position j, $j = 2, \ldots, l - k + 1$, we get the following new k-palindromes: $v_j \cdots v_{l+k}, v_j \cdots v_{l+2k}, \ldots v_j \cdots v_{l+mk}$ (m k-palindromes starting with a^{k-1}, of length $l-j+ik$ and containing i letters b, $i = 1, \ldots, m$)
- Starting from each position j, $j = l - k + 2, \ldots, l + 1$, we get the following new k-palindromes: $v_j \cdots v_{2l-j+2}, v_j \cdots v_{2l-j+2+k}, \ldots, v_j \cdots v_{2l-j+2+(m-1)k}$ (m k-palindromes starting with a^{l+1-j}, of length $2l - 2j + 3 + (i - 1)k$ and containing i letters b, $i = 1, \ldots, m$)

It is not hard to see that all the above k-palindromes are distinct up to k-abelian equivalence; in fact, they are abelian inequivalent. So, in total we have $(l + 1)(m + 1) = (n - mk + 1)(m + 1)$ distinct k-palindromes. Considering this as a function of m, we get that this function takes a maximal value when $m = \frac{n-k+1}{2k}$. Since all numbers are integer there, the actual maximal number of k-palindromes given by this construction is given by taking the closest integer value, i.e., $m = \lfloor \frac{n-k+1}{2k} \rceil$ (since the function is quadratic in m). Taking these values and taking into account the condition $n \geq k$, we derive that the number of k-palindromes is $(l + 1)(m + 1) \geq n^2/4k$. $\square$

We remark that in the $\Theta(n^2)$ number of k-palindromic factors the constant actually depends on k, so it makes sense when k is small relatively to n.

5   Conclusions and Open Problems

We have considered the numbers of k-abelian palindromes in finite and infinite words. These numbers are always between a constant and a quadratic bound, corresponding to so-called poor and rich words. In the case of poor words to avoid trivialities we always assumed that the words are closed under reversal. Our main result was a construction of infinite words containing only finitely many k-abelian palindromes. This construction could be modified for different pairs (k, l), where k was a constant in k-abelian equivalence and l was the size of the alphabet. For the remaining pairs to show that such an infinite poor word does not exist, we used a different approach, based on rewriting rules preserving k-abelian palindromicity. We also gave an example showing the existence of rich finite words, that is words containing the maximal number of k-abelian palindromes up to a constant multiplicative factor. The bound we found is $n^2/4k$, that is of order Cn^2, where C is a constant independent of n.

A few natural open problems remain. We built recurrent, but not uniformly recurrent k-abelian poor words. The problem is the following:

Problem 1. Does there exist an infinite uniformly recurrent word having the set of factors that is closed under reversal and containing only finitely many k-abelian palindromes?

The second problem asks for optimal constants for rich and poor words.

Problem 2. What is the exact minimal number of k-abelian palindromes an infinite word having a set of factors closed under reversal can contain? What is the exact maximal number of distinct k-abelian palindromes a word of length n can contain?

Some bounds for the constants can be found in this paper, although we did not try to find the best constants.

In the case of equality and classical palindromes the notion of a rich word can be extended to infinite word. The question is whether this is possible for k-abelian palindromes:

Problem 3. Does there exist an infinite k-abelian rich word? More precisely, does there exist an infinite word w, such that for some constant C each of its factors of length n contains at least Cn^2 distinct k-abelian palindromes?

References

1. Adamczewski, B.: Balances for fixed points of primitive substitutions. Theor. Comput. Sci. 307(1), 47–75 (2003)
2. Ambrož, P., Frougny, C., Masáková, Z., Pelantová, E.: Palindromic complexity of infinite words associated with simple Parry numbers. Ann. Inst. Fourier (Grenoble) 56, 2131–2160 (2006)
3. Avgustinovich, S., Karhumäki, J., Puzynina, S.: On abelian versions of Critical Factorization Theorem. RAIRO - Theoret. Inf. Appl. 46, 3–15 (2012)

4. Avgustinovich, S., Puzynina, S.: Weak Abelian Periodicity of Infinite Words. In: Bulatov, A.A., Shur, A.M. (eds.) CSR 2013. LNCS, vol. 7913, pp. 258–270. Springer, Heidelberg (2013)
5. Berstel, J., Boasson, L., Carton, O., Fagnot, I.: Infinite words without palindrome, Arxiv: 0903.2382 (2009)
6. Brlek, S., Hamel, S., Nivat, M., Reutenauer, C.: On The Palindromic Complexity of Infinite Words. Internat. J. Found. Comput. Sci. 15, 293–306 (2004)
7. Bucci, M., De Luca, A., Glen, A., Zamboni, L.Q.: A new characteristic property of rich words. Theoret. Comput. Sci. 410, 2860–2863 (2009)
8. Crochemore, M., Ilie, L., Tinta, L.: The "runs" conjecture. Theor. Comput. Sci. 412(27), 2931–2941 (2011)
9. de Luca, A.: Sturmian words: structures, combinatorics and their arithmetics. Theoret. Comput. Sci. 183, 45–82 (1997)
10. Droubay, X., Justin, J., Pirillo, G.: Episturmian words and some constructions of de Luca and Rauzy. Theoret. Comput. Sci. 255, 539–553 (2001)
11. Fici, G., Zamboni, L.Q.: On the least number of palindromes contained in an infinite word. Theor. Comput. Sci. 481, 1–8 (2013)
12. Glen, A., Justin, J., Widmer, S., Zamboni, L.Q.: Palindromic richness. European J. Combin. 30, 510–531 (2009)
13. Harju, T., Kärki, T., Nowotka, D.: The number of positions starting a square in binary words. Electron. J. Combin. 18(1) (2011)
14. Ilie, L.: A note on the number of squares in a word. Theor. Comput. Sci. 380(3), 373–376 (2007)
15. Karhumäki, J., Puzynina, S., Saarela, A.: Fine and Wilf's Theorem for k-Abelian Periods. Int. J. Found. Comput. Sci. 24(7), 1135–1152 (2013)
16. Karhumäki, J., Saarela, A., Zamboni, L.Q.: On a generalization of Abelian equivalence and complexity of infinite words. J. Comb. Theory, Ser. A 120(8), 2189–2206 (2013)
17. Keränen, V.: Abelian squares are avoidable on 4 letters. In: Kuich, W. (ed.) ICALP 1992. LNCS, vol. 623, pp. 41–52. Springer, Heidelberg (1992)
18. Kolpakov, R., Kucherov, G.: Finding Maximal Repetitions in a Word in Linear Time. In: FOCS 1999, pp. 596–604 (1999)
19. Kucherov, G., Ochem, P., Rao, M.: How Many Square Occurrences Must a Binary Sequence Contain? Electr. J. Comb. 10 (2003)
20. Lothaire, M.: Algebraic Combinatorics on Words. Cambridge University Press, Cambridge (2002)
21. Puzynina, S., Zamboni, L.Q.: Abelian returns in Sturmian words. J. Comb. Theory, Ser. A 120(2), 390–408 (2013)
22. Richomme, G., Saari, K., Zamboni, L.Q.: Abelian Complexity of Minimal Subshifts. J. London Math. Soc. 83, 79–95 (2011)
23. Rigo, M., Salimov, P.: Another Generalization of Abelian Equivalence: Binomial Complexity of Infinite Words. In: Karhumäki, J., Lepistö, A., Zamboni, L. (eds.) WORDS 2013. LNCS, vol. 8079, pp. 217–228. Springer, Heidelberg (2013)

Variations of the Morse-Hedlund Theorem
for k-Abelian Equivalence

Juhani Karhumäki[1,*], Aleksi Saarela[1,*], and Luca Q. Zamboni[1,2,**]

[1] Department of Mathematics and Statistics & FUNDIM, University of Turku,
FI-20014 Turku, Finland
[2] Université de Lyon, Université Lyon 1, CNRS UMR 5208, Institut Camille Jordan,
43 boulevard du 11 novembre 1918, F69622 Villeurbanne Cedex, France
{karhumak,amsaar}@utu.fi, lupastis@gmail.com

Abstract. In this paper we investigate local-to-global phenomena for
a new family of complexity functions of infinite words indexed by $k \in \mathbb{N}_1 \cup \{+\infty\}$ where $\mathbb{N}_1$ denotes the set of positive integers. Two finite words
u and v in A^* are said to be k-abelian equivalent if for all $x \in A^*$ of length
less than or equal to k, the number of occurrences of x in u is equal to
the number of occurrences of x in v. This defines a family of equivalence
relations $\sim_k$ on A^*, bridging the gap between the usual notion of abelian
equivalence (when $k = 1$) and equality (when $k = +\infty$). Given an infinite
word $w \in A^\omega$, we consider the associated complexity function $\mathcal{P}_w^{(k)}$:
$\mathbb{N}_1 \to \mathbb{N}_1$ which counts the number of k-abelian equivalence classes of
factors of w of length n. As a whole, these complexity functions have a
number of common features: Each gives a characterization of periodicity
in the context of bi-infinite words, and each can be used to characterize
Sturmian words in the framework of aperiodic one-sided infinite words.
Nevertheless, they also exhibit a number of striking differences, the study
of which is one of the main topics of our paper.

1 Introduction

A fundamental problem in both mathematics and computer science is to describe
local constraints which imply global regularities. A splendid example of this
phenomena may be found in the framework of combinatorics on words. In their
seminal papers [17, 18], G. A. Hedlund and M. Morse proved that a bi-infinite
word w is periodic if and only if for some positive integer n, the word w contains
at most n distinct factors of length n. In other words, it describes the exact
borderline between periodicity and aperiodicity of words in terms of the *factor
complexity function* which counts the number of distinct factors of each length
n. An analogous result was established some thirty years later by E. Coven and
G. A. Hedlund in the framework of abelian equivalence. They show that a bi-
infinite word is periodic if and only if for some positive integer n all factors of

---

* Partially supported by the Academy of Finland under grant 257857.
** Partially supported by a FiDiPro grant (137991) from the Academy of Finland and
by ANR grant *SUBTILE*.

A.M. Shur and M.V. Volkov (Eds.): DLT 2014, LNCS 8633, pp. 203–214, 2014.
© Springer International Publishing Switzerland 2014

w are abelian equivalent. Thus once again it is possible to distinguish between periodic and aperiodic words on a local level by counting the number of abelian equivalence classes of factors of length n.

In this paper we study the local-to-global behavior for a new family of complexity functions $\mathcal{P}_w^{(k)}$ of infinite words indexed by $k \in \mathbb{N}_1 \cup \{+\infty\}$ where $\mathbb{N}_1 = \{1, 2, 3, \ldots\}$ denotes the set of positive integers. Let $k \in \mathbb{N}_1 \cup \{+\infty\}$ and A be a finite non-empty set. Two finite words $u, v \in A^*$ are said to be k-abelian equivalent if for all $x \in A^*$ of length at most k, the number of occurrences of x in u is equal to the number of occurrences of x in v. This defines a family of equivalence relations $\sim_k$ on A^*, bridging the gap between the usual notion of abelian equivalence (when $k = 1$) and equality (when $k = +\infty$). Abelian equivalence of words has long been a subject of great interest (see, for instance, Erdős's problem, [4–6, 8, 15, 20–23]). Although the notion of k-abelian equivalence is quite new, there are already a number of papers on the topic [10–14, 16].

Given an infinite word $w \in A^\omega$, we consider the associated complexity function $\mathcal{P}_w^{(k)} : \mathbb{N}_1 \to \mathbb{N}_1$ which counts the number of k-abelian equivalence classes of factors of w of length n. Thus $\mathcal{P}_w^{(\infty)}$ corresponds to the usual factor complexity (sometimes called subword complexity in the literature) while $\mathcal{P}_w^{(1)}$ corresponds to abelian complexity. As it turns out, each intermediate complexity function $\mathcal{P}_w^{(k)}$ can be used to detect periodicity of words. As a starting point of our research, we list two classical results on factor and abelian complexity in connection with periodicity, and their k-abelian counterparts proved by the authors in [14]. We note that in each case, the first two items are included in the third.

Theorem 1. *Let w be a bi-infinite word over a finite alphabet. Then the following properties hold:*

- *(M. Morse, G. A. Hedlund, [17]) The word w is periodic if and only if $\mathcal{P}_w^{(\infty)}(n) < n + 1$ for some $n \geq 1$.*
- *(E. M. Coven, G. A. Hedlund, [5]) The word w is periodic if and only if $\mathcal{P}_w^{(1)}(n) < 2$ for some $n \geq 1$.*
- *The word w is periodic if and only if $\mathcal{P}_w^{(k)}(n) < \min\{n + 1, 2k\}$ for some $k \in \mathbb{N}_1 \cup \{+\infty\}$ and $n \geq 1$.*

Also, each complexity provides a characterization for an important class of binary words, the so-called *Sturmian* words:

Theorem 2. *Let w be an aperiodic one-sided infinite word. Then the following properties hold:*

- *(M. Morse, G. A. Hedlund, [18]). The word w is Sturmian if and only if $\mathcal{P}_w^{(\infty)}(n) = n + 1$ for all $n \geq 1$.*
- *(E. M. Coven, G. A. Hedlund, [5]). The word w is Sturmian if and only if $\mathcal{P}_w^{(1)}(n) = 2$ for all $n \geq 1$.*
- *The word w is Sturmian if and only if $\mathcal{P}_w^{(k)}(n) = \min\{n + 1, 2k\}$ for all $k \in \mathbb{N}_1 \cup \{+\infty\}$ and $n \geq 1$.*

However, in other respects, these various complexities exhibit radically different behaviors. For instance, in the context of one-sided infinite words, the first item in Theorem 1 gives rise to a characterization of ultimately periodic words, while for the other two, the result holds in only one direction: If $\mathcal{P}_w^{(k)}(n) < \min\{n + 1, 2k\}$ for some $k \in \mathbb{N}_1$ and $n \geq 1$ then w is ultimately periodic, but not conversely (see [14]). For instance in the simplest case when $k = 1$, it is easy to see that if w is the ultimately periodic word 01^ω, then for each positive integer n there are precisely two abelian classes of factors of w of length n. However, the same is true of the (aperiodic) *infinite Fibonacci word* $w = 0100101001001 \cdots$ defined as the fixed point of the morphism $0 \mapsto 01$, $1 \mapsto 0$. Analogously, in Theorem 2 the first item holds without the added assumption that w be aperiodic, while the other two items do not. Another striking difference between them is in their rate of growth. Consider for instance the binary Champernowne word $\mathcal{C} = 011011100101110111 \cdots$ obtained by concatenating the binary representation of the consecutive natural numbers. Let w denote the morphic image of $\mathcal{C}$ under the Thue-Morse morphism $0 \mapsto 01$, $1 \mapsto 10$. Then while $\mathcal{P}_w^{(\infty)}(n)$ has exponential growth, it can be shown that $\mathcal{P}_w^{(1)}(n) \leq 3$ for all n. Yet another fundamental disparity concerns the difference $\mathcal{P}_w^{(k)}(n+1) - \mathcal{P}_w^{(k)}(n)$. For factor complexity, one always has $\mathcal{P}_w^{(\infty)}(n + 1) - \mathcal{P}_w^{(\infty)}(n) \geq 0$, while for general k this inequality is far from being true.

A primary objective in this paper is to study the asymptotic lower and upper complexities defined by

$$\mathcal{L}_w^{(k)}(n) = \min_{m \geq n} \mathcal{P}_w^{(k)}(m) \quad \text{and} \quad \mathcal{U}_w^{(k)}(n) = \max_{m \leq n} \mathcal{P}_w^{(k)}(m).$$

Surprisingly these quantities can deviate from one another quite drastically. Indeed, one of our main results is to compute these values for the famous Thue-Morse word. We show that the upper limit is logarithmic, while the lower limit is just constant, in fact at most 8 in the case $k = 2$. This is quite unexpected considering the Thue-Morse word is both pure morphic and abelian periodic (of period 2). If we however allow more general words, then we obtain much stronger evidence of the non-existence of gaps in low k-abelian complexity classes. We construct uniformly recurrent infinite words having arbitrarily low upper limit and just constant lower limit. The concept of k-abelian complexity also leads to many interesting open questions. We conclude the paper in Section 6 by mentioning some of these problems.

2   Preliminaries

Let Σ be a finite non-empty set called the *alphabet*. The set of all finite words over Σ is denoted by Σ^* and the set of all (right) infinite words is denoted by Σ^ω. The set of positive integers is denoted by $\mathbb{N}_1$. A function $f : \mathbb{N}_1 \to \mathbb{R}$ is *increasing* if $f(m) \leq f(n)$ for all $m < n$, and *strictly increasing* if $f(m) < f(n)$ for all $m < n$.

Let $w \in \Sigma^\omega$. The word w is *periodic* if there is $u \in \Sigma^*$ such that $w = u^\omega$, and *ultimately periodic* if there are $u, v \in \Sigma^*$ such that $w = vu^\omega$. If w is not ultimately periodic, then it is *aperiodic*. Let $u = a_0 \cdots a_{m-1}$ and $a_0, \ldots, a_{m-1} \in \Sigma$. The *prefix* of length n of u is $\operatorname{pref}_n(u) = a_0 \cdots a_{n-1}$ and the *suffix* of length n of u is $\operatorname{suff}_n(u) = a_{m-n} \cdots a_{m-1}$. If $0 \le i \le m$, then the notation $\operatorname{rfact}_n^i(u) = a_i \cdots a_{i+n-1}$ is used. The length of a word u is denoted by $|u|$ and the number of occurrences of another word x as a factor of u by $|u|_x$. As a trivial boundary case, $|u|_\varepsilon = |u| + 1$. Two words $u, v \in \Sigma^*$ are *abelian equivalent* if $|u|_a = |v|_a$ for all $a \in \Sigma$.

Let $k \in \mathbb{N}_1$. Two words $u, v \in \Sigma^*$ are *k-abelian equivalent* if $|u|_x = |v|_x$ for all words x of length at most k. k-abelian equivalence is denoted by $\sim_k$. If the length of u and v is at least $k - 1$, then $u \sim_k v$ if and only if $|u|_x = |v|_x$ for all words x of length k and $\operatorname{pref}_{k-1}(u) = \operatorname{pref}_{k-1}(v)$ and $\operatorname{suff}_{k-1}(u) = \operatorname{suff}_{k-1}(v)$. This gives an alternative definition for k-abelian equivalence. A proof can be found in [14].

Let $w \in \Sigma^\omega$. The set of factors of w of length n is denoted by $\mathcal{F}_w(n)$. The *factor complexity* of w is the function $\mathcal{P}_w^{(\infty)} : \mathbb{N}_1 \to \mathbb{N}_1$ defined by

$$\mathcal{P}_w^{(\infty)}(n) = \#\mathcal{F}_w(n),$$

where $\#$ is used to denote the cardinality of a set. Let $k \in \mathbb{N}_1$. The *k-abelian complexity* of w is the function $\mathcal{P}_w^{(k)} : \mathbb{N}_1 \to \mathbb{N}_1$ defined by

$$\mathcal{P}_w^{(k)}(n) = \#(\mathcal{F}_w(n)/\sim_k).$$

Factor complexity functions are always increasing, and even strictly increasing for aperiodic words. For k-abelian complexity this is not true. This is why we define *upper k-abelian complexity* $\mathcal{U}_w^{(k)}$ and *lower k-abelian complexity* $\mathcal{L}_w^{(k)}$:

$$\mathcal{U}_w^{(k)}(n) = \max_{m \le n} \mathcal{P}_w^{(k)}(m) \qquad \text{and} \qquad \mathcal{L}_w^{(k)}(n) = \min_{m \ge n} \mathcal{P}_w^{(k)}(m).$$

These two functions can be significantly different. For example, if w is the Thue-Morse word and $k \ge 2$, then $\mathcal{U}_w^{(k)}(n) = \Theta(\log n)$ and $\mathcal{L}_w^{(k)}(n) = \Theta(1)$. This will be proved in Section 4.

The abelian complexity of a binary word $w \in \{0, 1\}^\omega$ can be determined by using the formula (see [22])

$$\mathcal{P}_w^{(1)}(n) = \max\{|u|_1 \mid u \in \mathcal{F}_n(w)\} - \min\{|u|_1 \mid u \in \mathcal{F}_n(w)\} + 1. \qquad (1)$$

For $k \in \mathbb{N}_1 \cup \{\infty\}$, we define $q^{(k)} : \mathbb{N}_1 \to \mathbb{N}_1, q^{(k)}(n) = \min\{n + 1, 2k\}$. The significance of this function is that if w is Sturmian, then $\mathcal{P}_w^{(k)} = q^{(k)}$. This is further discussed in Section 3.

There are large classes of words for which the k-abelian complexities are of the same order for many values of k. This is shown in the next two lemmas. Thus when analyzing the growth rate of the k-abelian complexity of a word, it may be sufficient to analyze the abelian or 2-abelian complexity.

Lemma 3. *Let $w \in \{0,1\}^\omega$ be such that every factor of w of length k contains at most one occurrence of 1. Then $\mathcal{P}_w^{(k)}(n) = \Theta(\mathcal{P}_w^{(1)}(n))$.*

Proof. Two factors of w are k-abelian equivalent if and only if they are abelian equivalent and have the same prefixes and suffixes of length $k - 1$.   □

Lemma 4. *Let $k, m \geq 2$ and let w be a fixed point of an m-uniform morphism h. Let i be such that $m^i \geq k - 1$. Then $\mathcal{P}_w^{(k)}(m^i(n + 1)) = O(\mathcal{P}_w^{(2)}(n))$.*

Proof. Every factor of w of length $m^i(n + 1)$ can be written as $ph^i(u)q$, where u is a factor of w of length n and $|pq| = m^i$. The k-abelian equivalence class of $ph^i(u)q$ is determined by p, q and the 2-abelian equivalence class of u.   □

In particular, Lemma 4 can be applied to the Thue-Morse word to analyze its k-abelian complexity once the behavior of its 2-abelian complexity is known.

It has been shown that there are many words for which the k-abelian and $(k+1)$-abelian complexities are similar, but there are also many words for which they are very different. For example, there are words having bounded k-abelian complexity but linear $(k + 1)$-abelian complexity. These words can even be assumed to be k-abelian periodic, meaning that they are of the form $u_1 u_2 \cdots$, where $u_1, u_2, \ldots$ are k-abelian equivalent. This is shown in the next lemma.

Lemma 5. *For every $k \geq 1$ there is a k-abelian periodic word w such that $\mathcal{P}_w^{(k+1)}(n) = \Theta(n)$.*

Proof. Let $w \in \{0,1\}^\omega$ be a word with linear abelian complexity (e.g., the Champernowne word) and let h be the morphism defined by $h(0) = 0^{k+1}10^{k-1}1$, $h(1) = 0^k 10^k 1$. Then the word $h(w)$ is k-abelian periodic and $\mathcal{P}_{h(w)}^{(k+1)}((2k+2)n) = \Theta(\mathcal{P}_w^{(1)}(n)) = \Theta(n)$. If m is the size of the alphabet, then $\mathcal{P}_{h(w)}^{(k+1)}(n + 1) \leq m\mathcal{P}_{h(w)}^{(k+1)}(n)$ for all n, so the claim follows.   □

3   Minimal k-Abelian Complexities

In this section classes of words with small k-abelian complexity are studied. Some well-known results about factor complexity are compared to results on k-abelian complexity proved in [14]. It should be expected that ultimately periodic words have low complexity, and this is indeed true for k-abelian complexity, although the k-abelian complexity of some ultimately periodic words is higher that the k-abelian complexity of some aperiodic words. For many complexity measures, Sturmian words have the lowest complexity among aperiodic words. This is also true for k-abelian complexity.

We recall the famous theorem of Morse and Hedlund [17] characterizing ultimately periodic words in terms of factor complexity. This theorem can be generalized for k-abelian complexity: If $\mathcal{P}_w^{(k)}(n) < q^{(k)}(n)$ for some n, then w is ultimately periodic, and if w is ultimately periodic, then $\mathcal{P}_w^{(\infty)}(n)$ is bounded. This was proved in [14].

If k is finite, then this generalization does not give a characterization of ultimately periodic words, because the function $q^{(k)}$ is bounded. In fact, it is impossible to characterize ultimately periodic words in terms of k-abelian complexity. For example, the word $0^{2k-1}1^\omega$ has the same k-abelian complexity as every Sturmian word. On the other hand, for every ultimately periodic word w there is a finite k such that $\mathcal{P}_w^{(k)}(n) < q^{(k)}(n)$ for all sufficiently large n.

The theorem of Morse and Hedlund has a couple of immediate consequences. The words w with $\mathcal{P}_w^{(\infty)}(n) = n+1$ for all n are, by definition, Sturmian words. Thus the following classification is obtained:

- w is ultimately periodic $\Leftrightarrow \mathcal{P}_w^{(\infty)}$ is bounded.
- w is Sturmian $\Leftrightarrow \mathcal{P}_w^{(\infty)}(n) = n+1$ for all n.
- w is aperiodic and not Sturmian $\Leftrightarrow \mathcal{P}_w^{(\infty)}(n) \geq n+1$ for all n and $\mathcal{P}_w^{(\infty)}(n) > n+1$ for some n.

This can be generalized for k-abelian complexity if the equivalences are replaced with implications:

- w is ultimately periodic $\Rightarrow \mathcal{P}_w^{(k)}$ is bounded.
- w is Sturmian $\Rightarrow \mathcal{P}_w^{(k)} = q^{(k)}$.
- w is aperiodic and not Sturmian $\Rightarrow \mathcal{P}_w^{(k)}(n) \geq q^{(k)}(n)$ for all n and $\mathcal{P}_w^{(k)}(n) > q^{(k)}(n)$ for some n.

For $k = 1$ this follows from the theorem of Coven and Hedlund [5]. For $k \geq 2$ it follows from a theorem in [14].

The above result means that one similarity between factor complexity and k-abelian complexity is that Sturmian words have the lowest complexity among aperiodic words. Another similarity between them is that ultimately periodic words have bounded complexity, and the largest values can be arbitrarily high: For every n, there is a finite word u having every possible factor of length n. Then $\mathcal{P}_{u^\omega}^{(k)}(n)$ is as high as it can be for any word, i.e., the number of k-abelian equivalence classes of words of length n.

Another direct consequence of the theorem of Morse and Hedlund is that there is a gap between constant complexity and the complexity of Sturmian words. For k-abelian complexity there cannot be a gap between bounded complexities and $q^{(k)}$, because the function $q^{(k)}$ itself is bounded. However, the question whether there is a gap above bounded complexity is more difficult. The answer is that there is no such gap, even if only uniformly recurrent words are considered. This is proved in Section 5.

4 k-Abelian Complexity of the Thue-Morse Word

In this section the k-abelian complexity of the Thue-Morse word is analyzed. Before that, the abelian complexity of a closely related word is determined.

Let σ be the morphism defined by $\sigma(0) = 01, \sigma(1) = 00$. Let $S = 01000101\cdots$ be the *period-doubling word*, which is the fixed point of σ; see, e.g., [7].

The abelian complexity of S is completely determined by the recurrence relations in the next lemma and by the first value $\mathcal{P}_S^{(1)}(1) = 2$. These relations were proved independently in [3]. It is an easy consequence that the abelian complexity of S is 2-regular (2-regular sequences were defined in [2]). The 2-abelian complexity of the Thue-Morse word has been conjectured to be 2-regular [19], and this is proved in [9].

Lemma 6. *For $n \geq 1$,*

$$\mathcal{P}_S^{(1)}(2n) = \mathcal{P}_S^{(1)}(n) \qquad and \qquad \mathcal{P}_S^{(1)}(4n \pm 1) = \mathcal{P}_S^{(1)}(n) + 1.$$

Proof. Let $p_n = \min\{|u|_1 \mid u \in \mathcal{F}_n(S)\}$ and $q_n = \max\{|u|_1 \mid u \in \mathcal{F}_n(S)\}$. Let $\overline{0} = 1$ and $\overline{1} = 0$. For $a \in \{0,1\}$, $\sigma(a) = 0\overline{a}$ and $\sigma^2(a) = 010a$. Because

$$\mathcal{F}_{2n}(S) = \{\sigma(u) \mid u \in \mathcal{F}_n(S)\} \cup \{\overline{a}\sigma(u)0 \mid au \in \mathcal{F}_n(S)\},$$
$$\mathcal{F}_{4n-1}(S) = \{\sigma^2(u)010 \mid u \in \mathcal{F}_{n-1}(S)\} \cup \{10a\sigma^2(u) \mid au \in \mathcal{F}_n(S)\} \cup$$
$$\{0a\sigma^2(u)0 \mid au \in \mathcal{F}_n(S)\} \cup \{a\sigma^2(u)01 \mid au \in \mathcal{F}_n(S)\},$$
$$\mathcal{F}_{4n+1}(S) = \{\sigma^2(u)0 \mid u \in \mathcal{F}_n(S)\} \cup \{10a\sigma^2(u)01 \mid au \in \mathcal{F}_n(S)\} \cup$$
$$\{0a\sigma^2(u)010 \mid au \in \mathcal{F}_n(S)\} \cup \{a\sigma^2(u) \mid au \in \mathcal{F}_{n+1}(S)\}$$

(here a always represents a letter), it can be seen that

$$p_{2n} = n - q_n, \qquad p_{4n-1} = p_n + n - 1, \qquad p_{4n+1} = p_n + n,$$
$$q_{2n} = n - p_n, \qquad q_{4n-1} = q_n + n, \qquad q_{4n+1} = q_n + n + 1.$$

The claim follows because $\mathcal{P}_S^{(1)}(n) = q_n - p_n + 1$ for all n. □

Theorem 7. *For $n \geq 1$ and $m \geq 0$,*

$$\mathcal{P}_S^{(1)}(n) = O(\log n), \quad \mathcal{P}_S^{(1)}((2 \cdot 4^m + 1)/3) = m + 2, \quad \mathcal{P}_S^{(1)}(2^m) = 2.$$

Proof. Follows from Lemma 6 by induction. □

The abelian complexity of S has a logarithmic upper bound and a constant lower bound. These bounds are the best possible increasing bounds.

Corollary 8. $\mathcal{U}_S^{(1)}(n) = \Theta(\log n)$ *and* $\mathcal{L}_S^{(1)}(n) = 2$.

Let τ be the Thue-Morse morphism defined by $\tau(0) = 01, \tau(1) = 10$. Let $T = 0110100110010110\cdots$ be the Thue-Morse word, which is a fixed point of τ. The first values of $\mathcal{P}_T^{(2)}$ are $2, 4, 6, 8, 6, 8, 10, 8, 6, 8, 8, 10, 10, 10, 8, 8, 6, 8, 10, 10$.

The 2-abelian equivalence of factors of T can be determined with the help of the following lemma.

Lemma 9. *Words $u, v \in \{0,1\}^*$ are 2-abelian equivalent if and only if*

$$|u| = |v|, \qquad |u|_{00} = |v|_{00}, \qquad |u|_{11} = |v|_{11} \qquad and \qquad \mathrm{pref}_1(u) = \mathrm{pref}_1(v).$$

Proof. The "only if" direction follows immediately from the alternative definition of 2-abelian equivalence. For the other direction, it follows from the assumptions that $|u|_{01} + |u|_{10} = |v|_{01} + |v|_{10}$. In any word $w \in \{0,1\}^*$, the numbers $|w|_{01}$ and $|w|_{10}$ can differ by at most one. If $|w|_{01} + |w|_{10}$ is even, then $|w|_{01} = |w|_{10}$. If it is odd and $\mathrm{pref}_1(w) = 0$, then $|w|_{01} = |w|_{10} + 1$. If it is odd and $\mathrm{pref}_1(w) = 1$, then $|w|_{01} + 1 = |w|_{10}$. This means that $|u|_{01} = |v|_{01}$ and $|u|_{10} = |v|_{10}$ and u and v are 2-abelian equivalent. ☐

The following lemma states that if u is a factor of T, then the numbers $|u|_{00}$ and $|u|_{11}$ can differ by at most one.

Lemma 10. *In the image of any word under τ, between any two occurrences of* 00 *there is an occurrence of* 11 *and vice versa.*

Proof. 00 can only occur in the middle of $\tau(10)$ and 11 can only occur in the middle of $\tau(01)$. The claim follows because 10's and 01's alternate in all binary words. ☐

Let u be a factor of T. If $|u|$ and $|u|_{00} + |u|_{11}$ are given, then there are at most 4 possibilities for the 2-abelian equivalence class of u. This is stated in a different way in the next lemma. First we define a function ϕ as follows. If $w = a_1 \cdots a_n$, then $\phi(w) = b_1 \cdots b_{n-1}$, where $b_i = 0$ if $a_i a_{i+1} \in \{01, 10\}$ and $b_i = 1$ if $a_i a_{i+1} \in \{00, 11\}$. If $w = a_1 a_2 \cdots$ is an infinite word, then $\phi(w) = b_1 b_2 \cdots$ is defined in an analogous way.

Lemma 11. *Let $u_1, \ldots, u_n$ be factors of T. If $\phi(u_1), \ldots, \phi(u_n)$ are abelian equivalent, then $u_1, \ldots, u_n$ are in at most 4 different 2-abelian equivalence classes.*

Proof. The numbers $|u_i|_{00} + |u_i|_{11} = |\phi(u_i)|_1$ are equal for all i; let this number be m. By Lemma 10, $\{|u_i|_{00}, |u_i|_{11}\} = \{\lfloor m/2 \rfloor, \lceil m/2 \rceil\}$. There are at most four different possible values for the triples $(|u_i|_{00}, |u_i|_{11}, \mathrm{pref}_1(u_i))$. The claim follows from Lemma 9. ☐

Now it can be proved that the 2-abelian complexity of T is of the same order as the abelian complexity of $\phi(T)$. It is known that $\phi(T)$ is actually the period-doubling word S [1].

Lemma 12. *For $n \geq 2$, $\mathcal{P}_S^{(1)}(n-1) \leq \mathcal{P}_T^{(2)}(n) \leq 4\mathcal{P}_S^{(1)}(n-1)$.*

Proof. If the factors of T of length n are $u_1, \ldots, u_m$, then the factors of $\phi(T)$ of length $n-1$ are $\phi(u_1), \ldots, \phi(u_m)$. If u_i and u_j are 2-abelian equivalent, then $\phi(u_i)$ and $\phi(u_j)$ are abelian equivalent, so the first inequality follows. The second inequality follows from Lemma 11 ☐

Theorem 13. *For $n \geq 1$ and $m \geq 0$,*

$$\mathcal{P}_T^{(2)}(n) = O(\log n), \quad \mathcal{P}_T^{(2)}((2 \cdot 4^m + 4)/3) = \Theta(m), \quad \mathcal{P}_T^{(2)}(2^m + 1) \leq 8.$$

Proof. Follows from Lemma 12 and Theorem 7. ☐

With the help of Lemma 4, we see that the k-abelian complexity of T behaves in a similar way as the abelian complexity of S.

Corollary 14. *Let $k \geq 2$. Then $\mathcal{U}_T^{(k)}(n) = \Theta(\log n)$ and $\mathcal{L}_T^{(k)}(n) = \Theta(1)$.*

5   Arbitrarily Slowly Growing k-Abelian Complexities

In this section we study whether there is a gap above bounded k-abelian complexity. This question can be formalized in two ways:

- Does there exist an increasing unbounded function $f : \mathbb{N}_1 \to \mathbb{N}_1$ such that for every infinite word w either $\mathcal{P}_w^{(k)}$ is bounded or $\mathcal{P}_w^{(k)} = \Omega(f)$?
- Does there exist an increasing unbounded function $f : \mathbb{N}_1 \to \mathbb{N}_1$ such that for every infinite word w either $\mathcal{P}_w^{(k)}$ is bounded or $\mathcal{P}_w^{(k)} \neq O(f)$?

The first question has already been answered negatively in Section 4. The answer to the second question is also negative, even if only uniformly recurrent words are considered. A construction proving this is given below.

Let $n_1, n_2, \ldots$ be a sequence of integers greater than 1. Let $m_j = \prod_{i=1}^{j} n_i$ for $j = 0, 1, 2, \ldots$. Let $a_i = 0$ if the greatest j such that $m_j | i$ is even and $a_i = 1$ otherwise. Let $U = a_1 a_2 a_3 \cdots$. The word U could also be described by a Toeplitz-type construction: Start with the word $(0^{n_1-1}\diamond)^\omega$, then replace the $\diamond$'s by the letters of $(1^{n_2-1}\diamond)^\omega$, then replace the remaining $\diamond$'s by the letters of $(0^{n_3-1}\diamond)^\omega$, and keep repeating this procedure so that U is obtained as a limit. It follows from the construction that $U \in (\text{pref}_{m_j-1}(U)\{0,1\})^\omega$ for all j.

Lemma 15. *The word U is uniformly recurrent.*

Proof. For every factor u of U, there is a j such that u is a factor of $\text{pref}_{m_j-1}(U)$. Because $U \in \{\text{pref}_{m_j-1}(U)0, \text{pref}_{m_j-1}(U)1\}^\omega$, every factor of U of length $m_j + |u| - 1$ contains u. $\qquad\square$

Lemma 16. *For every $n \geq 2$, let n' be such that $m_{n'-1} < n \leq m_{n'}$. Then*

$$\mathcal{P}_U^{(1)}(n) \leq n' + 1.$$

For all $J \geq 1$, if $n = 2\sum_{j=1}^{J}(m_{2j} - m_{2j-1})$, then

$$\mathcal{P}_U^{(1)}(n) \geq (n' + 1)/2.$$

For all $j \geq 1$,

$$\mathcal{P}_U^{(1)}(m_j) = 2.$$

Proof. Formula (1) will be used repeatedly in this proof. Another important simple fact is that if a, b, c are integers and c divides a, then $\lfloor (a+b)/c \rfloor = a/c + \lfloor b/c \rfloor$.

For all $n \geq 1$,

$$|\text{pref}_n(U)|_1 = \sum_{i=1}^{\infty}(-1)^{i+1}\left\lfloor \frac{n}{m_i} \right\rfloor,$$

and for all $n \geq 1$ and $l \geq 0$,

$$|\text{rfact}_n^l(U)|_1 = |\text{pref}_{n+l}(U)|_1 - |\text{pref}_l(U)|_1 = \sum_{i=1}^{\infty}(-1)^{i+1}\left(\left\lfloor \frac{n+l}{m_i} \right\rfloor - \left\lfloor \frac{l}{m_i} \right\rfloor\right).$$

For all i,

$$\lfloor (n+l)/m_i \rfloor - \lfloor l/m_i \rfloor \in \{\lfloor n/m_i \rfloor, \lceil n/m_i \rceil\}.$$

Moreover, for every n and l there is an i' such that, for $i \geq n'$,

$$\left\lfloor \frac{n+l}{m_i} \right\rfloor - \left\lfloor \frac{l}{m_i} \right\rfloor = \begin{cases} 1 & \text{if } n' \leq i < i', \\ 0 & \text{if } i \geq i' \end{cases},$$

so

$$\sum_{i=n'}^{\infty} (-1)^{i+1} \left(\left\lfloor \frac{n+l}{m_i} \right\rfloor - \left\lfloor \frac{l}{m_i} \right\rfloor \right) \in \left\{ 0, (-1)^{n'+1} \right\}.$$

Thus there are at most $n'+1$ possible values for $|\mathrm{rfact}_n^l(U)|_1$ and $\mathcal{P}_U^{(1)}(n) \leq n'+1$.

Consider the second claim. Let $n = 2 \sum_{j=1}^{J} (m_{2j} - m_{2j-1})$. The sequence (m_j) is increasing and, moreover, $m_{j+1} \geq 2m_j$ for all j, so by standard estimates for alternating sums,

$$m_{2J} \leq 2(m_{2J} - m_{2J-1}) < n < 2m_{2J} \leq m_{2J+1}.$$

Thus $n' = 2J + 1$. Let $l = m_{2J+1} - n/2$. Then

$$|\mathrm{rfact}_n^l(U)|_1 - |\mathrm{pref}_n(U)|_1 = \sum_{i=1}^{\infty} (-1)^{i+1} \left(\left\lfloor \frac{n+l}{m_i} \right\rfloor - \left\lfloor \frac{l}{m_i} \right\rfloor - \left\lfloor \frac{n}{m_i} \right\rfloor \right)$$

and for $i \leq 2J$ (recall that $m_i | m_j$ when $j \geq i$)

$$\lfloor (n+l)/m_i \rfloor - \lfloor l/m_i \rfloor - \lfloor n/m_i \rfloor$$
$$= \frac{m_{2J+1} + \sum_{(i+1)/2 \leq j \leq J} (m_{2j} - m_{2j-1})}{m_i} + \left\lfloor \frac{\sum_{1 \leq j < (i+1)/2} (m_{2j} - m_{2j-1})}{m_i} \right\rfloor$$
$$- \frac{m_{2J+1} - \sum_{(i+1)/2 \leq j \leq J} (m_{2j} - m_{2j-1})}{m_i} - \left\lfloor \frac{-\sum_{1 \leq j < (i+1)/2} (m_{2j} - m_{2j-1})}{m_i} \right\rfloor$$
$$- \frac{2 \sum_{(i+1)/2 \leq j \leq J} (m_{2j} - m_{2j-1})}{m_i} - \left\lfloor \frac{2 \sum_{1 \leq j < (i+1)/2} (m_{2j} - m_{2j-1})}{m_i} \right\rfloor$$
$$= \lfloor s/m_i \rfloor - \lfloor -s/m_i \rfloor - \lfloor 2s/m_i \rfloor,$$

where $s = \sum_{1 \leq j < (i+1)/2} (m_{2j} - m_{2j-1})$. If i is even, then $m_i/2 \leq s < m_i$, and if i is odd and $i > 1$, then $m_{i-1}/2 \leq s < m_{i-1}$. Thus

$$\left\lfloor \frac{s}{m_i} \right\rfloor - \left\lfloor \frac{s}{m_i} \right\rfloor - \left\lfloor \frac{2s}{m_i} \right\rfloor = \begin{cases} 0 & \text{if } i \text{ is even or } i = 1 \\ 1 & \text{if } i \text{ is odd and } i > 1 \end{cases}$$

and

$$\mathcal{P}_U^{(1)}(n) \geq |\mathrm{rfact}_n^l(U)|_1 - |\mathrm{pref}_n(U)|_1 + 1$$

$$= \sum_{i'=2}^{J} (-1)^{(2i'-1)+1} + \sum_{i=2J+1}^{\infty} (-1)^{i+1} \left(\left\lfloor \frac{n+l}{m_i} \right\rfloor - \left\lfloor \frac{l}{m_i} \right\rfloor - \left\lfloor \frac{n}{m_i} \right\rfloor \right) + 1$$

$$= J + 1 = \frac{n'+1}{2}.$$

Consider the third claim. Because $U \in \{\mathrm{pref}_{m_j-1}(U)0, \mathrm{pref}_{m_j-1}(U)1\}^\omega$, every factor of U of length m_j is abelian equivalent to either $\mathrm{pref}_{m_j-1}(U)0$ or $\mathrm{pref}_{m_j-1}(U)1$. Thus $\mathcal{P}_U^{(1)}(m_j) \leq 2$. Both $\mathrm{pref}_{m_j-1}(U)0$ and $\mathrm{pref}_{m_j-1}(U)1$ are factors of U, so $\mathcal{P}_U^{(1)}(m_j) = 2$. □

If $n_i = 2$ for all i, then the word U is the period-doubling word S. Thus Lemma 16 gives an alternative proof for Corollary 8.

Theorem 17. *For every increasing unbounded function $f : \mathbb{N}_1 \to \mathbb{N}_1$ there is a uniformly recurrent word $w \in \{0,1\}^\omega$ such that $\mathcal{P}_w^{(k)}(n) = O(f(n))$ but $\mathcal{P}_w^{(k)}(n)$ is not bounded.*

Proof. Follows from Lemmas 3, 15 and 16. □

6   Conclusion

In this paper we have investigated some generalizations of the results of Morse and Hedlund and those of Coven and Hedlund for k-abelian complexity. We have pointed out many similarities but also many differences. We have studied the k-abelian complexity of the Thue-Morse word and proved that there are uniformly recurrent words with arbitrarily slowly growing k-abelian complexities.

There are many open questions and possible directions for future work. One open problem related to Lemma 5 is to determine the maximal $(k+1)$-abelian complexity of a k-abelian periodic word. Another interesting topic would be the k-abelian complexities of morphic words. For example, for a morphic (or pure morphic) word w, how slowly can $\mathcal{U}_w^{(k)}(n)$ grow without being bounded? Can it grow slower than logarithmically? More generally, can the possible k-abelian complexities of some subclass of morphic words be classified?

References

1. Allouche, J.P., Arnold, A., Berstel, J., Brlek, S., Jockusch, W., Plouffe, S., Sagan, B.E.: A relative of the Thue-Morse sequence. Discrete Math. 139(1-3), 455–461 (1995)
2. Allouche, J.P., Shallit, J.: The ring of k-regular sequences. Theoret. Comput. Sci. 98(2), 163–197 (1992)

3. Blanchet-Sadri, F., Currie, J., Rampersad, N., Fox, N.: Abelian complexity of fixed point of morphism 0 ↦ 012, 1 ↦ 02, 2 ↦ 1. Integers 14, A11 (2014)
4. Cassaigne, J., Richomme, G., Saari, K., Zamboni, L.Q.: Avoiding Abelian powers in binary words with bounded Abelian complexity. Internat. J. Found. Comput. Sci. 22(4), 905–920 (2011)
5. Coven, E.M., Hedlund, G.A.: Sequences with minimal block growth. Math. Systems Theory 7, 138–153 (1973)
6. Currie, J., Rampersad, N.: Recurrent words with constant Abelian complexity. Adv. in Appl. Math. 47(1), 116–124 (2011)
7. Damanik, D.: Local symmetries in the period-doubling sequence. Discrete Appl. Math. 100(1-2), 115–121 (2000)
8. Dekking, M.: Strongly nonrepetitive sequences and progression-free sets. J. Combin. Theory Ser. A 27(2), 181–185 (1979)
9. Greinecker, F.: On the 2-abelian complexity of Thue-Morse subwords (Preprint), arXiv:1404.3906
10. Huova, M., Karhumäki, J.: Observations and problems on k-abelian avoidability. In: Combinatorial and Algorithmic Aspects of Sequence Processing (Dagstuhl Seminar 11081), pp. 2215–2219 (2011)
11. Huova, M., Karhumäki, J., Saarela, A.: Problems in between words and abelian words: k-abelian avoidability. Theoret. Comput. Sci. 454, 172–177 (2012)
12. Huova, M., Karhumäki, J., Saarela, A., Saari, K.: Local squares, periodicity and finite automata. In: Calude, C.S., Rozenberg, G., Salomaa, A. (eds.) Rainbow of Computer Science. LNCS, vol. 6570, pp. 90–101. Springer, Heidelberg (2011)
13. Karhumäki, J., Puzynina, S., Saarela, A.: Fine and Wilf's theorem for k-abelian periods. Internat. J. Found. Comput. Sci. 24(7), 1135–1152 (2013)
14. Karhumäki, J., Saarela, A., Zamboni, L.Q.: On a generalization of Abelian equivalence and complexity of infinite words. J. Combin. Theory Ser. A 120(8), 2189–2206 (2013)
15. Keränen, V.: Abelian squares are avoidable on 4 letters. In: Kuich, W. (ed.) ICALP 1992. LNCS, vol. 623, pp. 41–52. Springer, Heidelberg (1992)
16. Mercaş, R., Saarela, A.: 3-abelian cubes are avoidable on binary alphabets. In: Béal, M.-P., Carton, O. (eds.) DLT 2013. LNCS, vol. 7907, pp. 374–383. Springer, Heidelberg (2013)
17. Morse, M., Hedlund, G.A.: Symbolic dynamics. Amer. J. Math. 60(4), 815–866 (1938)
18. Morse, M., Hedlund, G.A.: Symbolic dynamics II: Sturmian trajectories. Amer. J. Math. 62(1), 1–42 (1940)
19. Parreau, A., Rigo, M., Vandomme, E.: A conjecture on the 2-abelian complexity of the Thue-Morse word. In: Representing Streams II (2014)
20. Puzynina, S., Zamboni, L.Q.: Abelian returns in Sturmian words. J. Combin. Theory Ser. A 120(2), 390–408 (2013)
21. Richomme, G., Saari, K., Zamboni, L.Q.: Balance and Abelian complexity of the Tribonacci word. Adv. in Appl. Math. 45(2), 212–231 (2010)
22. Richomme, G., Saari, K., Zamboni, L.Q.: Abelian complexity of minimal subshifts. J. Lond. Math. Soc (2) 83(1), 79–95 (2011)
23. Saarela, A.: Ultimately constant abelian complexity of infinite words. J. Autom. Lang. Comb. 14(3-4), 255–258 (2009)

Maximum Number of Distinct and Nonequivalent Nonstandard Squares in a Word*

Tomasz Kociumaka[1,**], Jakub Radoszewski[1,***],
Wojciech Rytter[1,2], and Tomasz Waleń[1,†]

[1] Faculty of Mathematics, Informatics and Mechanics,
University of Warsaw, Warsaw, Poland
{kociumaka,jrad,rytter,walen}@mimuw.edu.pl
[2] Faculty of Mathematics and Computer Science,
Copernicus University, Toruń, Poland

Abstract. The combinatorics of squares in a word depends on how the equivalence of halves of the square is defined. We consider Abelian squares, parameterized squares and order-preserving squares. The word uv is an Abelian (parameterized, order-preserving) square if u and v are equivalent in the Abelian (parameterized, order-preserving) sense. The maximum number of ordinary squares is known to be asymptotically linear, but the exact bound is still investigated. We present several results on the maximum number of distinct squares for nonstandard subword equivalence relations. Let $SQ_{\mathrm{Abel}}(n, k)$ and $SQ'_{\mathrm{Abel}}(n, k)$ denote the maximum number of Abelian squares in a word of length n over an alphabet of size k, which are distinct as words and which are nonequivalent in the Abelian sense, respectively. We prove that $SQ_{\mathrm{Abel}}(n, 2) = \Theta(n^2)$ and $SQ'_{\mathrm{Abel}}(n, 2) = \Omega(n^{1.5}/\log n)$. We also give linear bounds for parameterized and order-preserving squares for small alphabets: $SQ_{\mathrm{param}}(n, 2) = \Theta(n)$ and $SQ_{\mathrm{op}}(n, O(1)) = \Theta(n)$. As a side result we construct infinite words over the smallest alphabet which avoid nontrivial order-preserving squares and nontrivial parameterized cubes (nontrivial parameterized squares cannot be avoided in an infinite word).

1 Introduction

Repetitions in words are a fundamental topic in combinatorics on words [2]. They are widely used in many fields, such as pattern matching, automata theory, formal language theory, data compression, molecular biology, etc. Squares,

---

* The authors thank anonymous referees for many helpful comments.
** The author is supported by Polish budget funds for science in 2013-2017 as a research project under the 'Diamond Grant' program.
*** The author receives financial support of Foundation for Polish Science.
† The author is supported by Iuventus Plus grant (IP2011 058671) of the Polish Ministry of Science and Higher Education. This work is partially supported by ERC grant PAAl no. 259515.

A.M. Shur and M.V. Volkov (Eds.): DLT 2014, LNCS 8633, pp. 215–226, 2014.
© Springer International Publishing Switzerland 2014

that is, words of the form uu, are one of the most commonly studied types of repetitions. An example of an infinite square-free word over a ternary alphabet, given by Thue [24], is considered to be the foundation of combinatorics on words.

If we allow other equivalence relations on words, several generalizations of the notion of square can be obtained. One such generalization are Abelian squares, that is, words of the form uv where the multisets of symbols of u and v are the same. Abelian squares were first studied by Erdős [10], who posed a question on the smallest alphabet size for which there exists an infinite Abelian-square-free word. The first example of such a word over a finite alphabet was given by Evdokimov [11], later the alphabet size was improved to five by Pleasants [23] and finally an optimal example over four-letter alphabet was shown by Keränen [20].

In this paper we consider Abelian squares and introduce squares based on two other known equivalence relations on words. The first is parameterized equivalence [1], in which two words u, v of length n over alphabets $\mathrm{Alph}(u)$ and $\mathrm{Alph}(v)$ are considered equal if one can find a bijection $f : \mathrm{Alph}(u) \to \mathrm{Alph}(v)$ such that $v[i] = f(u[i])$ for all $i = 1, \ldots, n$. The second model, order-preserving equivalence [6], assumes that the alphabets are ordered. Two words u, v of the same length are considered equivalent in this model if they are equal in the parameterized sense with f being an strictly increasing bijection. We define a parameterized square and an order-preserving square as a concatenation of two words that are equivalent in the parameterized and in the order-preserving sense, respectively. Another recently studied model, lying in between standard equality and Abelian equivalence, is k-Abelian equivalence [17]. However, we do not consider this model here. The nonstandard types of squares can be viewed as a part of nonstandard stringology; see [21,22].

Example 1. Consider the alphabet $\Sigma = \{1, 2, 3, 4\}$ with the natural order. Then $1213\,1213$ is a square, $1213\,3112$ is an Abelian square, $1213\,4142$ is a parameterized square, and $1213\,1314$ is an order-preserving square over Σ.

An important combinatorial fact about ordinary squares is that the maximum number of distinct squares in a word of length n is linear in terms of n. Actually this number is smaller than $2n - \Theta(\log n)$ [14,18,19]. This bound has found applications in several text algorithms [5] including two different linear-time algorithms computing all distinct squares [15,8]. A recent result shows that the maximum number of distinct squares in a labeled tree is asymptotically $\Theta(n^{4/3})$ [7]. Also some facts about counting distinct squares in partial words are known [3,4]. In this paper we attempt the same type of combinatorial analysis for nonstandard squares. In turns out that the results that we obtain depend heavily on which squares we consider distinct.

Let $SQ_{\mathrm{Abel}}(n, k)$, $SQ_{\mathrm{param}}(n, k)$ and $SQ_{\mathrm{op}}(n, k)$ denote respectively the maximum number of Abelian, parameterized and order-preserving squares in a word of length n over an alphabet of size k which are *distinct* as words. Moreover let $SQ'_{\mathrm{Abel}}(n, k)$, $SQ'_{\mathrm{param}}(n, k)$ and $SQ'_{\mathrm{op}}(n, k)$ denote the maximum number of Abelian, parameterized and order-preserving squares in a word of length n over an alphabet of size k which are *nonequivalent* in the Abelian, parameterized and order-preserving sense, respectively. We also use analogous notation, e.g., $SQ_{\text{Abel}}(w)$, $SQ'_{\text{Abel}}(w)$, for any word w. Our main results are the following:

- $SQ_{\text{Abel}}(n,2) = \Theta(n^2)$, $SQ'_{\text{Abel}}(n,2) = \Omega(n^{1.5}/\log n)$;
- $SQ_{\text{op}}(n,k) = \Theta(n)$ and therefore $SQ'_{\text{op}}(n,k) = \Theta(n)$ for $k = O(1)$;
- $SQ_{\text{param}}(n,2) = \Theta(n)$ and therefore $SQ'_{\text{param}}(n,2) = \Theta(n)$.

Example 2. Consider a Fibonacci word $Fib_5 = 0100101001001$.[1] It contains 5 Abelian squares of length 6: $010\,010$, $001\,010$, $010\,100$, $100\,100$, and $001\,001$, which are all distinct as words but are Abelian-equivalent. In total Fib_5 contains 13 distinct subwords which are Abelian squares. Hence, $SQ_{\text{Abel}}(Fib_5) = 13$. On the other hand, Fib_5 contains only 5 Abelian-nonequivalent squares, with sample representatives: $0\,0$, $01\,01$, $001\,010$, $10010\,10010$, and $010010\,100100$. Hence, $SQ'_{\text{Abel}}(Fib_5) = 5$. The value SQ' is usually much smaller than SQ, e.g., for Fib_{14} of length 987, $SQ'_{\text{Abel}}(Fib_{14}) = 490$ and $SQ_{\text{Abel}}(Fib_{14}) = 57796$. In general one can show that $SQ'_{\text{Abel}}(Fib_k) = O(|Fib_k|)$. Abelian repetitions in Fibonacci words and Sturmian words were already studied in [13].

The second part of our paper can be viewed as an extension of the works of Thue [24], Evdokimov [11], Pleasants [23] and Keränen [20] on infinite square-free and Abelian-square-free words into the parameterized and order-preserving equivalence. As no square-free word of length larger than 1 exists, we consider words avoiding *nontrivial* nonstandard squares of length larger than 2. We present an infinite word over the minimum-size (ternary) alphabet avoiding nontrivial order-preserving squares. We also prove that there is no infinite word avoiding nontrivial parameterized squares, but there is one avoiding nontrivial parameterized cubes, that is, parameterized cubes of length greater than 3.

2   Bounds for Abelian Squares

For a word $w = w[1]\cdots w[n]$ we denote $|w| = n$. A *subword* of w is a word of the form $w[i]\cdots w[j]$ for $1 \le i \le j \le |w|$, which we denote by $w[i..j]$. A word is said to be *uniform* if all its letters are equal. A *block* (also known as a *run*) in a word is a maximal uniform subword.

In this section we restrict ourselves to the binary alphabet. First, we show a simple example which yields $SQ_{\text{Abel}}(n,2) = \Theta(n^2)$. Afterwards we attempt an analysis of $SQ'_{\text{Abel}}(n,2)$. Our main result is a lower bound of $\Omega(n^{1.5}/\log n)$. We also obtain an upper bound $O(nm)$ if the number of blocks is bounded by m.

A different proof of the following theorem was given independently by Fici [12].

Theorem 1. $SQ_{\text{Abel}}(n,2) = \Theta(n^2)$.

Proof. Consider the word $u_k = 0^k 10^k 10^{2k}$ of length $4k + 2$. It contains $\Theta(k^2)$ Abelian squares of the form $0^a 10^b\,0^{k-b}10^{a+2b-k}$ for all $0 \le a, b \le k$ and $a + 2b \ge k$. Thus we obtain $SQ_{\text{Abel}}(n,2) = \Theta(n^2)$ for $n = 4k + 2$. If $n \bmod 4 \ne 2$, we pick the longest word u_k such that $|u_k| \le n$ and extend it with $n - |u_k| \le 3$ zeros. $\square$

---

[1] Fibonacci words are defined as: $Fib_0 = 0$, $Fib_1 = 01$, $Fib_k = Fib_{k-1}Fib_{k-2}$ for $k \ge 2$.

2.1 Lower Bound for $SQ'_{Abel}(n, 2)$

For a word w and a letter c we denote the number of occurrences of c in w by $|w|_c$. The Parikh vector of a binary word w is $\mathcal{P}(w) = (|w|_0, |w|_1)$.

We say that (p, q) is *a square vector* in w if there exists an Abelian square $u_1 u_2$ in w such that $\mathcal{P}(u_1) = \mathcal{P}(u_2) = (p, q)$. Then $u_1 u_2$ is called a (p, q)-square. Let $SQV(w)$ denote the set of square vectors of w. Now $SQ'_{Abel}(n, 2)$ is the maximum number of different square vectors in a binary word of length n.

In the proof of the lower bound we require some number-theoretic tools. Erdős [9] investigated the problem of estimating the numbers:

$$P_k = |\{i \cdot j : 1 \leq i, j \leq k\}|.$$

It is known that $P_k = \Omega(k^2 / \log k)$. Our auxiliary problem is similar, but instead of the ordinary multiplication $i \cdot j$ we consider an operation

$$i \otimes j \;=\; \sum_{t=i}^{j} t = (i + j)(j - i + 1)/2.$$

We define

$$\mathrm{Sums}(a, b) = |\{i \otimes j \;:\; a \leq i \leq j \leq b\}|.$$

Example 3. $\mathrm{Sums}(2, 5) \;=\; \{2, 3, 4, 5, 7, 9, 12, 14\}$.

Lemma 1. $\mathrm{Sums}(\lceil \frac{3}{4}k \rceil, k) \;=\; \Omega(k^2 / \log k)$.

Proof. We use the following textbook fact:

Fact 1 ([16]). *Let $\pi(x)$ be the number of prime numbers in the range $[1..x]$. For any $\varepsilon > 0$ we have $\pi((1 + \varepsilon)x) - \pi(x) = \frac{\varepsilon x}{\log x} + o(\frac{x}{\log x})$.*

Let I_k denote the set of primes in the interval $[\lceil \frac{10}{12}k \rceil, \lfloor \frac{11}{12}k \rfloor]$ (this interval is a *middle third* of $[\lceil \frac{3}{4}k \rceil, k]$). Let

$$F_k = \{(i, j) \;:\; 0 \leq j - i < \lfloor \tfrac{k}{12} \rfloor - 1 \text{ and } \tfrac{i+j}{2} \in I_k\}.$$

Fact 1 implies that $|I_k| = \Omega(k/ \log k)$, and consequently $|F_k| = \Omega(k^2 / \log k)$. Note that $\{i \otimes j : (i, j) \in F_k\} \subseteq \mathrm{Sums}(\lceil \frac{3}{4}k \rceil, k)$. Therefore it suffices to prove the following:

Claim. If $(i, j), (i', j') \in F_k$, $(i, j) \neq (i', j')$, then $i \otimes j \neq i' \otimes j'$.

However $i \otimes j = p \cdot (j - i + 1)$, $i' \otimes j' = p' \cdot (j' - i' + 1)$, for $p, p' \in I_k$. The claim follows from the primality of p, p' and the inequalities $j - i + 1, j' - i' + 1 \leq \min(p, p')$. $\square$

In our construction a crucial role is played by *balanced* Abelian squares and *balanced* square vectors. A square vector (p, q) is called balanced if $p = q$, and a word w is called balanced if its Parikh vector is balanced. We define

$$\mathrm{neigh}^+((p, q), r) = \{(p, q + t) \;:\; 0 \leq t \leq r\},$$

$$\mathrm{neigh}^-((p, q), r) = \{(p, q - t) \;:\; 0 \leq t \leq r\},$$

$$\mathrm{neigh}((p, q), r) = \mathrm{neigh}^+((p, q), r) \cup \mathrm{neigh}^-((p, q), r).$$

For $i \leq j$ let us define the following word of length $2 \cdot (i \otimes j)$:

$$\mathbf{w}_{i,j} = 0^i 1^i 0^{i+1} 1^{i+1} \cdots 0^j 1^j.$$

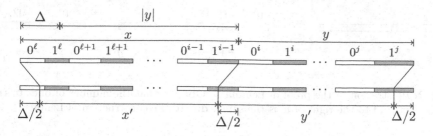

Fig. 1. Illustration of the proof of Lemma 2 — construction of balanced Abelian square $x'y'$

Observation 1. *Let* $w = \mathbf{w}_{i,j}$ *and* $\Delta \in \{0,\ldots,i\}$. *Then the subword* $w[1 + \Delta..|w| - \Delta]$ *is balanced.*

Let us take $\mathbf{w}_k = \mathbf{w}_{1,k}$. For example $\mathbf{w}_4 = 01001100011100001111$. Also, let $\mathbf{S}_k$ be a family of balanced vectors $\{(p,p) : p \in \mathrm{Sums}\left(\lceil \frac{3}{4}k \rceil, k\right)\}$.

Lemma 2. *If* $k > 16$, *then* $\mathbf{S}_k \subseteq SQV(\mathbf{w}_k)$.

Proof. Let $p = i \otimes j$ for i, j such that $\lceil \frac{3}{4}k \rceil \leq i, j \leq k$ and let $\ell < i$ be the largest index such that $\ell \otimes (i-1) \geq p$. Such an integer ℓ exists since for $k > 16$ we have $1 \otimes \left(\lceil \frac{3}{4}k \rceil - 1\right) \geq \lceil \frac{3}{4}k \rceil \otimes k$.

Consider the subwords $x = \mathbf{w}_{\ell,i-1}$, $y = \mathbf{w}_{i,j}$ of $\mathbf{w}_k$. If $|x| = |y|$, then we have just located a square xy with a square vector (p,p) and we are done. Otherwise, let $\Delta = |x| - |y| > 0$; see Fig. 1. Note that $0 < \Delta/2 < \ell$ and $|x|_0 = |x|_1 = p + \Delta/2$. We modify x into x' by cutting away the first $\Delta/2$ zeros and the last $\Delta/2$ ones: $x' = x[\Delta/2 + 1..|x| - \Delta/2]$. Then y' is obtained from y by adding $\Delta/2$ ones on the left side, and removing $\Delta/2$ ones from the right side. By Observation 1, $|x'|_0 = |x'|_1 = |y'|_0 = |y'|_1 = p$. $\qquad\square$

Lemma 3. *If* $k > 16$, *there exists* $r_k = \Omega\left(\sqrt{|\mathbf{w}_k|}\right)$ *such that for every* $\Gamma \in \mathbf{S}_k$

$$\mathrm{neigh}^+(\Gamma, r_k) \subseteq SQV(\mathbf{w}_k) \quad or \quad \mathrm{neigh}^-(\Gamma, r_k) \subseteq SQV(\mathbf{w}_k).$$

Proof. For $\Gamma \in \mathbf{S}_k$ we define i, j, and the Abelian square $x'y'$ corresponding to Γ as in the proof of Lemma 2. Let β and α be the distances from the right end of x' to the beginning and to the end of the block 1^{i-1}; see Fig. 2. Similarly we define δ as the distance of the right end of y' to the left endpoint of the block 1^j. One can easily check that the distance of the right end of y' to the end of the block 1^j equals α (see Fig. 2).

Note that $\alpha + \beta = i - 1 \geq \lceil \frac{3}{4}k \rceil - 1$ and $\delta \geq \beta$. There are two cases:

(a) If $\alpha \geq \beta$, then $\alpha \geq (i-1)/2$. Then $\mathrm{neigh}^+(\Gamma, \lfloor \alpha/2 \rfloor) \subseteq SQV(\mathbf{w}_k)$.
(b) If $\alpha < \beta$, then $\beta \geq (i-1)/2$. Then $\mathrm{neigh}^-(\Gamma, \lfloor \beta/2 \rfloor) \subseteq SQV(\mathbf{w}_k)$.

Thus we set $r_k = \lfloor (\lceil \frac{3}{4}k \rceil - 1)/4 \rfloor = \Omega\left(\sqrt{|\mathbf{w}_k|}\right)$ and the conclusion holds. $\qquad\square$

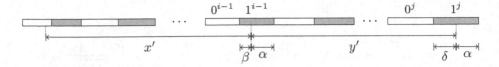

Fig. 2. Illustration of the proof of Lemma 3. Observe that the number of ones to the right of x' and to the right of y' is the same, due to the construction of Lemma 2.

Theorem 2. $SQ'_{\text{Abel}}(n) = \Omega(n^{1.5}/\log n)$.

Proof. We have constructed the family of words $\mathbf{w}_k$ together with the sets $\mathbf{S}_k$, which we show (Lemma 2) to be square vectors of $\mathbf{w}_k$. Due to Lemma 1 we have

$$|\mathbf{S}_k| = \Omega(|\mathbf{w}_k|/\log|\mathbf{w}_k|).$$

Note that for any $\Gamma_1, \Gamma_2 \in \mathbf{S}_k$, $\Gamma_1 \neq \Gamma_2$, and $r \geq 0$ we have $\text{neigh}(\Gamma_1, r) \cap \text{neigh}(\Gamma_2, r) = \emptyset$. Thus by Lemma 3 we obtain

$$SQ'_{\text{Abel}}(\mathbf{w}_k) = |SQV(\mathbf{w}_k)| \geq |\mathbf{S}_k|r_k = \Omega(|\mathbf{w}_k|^{1.5}/\log|\mathbf{w}_k|).$$

This completes the lower bound proof for $n = |\mathbf{w}_k|$. Otherwise we pick the longest word $\mathbf{w}_k$, $|\mathbf{w}_k| \leq n$, and append it with sufficiently many zeros.   □

2.2   An Upper Bound for $SQ'_{\text{Abel}}(n, 2)$

The number of blocks in a word w is defined as:

$$\#_{bl}(w) = 1 + |\{1 \leq i < |w| \ : \ w[i] \neq w[i+1]\}|.$$

For example $\#_{bl}(\mathbf{w}_k) = 2k$. We show a nontrivial upper bound for the number of nonequivalent Abelian squares in words with a given number m of blocks.

Lemma 4. *For a word w and a nonnegative integer δ suppose the following subwords are uniform but not all equal:*

$$w_1 = w[j..j+\delta], \quad w_2 = w[j+k..j+k+\delta], \quad w_3 = w[j+2k..j+2k+\delta].$$

If $w[j..j+2k-1]$ is an Abelian square, then no Abelian square of the same length starts at any position in the interval $[j+1..j+\delta]$.

Proof. Due to the binary alphabet we have exactly three cases: $w_1 = w_3$, $w_1 = w_2$ or $w_2 = w_3$. We prove the lemma only in the first case; see Fig. 3. The remaining cases admit similar proofs.

Let $w_1 = w_3 = (c_1)^\delta$ and $w_2 = (c_2)^\delta$ with $c_1 \neq c_2$. Denote $u_1 = w[j..j+k-1]$ and $u_2 = w[j+k..j+2k-1]$. Whenever we shift u_1 to the right (by at most δ positions), the number of occurrences of c_1 decreases and the number of occurrences of c_2 increases. However, when we shift u_2 to the right, the number of occurrences of c_1 increases and the number of occurrences of c_2 decreases. Therefore, we cannot obtain an Abelian square.   □

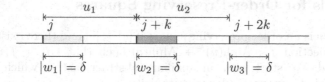

Fig. 3. Illustration of Lemma 4: the shaded areas correspond to uniform subwords (the first and the third one are composed of the same letter). An Abelian square u_1u_2 at position j excludes any Abelian square of the same length starting in the shaded area to the right of j.

Theorem 3. *If w is a binary word of length n with m blocks, then*

$$SQ'_{\text{Abel}}(w) \leq 3(m+1)\tfrac{n}{2}.$$

Proof. We call $\{0, 1, \ldots, n\}$ the set of *interpositions* of w. Intuitively, interpositions can be interpreted as locations between two consecutive letters, before the first letter or after the last letter. Let A, the set of *alternating* interpositions, contain 0, n and all interpositions i for which $w[i] \neq w[i+1]$. In other words if i is not alternating, then $w[i] = w[i+1]$, in particular both these letters are well defined. Note that $|A| = m+1$. To each Abelian square $w[i..i+2k-1]$ we assign three interpositions which we call *special*: the first interposition $i-1$ (F), the middle interposition $i+k-1$ (M), and the last interposition $i+2k-1$ (L).

For each square vector $\Gamma \in SQV(w)$ we consider only the rightmost occurrence of an Abelian square corresponding to Γ.

First, we consider Abelian squares for which one of the special interpositions is alternating. Let $v = w[i..i+2k-1]$ be such an Abelian square. We uniquely label v with a triple representing an alternating interposition, the type of this interposition (F/M/L) and the half of v's length: if $(i-1) \in A$, then the triple is $(i-1,\ \text{F},\ |v|/2)$, otherwise if $(i+k-1) \in A$, then it is $(i+k-1,\ \text{M},\ |v|/2)$, and otherwise it is $(i+2k-1,\ \text{L},\ |v|/2)$.

As a second group we consider the remaining (rightmost) Abelian squares. Let $v = w[i..i+2k-1]$ be such an Abelian square. Note that $w[i] = w[i+k] = w[i+2k]$ could not hold, otherwise v would not be the rightmost occurrence (v would be Abelian equivalent to $w[i+1..i+2k]$). Let ℓ_1 be the length of the maximal prefix of $w[i..n]$ of form $w[i]^*$, likewise ℓ_2 be the length of the maximal prefix of $w[i+k..n]$ of form $w[i+k]^*$, and ℓ_3 be the length of the maximal prefix of $w[i+2k..n]$ of form $w[i+2k]^*$. Let $\ell = \min(\ell_1, \ell_2, \ell_3) > 0$. We uniquely label v with a triple representing an alternating interposition, the type of this interposition and the half of v's length: if $\ell_1 = \ell$, then the triple is $(i+\ell-1,\ \text{F},\ |v|/2)$, otherwise if $\ell_2 = \ell$, then it is $(i+k+\ell-1,\ \text{M},\ |v|/2)$, and otherwise it is $(i+2k+\ell-1,\ \text{L},\ |v|/2)$.

Lemma 4 implies that each Abelian square receives a different label. Therefore there are at most $3(m+1)\tfrac{n}{2}$ Abelian squares in total. $\square$

In particular, Theorem 3 implies the following result:

Observation 2. $SQ'_{\text{Abel}}(\mathbf{w}_k) = O(|\mathbf{w}_k|^{1.5})$.

3   Bounds for Order-Preserving Squares

Recall that uv is an order-preserving square if $|u| = |v|$ and there exists a strictly increasing bijection $f : \mathrm{Alph}(u) \to \mathrm{Alph}(v)$ such that $v[i] = f(u[i])$ for all $i = 1, \ldots, |u|$. We start with an auxiliary abstract fact in which we do not require f to be of any particular monotonicity.

Lemma 5. *Let w be a word of length n over an alphabet Σ, and let Σ_1, Σ_2 be two distinct subsets of Σ of the same cardinality. Also, let f be a given bijection between Σ_1 and Σ_2. Then there are at most n distinct subwords of w of the form $xf(x)$, where $\mathrm{Alph}(x) = \Sigma_1$.*

Proof. Suppose a word $xf(x)$, where $\mathrm{Alph}(x) = \Sigma_1$, starts at position i in w. Let $j > i$ be the first occurrence of a letter in $\Sigma_2 - \Sigma_1$. Suppose it is the letter c. This letter is located in $f(x)$. Let $k \geq i$ be the first occurrence of $f^{-1}(c)$. Then $|x| = j - k$ and this determines the word $xf(x)$ as $w[i..i + 2(j - k) - 1]$.

Consequently there is at most one occurrence of a subword of the required form starting at a given position, so the number of such distinct subwords does not exceed n.   □

Theorem 4. *If $k = O(1)$, then $SQ_{\mathrm{op}}(n, k) = \Theta(n)$.*

Proof. Let w be a word of length n over a k-letter alphabet Σ. Each order-preserving square is of the form $xf(x)$ where $f : \mathrm{Alph}(x) \to \mathrm{Alph}(f(x))$ is a strictly increasing bijection. If $\mathrm{Alph}(x) = \mathrm{Alph}(f(x))$, then f must be the identity and thus $xf(x)$ is an ordinary square. However, there are at most $2n$ such squares in w [14]. Otherwise, $\mathrm{Alph}(x), \mathrm{Alph}(f(x))$ and f satisfy the assumptions of Lemma 5. The number of such triples is constant with respect to n, which, combined with Lemma 5, completes the proof.   □

4   Bounds for Parameterized Squares

In this section we consider words over the binary alphabet $\{0, 1\}$. An *antisquare* is a nonempty word of the form $x\bar{x}$, where $\bar{x}$ denotes bitwise negation of x. For example, $011\,100$ is an antisquare. Recall that uv is a parameterized square if $|u| = |v|$ and there exists a bijection $f : \mathrm{Alph}(u) \to \mathrm{Alph}(v)$ such that $v[i] = f(u[i])$ for all $i = 1, \ldots, |u|$. Observe that for binary alphabet each parameterized square is an ordinary square or an antisquare.

We also introduce *almost-squares*, which are the words of the form xax, where x is a word and $a \in \{0, 1\}$. Equivalently, an almost-square is an ordinary square with the last letter missing. The following words are examples of almost-squares: $011\,1\,011$, $11111\,0\,11111$, 0.

For a binary word w of length n we define the word $\hat{w}$ of length $n - 1$ so that $\hat{w}[i] = 1$ if $w[i] = w[i + 1]$ and $\hat{w}[i] = 0$ otherwise. For example, for $w = 00110101100010$ we have $\hat{w} = 1010000101100$.

For a word x of length ℓ we construct a rooted directed labeled tree $T(x)$ as follows. We start with a single path with edges labeled with the consecutive

Fig. 4. Rooted directed labeled tree $T(x)$, $|x| = \ell$

letters of x. Then we attach leaves to all nodes of the path so that each of them has two outgoing edges, one labeled with 0 and one labeled with 1; see Fig. 4. A square in a directed labeled tree is defined as a directed path such that the label of the path is an (ordinary) square.

Observation 3. *The following are equivalent:*
(a) the subword $w[i..j]$ is a parameterized square,
(b) the subword $\hat{w}[i..j-1]$ is an almost-square,
(c) the subword $\hat{w}[i..j-1]a$ for some $a \in \{0,1\}$ is a square in the tree $T(\hat{w})$.

The proof of the following lemma is an immediate generalization of the proof of the analogous upper bound on the number of ordinary squares in a word. It suffices to note that there are at most two topmost occurrences of distinct squares ending at each node of the tree; see [14,18].

Lemma 6. *In a labeled directed rooted tree with m nodes there are at most $2m$ distinct squares.*

Theorem 5. $SQ_{\mathrm{param}}(n, 2) \leq 8n$.

Proof. Let w be a word of length n. Observe that $T(\hat{w})$ has at most $2n$ nodes. It follows from Observation 3 and Lemma 6 that the number of distinct almost-squares in w is at most $4n$. For each almost-square v there are exactly two parameterized squares u_1, u_2 such that $\hat{u}_1 = \hat{u}_2 = v$ ($\bar{u}_1 = u_2$ and u_1, u_2 are both ordinary squares or both antisquares). Consequently $SQ_{\mathrm{param}}(w)$ does not exceed twice the number of distinct almost-squares in $\hat{w}$. Hence $SQ_{\mathrm{param}}(w) \leq 8n$. □

5   Infinite Words Avoiding Nonstandard Squares/Cubes

It is known that there are infinite words over a 4-letter alphabet avoiding Abelian squares while over 3-letter alphabets such words do not exist [20]. Here, we investigate an analogous problem for other nonstandard repetitions.

We say that a word is op-square-free if it does not contain an order-preserving square of length greater than 2. Let $\Sigma_3 = \{0, 1, 2\}$ ordered in the natural way. Consider the morphism:

$$\psi : 0 \mapsto 10,\ 1 \mapsto 11,\ 2 \mapsto 12.$$

Lemma 7. *If a word $w \in \Sigma_3^*$ is square free, then $\psi(w)$ is op-square-free.*

Proof. Let $\approx$ denote the order-preserving equivalence (i.e., $u \approx v$ if $|u| = |v|$ and uv is an order-preserving square). We have the following simple observation.

Observation 4. *For any symbols* $a, b, c \in \Sigma_3$ *we have:*
(a) $1\,a \approx 1\,b \Leftrightarrow a = b$;
(b) $a\,1\,b \approx 1\,c\,1 \Rightarrow a = b$.

Suppose to the contrary that $w' = \psi(w)$ contains an order-preserving square $u'v' = w'[i..i+2k-1]$, with $|u'| = |v'| = k \geq 2$. We consider four cases depending on the parity of i and k.

If $2 \mid k$ and $2 \nmid i$, then u' and v' start with a 1 and every second symbol of each of them is a 1. Consequently, by Observation 4(a), $u' = v'$. Moreover, in this case we have $u' = \psi(u)$ and $v' = \psi(v)$ for some subword uv of w. Hence, uv is a square in w, a contradiction.

If $2 \mid k$ and $2 \mid i$, then $w'[i-1...i+2k-2]$ is also an order-preserving square. The conclusion follows from the previous case.

If $2 \nmid k$ and $2 \nmid i$, then u' and v' start with $1c1$ and $a1b$ for some $a, b, c \in \Sigma_3$, respectively. By Observation 4(b) we conclude that $a = b$, which implies a square ab in w, a contradiction.

The final case, $2 \nmid k$ and $2 \mid i$, also implies a 2-letter square in w just as in the previous case. This completes the proof that w' is op-square-free. ☐

We apply Lemma 7 to all prefixes of an infinite square-free word [24] over a ternary alphabet and obtain the following result.

Theorem 6. *There exists an infinite op-square-free word over 3-letter alphabet.*

A parameterized cube is a word uvw such that both uv and vw are parameterized squares. A word is called parameterized-square-free (parameterized-cube-free) if it does not contain parameterized squares (parameterized cubes) of length greater than 3. We show that there is no infinite parameterized-square-free word and construct a binary parameterized-cube-free word.

Theorem 7. *There is no infinite parameterized-square-free word.*

Proof. Suppose to the contrary that such an infinite word x exists. In the proof we denote symbols of $\mathrm{Alph}(x)$ by a, b, c, d. Note that every suffix of x has to contain two adjacent equal symbols. This is because $abcd$ for $a \neq b$ and $c \neq d$ is a parameterized square. Moreover, x has to contain some three adjacent equal symbols. The reason is that $abbd$ for $a \neq b \neq d$ is a parameterized square.

We can therefore assume that x contains a subword aaa. To avoid a parameterized square of length 4, this subword must be followed in x by some letter $b \neq a$. For the same reason the next letter c must satisfy $c \neq b$, and afterwards the subword $aaabc$ must be followed by two more occurrences of c. Finally the next letter must be $d \neq c$ to avoid a parameterized square $cccc$. We conclude that x contains a subword $aaabcccd$ for $b \neq a$ and $d \neq c$, which turns out to be a parameterized square. This contradiction completes the proof. ☐

Let τ be the infinite Thue-Morse word. Recall that τ is cube-free [25]. Also recall the morphism ψ defined just before Lemma 7.

Theorem 8. *The word* $\psi(\tau)$ *is parameterized-cube-free.*

Proof. Suppose to the contrary that $u_1 u_2 u_3$ is a parameterized cube in $\psi(\tau)$, with $|u_1| = |u_2| = |u_3| = k > 1$. Note that $\psi(\tau)$ does not contain 6 ones in a row. Hence, at least one of the words u_1, u_2, u_3 contains 0, therefore each of them contains 0. Moreover every second symbol of u_1, u_2, u_3 is 1.

Recall from Section 4 that a binary parameterized square is either an ordinary square or an antisquare. If $2 \mid k$, then the ones of every second position of u_1, u_2, u_3 align and $u_1 u_2$, $u_2 u_3$ must be ordinary squares. Therefore $u_1 u_2 u_3$ is an ordinary cube in $\psi(\tau)$ which induces a cube in τ.

If $2 \nmid k$, the same argument implies that both $u_1 u_2$ and $u_2 u_3$ are antisquares. Because of the ones on every second position of u_1, u_2, u_3 we actually have $u_1 = 0101\cdots$, $u_2 = 1010\cdots$, $u_3 = 0101\cdots$ or $u_1 = 1010\cdots$, $u_2 = 0101\cdots$, $u_3 = 1010\cdots$. In both cases we obtain a cube $(10)^3$ in $\psi(\tau)$ which induces 0^3 in τ. $\square$

6   Final Remarks

We have presented several combinatorial results related to the maximum number of nonstandard squares in a word of length n. For Abelian squares we have shown that $SQ_{\text{Abel}}(n,2) = \Theta(n^2)$ and $SQ'_{\text{Abel}}(n,2) = \Omega(n^{1.5}/\log n)$. The latter bound, although reached by a simple family of words, required a rather involved proof.

For squares in order-preserving and parameterized setting we show that their maximum number is linear of n for a constant and a binary alphabet, respectively. We have also presented examples of infinite words over a minimal alphabet that avoid squares in order-preserving setting and cubes in parameterized setting, respectively.

The main open question that arises from our work is to provide an upper bound for $SQ'_{\text{Abel}}(n,2)$. We have made a step towards this bound by showing that the maximum number of distinct Abelian squares in a word of length n containing m blocks is $O(nm)$. The remaining open questions are connected to $SQ'_{\text{op}}(n,k)$ and $SQ'_{\text{param}}(n,k)$ for arbitrary k (not necessarily a constant). Based on experimental results, we state the following conjecture:

Conjecture 1. $SQ'_{\text{Abel}}(n,2) = O(n^{1.5})$, $SQ'_{\text{op}}(n,k) = SQ'_{\text{param}}(n,k) = \Theta(n)$ for any $k \geq 2$.

References

1. Baker, B.S.: Parameterized pattern matching: Algorithms and applications. Journal of Computer and System Sciences 52(1), 28–42 (1996)
2. Berstel, J., Karhumäki, J.: Combinatorics on words: A tutorial. Bulletin of the EATCS 79, 178–228 (2003)
3. Blanchet-Sadri, F., Jiao, Y., Machacek, J.M., Quigley, J., Zhang, X.: Squares in partial words. Theoretical Computer Science 530, 42–57 (2014)
4. Blanchet-Sadri, F., Merca, R., Scott, G.: Counting distinct squares in partial words. Acta Cybernetica 19(2), 465–477 (2009)
5. Crochemore, M., Ilie, L., Rytter, W.: Repetitions in strings: Algorithms and combinatorics. Theoretical Computer Science 410(50), 5227–5235 (2009)

6. Crochemore, M., et al.: Order-preserving incomplete suffix trees and order-preserving indexes. In: Kurland, O., Lewenstein, M., Porat, E. (eds.) SPIRE 2013. LNCS, vol. 8214, pp. 84–95. Springer, Heidelberg (2013)
7. Crochemore, M., Iliopoulos, C.S., Kociumaka, T., Kubica, M., Radoszewski, J., Rytter, W., Tyczyński, W., Waleń, T.: The maximum number of squares in a tree. In: Kärkkäinen, J., Stoye, J. (eds.) CPM 2012. LNCS, vol. 7354, pp. 27–40. Springer, Heidelberg (2012)
8. Crochemore, M., Iliopoulos, C.S., Kubica, M., Radoszewski, J., Rytter, W., Waleń, T.: Extracting powers and periods in a word from its runs structure. Theoretical Computer Science 521, 29–41 (2014)
9. Erdős, P.: An asymptotic inequality in the theory of numbers (in Russian). Vestnik Leningrad University: Mathematics 15, 41–49 (1960)
10. Erdős, P.: Some unsolved problems, Hungarian Academy of Sciences Mat. Kutató Intézet Közl. 6, 221–254 (1961)
11. Evdokimov, A.A.: Strongly asymmetric sequences generated by a finite number of symbols. Doklady Akademii Nauk SSSR 179(6), 1268–1271 (1968)
12. Fici, G.: Personal communication
13. Fici, G., Langiu, A., Lecroq, T., Lefebvre, A., Mignosi, F., Prieur-Gaston, É.: Abelian repetitions in Sturmian words. In: Béal, M.-P., Carton, O. (eds.) DLT 2013. LNCS, vol. 7907, pp. 227–238. Springer, Heidelberg (2013)
14. Fraenkel, A.S., Simpson, J.: How many squares can a string contain? Journal of Combinatorial Theory, Series A 82, 112–120 (1998)
15. Gusfield, D., Stoye, J.: Linear time algorithms for finding and representing all the tandem repeats in a string. Journal of Computer and System Sciences 69(4), 525–546 (2004)
16. Hardy, G., Wright, E., Heath-Brown, D., Silverman, J.: An Introduction to the Theory of Numbers. Oxford mathematics. OUP, Oxford (2008)
17. Huova, M., Karhumäki, J., Saarela, A.: Problems in between words and abelian words: k-abelian avoidability. Theor. Comput. Sci. 454, 172–177 (2012)
18. Ilie, L.: A simple proof that a word of length n has at most 2n distinct squares. Journal of Combinatorial Theory, Series A 112(1), 163–164 (2005)
19. Ilie, L.: A note on the number of squares in a word. Theoretical Computer Science 380(3), 373–376 (2007)
20. Keränen, V.: Abelian squares are avoidable on 4 letters. In: Kuich, W. (ed.) ICALP 1992. LNCS, vol. 623, pp. 41–52. Springer, Heidelberg (1992)
21. Muthukrishnan, S.: New results and open problems related to non-standard stringology. In: Galil, Z., Ukkonen, E. (eds.) CPM 1995. LNCS, vol. 937, pp. 298–317. Springer, Heidelberg (1995)
22. Muthukrishnan, S., Palem, K.: Non-standard stringology: Algorithms and complexity. In: Proceedings of the Twenty-sixth Annual ACM Symposium on Theory of Computing, STOC 1994, pp. 770–779. ACM, New York (1994)
23. Pleasants, P.A.: Non-repetitive sequences. Mathematical Proceedings of the Cambridge Philosophical Society 68, 267–274 (1970)
24. Thue, A.: Über unendliche Zeichenreihen. Norske Videnskabers Selskabs Skrifter Mat.-Nat. Kl. 7, 1–22 (1906)
25. Thue, A.: Über die gegenseitige Lage gleicher Teile gewisser Zeichenreihen. Norske Videnskabers Selskabs Skrifter Mat.-Nat. Kl. 10, 1–67 (1912)

Knight Tiles: Particles and Collisions in the Realm of 4-Way Deterministic Tilings

Bastien Le Gloannec and Nicolas Ollinger

Univ. Orléans, INSA Centre Val de Loire, LIFO EA 4022, FR-45067 Orléans, France
{Bastien.Le-Gloannec,Nicolas.Ollinger}@univ-orleans.fr

Abstract. Particles and collisions are convenient construction tools to compute inside tilings and enforce complex sets of tilings with simple tilesets. Locally enforceable particles being incompatible with expansivity in the orthogonal direction, a compromise has to be found to combine both notions in a same tileset. This paper introduces knight tiles: a framework to construct 4-way deterministic tilings, that is tilings completely determined by any infinite diagonal of tiles, for which local particles and collisions with many slopes can still be constructed while being expansive in infinitely many directions. The framework is then illustrated by an elegant yet simple construction to mark a diagonal with a 4-way deterministic knight tileset.

Keywords: Deterministic tiles, domino problem, tiling problem, expansive subdynamics, particles, Wang tiles.

A tiling is a coloring of the discrete plane $\mathbb{Z}^2$ assigning a tile from a finite tileset to each position so that the local tiling rules associated to the tileset are satisfied for every tile in its neighborhood. Wang tiles provide a convenient and universal syntactic description of tiling rules: a Wang tile is a unit square with colored edges and the tiling rule requires adjacent squares to share a same color along their common edge – tiling constraints are readable directly on the tiling. Starting with the study of the Domino Problem [2,11], computations have been successfully embedded into tilings to enforce computational phenomena and prove undecidability and complexity results. Following what is done in the case of cellular automata, particles and collisions have been successfully used as a tool to transmit information quanta through tilings and mark positions.

A tiling is 4-way deterministic if any infinite diagonal strip of tiles uniquely determines the whole tiling. A syntactic way to enforce such a property on a tileset is to require any consecutive pair of colors along its edges to uniquely determine a tile. Kari and Papasoglu [4] were able to prove that it is still possible to construct aperiodic tilesets under such constraint as are every family of tilings generated by substitution systems [6]. Ten years later, Lukkarila [8] showed that Turing machines can still be simulated in this setting and proved that the Domino Problem is still undecidable for 4-way deterministic tilesets. The construction is subtle and quite involved, building on the previous construction by Kari and Papasoglu [4]. The construction has to be quite involved

A.M. Shur and M.V. Volkov (Eds.): DLT 2014, LNCS 8633, pp. 227–238, 2014.
© Springer International Publishing Switzerland 2014

in particular because such a syntactic constraint enforces the expansivity [3] of the tilings in every directions but two, prohibiting the use of locally defined particles that are neither horizontal nor vertical. In this paper, inspired by the partitioned cellular automata [9], we propose to replace Wang tiles with knight tiles, keeping a syntactic condition to enforce 4-way determinism and gaining back particles and collisions.

Similar to a Wang tile, a knight tile is a colored unit square sharing colors not only with its 4 direct neighbors but also with the 8 neighbors located at a chess knight move. Determinism along one diagonal is then simply expressed as a syntactic condition on tuples of colors. While keeping expansive in infinitely many directions, knight tiles can be combined to construct particles along slopes that are not too steep. The interest of this family of tilesets is illustrated by considering two problems handled with complicated constructions in Lukkarila [8].

The first construction is the key ingredient to the proof of the undecidability of the Domino Problem: solving the seeded Domino Problem by simulating a Turing machine.

The second construction is a key ingredient for the previous construction in the classical case: marking a diagonal — a slope that cannot be obtained directly by knight tiles using a particle. Indeed, Lukkarila [8] asks "Could there be a significantly simpler tile set for drawing a single diagonal line 4-way deterministically?" The construction provided here is significantly simpler, combining time-symmetry and the infinite Firing Squad technique by Kari [5] with properties of the Thue-Morse substitution to enforce 4-way determinism.

1   Preliminary Definitions

Wang Tiles and Deterministic Tilesets. Given a finite alphabet of *colors* $\mathcal{C}$, a *Wang tile* is an oriented unit square tile with one color on each side. Formally, it is a quadruple $t \in \mathcal{C}^4$ whose four components are identified with the four directions $\{\text{W}, \text{S}, \text{E}, \text{N}\}$ and, for convenience, denoted as t_W, t_S, t_E and t_N. A *Wang tileset* τ is a finite set of Wang tiles. A *tiling* by τ is a map $T : \mathbb{Z}^2 \to \tau$ associating a tile of τ to each cell of the discrete plane $\mathbb{Z}^2$ such that two adjacent tiles (for the 4-connectedness) share the same color on their common edge. More formally, a tiling T satisfies the following constraints, for all $(x, y) \in \mathbb{Z}^2$: $T(x, y)_\text{W} = T(x - 1, y)_\text{E}$, $T(x, y)_\text{S} = T(x, y - 1)_\text{N}$, $T(x, y)_\text{E} = T(x + 1, y)_\text{W}$, $T(x, y)_\text{N} = T(x, y + 1)_\text{S}$. The set of all tilings by a tileset τ is denoted as $\mathcal{X}_\tau$.

The tileset τ is NE-*deterministic* if for all couples of tiles $(w, s) \in \tau^2$, there exists at most one tile $t \in \tau$ simultaneously compatible to the west with w and to the south with s: $w_\text{E} = t_\text{W}$ and $s_\text{N} = t_\text{S}$. In the case of Wang tiles, one can equivalently say that any tile $t \in \tau$ is uniquely identified by its couple of colors (t_W, t_S). $\{\text{SW}, \text{SE}, \text{NW}\}$-*determinism* is defined symmetrically. A tileset is *4-way deterministic* if it is simultaneously SW, SE, NE and NW-deterministic.

Subshifts and Expansiveness. Given a finite alphabet Σ, a Σ-*coloring* (or simply *coloring* in the absence of ambiguity) of the discrete plane is a map

$c : \mathbb{Z}^2 \to \Sigma$. For all $u \in \mathbb{Z}^2$, we define the translation σ_u over colorings by $\sigma_u(c)(x) = c(x - u)$ for all $c \in \Sigma^{\mathbb{Z}^2}$ and $x \in \mathbb{Z}^2$. A coloring $c \in \Sigma^{\mathbb{Z}^2}$ is *periodic* of period $p \in \mathbb{Z}^2$ if $\sigma_p(c) = c$. The set $\Sigma^{\mathbb{Z}^2}$ of all Σ-colorings is endowed with the product (over $\mathbb{Z}^2$) topology of the discrete topology (over Σ). A *subshift* $\mathcal{X}$ is a (topologically) closed and translation-invariant ($\forall u \in \mathbb{Z}^2$, $\sigma_u(\mathcal{X}) = \mathcal{X}$) subset of $\Sigma^{\mathbb{Z}^2}$. In particular, the set of tilings by a tileset τ is a subshift of $\tau^{\mathbb{Z}^2}$.

For all *slope* $\alpha \in \mathbb{R} \cup \{\infty\}$, let us denote as l_α the real line of slope α going through the origin. For all *radius* $\rho > 0$, let us define $L_\alpha(\rho) = (l_\alpha + [-\rho, \rho]u) \cap \mathbb{Z}^2$ for $u \in \mathbb{R}^2$ a unit vector orthogonal to l_α. $L_\alpha(\rho)$ is the discrete thick line of slope α and width 2ρ centered on the origin. A slope α is a *direction of expansiveness* of a subshift $\mathcal{X} \subseteq \Sigma^{\mathbb{Z}^2}$ if there exists a radius $\rho > 0$ such that for all $x, y \in \mathcal{X}$, $x|_{L_\alpha(\rho)} = y|_{L_\alpha(\rho)} \implies x = y$. In particular, if τ is a 4-way deterministic Wang tileset, $\mathcal{X}_\tau$ is expansive in (at least) all directions of $\mathbb{R} \backslash \{0\}$.

2   From Wang Tiles to Knight Tiles

The notion of 4-way determinism introduces a very strong constraint on the tileset. Indeed, it strongly limits its capacity to construct particles locally. To mix particles and determinism, one might loosen the constraints by generalizing the notion of local determinism to a broader *radius* $r \geq 1$ of determinism.

Radius of Determinism. A tileset τ is NE *deterministic with radius* r if for all *valid* (i.e. containing no tiling error along its inner edges) $(2r + 1) \times (2r + 1)$ square pattern by τ, the center tile is perfectly determined by the shifted diagonal formed by the $2r$ tiles at positions $(1, 2r), (2, 2r - 1), \ldots, (2r - 1, 1)$. Determinism with radius r in the three other diagonal directions (SW, SE and NW) is defined symmetrically. The tileset is *4-way deterministic with radius* r if it is simultaneously deterministic with radius r in the four diagonal directions.

The different radiuses can be compared through the expansiveness of their tilings as pointed out by the following proposition.

Proposition 1. *If τ is a 4-way deterministic tileset with radius r, then $\mathcal{X}_\tau$ is (at least) expansive in directions* $\left]-\frac{r}{r-1}, -\frac{r-1}{r}\right[\cup \left]\frac{r-1}{r}, \frac{r}{r-1}\right[$.

The expansiveness directions are actually tightly related to the *particles* that can be locally realized. Expansiveness is, up to a change in the radius ρ, independent of the profile chosen for the slope. At a fixed radius of determinism, the ability to mark a line however depends on its profile. Let us denote the horizontal and vertical unit vectors of $\mathbb{Z}^2$ as $e_1 = (1, 0)$ and $e_2 = (0, 1)$. For any slope $\alpha \in \mathbb{Q} \cup \{\infty\}$, a *(periodic) profile* for α is a finite sequence $(u_0, \ldots, u_{k-1}) \in \{e_1, e_2\}^k$ such that $\alpha = \frac{\sum_i u_i \cdot e_2}{\sum_i u_i \cdot e_1}$. Given a profile $P = (u_0, \ldots, u_{k-1})$ for α, we define the associated subshift $\mathcal{L}_P \subseteq \{0, 1\}^{\mathbb{Z}^2}$ as the subshift generated by (i.e. the smallest subshift containing) the configuration c_P containing exactly one discrete 4-connected line of slope α drawn following the profile P, i.e. formally defined by $c_P(x) = 1$ if $x = \sum_{0 \leq i \leq n} u_{i \bmod k}$ or $x = -\sum_{-n \leq i < 0} u_{i \bmod k}$ for some $n \geq 0$, and $c_P(x) = 0$ for all other $x \in \mathbb{Z}^2$. Note that $\mathcal{L}_P$ exactly contains all translated versions of c_P, plus the blank coloring (all 0s) for compactness reasons. A coloring $c \in \Sigma^{\mathbb{Z}^2}$, where Σ contains an identified blank color $c_0 \in \Sigma$, is a *particle* of direction $\alpha \in \mathbb{Q} \cup \{\infty\}$ if it is periodic in direction α (i.e. c admits a periodicity vector $(p_x, p_y) \in \mathbb{Z}^2$ such that $\alpha = \frac{p_y}{p_x}$) and ultimately constant[1] equal to c_0 in any other direction. This equivalently means that all non-blank cells of c are contained in $L_\alpha(\rho) + \Delta$ for some $\rho > 0$ and $\Delta \in \mathbb{Z}^2$. We say that a slope $\alpha \in \mathbb{Q} \cup \{\infty\}$ is *locally realized* by a tileset τ if there exists a projection $\pi : \tau \rightarrow \{0, 1\}$ (naturally extended to colorings $\pi : \tau^{\mathbb{Z}^2} \rightarrow \{0, 1\}^{\mathbb{Z}^2}$) such that there exists a profile P for the slope α such that $\pi(\mathcal{X}_\tau) = \mathcal{L}_P$ and every tiling of $\mathcal{X}_\tau$ is a particle. The previous definition introduces a reinforced notion of *soficity* for $\mathcal{L}_P$.

The following results conclude our remarks on radiuses of determinism.

Proposition 2. *If a slope α is locally realizable by a tileset τ then α is not a direction of expansiveness for $\mathcal{X}_\tau$.*

Proposition 3. *The slopes that are locally realizable by 4-way deterministic tilesets at radius r are exactly $\mathbb{Q} \cap \left(\left[-\infty, -\frac{r}{r-1} \right] \cup \left[-\frac{r-1}{r}, \frac{r-1}{r} \right] \cup \left[\frac{r}{r-1}, +\infty \right] \right)$.*

Knight Tiles. Whereas classical 4-way determinism is a purely syntactic property of the tileset, to check determinism at radius r one has to consider all tilings of $(2r+1) \times (2r+1)$ squares. We introduce deterministic knight tiles as a convenient and purely syntactic notion of determinism at radius 2 that can be checked directly on the tiles.

Given a finite alphabet $\mathcal{C}$ of *colors*, a *knight tile* is formally a 12-tuple of $\mathcal{C}^{12}$ and a *knight tileset* is a finite set of knight tiles. For convenience, each of the twelve components of a knight tile will be identified by a direction among $\{W, S, E, N, WS, SW, SE, ES, EN, NE, NW, WN\}$ and for a knight tile $T \in \mathcal{C}^{12}$, we will for instance denote by $T_W \in \mathcal{C}$ the corresponding W color. A *tiling* of the discrete plane by a knight tileset τ is a map $t : \mathbb{Z}^2 \rightarrow \tau$ satisfying the following constraints, for all $(x, y) \in \mathbb{Z}^2$: $t(x,y)_W = t(x-1,y)_E$, $t(x,y)_S = t(x,y-1)_N$, $t(x,y)_E = t(x+1,y)_W$, $t(x,y)_N = t(x,y+1)_S$, $t(x,y)_{WS} = t(x-2,y-1)_{EN}$, $t(x,y)_{SW} = t(x-1,y-2)_{NE}$, $t(x,y)_{SE} = t(x+1,y-2)_{NW}$, $t(x,y)_{ES} = t(x+2,y-1)_{WN}$, $t(x,y)_{EN} = t(x+2,y+1)_{WS}$, $t(x,y)_{NE} = t(x+1,y+2)_{SW}$, $t(x,y)_{NW} = t(x-1,y+2)_{SE}$, and $t(x,y)_{WN} = t(x-2,y+1)_{ES}$. This means that each tile of a tiling shares each of its different color components with one of the twelve neighboring tiles represented on the Fig. 1a.

A knight tileset τ is NE-*deterministic* if any tile $t \in \tau$ is uniquely identified by its quadruple of colors $(t_{WN}, t_W, t_S, t_{SE})$, i.e. there is at most one tile in the set that is compatible with the four colors pointed out on the Fig. 1b. Determinism in the three other diagonal directions (SW, SE and NW) is defined symmetrically.

---

[1] Note that all our results would remain true replacing "ultimately constant" by "ultimately periodic" in the definition of a particle.

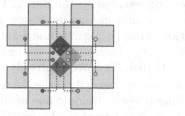
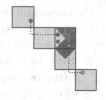

(a) Neighborhood of a knight tile   (b) NE determinism

Fig. 1. Knight tiles and determinism

A knight tileset is *4-way deterministic* if it is simultaneously deterministic in the four diagonal directions SW, SE, NE and NW.

The rest of the paper is dedicated to advocate for the use of this notion of radius-2 determinism by showing that constructions that are painful to handle in the classical 4-way deterministic setting can be treated with particles and collisions in the 4-way deterministic knight setting.

3   The Seeded Knight Domino Problem Is Undecidable

Simulating a Turing machine with classical 4-way tiles requires a complicated machinery [8] involving particular aperiodic tilesets. Knight colors provide enough flexibility to handle it directly and prove the undecidability of the Domino Problem with a seed tile in a classical way.

Problem 1 (Domino Problem with a seed tile). Given a tileset τ and a specified *seed tile* $t_0 \in \tau$, does there exist a tiling of $\mathbb{Z}^2$ by τ using (at least once) the tile t_0?

The reader is assumed familiar with Turing machines. As every Turing computation can be made reversible [7,1], we directly work with reversible Turing machines. A *reversible Turing machine* (RTM) is a 5-tuple $(\Sigma, Q, \overleftrightarrow{q_i}, F, \delta)$ where Σ is the tape (finite) alphabet, Q the finite set of states. Before defining the remaining elements of the tuple, let us state that the head always moves at each transition and we denote its two possible moves by $\{\leftarrow, \rightarrow\}$. Let us also define $\overleftrightarrow{Q} = Q \times \{\leftarrow, \rightarrow\}$ whose elements will be written $\overrightarrow{q}$ (resp. $\overleftarrow{q}$) to denote $(q, \rightarrow)$ (resp. $(q, \leftarrow)$). Then we define the **partial injective** transition map $\delta : \overleftrightarrow{Q} \times \Sigma \rightarrow \overleftrightarrow{Q} \times \Sigma$. Finally $\overleftrightarrow{q_i} \in \overleftrightarrow{Q}$ is the initial (oriented) state and $F \subset \overleftrightarrow{Q}$ is the set of final (oriented) states (and we will assume that δ is not defined on states of F). This is an acceptable model for reversible Turing machines that will be furthermore well-fitted to our purpose.

The tileset described on Fig. 2, where colors are conveniently represented by arrows that must go uninterrupted across tile edges, is a 4-way deterministic knight tileset that simulates a given RTM $(\Sigma, Q, \overleftrightarrow{q_i}, F, \delta)$. First note that the only knight color components used in this simulation are $\{SW, SE, NE, NW\}$ hence there is no ambiguity on the knight colors represented on the Fig. 2. To be able to use efficiently these knight colors for determinism, one must slow down the simulation. To that purpose, each transition of the Turing machine is decomposed into three steps in this fixed order: 1. a state transition step where the transition is done but the head does not yet move, though its move is already contained in the state of $\overleftrightarrow{Q}$ (the corresponding tiles are represented on the bottom line of the figure); 2. a waiting step where nothing happens (middle line of the figure); 3. a move step where the head finally moves (top line of the figure). The Wang constraints to enforce this order for the steps of the computation process are not explicitly represented on the tiles of Fig. 2, but it is a simple counting modulo 3 in the vertical colors. Assuming that a Turing computation is correctly initialized in a tiling, then each line of the space-time diagram of the Turing machine can be read once every three lines of the tiling.

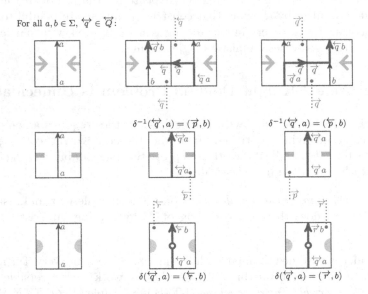

Fig. 2. Deterministic knight tiles simulating a RTM

One can enforce the initialization of the Turing computation using a seed tile and the tiles of the Fig. 3a that force a blank tape containing a unique Turing machine head to appear in the tilings. The labels "≤ -2", "-1", "0", "1" and "≥ 2" appearing on the tiles denote some specifications for the horizontal colors: the tile with index 0 is only compatible to the left (resp. right) with the tile with index -1 (resp. 1) and, to the left (resp. right) of the -1 (resp. 1) tile, only the "≤ -2" (resp. "≥ 2") tile can appear. However, these tiles cannot directly be added to the previous simulation tile set without losing the 4-way determinism. Indeed, it not difficult to see that, as is, predicting the tiles with index -1 or 1 is not possible in every direction. To solve that problem, one can add a very simple layer of information represented by the Wang tiles of the

Fig. 3b: a blank tile g_0 and a tile g_1 carrying a vertical "ghost" signal (that must be vertically preserved along columns of the tilings). Denoting as τ_T the Turing simulation tiles of the Fig. 2, τ_i^1 the three initialization tiles with indices $\{-1, 0, 1\}$ of the Fig. 3a, τ_i^2 the two initialization tiles with labels "≤ -2" and "≥ 2" of the Fig. 3a and b the blank tile of the Fig. 3a, we define the two-layered knight tileset $\tau = ((\tau_T \cup \{b\}) \times \{g_0, g_1\}) \cup (\tau_i^1 \times \{g_1\}) \cup (\tau_i^2 \times \{g_0\})$, which can be interpreted as a standard knight tileset on couples of colors of each layer. In a tiling, the three consecutive columns containing the initialization tiles with indices $\{-1, 0, 1\}$ are the only ones carrying a ghost signal (tile g_1) on the second layer. This allows to predict these tiles in every direction, hence the knight tileset τ is 4-way deterministic. As the transition tiles are not defined on final states, τ tiles the plane if and only if the simulated Turing machine does not halt on the blank tape. It is then straightforward to derive the following result.

Theorem 1. *The Domino Problem with a seed tile is undecidable for 4-way deterministic knight tilesets.*

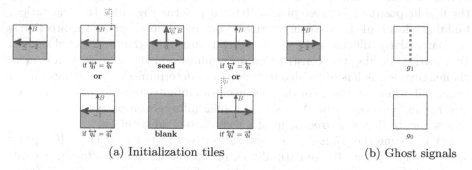

(a) Initialization tiles   (b) Ghost signals

Fig. 3. Additional knight tiles to initialize the Turing computation

4   Marking a Diagonal

Considering deterministic knight tiles allows to easily build particles that are not realizable in the usual deterministic Wang framework (where the only realizable slopes are 0 and ∞ by proposition 3) as illustrated by the following result.

Proposition 4. *The slopes that are locally realizable using 4-way deterministic knight tiles are exactly* $\mathbb{Q} \cap ([-\infty, -2] \cup [-\frac{1}{2}, \frac{1}{2}] \cup [2, +\infty])$.

Our main construction will rely on these realizable slopes to mark a diagonal in a lighter way than the construction of [8] in the Wang case. The general idea is to use a fundamentaly one-dimensional hierarchical structure of signals similar to an infinite version of Minsky's classical solution to the *Firing Squad* synchronization problem on cellular automata, as what is done for instance in [5]. This structure has moreover to be reversible so that the tileset can be made deterministic in two opposite directions, say NE and SW. The structure also has to be compatible with determinism for knight tiles in the two orthogonal directions NW and SE. The reader might keep in mind that only knight colors {WS, SW, EN, NE} will be used. Thus the tileset will be deterministic at radius 1 (classical Wang tiles determinism) in directions NE and SW, while it will be deterministic at radius 2 (using knights) in directions NW and SE.

The general structure is described on Fig. 4 where the red dots represent the diagonal to be marked, referred to as the *fire line*. For convenience, the four kinds of signals are named H, H', V and V' according to the figure. Their respective slopes are 0, $\frac{1}{3}$, $\frac{2}{3}$ and ∞. A binary hierarchical structure is used to mark some points along successive *front lines* until the granularity of space allows all the marked points of a front to be sufficiently close to decide by a local rule to *fire* at that point (and consequently mark the diagonal). Let us enumerate the front lines by their *rank* starting at 0 for the fire line. The points of the front of rank k are regularly arranged and two consecutive points are separated by $2^k - 1$ non-marked positions. The marked points of the front of rank k will be referred to as *pillars*, which correspond to the points that are marked on the front of rank $k + 1$ (i.e. points that were already marked seen things as a firing squad with time running towards NE), and *middles*, which exactly correspond to the middle positions between pillars. Referring to the Fig. 4 for the orientation, building a local rule for the construction to be deterministic in the NE direction is not particularly difficult and goes straightforward using the represented signals. By contrast, making this construction deterministic in the SW direction is more challenging as one has to be able to locally and deterministically distinguish between pillars and middles. For this, one needs an infinite source of well-structured alternating bits sequences. A way to get such information properly is to resort to a sequence that is a fixed-point of a well-chosen substitution.

Let us define the *Thue-Morse substitution* (see [10]) $s : \{0,1\} \rightarrow \{0,1\}^*$ by $s(0) = 01$ and $(1) = 10$. We naturally extend the definition of s to finite words of $\{0,1\}^*$ (resp. infinite words of $\{0,1\}^{\mathbb{N}}$) defining, for all $u \in \{0,1\}^*$ (resp. $u \in \{0,1\}^{\mathbb{N}}$), $s(u)_{2i+k} = s(u_i)_k$ for all $i \in \{0, \ldots, |u| - 1\}$ (resp. $i \in \mathbb{N}$) and $k \in \{0,1\}$. The *Thue-Morse word* $\mathcal{T} \in \{0,1\}^{\mathbb{N}}$ is the unique fixed-point of s starting with the letter 0: $\mathcal{T} = \lim_{n \rightarrow +\infty} s^n(0)$ that exists as $s(0)_1 = 0$ and is infinite as $|s(0)| > 1$. As the Thue-Morse word is a fixed-point for s, it can be unsubstituted into itself by s placing bars every two letters starting with a bar at position 0. A given factor v of the Thue-Morse word is said *even* if there exists an even position $p \in \mathbb{N}$ such that $\mathcal{T}_p \cdots \mathcal{T}_{p+|v|} = v$, which means that there is a bar just before its first letter v_0 (i.e. between $\mathcal{T}_{p-1}$ and $\mathcal{T}_p$ in the Thue-Morse word) in one of the possible decompositions of v. Complementarily, a factor is said *odd* if it appears at an odd position p in the Thus-Morse word, which means that the bar is between v_0 and v_1 in a decomposition of v. A factor can simultaneously be even and odd if it appears in the Thue-Morse word at both even and odd positions: for instance 101 appears at positions 2 and 11. However, the following lemma points out the fact that a sufficiently large factor cannot be simultaneously even and odd.

Lemma 1. *Any factor of $\mathcal{T}$ of length at least four is either even or odd, which means that it admits a unique decomposition.*

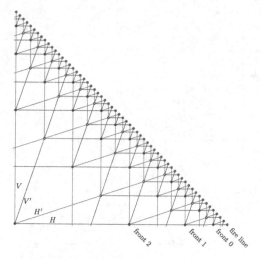

Fig. 4. General scheme of the structure

Hence it is sufficient to look at factors of size four to be able to determine unambiguously their alignment. Observe that moving letter by letter a window of size four over the Thue-Morse word, one sees an alternating sequence of even/odd factors. This will constitute the alternating bits sequence we require.

Now that all the required objects are defined, each signal of the construction will henceforth carry a factor of size four of the Thue-Morse sequence. The signals H and V only carry the even words while H' and V' only carry odd words. When two signals meet on a marked point, we require them to carry the same factor u ($|u| = 4$) and we substitute it by s ($|s(u)| = 8$) to derive the four factors carried by the four "outgoing" (seeing time going NE) signals: the produced type V signal must carry $s(u)_0 \cdots s(u)_3$, the type V' must carry $s(u)_1 \cdots s(u)_4$, the type H' must carry $s(u)_3 \cdots s(u)_6$ and the type H must carry $s(u)_4 \cdots s(u)_7$. First remark that knowing $s(u)$, one uniquely deduces u. Although there is no "middle signal" to carry $s(u)_2 \cdots s(u)_5$, it can easily be deduced from the four other derived words we dispose of.

This defines a coherent structure. Indeed, let us consider any bi-infinite word of $\{0,1\}^{\mathbb{Z}}$ such that every factor appears in the Thue-Morse sequence[2], "written" along a front line of rank $k > 1$ in the following sense: two consecutive marked points hold two consecutive factors of size 4 with a one-letter shift. The previously described local mecanisms of substitution/unsubstitution on marked points enforce the same word (as it is a fixed-point) to appear on all the other front lines and to be used as a ressource to properly alternate between middles and pillars on a front line, coherently with other lines so that a proper global binary hierarchical structure is enforced.

---

[2] For topological reasons, such a word exists. The set of all these words is actually a non-empty subshift of $\{0,1\}^{\mathbb{Z}}$. All of them are fixed-point for the Thue-Morse substitution.

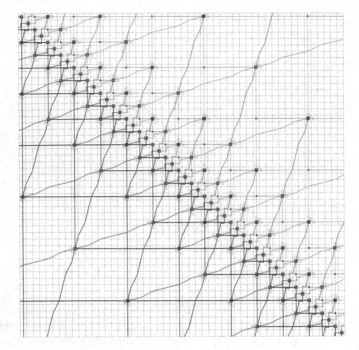

Fig. 5. Marking a diagonal

A realization of this structure using knight tiles is given on Fig. 5 where the colors are represented using some convenient signal notations. The knight colors are not represented on this figure but are all illustrated on Fig. 6. H and V signals, of respective slopes 0 and ∞, carry even words and meet on pillars (solid black squares on the figures). The pilar tiles force their two incoming H and V signals to carry the same factor u. As the four outcoming factors are derived from this common word, a pillar tile is perfectly determined by u, which is explicitly written next to each pillar on Fig. 5. Similarly, H' and V' signals, of respective slopes $\frac{1}{3}$ and $\frac{2}{3}$ realized using particular profiles for reasons that will be exposed later and two different types of knight colors to distinguish the two different types of angles that appear in that profile as illustrated by figures 6a and 6b, carry odd factors. These signals meet on middles, that are represented by gray squares on the figures. The middle tiles force their two incoming H' and V' signals to carry the same factor u. As the four outcoming factors are derived from this common word, a middle tile is perfectly determined by u, which is explicitly written next to each middle on Fig. 5.

We do not dispose of enough space to formally describe the whole tileset. Although most of it is a rather straightforward translation of the previously exposed scheme (Fig. 4 and its symmetric along the fire line) into a tileset, several points certainly require explanations. This is what the remaining paragraphs are dedicated to.

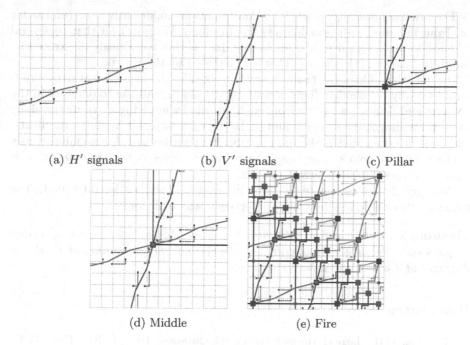

(a) H' signals   (b) V' signals   (c) Pillar

(d) Middle   (e) Fire

Fig. 6. Knight colors of the tileset

First, when a H signal meets a V, they do not simply cross each other, there is actually a *collision* occuring (grey dot on the figures) and there are actually two kinds of H/V signals. When two signals of the first kind meet, they cross each other and by the way change into the second kind (darker color on the figures). When two signals of the second kind meet, they produce a pillar that will in turn produce two signals of the first kind. This collision has to be treated particularly when it is the last collision before the fire line, which can be detected as the collision tile also has to contain some H'/V' signals when it is near the fire line (see Fig. 6e). In that case, the first kind of H/V signals is changed into a third kind that produces a fire tile (instead of a pillar) when meeting.

The case of H' and V' signals requires some careful attention as we need to use knight colors for them to be deterministic in the NW and SE directions. The knight colors used basically contain the type of signal and the word it carries. However a knight must absolutely not cross a front line as in this case the determinism in one of the orthogonal NE or SW directions would fail. Without entering into details, our solution consists in using the $(e_1, e_1, e_2, e_1, e_1, e_1, e_1, e_2)$ profile (alternating between 2 and 4 horizontal steps) such that one can make sure that the tiles that appear at both end of a signal are tiles with blank knight colors (hence no knight crosses the front line). We use the symmetrical profile for V' signals. All situations are illustrated by Fig. 6.

The fire line must appear when marked points are sufficiently close (and as for the front line, no knight should cross the fire line). We resort for that to several *ad-hoc* tiles that appear only near the fire line, as this was already illustrated by the third kind of H/V signals. This is illustrated by the Fig. 6e.

On the other side of the fire line, say the NE side, the hierarchical structure is symmetrically dismantled. This is done by duplicating all the colors used in the SW part and defining a NE tile from each SW tile by applying a symmetry along the NW–SW diagonal on the tile and replacing each SW color by its NE duplicate. Only the fire tiles, that make the junction between boths sides, simultaneously hold SW and NE colors. That way, the resulting tileset is *time-symmetric*: it is its own "inverse" up to a swap in colors.

To conclude, let us denote as κ the previously described knight tileset. The following theorem underlines the properties κ was built for.

Theorem 2. κ *is a 4-way deterministic knight tileset and its set of tilings projects onto the diagonal subshift, i.e.* $\{0, 1\}$*-colorings containing at most one diagonal of 1 and 0 everywhere else.*

References

1. Bennett, C.H.: Logical reversibility of computation. IBM J. Res. Dev. 17(6), 525–532 (1973), http://dx.doi.org/10.1147/rd.176.0525
2. Berger, R.: The Undecidability of the Domino Problem. Ph.D. thesis. Harvard University (1964)
3. Boyle, M., Lind, D.: Expansive subdynamics. Transactions of the American Mathematical Society 349(1), 55–102 (1997)
4. Kari, J., Papasoglu, P.: Deterministic aperiodic tile sets. Geometric and Functional Analysis 9, 353–369 (1999), http://dx.doi.org/10.1007/s000390050090
5. Kari, J.: Rice's theorem for the limit sets of cellular automata. Theor. Comput. Sci. 127(2), 229–254 (1994)
6. Le Gloannec, B., Ollinger, N.: Substitutions and strongly deterministic tilesets. In: Cooper, S.B., Dawar, A., Löwe, B. (eds.) CiE 2012. LNCS, vol. 7318, pp. 462–471. Springer, Heidelberg (2012)
7. Lecerf, Y.: Machines de Turing réversibles. Comptes rendus de l'Académie française des Sciences 257, 2597–2600 (1963)
8. Lukkarila, V.: The 4-way deterministic tiling problem is undecidable. Theor. Comput. Sci. 410(16), 1516–1533 (2009)
9. Morita, K.: Reversible simulation of one-dimensional irreversible cellular automata. Theoretical Computer Science 148(1), 157–163 (1995)
10. Pytheas Fogg, N.: Substitutions in Dynamics, Arithmetics and Combinatorics. Lecture Notes in Mathematics. Springer (2002), http://books.google.fr/books?id=Cmogpq-1SnoC
11. Robinson, R.: Undecidability and nonperiodicity for tilings of the plane. Inventiones Mathematicae 12(3), 177–209 (1971)

Eigenvalues and Transduction
of Morphic Sequences

David Sprunger[1], William Tune[1], Jörg Endrullis[1,2], and Lawrence S. Moss[1,*]

[1] Department of Mathematics, Indiana University, Bloomington IU 47405, USA
[2] Vrije Universiteit Amsterdam, Department of Computer Science, 1081 HV
Amsterdam, The Netherlands

Abstract. We study finite state transduction of automatic and morphic sequences. Dekking [4] proved that morphic sequences are closed
under transduction and in particular morphic images. We present a simple proof of this fact, and use the construction in the proof to show that
non-erasing transductions preserve a condition called α-substitutivity.
Roughly, a sequence is α-substitutive if the sequence can be obtained
as the limit of iterating a substitution with dominant eigenvalue α. Our
results culminate in the following fact: for multiplicatively independent
real numbers α and β, if v is a α-substitutive sequence and w is an β-
substitutive sequence, then v and w have no common non-erasing trans-
ducts except for the ultimately periodic sequences. We rely on Cobham's
theorem for substitutions, a recent result of Durand [5].

1 Introduction

Infinite sequences of symbols are of paramount importance in a wide range of
fields, ranging from formal languages to pure mathematics and physics. A land-
mark was the discovery in 1912 by Axel Thue, founding father of formal language
theory, of the famous sequence 0110 1001 1001 0110 1001 0110 $\cdots$.Thue was interested in infinite words which avoid certain patterns, like squares ww or cubes
www, when w is a non-empty word. Indeed, the sequence shown above, called
the Thue–Morse sequence, is cube-free. It is perhaps the most natural cube-free
infinite word.

A common way to transform infinite sequences is by using *finite state transducers*.
These transducers are deterministic finite automata with input letters and output words
for each transition; an example is shown in
Figure 1. Usually we omit the words "finite
state" and refer to *transducers*. A transducer
maps infinite sequences to infinite sequences
by reading the input sequence letter by let-

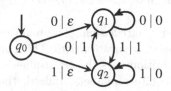

Fig. 1. A transducer computing the
difference (exclusive or) of consecutive bits

ter. Each of these transitions produces an output word, and the sequence formed

* Partially supported by a grant #245591 from the Simons Foundation.

by concatenating each of these output words in the order they were produced is the output sequence. In particular, since this transducer runs for infinite time to read its entire input, this model of transduction does not have final states. A transducer is called *k-uniform* if each step produces *k*-letter words. For example, *Mealy machines* are 1-uniform transducers. A transducer is *non-erasing* if each step produces a non-empty word; this condition is prominent in this paper.

Although transducers are a natural machine model, hardly anything is known about their capabilities of transforming infinite sequences. To state the issues more clearly, let us write $x \trianglelefteq y$ if there is a transducer taking y to x. This transducibility gives rise to a partial order of *stream degrees* [6] that is analogous to, but more fine-grained than, recursion-theoretic orderings such as *Turing reducibility* $\leq_T$ and *many-one reducibility* $\leq_m$. We find it surprising that so little is known about $\trianglelefteq$. As of now, the structure of this order is vastly unexplored territory with many open questions. To answer these questions, we need a better understanding of transducers.

The main things that are known at this point concern two particularly well-known sets of streams, namely the *morphic* and *automatic* sequences. Morphic sequences are obtained as the limit of iterating a morphism on a starting word (and perhaps applying a coding to the limit word). Automatic sequences have a number of independent characterizations (see [1]); we shall not repeat these here. There are two seminal closure results concerning the transduction of morphic and automatic sequences:

(1) The class of morphic sequences is closed under transduction (Dekking [4]).
(2) For all k, the class of k-automatic sequences is closed under uniform transduction (Cobham [3]).

In this paper, we do not attack the central problems concerning the stream degrees. Instead, we are interested in a closure result for non-erasing transductions. Our interest comes from the following easy observation:

(3) For every morphic sequence $w \in \Sigma^\omega$ there is a 2-automatic sequence $w' \in (\Sigma \cup \{a\})^\omega$ such that w is obtained from w' by erasing all occurrences of a. (See Allouche and Shallit [1, Theorem 7.7.1])

This motivates the question: how powerful is non-erasing transduction?

Our Contribution. The main result of this paper is stated in terms of the notion of α-*substitutivity*. This condition is defined in Definition 5 below, and the definition uses the eigenvalues of matrices naturally associated with morphisms on finite alphabets. Indeed, the core of our work is a collection of results on eigenvalues of these matrices.

We prove that the set of α-substitutive words is closed under non-erasing finite state transduction. We follow Allouche and Shallit [1] in obtaining transducts of a given morphic sequence w by *annotating* an iteration morphism, and then taking a morphic image of the annotated limit sequence. For the first part of this transformation, we show that a morphism and its annotation have the same eigenvalues with non-negative eigenvectors. For the second part, we revisit the proof given in Allouche and Shallit [1] of Dekking's theorem that morphic images of morphic sequences are morphic. We simplify the construction in the proof to make it amenable for an analysis of the eigenvalues of the resulting morphism.

For an extended version of this paper with examples we refer to [9].

Related Work. Durand [5] proved that if w is an α-substitutive sequence and h is a non-erasing morphism, then $h(w)$ is α^k-substitutive for some $k \in \mathbb{N}$. We strengthen this result in two directions. First, we show that k may be taken to be 1; hence $h(w)$ is α^k-substitutive for every $k \in \mathbb{N}$. Second, we show that Durand's result also holds for non-erasing transductions.

2 Preliminaries

We recall some of the main concepts that we use in the paper. For a thorough introduction to morphic sequences, automatic sequences and finite state transducers, we refer to [1,8].

We are concerned with infinite sequences Σ^ω over a finite alphabet Σ. We write Σ^* for the set of finite words, Σ^+ for the finite, non-empty words, Σ^ω for the infinite words, and $\Sigma^\infty = \Sigma^* \cup \Sigma^\omega$ for all finite or infinite words over Σ.

2.1 Morphic Sequences and Automatic Sequences

Definition 1. *A* morphism *is a map* $h : \Sigma \to \Gamma^*$. *This map extends by concatenation to* $h : \Sigma^* \to \Gamma^*$, *and we do not distinguish the two notationally. Notice also that* $h(vu) = h(v)h(u)$ *for all* $u,v \in \Sigma^*$. *If* $h_1, h_2 : \Sigma \to \Sigma^*$, *we have a composition* $h_2 \circ h_1 : \Sigma \to \Sigma^*$.

An erased letter *(with respect to* h*) is some* $a \in \Sigma$ *such that* $h(a) = \varepsilon$. *A morphism* $h : \Sigma^* \to \Gamma^*$ *is called* erasing *if has an erased letter. A morphism is* k-uniform *(for* $k \in \mathbb{N}$*) if* $|h(a)| = k$ *for all* $a \in \Sigma$. *A* coding *is a 1-uniform morphism* $c : \Sigma \to \Gamma$.

A morphic sequence is obtained by iterating a morphism, and applying a coding to the limit word.

Definition 2. *Let* $s \in \Sigma^+$ *be a word,* $h : \Sigma \to \Sigma^*$ *a morphism, and* $c : \Sigma \to \Gamma$ *a coding. If the limit* $h^\omega(s) = \lim_{n \to \infty} h^n(s)$ *exists and is infinite, then* $h^\omega(s)$ *is a* pure morphic sequence, *and* $c(h^\omega(s))$ *a* morphic sequence.

If $h(x_1) = x_1 z$ *for some* $z \in \Sigma^+$, *we say that* h *is* prolongable *on* x_1. *In this case,* $h^\omega(x_1)$ *is a pure morphic sequence. If additionally, the morphism* h *is* k-uniform, *then* $c(h^\omega(s))$ *is a* k-automatic *sequence. A sequence* $w \in \Sigma^\omega$ *is called* automatic *if* w *is* k-automatic *for some* $k \in \mathbb{N}$.

2.2 Cobham's Theorem for Morphic Words

Definition 3. *For* $a \in \Sigma$ *and* $w \in \Sigma^*$ *we write* $|w|_a$ *for the number of occurrences of* a *in* w. *Let* h *be a morphism over* Σ. *The* incidence matrix *of* h *is the* *matrix* $M_h = (m_{i,j})_{i \in \Sigma, j \in \Sigma}$ *where* $m_{i,j} = |h(j)|_i$ *is the number of occurrences of the letter* i *in the word* $h(j)$.

Theorem 4 (Perron-Frobenius). *Every non-negative square matrix* M *has a real eigenvalue* $\alpha \geq 0$ *that is greater than or equal to the absolute value of any other eigenvalue of* M *and the corresponding eigenvector is non-negative. We refer to* α *as the* dominating eigenvalue *of* M.

Definition 5. *The* dominating eigenvalue *of a morphism* h *is the dominating eigenvalue of* M_h. *An infinite sequence* $w \in \Sigma^\omega$ *over a finite alphabet* Σ *is said to be* α-substitutive *($\alpha \in \mathbb{R}$) if there exist a morphism* $h : \Sigma \to \Sigma^*$ *with dominating eigenvalue* α, *a coding* $c : \Sigma \to \Sigma$ *and a letter* $a \in \Sigma$ *such that (i)* $w = c(h^\omega(a))$, *and (ii) every letter of* Σ *occurs in* $h^\omega(a)$.

Two complex numbers x, y are called *multiplicatively independent* if for all $k, \ell \in \mathbb{Z}$ it holds that $x^k = y^\ell$ implies $k = \ell = 0$. We shall use the following version of Cobham's theorem due to Durand [5].

Theorem 6. *Let* α *and* β *be multiplicatively independent Perron numbers. If a sequence* w *is both* α-substitutive *and* β-substitutive, *then* w *is eventually periodic.* $\qquad\square$

2.3 Transducers

Definition 7. *A* sequential finite-state transducer (FST) $M = (\Sigma, \Delta, Q, q_0, \delta, \lambda)$ *consists of (i) a finite input alphabet* Σ, *(ii) a finite output alphabet* Δ, *(iii) a finite set of states* Q, *(iv) an initial state* $q_0 \in Q$, *(v) a transition function* $\delta : Q \times \Sigma \to Q$, *and (vi) an output function* $\lambda : Q \times \Sigma \to \Delta^*$.

We use transducers to transform infinite words. The transducer reads the input word letter by letter, and the transformation result is the concatenation of the output words encountered along the edges.

Definition 8. *Let* $M = (\Sigma, \Delta, Q, q_0, \delta, \lambda)$ *be a transducer. We extend the state transition function* δ *from letters* Σ *to finite words* Σ^* *as follows:* $\delta(q, \varepsilon) = q$ *and* $\delta(q, aw) = \delta(\delta(q, a), w)$ *for* $q \in Q$, $a \in \Sigma$, $w \in \Sigma^*$.

The output function λ is extended to the set of all words $\Sigma^\infty = \Sigma^\omega \cup \Sigma^*$ by the following definition: $\lambda(q, \varepsilon) = \varepsilon$ and $\lambda(q, aw) = \lambda(q, a)\lambda(\delta(q, a), w)$ for $q \in Q$, $a \in \Sigma$, $w \in \Sigma^\infty$.

We introduce $\delta(w)$ and $\lambda(w)$ as shorthand for $\delta(q_0, w)$ and $\lambda(q_0, w)$, respectively. Moreover, we define $M(w) = \lambda(w)$, the *output of* M *on* $w \in \Sigma^\omega$. In this way, we think of M as a function from (finite or infinite) words on its input alphabet to infinite words on its output alphabet $M : \Sigma^\infty \to \Delta^\infty$. If $x \in \Sigma^\omega$ and $y \in \Delta^\omega$, we write $y \trianglelefteq x$ if for some transducer M, we have $M(x) = y$.

Notice that every morphism is computable by a transducer (with one state). In particular, every coding is computable by a transducer.

3   Closure of Morphic Sequences under Morphic Images

Definition 9. Let $h : \Sigma^* \to \Sigma^*$ be morphisms, and let $\Gamma \subseteq \Sigma$ be a set of letters. We call a letter $a \in \Sigma$

  (i) *dead* if $h^n(a) \in \Gamma^*$ for all $n \geq 0$,
 (ii) *near dead* if $a \notin \Gamma$, and for all $n > 0$, $h^n(a)$ consists of dead letters,
(iii) *resilient* if $h^n(a) \notin \Gamma^*$ for all $n \geq 0$,
 (iv) *resurrecting* if $a \in \Gamma$ and $h^n(a) \notin \Gamma^*$ for all $n > 0$

with respect to h and Γ. We say that the morphism h *respects* Γ if every letter $a \in \Sigma$ is either dead, near dead, resilient, or resurrecting. (Note that all of these definitions are with respect to some fixed h and Γ.)

Lemma 10. *Let $g : \Sigma^* \to \Sigma^*$ be a morphism, and let $\Gamma \subseteq \Sigma$. Then g^r respects Γ for some natural number $r > 0$.*

Proof. See Lemma 7.7.3 in Allouche and Shallit [1].   □

Definition 11. *For a set of letters $\Gamma \subseteq \Sigma$ and a word $w \in \Sigma^\infty$, we write $\gamma_\Gamma(w)$ for the word obtained from w by erasing all occurrences of letters in Γ.*

Definition 12. Let $g : \Sigma^* \to \Sigma^*$ be a morphism, and $\Gamma \subseteq \Sigma$ a set of letters. We construct an alphabet Δ, a morphism $\xi : \Delta^* \to \Delta^*$ and a coding $\rho : \Delta \to \Sigma$ as follows. We refer to Δ, ξ, ρ as the *morphic system associated with the erasure of Γ from g^ω*.

Let $r \in \mathbb{N}_{>0}$ be minimal such that g^r respects Γ (r exists by Lemma 10). Let $\mathcal{D}$ be the set of dead letters with respect to g^r and Γ. For $x \in \Sigma^*$ we use brackets $[x]$ to denote a new letter. For words $w \in \{g^r(a) \mid a \in \Sigma\}$, whenever $\gamma_\mathcal{D}(w) = w_0\, a_1 w_1\, a_2 w_2\, \cdots\, a_{k-1} w_{k-1}\, a_k w_k$ with $a_1, \ldots, a_k \notin \Gamma$ and $w_0, \ldots, w_k \in \Gamma^*$, we define $\mathsf{blocks}(w) = [w_0 a_1 w_1]\, [a_2 w_2]\, \cdots\, [a_{k-1} w_{k-1}]\, [a_k w_k]$. Here it is to be understood that $\mathsf{blocks}(w) = \varepsilon$ if $\gamma_\mathcal{D}(w) = \varepsilon$, and $\mathsf{blocks}(w)$ is *undefined* if $\gamma_\mathcal{D}(w) \in \Gamma^+$.

Let the alphabet Δ consist of all letters $[a]$ and all bracketed letters $[w]$ occurring in words $\mathsf{blocks}(g^r(a))$ for $a \in \Sigma$. We define the morphism $\xi : \Delta \to \Delta^*$ and the coding $\rho : \Delta \to \Sigma$ by $\xi([a_1 \cdots a_k]) = \mathsf{blocks}(g^r(a_1)) \cdots \mathsf{blocks}(g^r(a_k))$ and $\rho([w\,a\,u]) = a$ for $[a_1 \cdots a_k] \in \Delta$ and $a \notin \Gamma$, $w, u \in \Gamma^*$. For $a \in \Gamma$ we can define $\rho([a])$ arbitrarily, for example, $\rho(a) = a$.

Proposition 13. *Let $g : \Sigma^* \to \Sigma^*$ be a morphism, $a \in \Sigma$ such that $g^\omega(a) \in \Sigma^\omega$, and $\Gamma \subseteq \Sigma$ a set of letters. Let Δ, ξ and ρ be the morphic system associated to the erasure of Γ from g^ω in Definition 12. Then $\rho(\xi^\omega([a])) = \gamma_\Gamma(g^\omega(a))$.*

Proof. For $\ell \in \mathbb{N}$ and $[w_1], \ldots, [w_\ell] \in \Delta$ we define $\mathsf{cat}([w_1] \cdots [w_\ell]) = w_1 \cdots w_\ell$. We prove by induction on n that for all words $w \in \Delta^*$, and for all $n \in \mathbb{N}$, $\mathsf{cat}(\xi^n(w)) = g^{nr}(\mathsf{cat}(w))$. The base case is immediate. For the induction step, assume that we have $n \in \mathbb{N}$ such that for all words $w \in \Delta^*$, $\mathsf{cat}(\xi^n(w)) = g^{nr}(\mathsf{cat}(w))$. Let $w \in \Delta^*$, $w = [a_{1,1} \cdots a_{1,\ell_1}] \cdots [a_{k,1} \cdots a_{k,\ell_k}]$. Then

$$\mathsf{cat}(\xi(w)) = \mathsf{cat}(\xi([a_{1,1} \cdots a_{1,\ell_1}]) \cdots \xi([a_{k,1} \cdots a_{k,\ell_k}]))$$
$$= \mathsf{cat}(\mathsf{blocks}(g^r(a_{1,1})) \cdots \mathsf{blocks}(g^r(a_{k,\ell_k}))) = g^r(\mathsf{cat}(w))$$

By the induction hypothesis, $\mathsf{cat}(\xi^{n+1}(w)) = g^{nr}(\mathsf{cat}(\xi(w))) = g^{nr}(g^r(\mathsf{cat}(w))) = g^{(n+1)r}(\mathsf{cat}(w))$. To complete the proof, note that by definition $\rho([w\,a\,u]) = \gamma_\Gamma(w\,a\,u)$ and thus $\rho(w) = \gamma_\Gamma(\mathsf{cat}(w))$ for every $w \in \Delta^*$. Hence, for all $n \geq 1$, $\rho(\xi^n([a])) = \gamma_\Gamma(\mathsf{cat}(\xi^n([a]))) = \gamma_\Gamma(g^{nr}(a))$. Taking limits, we obtain $\rho(\xi^\omega([a])) = \gamma_\Gamma(g^\omega(a))$. $\qquad \square$

Definition 14. Let $g, h : \Sigma^* \to \Sigma^*$ be morphisms such that h is non-erasing. We construct an alphabet Δ, a morphism $\xi : \Delta^* \to \Delta^*$ and a coding $\rho : \Delta \to \Sigma$ as follows. We refer to Δ, ξ, ρ as the *morphic system associated with the morphic image of g^ω under h*.

Let $\Delta = \Sigma \cup \{ [a] \mid a \in \Sigma \}$. For nonempty words $w = a_1 a_2 \cdots a_k \in \Sigma^*$ we define $\mathsf{head}(w) = a_1$, $\mathsf{tail}(w) = a_2 \cdots a_k$ and $\mathsf{img}(w) = [a_1]u_1 [a_2]u_2 \cdots [a_k]u_k$ where $u_i = \mathsf{tail}(h(a_i)) \in \Sigma^*$. We define the morphism $\xi : \Delta^* \to \Delta^*$ and the coding $\rho : \Delta \to \Sigma$ by $\xi([a]) = \mathsf{img}(g(a)))$ and $\xi(a) = \varepsilon$, and $\rho([a]) = \mathsf{head}(h(a))$ and $\rho(a) = a$ for $a \in \Sigma$.

Notice here the $\rho([a])$ and u_i, defined using $\mathsf{head}()$ and $\mathsf{tail}()$, are well-defined since h is non-erasing and hence $h(a_i)$ will be nonempty.

Proposition 15. *Let $g, h : \Sigma^* \to \Sigma^*$ be morphisms such that h is non-erasing, and $a \in \Sigma$ such that $g^\omega(a) \in \Sigma^\omega$. Let Δ, ξ and ρ be as in Definition 12. Then $\rho(\xi^\omega([a])) = h(g^\omega(a))$.*

Proof. We define $z : \Delta \to \Sigma^*$ by $z(a) = \varepsilon$ and $z([a]) = a$ for all $a \in \Sigma$. By induction on $n > 0$ we show $\rho(\xi^n(w)) = h(g^n(z(w)))$ and $z(\xi^n(w)) = g^n(z(w))$ for all $w \in \Delta^*$.

We start with the base case. Note that $\rho(\xi([a])) = h(g(a)) = h(g(z([a])))$ and $\rho(\xi(a)) = \varepsilon = h(g(z(a)))$ for all $a \in \Sigma$, and thus $\rho(\xi(w)) = h(g(z(w)))$ for all $w \in \Delta^*$. Moreover, we have $z(\xi([a])) = g(a) = g(z([a]))$ and $z(\xi(a)) = \varepsilon = g(z(a))$ for all $a \in \Sigma$, and hence $z(\xi(w)) = g(z(w))$ for all $w \in \Delta^*$.

Let us consider the induction step. By the base case and induction hypothesis $\rho(\xi^{n+1}(w)) = \rho(\xi(\xi^n(w))) = h(g(z(\xi^n(w)))) = h(g(g^n(z(w)))) = h(g^{n+1}(z(w)))$ and $z(\xi^{n+1}(w)) = z(\xi(\xi^n(w))) = g(z(\xi^n(w))) = g(g^n(z(w))) = g^{n+1}(z(w))$. Thus $\rho(\xi^n([a])) = h(g^n(a))$ for all $n \in \mathbb{N}$, and taking limits: $\rho(\xi^\omega([a])) = h(g^\omega(a))$. $\qquad \square$

Every morphic image of a word can be obtained by erasing letters, followed by the application of a non-erasing morphism. As a consequence we obtain:

Corollary 16. *The morphic image of a pure morphic word is morphic or finite.*

Proof. Let $w \in \Sigma^\omega$ be a word and $h : \Sigma \to \Sigma^*$ a morphism. Let $\Gamma = \{ a \mid h(a) = \varepsilon \}$ be the set of letters erased by h, and $\Delta = \Sigma \setminus \Gamma$. Then $h(w) = g(\gamma_\Gamma(w))$ where g is the non-erasing morphism obtained by restricting h to Δ. Hence for purely morphic w, the result follows from Propositions 13 and 15.   □

Theorem 17 (Cobham [2], Pansiot [7]). *The morphic image of a morphic word is morphic.*

Proof. Follows from Corollary 16 since the coding can be absorbed into the morphic image.   □

Eigenvalue Analysis

The following lemma states that if a square matrix N is an extension of a square matrix M, and all added columns contain only zeros, then M and N have the same non-zero eigenvalues.

Lemma 18. *Let Σ, Δ be disjoint, finite alphabets. Let $M = (m_{i,j})_{i,j \in \Sigma}$ and $N = (n_{i,j})_{i,j \in \Sigma \cup \Delta}$ be matrices such that (i) $n_{i,j} = m_{i,j}$ for all $i, j \in \Sigma$ and (ii) $n_{i,j} = 0$ for all $i \in \Sigma \cup \Delta$, $j \in \Delta$. Then M and N have the same non-zero eigenvalues.*

$$\begin{pmatrix} M & \begin{matrix} 0 \cdots 0 \\ \end{matrix} \\ \hline \begin{matrix} 0 \cdots 0 \\ 0 \cdots 0 \end{matrix} \end{pmatrix}$$

Proof. N is a block lower triangular matrix with M and 0 as the matrices on the diagonal. Hence the eigenvalues of N are the combined eigenvalues of M and 0. Therefore M and N have the same non-zero eigenvalues.   □

We now show that morphic images with respect to non-erasing morphisms preserve α-substitutivity. This strengthens a result obtained in [5] where it has been shown that the non-erasing morphic image of an α-substitutive sequence is α^k-substitutive for some $k \in \mathbb{N}$. We show that one can always take $k = 1$. Note that every α-substitutive sequence is also α^k-substitutive for all $k \in \mathbb{N}, k > 0$.

Theorem 19. *Let Σ be a finite alphabet, $w \in \Sigma^\omega$ be an α-substitutive sequence and $h : \Sigma \to \Sigma^*$ a non-erasing morphism. Then the morphic image of w under h, that is $h(w)$, is α-substitutive.*

Proof. Let $\Sigma = \{ a_1, \ldots, a_k \}$ be a finite alphabet, $w \in \Sigma^\omega$ be an α-substitutive sequence and $h : \Sigma \to \Sigma^*$ a non-erasing morphism. As the sequence w is α-substitutive, there exist a morphism $g : \Sigma \to \Sigma^*$ with dominant eigenvalue α, a coding $c : \Sigma \to \Sigma$ and a letter $a \in \Sigma$ such that $w = c(g^\omega(a))$ and all letters from Σ occur in $g^\omega(a)$. Then $h(w) = h(c(g^\omega(a))) = (h \circ c)(g^\omega(a)))$, and $h \circ c$ is a non-erasing morphism. Without loss of generality, by absorbing c into h, we may assume that c is the identity.

From h and g, we obtain an alphabet Δ, a morphism ξ, and a coding ρ as in Definition 14. Then by Proposition 15, we have $\rho(\xi^\omega([a])) = h(g^\omega(a))$. As a consequence, it suffices to show that $\rho(\xi^\omega([a]))$ is α-substitutive. Let $M = (M_{i,j})_{i,j \in \Sigma}$ and $N = (N_{i,j})_{i,j \in \Delta}$ be the incidence matrices of g and ξ, respectively. By Definition 14 we have for all $a, b \in \Sigma$: $|\xi([a])|_{[b]} = |g(a)|_b$ and

$|\xi(a)|_b = |\xi(a)|_{[b]} = 0$. Hence we obtain $N_{[b],[a]} = M_{b,a}$, $N_{b,a} = 0$ and $N_{[b],a} = 0$ for all $a,b \in \Sigma$. After changing the names (swapping a with $[a]$) in N, we obtain from Lemma 18 that N and M have the same non-zero eigenvalues, and thus the same dominant eigenvalue.   □

4   Closure of Morphic Sequences under Transduction

In this section, we give a proof of the following theorem due to Dekking [4].

Theorem 20 (Transducts of morphic sequences are morphic). *If M is a transducer with input alphabet Σ and $x \in \Sigma^\omega$ is a morphic sequence, then $M(x)$ is morphic or finite.*

This proof will proceed by *annotating* entries in the original sequence x with information about what state the transducer is in upon reaching that entry. This allows us to construct a new morphism which produces the transduced sequence $M(x)$ as output. After proving this theorem, we will show that this process of annotation preserves α-substitutivity.

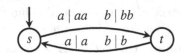

Fig. 2. A transducer that doubles every other letter

4.1   Transducts of Morphic Sequences Are Morphic

We show in Lemma 27 that transducts of morphic sequences are morphic. In order to prove this, we also need several lemmas about transducers which are of independent interest. The approach here is adapted from a result in Allouche and Shallit [1]; it is attributed in that book to Dekking. We repeat it here partly for the convenience of the reader, but mostly because there are some details of the proof which are used in the analysis of the substitutivity property.

Definition 21 (τ_w, $\Xi(w)$). *Given a transducer $M = (\Sigma, \Delta, Q, q_0, \delta, \lambda)$ and a word $w \in \Sigma^*$, we define $\tau_w \in Q^Q$ to be $\tau_w(q) = \delta(q, w)$. Note that $\tau_{wv} = \tau_v \circ \tau_w$. Further, we define $\Xi : \Sigma^* \to (Q^Q)^\omega$ by $\Xi(w) = (\tau_w, \tau_{h(w)}, \tau_{h^2(w)}, \ldots, \tau_{h^n(w)}, \ldots)$.*

Next, we show that $\{\, \Xi(w) : w \in \Sigma^* \,\}$ is finite.

Lemma 22. *For any transducer M and any morphism $h : \Sigma \to \Sigma^*$, there are natural numbers $p \geq 1$ and $n \geq 0$ so that for all $w \in \Sigma^*$, $\tau_{h^i(w)} = \tau_{h^{i+p}(w)}$ for all $i \geq n$.*

Proof. Let $\Sigma = \{1, 2, \ldots, s\}$. Define $H : (Q^Q)^s \to (Q^Q)^s$ by $H(f_1, f_2, \ldots, f_s) = (f_{h(1)}, f_{h(2)}, \ldots, f_{h(s)})$. When we write $f_{h(i)}$ on the right, here is what we mean. Suppose that $h(i) = v_0 \cdots v_j$. Then $f_{h(i)}$ is short for the composition $f_{v_j} \circ f_{v_{j-1}} \circ \cdots \circ f_{v_1} \circ f_{v_0}$. Recall the notation τ_w from Definition 21; we thus have τ_i for the individual letters $i \in \Sigma$. Consider $T_0 = (\tau_1, \tau_2, \ldots, \tau_s)$. We define its *orbit* as the infinite sequence $(T_i)_{i \in \omega}$ of elements of $(Q^Q)^s$ given by $T_i = H^i(T_0) = H^i(\tau_1, \ldots \tau_s) = (\tau_{h^i(1)}, \ldots, \tau_{h^i(s)})$. Since each of the T_i belongs to the finite set $(Q^Q)^s$, the orbit of T_0 is eventually periodic. Let n be the preperiod length and p be the period length. The periodicity implies that $(*)$ $\tau_{h^i(j)} = \tau_{h^{i+p}(j)}$ for each $j \in \Sigma$ and for all $i \geq n$.

Let $w \in \Sigma^*$ and $i \geq n$. Since $w \in \Sigma^*$, we can write it as $w = \sigma_1 \sigma_2 \cdots \sigma_m$. We prove that $\tau_{h^i(w)} = \tau_{h^{i+p}(w)}$. Note that $\tau_{h^i(w)} = \tau_{h^i(\sigma_1 \cdots \sigma_m)} = \tau_{h^i(\sigma_1) \cdots h^i(\sigma_m)} = \tau_{h^i(\sigma_n)} \circ \cdots \circ \tau_{h^i(\sigma_1)}$. We got this by breaking w into individual letters, then using the fact that h is a morphism, and finally using the fact that $\tau_{uv} = \tau_u \circ \tau_v$. Finally we know by $(*)$ that for individual letters, $\tau_{h^i(\sigma_j)} = \tau_{h^{i+p}(\sigma_j)}$. So $\tau_{h^i(w)} = \tau_{h^{i+p}(w)}$, as desired. $\qquad \square$

Definition 23 ($\Theta(w)$). *Given a transducer M and a morphism h, we find p and n as in Lemma 22 just above and define $\Theta(w) = (\tau_w, \tau_{h(w)}, \ldots, \tau_{h^{n+p-1}(w)})$.*

Lemma 24.   *(i) Given M and h, the set $A = \{\Theta(w) : w \in \Sigma^*\}$ is finite.*
(ii) If $\Theta(w) = \Theta(y)$, then $\Theta(h(w)) = \Theta(h(y))$.
(iii) If $\Theta(w) = \Theta(y)$, then for all $u \in \Sigma^*$, $\Theta(wu) = \Theta(yu)$.

Proof. Part (i) comes from the fact that each of the $n + p$ coordinates of $\Theta(w)$ comes from the finite set Q^Q. For (ii), we calculate:

$$\Theta(h(w)) = (\tau_{h(w)}, \tau_{h^2(w)}, \ldots, \tau_{h^{n+p}(w)}) = (\tau_{h(w)}, \tau_{h^2(w)}, \ldots, \tau_{h^{n+p-1}(w)}, \tau_{h^n(w)})$$
$$= (\tau_{h(y)}, \tau_{h^2(y)}, \ldots, \tau_{h^{n+p-1}(y)}, \tau_{h^n(y)}) = \Theta(h(y))$$

using by Lemma 22 and since $\Theta(w) = \Theta(y)$. Part (iii) uses $\Theta(w) = \Theta(y)$ as follows:

$$\Theta(wu) = (\tau_u \circ \tau_w, \tau_{h(u)} \circ \tau_{h(w)}, \tau_{h^2(u)} \circ \tau_{h^2(w)}, \ldots, \tau_{h^{n+p-1}(u)} \circ \tau_{h^{n+p-1}(w)})$$
$$= (\tau_u \circ \tau_y, \tau_{h(u)} \circ \tau_{h(y)}, \tau_{h^2(u)} \circ \tau_{h^2(y)}, \ldots, \tau_{h^{n+p-1}(u)} \circ \tau_{h^{n+p-1}(y)}) = \Theta(yu) \quad \square$$

Definition 25 ($\overline{h}$). *Given a transducer M and a morphism h, let A be as in Lemma 24(i). Define the morphism $\overline{h} : \Sigma \times A \to (\Sigma \times A)^*$ as follows. For for all $\sigma \in \Sigma$, whenever $h(\sigma) = s_1 s_2 s_3 \cdots s_\ell$, let $\overline{h}((\sigma, \Theta(w)))$ be defined as*

$$(s_1, \Theta(hw)) \, (s_2, \Theta((hw)s_1)) \, (s_3, \Theta((hw)s_1 s_2)) \, \cdots \, (s_\ell, \Theta((hw)s_1 s_2 \cdots s_{\ell-1}))$$

By Lemma 24, $\overline{h}$ is well-defined. Notice that $|\overline{h}(\sigma, a)| = |h(\sigma)|$ for all σ.

Lemma 26. *For all $\sigma \in \Sigma$, all $w \in \Sigma^*$ and all $n \in \mathbb{N}$, if $h^n(\sigma) = s_1 s_2 \cdots s_\ell$, then*

$$\overline{h}^n((\sigma, \Theta(w))) = (s_1, \Theta(h^n w)) \, (s_2, \Theta((h^n w)s_1)) \, \cdots \, (s_\ell, \Theta((h^n w)s_1 \cdots s_{\ell-1})) \, .$$

In particular, for $1 \leq i \leq \ell$, the first component of the ith term in $h^n(\sigma, \Theta(w))$ is s_i.

Proof. By induction on n. For $n = 0$, the claim is trivial. Assume that it holds for n. Let $h^n(\sigma) = s_1 s_2 \cdots s_\ell$, and for $1 \le i \le \ell$, let $h(s_i) = t_1^i t_2^i \cdots t_{k_i}^i$. Thus $h^{n+1}(\sigma) = h(s_1 s_2 \cdots s_\ell) = t_1^1 t_2^1 \cdots t_{k_1}^1 t_1^2 t_2^2 \cdots t_{k_2}^2 t_1^\ell t_2^\ell \cdots t_{k_\ell}^\ell$. Then:

$$\overline{h}(\overline{h}^n(\sigma, \Theta(w))) = \overline{h}(s_1, \Theta((h^n w))) \cdots \overline{h}(s_\ell, \Theta((h^n w)s_1 s_2 \cdots s_{\ell-1}))$$

For $1 \le i \le \ell$, we have

$$\begin{aligned}
&\overline{h}(s_i, \Theta((h^n w)s_1 \cdots s_{i-1})) \\
&= (t_1^i, \Theta((hh^n w)h(s_1 \cdots s_{i-1}))) \quad (t_2^i, \Theta((hh^n w)h(s_1 \cdots s_{i-1})t_1^i)) \\
&\quad \cdots \quad (t_{k_i}^i, \Theta(hh^n w)h(s_1 \cdots s_{i-1})t_1^i t_2^i \cdots t_{k_i-1}^i)) \\
&= (t_1^i, \Theta((h^{n+1} w)t_1^1 t_2^1 \cdots t_{k_1}^1 \cdots t_1^{i-1} t_2^{i-1} \cdots t_{k_i-1}^{i-1})) \quad (t_2^i, \Theta((h^{n+1} w)t_1^1 t_2^1 \\
&\quad \cdots \quad t_{k_1}^1 \cdots t_1^{i-1} t_2^{i-1} \cdots t_{k_i-1}^{i-1} t_1^i)) \\
&\quad \cdots \quad (t_{k_i}^i, \Theta((h^{n+1} w)t_1^1 t_2^1 \cdots t_{k_i}^1 \cdots t_1^{i-1} t_2^{i-1} \cdots t_{k_i-1}^{i-1} t_1^i \cdots t_{k_i-1}^i))
\end{aligned}$$

Concatenating the sequences $\overline{h}(s_i, \Theta((h^n w)s_1 \cdots s_{i-1}))$ for $i = 1, \ldots, \ell$ completes our induction step. $\square$

Lemma 27. *Let $M = (\Sigma, \Delta, Q, q_0, \delta, \lambda)$ be a transducer, let h be a morphism prolongable on the letter x_1, and write $h^\omega(x_1)$ as $x = x_1 x_2 x_3 \cdots x_n \cdots$. Let Θ be from Definition 23. Using this, let A be from Lemma 24(i), and $\overline{h}$ from Definition 25. Then*

(i) $\overline{h}$ is prolongable on $(x_1, \Theta(\varepsilon))$.

(ii) Let $c : \Sigma \times A \to \Sigma \times Q$ be the coding $c(\sigma, \Theta(w)) = (\sigma, \tau_w(q_0))$. Then c is well-defined.

(iii) The image under c of $\overline{h}^\omega((x_1, \Theta(\varepsilon))$ is

$$z = (x_1, \delta(q_0, \varepsilon)) \; (x_2, \delta(q_0, x_1)) \; \cdots \; (x_n, \delta(q_0, x_1 x_2 \cdots x_{n-1})) \; \cdots \quad (1)$$

This sequence z is morphic in the alphabet $\Sigma \times Q$.

Proof. For (i), write $h(x_1)$ as $x_1 x_2 \cdots x_\ell$. Using the fact that $h^i(\varepsilon) = \varepsilon$ for all i, we see that $\overline{h}((x_1, \Theta(\varepsilon))) = (x_1, \Theta(\varepsilon)) \; (x_2, \Theta(x_1)) \; \cdots \; (x_\ell, \Theta(x_1, \ldots, x_{\ell-1}))$. This verifies the prolongability. For (ii): if $\Theta(w) = \Theta(u)$, then τ_w and τ_u are the first component of $\Theta(w)$ and are thus equal. We turn to (iii). Taking $w = \varepsilon$ in Lemma 26 shows that $\overline{h}^\omega((x_1, \Theta(\varepsilon))$ is

$$(x_1, \Theta(\varepsilon)) \; (x_2, \Theta(x_1)) \; (x_3, \Theta(x_1 x_2)) \; \cdots \; (x_m, \Theta(x_1 x_2 \cdots x_{m-1})) \; \cdots .$$

The image of this sequence under the coding c is

$$(x_1, \tau_\varepsilon(q_0)) \; (x_2, \tau_{x_1}(q_0)) \; (x_3, \tau_{x_1 x_2}(q_0)) \; \cdots \; (x_m, \tau_{x_1 x_2 \cdots x_{m-1}}(q_0)) \; \cdots .$$

In view of the τ functions' definition (Def. 21), we obtain z in (1). By definition, z is morphic. $\square$

This is most of the work required to prove Theorem 20, the main result of this section.

Proof (Theorem 20). Since x is morphic, there is a morphism $h : \Sigma' \to (\Sigma')^*$, a coding $c : \Sigma' \to \Sigma$, and a letter $x_1 \in \Sigma'$ so that $x = c(h^\omega(x_1))$. We are to show that $M(c(h^\omega(x_1)))$ is morphic. Since c is computable by a transducer, we have $x = (M \circ c)(h^\omega(x_1))$, where $\circ$ is the wreath product of transducers. It is thus sufficient to show that given a transducer M, the sequence $M(h^\omega(x_1))$ is morphic.

The sequence $z = (x_1, \delta(q_0, \varepsilon))\ (x_2, \delta(q_0, x_1))\ (x_3, \delta(q_0, x_1 x_2))\ \cdots$ is morphic by Lemma 27. The output function of M is a morphism $\lambda : \Sigma \times Q \to \Delta^*$. By Corollary 16, $\lambda(z)$ is morphic or finite. But $\lambda(z)$ is exactly $M(x)$. $\qquad\square$

4.2 Substitutivity of Transducts

We are also interested in analyzing the α-substitutivity of transducts. We claim that if a sequence x is α-substitutive, then $M(x)$ is also α-substitutive for all M.

As a first step, we show that annotating a morphism does not change α-substitutivity.

Definition 28. *Let Σ be an alphabet, A any set and $w = (b_1, a_1) \ldots (b_k, a_k) \in (\Sigma \times A)^*$ be a word. We call A the* set of annotations. *We write $\lfloor w \rfloor$ for the word $b_1 b_2 \ldots b_k$, that is, the word obtained by dropping the annotations.*

A morphism $\overline{h} : (\Sigma \times A) \to (\Sigma \times A)^*$ is an annotation *of $h : \Sigma \to \Sigma^*$ if $h(b) = \lfloor \overline{h}(b, a) \rfloor$ for all $b \in \Sigma$, $a \in A$.*

Note that the morphism $\overline{h}$ from Definition 25 is an annotation of h in this sense. Then from the following proposition it follows that if x is α-substitutive, then the sequence z in Lemma 27 is also α-substitutive.

Proposition 29. *If $x = h^\omega(\sigma)$ is an α-substitutive morphic sequence with morphism $h : \Sigma \to \Sigma^*$ and A is any set of annotations, then any annotated morphism $\overline{h} : \Sigma \times A \to (\Sigma \times A)^*$ also has an infinite fixpoint $\overline{h}^\omega((\sigma, a))$ which is also α-substitutive.*

The proof of this proposition is in two lemmas: first that the eigenvalues of the morphism are preserved by the annotation process, and second that if α is the dominant eigenvalue for h, then no greater eigenvalues are introduced for $\overline{h}$.

Lemma 30. *All eigenvalues for h are also eigenvalues for any annotated version $\overline{h}$ of h.*

Proof. Let $M = (m_{i,j})_{i,j \in \Sigma}$ be the incidence matrix of h. Order the elements of the annotated alphabet $\Sigma \times A$ lexicographically. Then the incidence matrix of $\overline{h}$, call it $N = (n_{i,j})_{i,j \in \Sigma \times A}$, can be thought of as a block matrix where the blocks have size $|A| \times |A|$ and there are $|\Sigma| \times |\Sigma|$ such blocks in N. Note that by the definition of annotation, the row sum in each row of the (a, b) block of N is $m_{a,b}$. To simplify the notation, for the rest of this proof we write J for $|\Sigma|$ and K for $|A|$. Suppose $v = (v_1, v_2, \ldots, v_J)$ is a column eigenvector for M with eigenvalue α. Consider $\overline{v} = (v_1, \ldots, v_1, v_2, \ldots, v_2, \ldots, v_n, \ldots v_n)$. This is a "block vector":

the first K entries are v_1, the second K entries are v_2, and so on, for a total of $K \cdot J$ entries. We claim that $\overline{v}$ is a column eigenvector for N with eigenvalue α.

Consider the product of row k of N with $\overline{v}$. This is $\sum_{j=1}^{K \cdot J} n_{k,j} \overline{v}_j = \sum_{b=1}^{J} v_b \cdot (\sum_{j=1}^{K} n_{k,Kb+j})$. Now $k = Ka + r$. So $\sum_{j=1}^{K} n_{k,Kb+j}$ is the row sum of the (a,b) block of N and hence is $m_{a,b}$. Therefore, row k of N times $\overline{v}$ is $\sum_{b=1}^{J} v_b m_{a,b} = \alpha v_a$, since v is an eigenvector of M. Finally we note that the kth entry of $\overline{v}$ is v_a by its definition. Hence multiplying $\overline{v}$ by N multiplies the kth entry of $\overline{v}$ by α for all k.

We have shown that $\overline{v}$ is a column eigenvector of N with eigenvalue α, so the (column) eigenvalues of M are all present in N. However, since a matrix and its transpose have the same eigenvalues, the (column) qualification on the eigenvalues is unnecessary. □

If $\overline{h}$ is an annotation of h, then we have

$$|h(b)|_{b'} = \sum_{a' \in A} |\overline{h}((b,a))|_{(b',a')} \qquad \text{for all } b, b' \in \Sigma \text{ and } a \in A \qquad (2)$$

Lemma 31. *Let $h, \overline{h}$ be morphisms such that $\overline{h} : (\Sigma \times A) \to (\Sigma \times A)^*$ is an annotation of $h : \Sigma \to \Sigma^*$. Then every eigenvalue of $\overline{h}$ with a non-negative eigenvector is also an eigenvalue for h.*

Proof. Let $M = (m_{i,j})_{i,j \in \Sigma}$ be the incidence matrix of h and $N = (n_{i,j})_{i,j \in \Sigma \times A}$ be the incidence matrix of $\overline{h}$. Let r be an eigenvalue of N with corresponding eigenvector $v = (v_{(b,a)})_{(b,a) \in \Sigma \times A}$, that is, $Nv = rv$ and $v \neq 0$. We define a vector $w = (w_b)_{b \in \Sigma}$ as follows: $w_b = \sum_{a \in A} v_{(b,a)}$. We show that $Mw = rw$. Let $b' \in \Sigma$, then:

$$(Mw)_{b'} = \sum_{b \in \Sigma} M_{b',b} w_b = \sum_{b \in \Sigma} \left(M_{b',b} \sum_{a \in A} v_{(b,a)} \right) = \sum_{b \in \Sigma} \sum_{a \in A} M_{b',b} v_{(b,a)}$$

$$\overset{\text{by (2)}}{=} \sum_{b \in \Sigma} \sum_{a \in A} \left(\sum_{a' \in A} N_{(b',a'),(b,a)} \right) v_{(b,a)} = \sum_{a' \in A} \sum_{b \in \Sigma} \sum_{a \in A} N_{(b',a'),(b,a)} v_{(b,a)}$$

$$\overset{Nv=rv}{=} \sum_{a' \in A} r v_{(b',a')} = r \sum_{a' \in A} v_{(b',a')} = r w_{b'}$$

Hence $Mw = rw$. If $w \neq 0$ it follows that r is an eigenvalue of M. Note that if v is non-negative, then $w \neq 0$. This proves the claim. □

Corollary 32. *Let $h, \overline{h}$ be morphisms such that $\overline{h} : (\Sigma \times A) \to (\Sigma \times A)^*$ is an annotation of $h : \Sigma \to \Sigma^*$. Then the dominant eigenvalue for h coincides with the dominant eigenvalue for $\overline{h}$.*

Proof. By Lemma 30 every eigenvalue of h is an eigenvalue of $\overline{h}$. Thus the dominant eigenvalue of $\overline{h}$ is greater or equal to that of h. By Theorem 4, the dominant eigenvalue of a non-negative matrix is a real number $\alpha > 1$ and its corresponding eigenvector is non-negative. By Lemma 30, every eigenvalue of $\overline{h}$ with a non-negative eigenvector is also an eigenvalue of h. Thus the dominant eigenvalue of h is also greater or equal to that of $\overline{h}$. Hence the dominant eigenvalues of h and $\overline{h}$ must be equal. □

Theorem 33. *Let α and β be multiplicatively independent real numbers. If v is a α-substitutive sequence and w is an β-substitutive sequence, then v and w have no common non-erasing transducts except for the ultimately periodic sequences.*

Proof. Let h_v and h_w be morphisms whose fixed points are v and w, respectively. By the proof of Theorem 20, x is a morphic image of an annotation $\overline{h}_v$ of h_v, and also of an annotation $\overline{h}_w$ of h_w. The morphisms must be non-erasing, by the assumption in this theorem. By Corollary 32 and Theorem 19, x is both α- and β-substitutive. By Durand's Theorem 6, x is eventually periodic. □

5   Conclusion

We have re-proven some of the central results in the area of morphic sequences, the closure of the morphic sequences under morphic images and transduction. However, the main results in this paper come from the eigenvalue analyses which followed our proofs in Sections 3 and 4. These are some of the only results known to us which enable one to prove negative results on the transducibility relation $\trianglelefteq$. One such result is in Theorem 33; this is perhaps the culmination of this paper. The next step in this line of work is to weaken the hypothesis in some of results that the transducers be *non-erasing*.

References

1. Allouche, J.P., Shallit, J.: Automatic Sequences: Theory, Applications, Generalizations. Cambridge University Press, New York (2003)
2. Cobham, A.: On the Hartmanis-Stearns problem for a class of tag machines. In: IEEE Conference Record of 1968 Ninth Annual Symposium on Switching and Automata Theory, pp. 51–60 (1968)
3. Cobham, A.: Uniform tag sequences. Math. Systems Theory 6, 164–192 (1972)
4. Dekking, F.M.: Iteration of maps by an automaton. Discrete Math. 126, 81–86 (1994)
5. Durand, F.: Cobham's theorem for substitutions. Journal of the European Mathematical Society 13, 1797–1812 (2011)
6. Endrullis, J., Hendriks, D., Klop, J.W.: Degrees of Streams. Integers 11B(A6), 1–40 (2011), proceedings of the Leiden Numeration Conference 2010
7. Pansiot, J.J.: Hiérarchie et fermeture de certaines classes de tag-systèmes. Acta Inform. 20(2), 179–196 (1983)
8. Sakarovitch, J.: Elements of Automata Theory. Cambridge University Press (2009)
9. Sprunger, D., Tune, W., Endrullis, J., Moss, L.S.: Eigenvalues and transduction of morphic sequences: Extended version. Tech. rep., CoRR (2014), http://arxiv.org/

Breadth-First Serialisation
of Trees and Rational Languages
(Short Paper)

Victor Marsault* and Jacques Sakarovitch

Telecom-ParisTech and CNRS, 46 rue Barrault 75013 Paris, France
victor.marsault@telecom-paristech.fr

Abstract. We present here the notion of *breadth-first signature* of trees and of prefix-closed languages; and its relationship with numeration system theory. A signature is the serialisation into an *infinite word* of an ordered infinite tree of finite degree. Using a known construction from numeration system theory, we prove that the signature of (prefix-closed) rational languages are substitutive words and conversely that a special subclass of substitutive words define (prefix-closed) rational languages.

1 Introduction

This work introduces a new notion: the breadth-first signature of a tree (or of a language). It consists of an infinite word describing the tree (or the language). Depending on the direction (from tree to word, or conversely), it is either a *serialisation* of the tree into an infinite word or a *generation* of the tree by the word. We study here the serialisation of rational, or regular, languages.

The (breath-first) signature of an ordered tree of finite degree is a sequence of integers, the sequence of the degrees of the nodes visited by a breadth-first traversal of the tree. Since the tree is ordered, there is a *canonical* breadth-first traversal; hence the signature is uniquely defined and characteristic of the tree.

Similarly, we call *labelling* the infinite sequence of the labels of the edges visited by the breadth-first traversal of a labelled tree. The pair signature/labelling is once again characteristic of the labelled tree. It provides an effective serialisation of labelled trees, hence of prefix-closed languages.

The serialisation of a (prefix-closed) language is very close, and in some sense, equivalent to the enumeration of the words of the language in the radix order. It makes then this notion particularly fit to describing the languages of integer representations in various numeration systems. It is of course the case for the representations in an integer base p which corresponds to the signature p^ω, the constant sequence. But it is also the case for non-standard numeration systems such as the Fibonacci numeration system whose representation language has for signature the Fibonacci word (*cf.* Section 4); and the rational base numeration systems as defined in [1] and whose representation languages have periodic signatures, that is, signatures that are infinite periodic words. To tell the truth, it is the latter case that first motivated our study of signatures. In another work still in preparation [2], we study trees and languages that have periodic signatures.

---

* Corresponding author.

A.M. Shur and M.V. Volkov (Eds.): DLT 2014, LNCS 8633, pp. 252–260, 2014.

In the present work, we first introduce the notion of signature of trees (Section 2) and of languages (Section 3). Then, in Section 4, we give with Theorem 1 a characterisation of the signatures of (prefix-closed) rational languages as those whose signature is substitutive. The proof of this result relies on a correspondence between substitutive words and automata due to Maes and Rigo [3] or Dumont and Thomas [4] and whose principle goes back to the work of Cobham [5].

2   Signatures of Trees

Classically, trees are undirected graphs in which any two vertices are connected by exactly one path (*cf.* [6], for instance). Our view differs in two respects.

First, a tree is a *directed* graph $\mathcal{T} = (V, \Gamma)$ such that there exists a *unique* vertex, called *root*, which has no incoming arc, and there is a *unique (oriented) path* from the root to every other vertex. Elements of a tree get particular names: vertices are called *nodes*; if (x, y) is an arc, y is called *a child* of x and x *the father* of y. We draw trees with the root on the left, and arcs rightwards.

Second, our trees are *ordered*, that is, that there is a total order on the set of children of every node. The order will be implicit in the figures, with the convention that lower children are smaller (according to this order).

The *degree* $\mathsf{d}(x)$ of a node is the number of children of x. A breadth-first traversal of a tree $\mathcal{T}$ eventually meets every node of $\mathcal{T}$ if and only if all degrees are finite. In the following, and as we are interested in trees in relation with infinite languages (over finite alphabets), we deal with infinite trees of bounded degree only. Since trees are ordered, there is a canonical breadth-first traversal for every tree. We may then consider that the set of nodes of a tree is always the set of integers $\mathbb{N}$: 0 is the root and the integer i is the $(i+1)$-th node visited by the breadth-first traversal of the tree.

It will prove to be extremely convenient to have a slightly different look at trees and to consider that the root of a tree is also a *child of itself*, that is, bears a loop onto itself. This convention is sometimes taken when implementing tree-like structures (*e.g.* file systems): it makes the father function total. We call such a structure an *i-tree*. It is so close to a tree that we pass from tree to i-tree (or conversely) with no further ado.

We call *signature* any infinite sequence s of non-negative integers. The signature $s = s_0 s_1 s_2 \cdots$ is *valid* if the following holds:

$$\forall n \in \mathbb{N} \qquad \sum_{i=0}^{n} s_i > j + 1 \ . \tag{1}$$

Definition 1. *The* breadth-first signature *or, for short, the* signature, *of a tree, or an i-tree, $\mathcal{T}$ is the sequence of the degrees of the nodes of the i-tree $\mathcal{T}$ in the order given by the breadth-first traversal of $\mathcal{T}$.*

In other words, $s = s_0 s_1 s_2 \cdots$ is the signature of a tree $\mathcal{T}$ if $s_0 = \mathsf{d}(0) + 1$ and $s_i = \mathsf{d}(i)$ for every node i of $\mathcal{T}$. Note that the definition implies that the signatures of a tree and of the corresponding i-tree are the same.

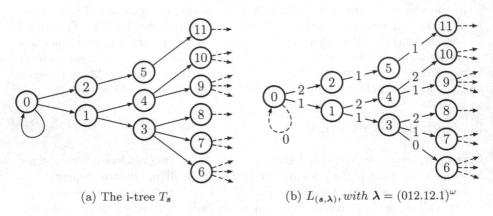

(a) The i-tree T_s (b) $L_{(s,\lambda)}$, with $\lambda = (012.12.1)^\omega$

Fig. 1. The i-tree and a language whose signature is $s = (321)^\omega$

Proposition 1. *A tree has a valid signature and conversely a valid signature s uniquely defines a tree T_s whose signature is s.*

The proof of Proposition 1 takes essentially the form of a procedure that generates an i-tree from a valid signature $s = s_0 s_1 s_2 \cdots$. It maintains two integers: the starting point n and the end point m of the transition, both initially set to 0. In one step of the procedure, s_n nodes are created, corresponding to the integers $m, m+1, \ldots, (m + s_n - 1)$, and s_n edges are created (all from n, and one to each of these new nodes). Then n is incremented by 1, and m by s_n.

The validity of s ensures that at each step of the procedure $n < m$, with the exception of the first step where $n = m = 0$. It follows that every node is strictly larger than its father, excepted for the root, whose father is itself. Figure 1a shows the i-tree whose signature is $(321)^\omega$.

3   Labelled Signatures of Languages

In the sequel, alphabets are totally ordered; and we use implicitly the natural order on digit alphabets (that is $0 < 1 < 2 < \cdots$). A word $w = a_0 a_1 \cdots a_{k-1}$ is *increasing* if $a_0 < a_1 < \cdots < a_{k-1}$. The length of a finite word w is denoted by $|w|$.

A labelled (i-)tree $\mathcal{T}$ is an (i-)tree whose arcs hold a label taken in an alphabet A. Since both $\mathcal{T}$ and A are ordered, the labels on arcs have to be *consistent*, that is, the labels of the arcs to the children of a same node are in the same order as the children: an arc to a smaller child is labelled by a smaller letter.

A labelled (i-)tree $\mathcal{T}$ defines the language of the branch labels. Conversely, a prefix-closed language L (over an ordered alphabet) uniquely defines a labelled (ordered) tree.

The labelling λ of a labelled tree $\mathcal{T}$ (labelled in A) is the infinite word in A^ω obtained as the sequence of the arc labels of $\mathcal{T}$ visited in a breadth-first search.

Definition 2. *Let* s *be a signature. An infinite word* $\boldsymbol{\lambda}$ *in* A^ω *is* consistent *with* s *if the factorisation of* $\boldsymbol{\lambda}$ *in the infinite sequence* $(w_n)_{n\in\mathbb{N}}$ *of words in* A^*:
$\boldsymbol{\lambda} = w_0\,w_1\,w_2\cdots$ *induced by the condition that for every* n *in* $\mathbb{N}$, $|w_n| = s_n$, *has the property that for every* n *in* $\mathbb{N}$, w_n *is an* increasing *word.*

A pair $(s, \boldsymbol{\lambda})$ *is a* valid labelled signature *if* s *is a valid signature and if* $\boldsymbol{\lambda}$ *is an infinite word consistent with* s.

A simple and formal verification yields the following.

Proposition 2. *A prefix-closed language* L *uniquely determines a labelled tree and hence a valid labelled signature, the labelled signature of* L *and conversely any valid labelled signature* $(s, \boldsymbol{\lambda})$ *uniquely determines a labelled tree* $\mathcal{T}_{(s,\lambda)}$ *and hence a prefix-closed language* $L_{(s,\lambda)}$, *whose signature is precisely* $(s, \boldsymbol{\lambda})$.

Figure 1b shows the labelling of the i-tree whose signature is $s = (321)^\omega$ by the infinite word $\boldsymbol{\lambda} = (012.12.1)^\omega$. This is of course a very special labelling: labellings consistent with s need not be periodic.

The identification between a prefix-closed language L and the tree $\mathcal{T}_L$ whose branch language is L (and whose set of nodes is $\mathbb{N}$) is very similar to the processes proposed in the works of Lecomte et Rigo [7,8] for the definition of the *Abstract Numeration Systems* (ANS) — without the assumption that L is rational, and with the restriction that L is prefix-closed. Indeed, the $(n+1)$-th word of L in the radix order is the label of the path from the root 0 to the node n in $\mathcal{T}_L$. (The first word of L is always ε and labels the empty path from the root to itself.)

Remark 1. A very simple tree paired with the appropriate labelling may produce an artificially complex language. For instance, the infinite unary tree may be labelled by a non-recursive word. This explains why a result relative to the regularity of languages defined by signatures will always require some restriction on the labelling. The notion of *substitutive labelled signature* defined in the next Section 4 is an example of such a restriction.

4   Substitutive Signature and Rational Languages

We follow [9] for the terminology and basic definitions on *substitutions*. Let A be an alphabet. A morphism $\sigma : A^* \to A^*$ is *prolongable* on a letter a in A if $\sigma(a) = au$ for some word u and moreover $\lim_{n\to+\infty}|\sigma^n(a)| = +\infty$. Then, the sequence $(\sigma^n(a))_{n\in\mathbb{N}}$ converges to an infinite word denoted by $\sigma^\omega(a)$; any such word is called *purely substitutive*. The image $f(w)$ of a purely substitutive word w by a letter-to-letter morphism f is called a *substitutive word*.

Definition 3. *Let* $\sigma : A^* \to A^*$ *be a morphism prolongable on* a *in* A *and let* $f_\sigma : A^* \to D^*$ *be the letter-to-letter morphism defined by* $\forall b \in A,\, f_\sigma(b) = |\sigma(b)|$. *The substitutive word* $f_\sigma(\sigma^\omega(a))$ *is called a* substitutive signature.

Furthermore, let $g : A^* \to B^*$ *be a morphism satisfying the following condition:* $\forall b \in A,\, |g(b)| = f_\sigma(b)$. *The pair* $(f_\sigma(\sigma^\omega(a)),\, g(\sigma^\omega(a)))$ *is called a* substitutive labelled signature.

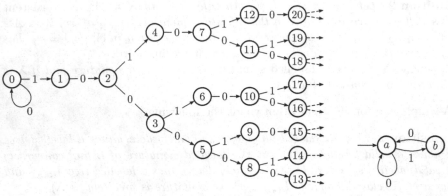

(a) $L_{(r,\mu)}$: integer representations in the Fibonacci numeration system.

(b) $\mathcal{A}_{(r,\mu)}$: automaton accepting $L_{(r,\mu)}$.

Fig. 2. The Fibonacci signature $r = \sigma^\omega(a)$ with $\sigma(a) = ab$ and $\sigma(b) = a$

The next lemma is a direct consequence of the fact that if σ denotes a morphism prolongable on a and if w denotes a prefix of $\sigma^\omega(a)$, then $|\sigma(w)| > |w|$.

Lemma 1. *A substitutive signature is valid.*

Example 1 (The Fibonacci signature). The Fibonacci word is the purely substitutive word $\sigma^\omega(a)$ defined by $\sigma(a) = ab$ and $\sigma(b) = a$:

$$\sigma^\omega(a) \quad = \quad abaababaabaab\cdots$$

Hence the substitutive signature defined by σ is

$$r \quad = \quad f_\sigma(\sigma^\omega(a)) \quad = \quad 2122121221221\cdots$$

Let g be the morphism defined by $g(a) = 01$ and $g(b) = 1$ defining the labelling $\mu = g(\sigma^\omega(a))$ (which is consistent with r):

$$\mu \quad = \quad g(\sigma^\omega(a)) \quad = \quad 01.0.01.01.0.01.0.01.01.0.01.01.0\ldots$$

The language $L_{(r,\mu)}$, as shown at Figure 2a, is the language of integer representations in the Fibonacci numeration system.

Theorem 1. *A prefix-closed language is rational if and only if its labelled signature is substitutive.*

The proof of this theorem relies on a correspondence between finite automata and substitutive words used by Rigo and Maes in [3] (*cf.* also [8, Section 3.4]) to prove the equivalence between two decision problems. A similar construction was used by Dumont and Thomas in [4] to define the prefix-suffix graph.

We give here the proof of one direction in detail, reformulated into the next proposition. The other direction is analogous.

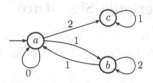

Fig. 3. The automaton $\mathcal{A}_{(\theta,h)}$ accepting the language $L_{(s,\lambda)}$ shown at Figure 1b with $\theta(a) = abc$, $\theta(b) = ab$, $\theta(c) = c$, $h(a) = 012$, $h(b) = 12$, $h(c) = 1$

Proposition 3. *If (s,λ) is a valid substitutive labelled signature, then $L_{(s,\lambda)}$ is a rational language.*

Proof. Let $\sigma : A^* \to A^*$ such that $s = f_\sigma(\sigma^\omega(a))$ and $g : A^* \to B^*$ such that $\lambda = g(\sigma^\omega(a))$. Since we are using two alphabets at the same time, a, b, c will denote letters of A and x, y letters of B.

Let $\mathcal{A}_{(s,\lambda)} = \langle A, B, \delta, a, A \rangle$ be the automaton whose set of states is A; the alphabet is B; the initial state is a; all states are final; and the transition function is defined as follows. For every b in A, let $k = |\sigma(b)| = |g(b)|$. From b, there are k outgoings transitions and for every i, $1 \leqslant i \leqslant k$, $b \xrightarrow{y} c$, where c is the i-th letter of $\sigma(b)$ and y is the i-th letter of $g(b)$. Figure 2b shows the automaton computed from the Fibonacci signature; see Figure 3 for the signature of Example 2 below.

Note that since the morphism σ is prolongable on a, the automaton $\mathcal{A}_{(s,\lambda)}$ features a loop $a \xrightarrow{x} a$ on the initial state whose label x is the first letter of $g(a)$. This loop induces that $L(\mathcal{A}_{(s,\lambda)})$ is of the form x^*L and leading x's serve the same role as leading 0's in usual numeration systems. We denote by L the language containing the words of $L(\mathcal{A}_{(s,\lambda)})$ that does not start with an x. Proving that $L(\mathcal{A}_{(s,\lambda)})$ has (s,λ) for signature amounts to prove that if w_i denotes the $(i+1)$-th word of L in the radix order, then w_i reaches the state corresponding to the $(i+1)$ letter of $\sigma^\omega(a)$.

Let b be a letter of A, hence a state of $\mathcal{A}_{(s,\lambda)}$. The word $\sigma(b)$ is exactly the sequence of the states that are direct successors of b in $\mathcal{A}_{(s,\lambda)}$ in the right order that is, a successor by a smaller label is before a successor by a larger label. It follows that the word $\sigma(\sigma(b))$ is the sequence of the states that are reachable from b in two steps and once again, in the right order. An easy induction yields that $\sigma^i(b)$ is the sequence of the states reachable in exactly i steps.

If $\sigma(a) = au$, then the words of length 1 of L reach the states of u (and the empty word reaches the state a). An easy induction yields that the words of length i belonging to L reach the sequence of states $\sigma^i(u)$. Hence the words of L taken in the radix order reach the state sequence $a\,u\,\sigma(u)\sigma^2(u)\sigma^3(u)\cdots$ which is equal to $\sigma^\omega(a)$.

Remark 2. As we said, our view on a prefix-closed rational language L essentially amounts to considering L as an ANS, in the sense of [7,8]. The consequence of Theorem 1 is to associate with L a substitution σ_L. In [4], Dumont and Thomas described the numeration system associated with a substitution σ. It can be derived from the construction of Theorem 1 that the ANS L may be mapped onto the Dumont-Thomas numeration system associated with σ_L.

5   On Ultimately Periodic Signature

Let $s = uv^\omega$ be an ultimately periodic word over the alphabet $\{0, 1, \ldots, k\}$; we call *growth ratio* of v, denoted by $gr(v)$, the average of the letters of v:

$$gr(v) \quad = \quad \frac{\sum_{i=0}^{|v|-1} v[i]}{|v|} \ .$$

We treat here the case where $gr(v)$ is an integer that is, when the sum of the letters of v is a multiple its length. In this case, uv^ω is a substitutive signature.

Proposition 4. *If s denotes an ultimately periodic valid signature whose growth ratio is an integer, then s is a substitutive signature.*

Proof. Let $s = uv^\omega$ be an ultimately periodic signature. We write $k = |u|$, $n = |v|$ and denote by A an alphabet whose $(k + n)$ letters are denoted as follows.

$$A = B \uplus C \quad \text{where} \quad B = \{b_0, b_1, \ldots, b_{(k-1)}\} \quad \text{and} \quad C = \{c_0, c_1, \ldots, c_{(n-1)}\} \ .$$

The letters of B correspond to positions of u and those of C to positions of v. Let $\sigma : A^* \to A^*$ be a morphism defined implicitly by

$$\sigma(b_0 b_1 \cdots b_{(k-1)} c_0 c_1 \cdots c_{(n-1)}) \text{ is prefix of } b_0 b_1 \cdots b_{(k-1)} (c_0 c_1 \cdots c_{(n-1)})^\omega \quad \text{(2a)}$$
$$\forall i < k \quad |\sigma(b_i)| = u_i \quad \text{(2b)}$$
$$\forall i < n \quad |\sigma(c_i)| = v_i \quad \text{(2c)}$$

Let us denote by $\overline{u} = b_0 b_1 \cdots b_{(k-1)}$ and by $\overline{v} = c_0 c_1 \cdots c_{(n-1)}$, hence, respectively from Equations 2b and 2c, $f_\sigma(\overline{u}) = u$ and $f_\sigma(\overline{v}) = v$. Let i and j be the two integers such that $\sigma(\overline{u}) = \overline{u}(\overline{v})^i c_0 \cdots c_{(j-1)}$. Equation 2c implies that $|\sigma(\overline{v})| = n \times gr(v)$ hence, from Equation 2a,

$$\sigma(\overline{v}) = c_j \cdots c_{(n-1)} (\overline{v})^{gr(v)-1} c_0 \cdots c_{(j-1)} \ .$$

It follows that $\overline{u}(\overline{v})^\omega$ is a fixed point of σ.

It remains to prove that the morphism σ is prolongable on b_0 or, more precisely, that $\lim_{n\to+\infty} |\sigma^n(b_0)| = +\infty$. Let us denote by w any prefix of $\overline{u}(\overline{v})^\omega$ and prove that $|\sigma(w)| > |w|$. Since w is a prefix of $\overline{u}(\overline{v})^\omega$, $f_\sigma(w)$ is a prefix of s, and since s is valid, the sum of the letters of $f_\sigma(w)$ is strictly greater than $|w|$. From the definition of f_σ, $|\sigma(w)|$ is equal to the sum of the letters of $f_\sigma(w)$, hence is strictly greater than $|w|$.

Example 2. The purely periodic signature $s = (321)^\omega$ is the substitutive signature $f_\theta(\theta^\omega(b_0))$ where θ is defined by $\theta(c_0) = c_0 c_1 c_2$, $\theta(c_1) = c_0 c_1$ and $\theta(c_2) = c_2$. Figure 3 shows the automaton $\mathcal{A}_{(\theta,h)}$ accepting $L_{(s,\lambda)}$ (shown at Figure 1b) where $\lambda = h(\theta^\omega(c_0))$ with $h(c_0) = 012$, $h(c_1) = 12$ and $h(c_2) = 1$. This language consists of non-canonical representations of the integers in base 2 (that is, the growth ratio of s): the $(n+1)$-th word[1] of $L_{(s,\lambda)}$ in the radix order is a word $d_n d_{n-1} \cdots d_0$ over the alphabet $\{0, 1, 2\}$ and its binary value $\sum_{i=0}^n d_i 2^i$ is equal to n.

---

[1] Recall that we are ignoring leading 0's, hence the $(n + 1)$-th word of $L_{(s,\lambda)}$ is the one labelling the path $0 \to n$ in Figure 1b.

6   Conclusion and Future Work

In this work, we introduced a way of effectively describing infinite trees and languages by infinite words using a simple breadth-first traversal. Since this transformation is essentially one-to-one, it is natural to wonder which class of words is associated with which class of languages.

In this first work on the subject, we have proved that rational languages are associated with (a particular subclass of) substitutive words. We also proved that ultimately periodic signatures whose growth ratio is an integer are substitutive, and hinted their link to integer base numeration systems.

In a forthcoming paper [2], we study the class of languages associated with periodic signatures whose growth ratio is not an integer and how they are related to the representation language in rational base numeration systems. In the future, our aim is to further explore this relationship by means of the notion of direction, that extends the notion of growth ratio to aperiodic signatures.

Acknowledgements. The authors are grateful to the referee who drew their attention to the work of Dumont and Thomas.

References

1. Akiyama, S., Frougny, C., Sakarovitch, J.: Powers of rationals modulo 1 and rational base number systems. Israel J. Math. 168, 53–91 (2008)
2. Marsault, V., Sakarovitch, J.: Rhythmic generation of infinite trees and languages (2014), In preparation, early version accessible at arXiv:1403.5190
3. Rigo, M., Maes, A.: More on generalized automatic sequences. Journal of Automata, Languages and Combinatorics 7(3), 351–376 (2002)
4. Dumont, J.M., Thomas, A.: Systèmes de numŕation et fonctions fractales relatifs aux substitutions. Theor. Comput. Sci. 65(2), 153–169 (1989)
5. Cobham, A.: Uniform tag sequences. Math. Systems Theory 6, 164–192 (1972)
6. Diestel, R.: Graph Theory. Springer (1997)
7. Lecomte, P., Rigo, M.: Numeration systems on a regular language. Theory Comput. Syst. 34, 27–44 (2001)
8. Lecomte, P., Rigo, M.: Abstract numeration systems. In: Berthé, V., Rigo, M. (eds.) Combinatorics, Automata and Number Theory, pp. 108–162. Cambridge Univ. Press (2010)
9. Berthé, V., Rigo, M.: Combinatorics, Automata and Number Theory. Cambridge University Press (2010)

Measuring Communication
in Automata Systems
(Invited Paper)

Martin Kutrib and Andreas Malcher

Institut für Informatik, Universität Giessen
Arndtstr. 2, 35392 Giessen, Germany
{kutrib,malcher}@informatik.uni-giessen.de

Abstract. We consider systems of interacting finite automata. On the one hand, we look at automata systems consisting of a small constant number of synchronous and autonomous finite automata that share a common input and communicate with each other as weakly parallel models. On the other hand, we consider cellular automata consisting of a huge number of interacting automata as massively parallel systems. The communication in both types of automata systems is quantitatively measured by the number of messages sent by the components. In cellular automata it is also qualitatively measured by the bandwidth of the communication links. We address several aspects concerning the complexity of such systems. In particular, fundamental types of communication are considered and the questions of how much communication is necessary to accept a certain language and whether there are communication hierarchies are investigated. Since even for systems with few communication many properties are undecidable, another question is to what extent the communication has to be limited in order to obtain systems with decidable properties again. We present some selected results on these topics and want to draw attention to the overall picture and to some of the main ideas involved.

1 Introduction

Parallel computational models are appealing and widely used in order to describe, understand, and manage parallel processes occurring in real life. One principal task in order to employ a parallel computational model in an optimal way is to understand how cooperation of several processors is organized optimally. To this end, it is essential to know which communication and which amount of communication must or should take place between several processors. From the viewpoint of energy and the costs of communication links, it would be desirable to communicate a minimal number of times with a minimum amount of information transmitted. On the other hand, it would be interesting to know how much communication is necessary in a certain parallel model to accomplish a certain task.

Whenever several heads of a device are controlled by a common finite-state control, one may suppose that the heads are synchronous and autonomous finite

A.M. Shur and M.V. Volkov (Eds.): DLT 2014, LNCS 8633, pp. 260–274, 2014.

automata that communicate their states in every time step. Similarly, for the massively parallel model of cellular automata, one may assume that every cell communicates its current state to its neighbors in every time step. From this viewpoint it is almost self-evident to limit the communication allowed to address the problems mentioned above.

Here we measure the communication in automata systems in two ways, namely, quantitatively and qualitatively. First, the number of messages allowed to be sent by the components is limited depending on the length of the input, where we can distinguish between the total number of messages and the number of messages between each two components. Clearly, the communication mode plays a crucial role when posing these restrictions. Second, the bandwidth of the communication links between the components can be restricted as well, that is, the number of different messages is limited. In the following we address several aspects concerning the complexity of such communication restricted devices.

In Section 2, we consider automata systems consisting of a few number of interacting finite automata. We distinguish two types of communication. First, we look at systems which communicate by broadcasting messages and we can present a finite hierarchy with regard to the number of messages sent. Concerning decidability questions, we obtain undecidability if the number of messages broadcast is at least logarithmic with respect to the length of the input, but we have decidability in case of a constant number of messages. In the second part, we consider systems which communicate by requesting information from other components. For such systems we yield an infinite hierarchy concerning the number of messages sent and obtain similar undecidability results.

In Section 3, we restrict communication in the massively parallel model of cellular automata. First, we investigate the qualitative restriction of a limited bandwidth for the messages transmitted. The main result here is that cellular automata communicating k different messages are more powerful than those which are allowed to communicate at most $k-1$ different messages only. Second, we consider the quantitative restriction to the number of messages sent and distinguish between the maximum of messages sent between two cells and the sum of all such messages. Again, it is possible to establish proper hierarchies with respect to the number of messages sent. Finally, we look at cellular automata which are qualitatively and quantitatively communication restricted. It turns out that even such automata with minimal communication are still very powerful which is demonstrated by the fact that almost all commonly studied decidability question are undecidable.

The reader is assumed to be familiar with the basic notions of automata theory as contained, for example, in [7, 9]. In the present paper we will use the following notational conventions. We denote the set of *nonnegative integers* by $\mathbb{N}$. An *alphabet* Σ is a non-empty finite set, its elements are called *letters* or *symbols*. We write Σ^* for the set of all words over the finite alphabet Σ. The *empty word* is denoted by λ, and $\Sigma^+ = \Sigma^* \setminus \{\lambda\}$. The *reversal* of a word w is denoted by w^R and for the *length* of w we write $|w|$. We use $\subseteq$ for *inclusions* and $\subset$ for *strict* *inclusions.* The *family of languages accepted* by devices of type X is denoted by $\mathscr{L}(X)$.

2 Weakly Parallel Systems

One of the simplest systems that could be called *parallel* is a device consisting of some constant number of synchronous and autonomous finite automata that may communicate with each other. Basically, here we distinguish two fundamental types of communication. On the one hand, the single automata can broadcast messages to other components and, on the other hand, the single automata can request information of other components.

2.1 Broadcasting Messages

When systems of synchronous and autonomous finite automata communicate by broadcasting messages, several technical aspects have to be considered. For example, is it allowed that more than one component broadcast messages at the same time? If yes, what happens at the recipients? Is only one message processed? If yes, which one? Or else are all messages processed? If yes, in which order or in parallel? In this subsection we dodge these problems by starting our considerations with very simple systems consisting of two one-way components only.

A one-way two-party finite automata system (see Figure 1) is a device of two finite automata working independently on a common read-only one-way input tape. The automata communicate by broadcasting messages. The transition function of a single automaton depends on its current state, the currently scanned input symbol, and the message currently received from the other automaton. Both automata work synchronously and the messages are delivered instantly.

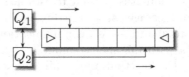

Fig. 1. A two-party finite automata system

Formally, we define a *deterministic one-way two-party finite automata system* (DFAS(2)) as a construct $M = \langle \Sigma, B, \triangleright, \triangleleft, A_1, A_2, \rangle$, where Σ is the finite set of *input symbols*, B is the set of possible *messages*, $\triangleright \notin \Sigma$ and $\triangleleft \notin \Sigma$ are the *left and right endmarkers*, and each $A_i = \langle Q_i, \Sigma, \delta_i, \mu_i, q_{0,i}, F_i \rangle$, $i \in \{1, 2\}$, is basically a *deterministic finite automaton* with state set Q_i, initial state $q_{0,i} \in Q_i$, and set of *accepting states* $F_i \subseteq Q_i$. Additionally, each A_i has a *broadcast function* $\mu_i : Q_i \times (\Sigma \cup \{\triangleright, \triangleleft\}) \to B \cup \{\bot\}$ which determines the message *to be sent*, where $\perp \notin B$ means *nothing to send*, and a *(partial) transition function* $\delta_i : Q_i \times (\Sigma \cup \{\triangleright, \triangleleft\}) \times (B \cup \{\perp\}) \to Q_i \times \{0, 1\}$, where 1 means to move the head one square to the right and 0 means to keep the head on the current square.

The automata A_1 and A_2 are called *components* of the system M. Initially, the input is presented on the tape in between the endmarkers, and both components are in their initial states at the left endmarker. Whenever the transition function of (at least) one of the single automata is undefined the whole systems *halts*. The input is accepted if at least one of the automata is in an accepting state at this moment. To study the impact of communication in DFAS(2), the communication is measured by the total number of messages sent during a computation, where it is understood that $\perp$ means no message and, thus, is not counted. Let $f : \mathbb{N} \to \mathbb{N}$ be a mapping. If all $w \in L(M)$ are accepted with computations where the total number of messages sent is bounded by $f(|w|)$, then M is said to be *communication bounded by f*. We denote the class of DFAS(2) that are communication bounded by f by (f)-DFAS(2).

The next example reveals that devices allowed to send just one message during a computation can accept non-regular languages.

Example 1. The language $\{a^n b^n \mid n \geq 1\}$ belongs to $\mathscr{L}((1)\text{-DFAS}(2))$.

The principal idea of the construction is that both components move their heads with different speeds. The first component moves in every time step, while the second component moves in every other time step. The first component sends a message when it arrives at the right endmarker. Now the second component halts and accepts if and only if it receives the message at the first symbol b.   □

A straightforward generalization of Example 1 shows that another single communication step strengthens the computational capacity so that even non-context-free languages are accepted. In particular, the language $\{a^n b^n c^n \mid n \geq 1\}$ belongs to $\mathscr{L}((2)\text{-DFAS}(2))$. More general, any DFAS(2) can be simulated by a one-way two-head finite automaton in a straightforward manner. Therefore, the family $\mathscr{L}(\text{DFAS}(2))$ is a proper subclass of the complexity class L. However, the witness languages from above are semilinear. In [8] it has been shown that DFAS(2) that communicate in every time step accept non-semilinear languages. So, the question arises how much communication is necessary at all to accept a non-semilinear language. An upper bound can be derived from a result in [23]. The language $L_{expo} = \{a^{2^0} b a^{2^1} b \cdots b a^{2^m} \mid m \geq 1\}$ is accepted by some $(O(\log(n)))\text{-DFAS}(2)$, where the components communicate only when a b or the right endmarker is reached. Similarly, $L_{poly} = \{aba^3 b a^5 b \cdots b a^{2m+1} \mid m \geq 0\}$ belongs to $\mathscr{L}((O(\sqrt{n}))\text{-DFAS}(2))$. By means of Kolmogorov arguments, variants of these languages and the copy language $\{wcw \mid w \in \{0, 1\}^*\}$ can be used to separate the upper three levels of the finite hierarchy

$$\mathscr{L}((O(1))\text{-DFAS}(2)) \subset \mathscr{L}((O(\log(n)))\text{-DFAS}(2)) \subset$$

$$\mathscr{L}((O(\sqrt{n}))\text{-DFAS}(2)) \subset \mathscr{L}((O(n))\text{-DFAS}(2)).$$

Concerning the separation of the first two levels and a lower bound for the communication necessary to accept non-semilinear languages, a simulation result from [23] is of tangible advantage. In particular, it reveals that any (k)-DFAS(2) can effectively be simulated by a *deterministic (r, s)-reversal bounded two-way k-counter machine*, where k, r, and s are constant. Basically, such a machine is a device with a finite-state control, a two-way read-only input head, and k counters. A move consists of moving the input head a position to the right, to the left, or to keep it at its current position, adding -1, 0, or $+1$ to each counter, and changing the state. The machine can test the counters for zero. The input is accepted if the device eventually halts in an accepting state. The machine is said to be (r, s)-*reversal bounded* if in every accepting computation the input head reverses its direction at most r times and the content in each counter alternately increases and decreases by at most s times. In [10] these machines are formally defined and studied. In particular, it is shown that any language which is accepted by a deterministic reversal-bounded two-way finite-counter machine has a semilinear Parikh image. Together with the witness languages from above, this separates the first two levels of the hierarchy. Moreover, any *unary* language from $\mathscr{L}((O(1))\text{-DFAS}(2))$ is regular.

We continue this subsection with a focus on the edge between decidability and undecidability. The question is to what extent communication has to be reduced in order to regain the decidability of certain problems. By reduction of the corresponding problems for two-way cellular automata [14] one can show that the problems of testing emptiness, finiteness, and infiniteness are not semidecidable for a given $(O(\log(n)))$-DFAS(2). The undecidability results for cellular automata have been shown in [24]. A further reduction of these non-semidecidable problems proves that also the problems of testing inclusion, equivalence, regularity, and context-freeness are not semidecidable for a given $(O(\log(n)))$-DFAS(2). However, the edge between decidability and undecidability is crossed when the communication is reduced to be constant. Another result in [10] says that the properties emptiness, finiteness, inclusion, and equivalence are decidable for deterministic reversal-bounded two-way finite-counter machines. Due to the effectiveness of the simulation result mentioned above they are decidable for $(O(1))$-DFAS(2) as well.

Finally, we conclude the subsection with some deep results from [11, 12] dealing mostly with deterministic two-way multi-party finite automata systems (2DFAS(k)). In these models each of the k components may also move to the left. If more than one component broadcasts a message at the same time, all components receive all messages in parallel. In [11] it is shown that there are gaps in the hierarchy of message complexity. For one-way devices there is no language accepted by some DFAS(k) that requires more than constant, that is $\omega(1)$, and less than logarithmic, that is $o(\log(n))$, messages to be sent. For two-way devices there is a gap between $\omega(1)$ and $o((\log\log\log(n))^c)$, where c is a constant. Solely for two-way devices, in [12] infinite communication hierarchies are shown. On the one hand, increasing a constant number of messages allowed by two gives a strictly stronger device: $\mathscr{L}((m-1)\text{-2DFAS}(k)) \subset \mathscr{L}((m+1)\text{-2DFAS}(k))$, for all $m \geq 1$. On the other hand, there is such a hierarchy between the communication bounds $\log\log(n)$ and n. Furthermore, it is proved that the copy language $\{ wcw \mid w \in \{0,1\}^* \}$ is not accepted by any two-way system with an arbitrary number of components if the communication is bounded by $o(n)$. On the other hand, there are languages that require a superlinear number of communications.

2.2   Requesting Messages

When systems of synchronous and autonomous finite automata communicate by requesting messages, some of the technical issues for broadcasting messages can be avoided. For example, if more than one component requests information from other components, there is no conflict a priori. Particular models based on this type of communication are the so-called *parallel communicating finite automata systems* which were introduced in [25] as a simple automaton model of parallel processes and cooperating systems, see also [1, 2, 5]. As before, the input is read and processed in parallel by several one-way finite automata. The communication takes place in such a way that a component can request the current *state* from another component, and *is set to that state* after receiving it whereby its former state is lost. If there is just one component that is allowed to query for states, the system is said to be *centralized*. A further distinction of the models can be made dependent on whether each automaton sending its current state is reset to its initial state (*returning mode*) or stays in its state (*non-returning mode*). Recently, these communication protocols have been refined in [35] and further investigated for the case of parallel communicating systems of pushdown automata [28]. There, the communication process is performed in an asynchronous manner, reflecting the technical features of many real communication processes. Before we turn to discuss results on the degree of communication in such devices, we present more details of the informal definition of the automata systems.

A deterministic parallel communicating finite automata system of degree k (DPCFA(k)) is a device of k deterministic finite automata working independently on a common read-only one-way input tape. The transition function of a single automaton depends on its current state and the currently scanned input symbol. All automata work synchronously. Each computation step of the whole system consists of two phases. Let $\{q_1, q_2, \ldots, q_k\}$ be a set of distinguished query states. In a first phase, all components are in non-query states and perform an ordinary (non-communicating) step independently. The second phase is the communication phase during which components in query states receive the requested states as long as the sender is not in a query state itself. That is, if a component i is in query state q_j, then component i is set to the current state of component j. This process is repeated until all requests are resolved, if possible. If the requests are cyclic the computation halts. Moreover, whenever the transition function of (at least) one of the single automata is undefined the whole systems halts as well. As before, the input is accepted if at least one of the components is in an accepting state when the system halts.

Communication in nondeterministic PCFA was studied in [27], where the amount of communication necessary to accept a language is considered as a dynamical measure of descriptional complexity as follows: The degree of communication of an accepting computation is defined as the number of queries posed. The degree of communication $Comm(w)$ of a nondeterministic PCFA M

on input w is defined as the minimal number of queries posed in accepting computations on w. The degree of communication $Comm(M)$ of a PCFA M is then defined as $\sup\{\,Comm(x)\mid w\in L(M)\,\}$, and the degree of communication of a language (with respect to a PCFA of type X) is the infimum of the degrees of communication taken over all PCFA of type X that accept the language. It is shown in [27] that this measure cannot be algorithmically computed for languages accepted by nondeterministic centralized or non-centralized non-returning PCFA. The computability status of the degree of communication for the other types of PCFA languages as well as for all types of PCFA is stated as open question.

In [3] deterministic centralized returning PCFA are studied, where the degree of communication is bounded by a function f in the length of the input word $((f)$-DRCPCFA$(k))$. The example languages L_{expo} and L_{poly} from the previous subsection are similarly accepted as before. We have the containment $L_{expo}\in\mathscr{L}((O(\log(n)))\text{-DRCPCFA}(2))$ and $L_{poly}\in\mathscr{L}((O(\sqrt{n}))\text{-DRCPCFA}(2))$. So, again, two components are sufficient to accept non-semilinear languages. However, in order to separate levels in a communication hierarchy the witnesses are more sophisticated. The language of the next example combines the well-known non-context-free copy language with L_{expo}.

Example 2. The language

$$L_{expo,wbw}=\{\,\$w_1w_2\cdots w_mba^{2^0}w_1w_1a^{2^1}w_2w_2\cdots a^{2^{m-1}}w_mw_m\&\mid$$
$$m\geq 1, w_i\in\{0,1\}, 1\leq i\leq m\,\}$$

is accepted by some $(O(\log(n)))$-DRCPCFA(3).

We call the sole component allowed to request communication *master*. The idea of the construction is as follows. The second non-master component initially passes over the $\$$ and, then, it reads a symbol, remembers it in its state, and loops without moving. Whenever the component is reset into its initial state after a communication, it reads the next symbol, remembers it, and loops without moving. This component is used by the master to match the w_i from the prefix with the w_i from the suffix.

The first non-master component initially passes over the prefix $\$w_1w_2\cdots w_m$, the b, and the adjacent infix $aw_1w_1aaw_2w_2$. On its way it checks whether the neighboring symbols w_i are in fact the same. If the second check is successful the component enters some state s_{ww}. Exactly at that time it has to be queried by the master, otherwise it blocks the computation. Subsequently, it repeatedly continues to read the input, where each occurrence of neighboring symbols w_i is checked for equality, which is indicated by entering state s_{ww} again. This component is used to verify that all neighboring symbols w_i in the suffix are equal and, by the master, to check the lengths of the a-blocks. After being reset to its initial state, the component takes a number of time steps equal to the length of the next a-block plus 2 to get on the first symbol after the next w_iw_i.

The master initially passes over the prefix $\$w_1w_2\cdots w_m$, the b, and the first a. Then it reads the first of two adjacent symbols w_i and requests the state of the second non-master component (the equality of the symbols w_i has already been checked by the first non-master component). It receives the information about the matching symbol w_i from the prefix. If this symbol is the same as the next input symbol, then the computation continues by requesting the state of the first non-master component. If the master receives state s_{ww} the length of the first two a-blocks are verified. Now the master repeatedly continues to read the input, where on each occurrence of neighboring symbols w_i the equality with the corresponding symbol in the prefix is checked by querying the second non-master component and the lengths of the a-blocks are compared by querying the other non-master component. After that, it takes a number of time steps equal to the length of the adjacent a-block plus 2 to request the state of the first non-master component again. Finally, when the master has checked the last symbol w_m and gets the information that the first non-master component has read symbol &, it queries the second non-master component. If it receives a b, the input is accepted. In all other cases it is rejected.

The length of a word $w \in L_{expo,wbw}$ is $|w| = 3m+3+\sum_{i=0}^{m-1} 2^i = 2^m+3m+2$, for some $m \geq 1$. In its accepting computation, two communications take place for every w_iw_i and one more communication on the endmarker. So there are $2m+1$ communications which is of order $O(\log(|w|))$. $\qquad\square$

From the combination of the copy language with L_{poly} one gets another witness language. Again, by Kolmogorov arguments lower bounds on the communication necessary to accept these languages can be established. So, basically, a finite hierarchy as in the case of broadcasting messages has been derived [3]:

$$\mathscr{L}((O(\log(n)))\text{-DRCPCFA}(k)) \subset \mathscr{L}((O(\sqrt{n}))\text{-DRCPCFA}(k))$$

for every $k \geq 3$, and

$$\mathscr{L}((O(\sqrt{n}))\text{-DRCPCFA}(k)) \subset \mathscr{L}((O(n))\text{-DRCPCFA}(k))$$

for every $k \geq 2$. However, for at least four components, an infinite strict hierarchy of language classes in between the logarithm and the square root can be shown. The ingredients of the proof are functions $f : \mathbb{N} \to \mathbb{N}$ with $f \in \Theta(n^r)$, $r \geq 1$, that are time-computable by *one-way cellular automata* [4] and *valid computations* of such devices (see, for example, [13, 14]). So, let M_r be a one-way cellular automaton that time-computes $f \in \Theta(n^r)$. Then the language L_r is designed as

$$L_r = \{\ \$_1x_1x_2\cdots x_\ell\$_2w_1'w_2'\cdots w_m'w_{m+1}\cdots w_\ell\$_3$$
$$w_1'w_2'\cdots w_m'w_{m+1}\cdots w_\ell\$_4a^{2^0}bba^{2^1}bb\cdots a^{2^{m-1}}bb\& \mid m \geq 1,$$
$$x_1x_2\cdots x_\ell \text{ is the valid computation of } M_r \text{ on input } a^m,$$
$$w_i' \in \{0',1'\}, 1 \leq i \leq m, w_i \in \{0,1\}, m+1 \leq i \leq \ell\}.$$

Let $r \geq 1$ be an integer. Then L_r belongs to $\mathscr{L}((O(\log(n)^{r+2}))\text{-DRCPCFA}(4))$ but does not belong to the family $\mathscr{L}((O(\log(n)^{r}))\text{-DRCPCFA}(4))$. The proofs of these results do not rely on a specific number of components as long as at least four components are provided (see [3] for details). Therefore, the hierarchy follows for any number of components $k \geq 4$:

$$\mathscr{L}((O(\log(n)^{r}))\text{-DRCPCFA}(k)) \subset \mathscr{L}((O(\log(n)^{r+2}))\text{-DRCPCFA}(k)).$$

Next we come back to decidability issues. The fact that emptiness, finiteness, infiniteness, universality, inclusion, equivalence, regularity, and context-freeness are not semi-decidable for $(O(\log(n)))\text{-DRCPCFA}(k)$ if $k \geq 4$, nicely complements the situation for devices that broadcast messages. So, let us turn to questions concerning the decidability or computability of the communication bounds. In principle, we deal with three different types of problems. The first type is to decide for a given DRCPCFA(k) M and a given function f whether or not M is communication bounded by f. The problem has been solved negatively for all non-trivial communication bounds and all degrees $k \geq 3$:

Let $k \geq 3$ be any degree, $f \in o(n)$, and M be a DRCPCFA(k). Then it is not semi-decidable whether M is communication bounded by f.

The second type of problems we are dealing with has been raised in [27]. The question is whether the degree of communication $Comm(M)$ is computable for a given nondeterministic PCFA(k) M. Since $Comm(M)$ is either finite or infinite, in our terms the question is to decide whether or not M is communication bounded by some function $f \in O(1)$ and, if it is, to compute the precise constant. Again, the problem has been solved negatively for all degrees $k \geq 3$ in [3]:

Let $k \geq 3$ be an integer. Then the degree of communication $Comm(M)$ is not computable for DRCPCFA(k).

The third type of problems we are dealing with is now the question whether the degree of communication is computable for the *language* accepted by a given nondeterministic PCFA(k) M. In [27] it is shown that $Comm_{\text{CPCFA}}(L(M))$ for some nondeterministic non-returning centralized PCFA M is not computable. However, the degree is not even computable for DRCPCFA [3]:

Let $k \geq 3$. Then the degree of communication $Comm_{\text{DRCPCFA}(k)}(L(M))$ is not computable.

3   Massively Parallel Systems

In this section we turn to parallel automata systems at the opposite end of the parallelism scale. While in the previous section a finite number of components process inputs of arbitrary lengths, here so-called *cellular automata* are considered, where the number of available processing elements, the cells, is given by the length of the input. So, for every symbol of the input string we have one cell. For this reason such devices are often called massively parallel.

Before we discuss communication issues, we present more details of the devices in question. A two-way cellular automaton (CA) is a linear array of identical deterministic finite automata, called cells, that are identified by natural numbers.

Except for the outermost cells each one is connected to its both nearest neighbors. In general, the state transition depends on the current state of each cell and on the current states of its neighbors. With an eye toward restrictions of the communication, we have the viewpoint that the state transition depends on the current state of each cell and the messages that are currently sent by its neighbors. The possible messages are formalized as a set of possible communication symbols. If this set is equal to the state set, the general case applies. The messages to be sent by a cell depend on its current state and are determined by so-called communication functions. The two outermost cells receive a boundary symbol # on their free input lines once during the first time step from the outside world. Subsequently, these input lines are never used again. The state changes take place simultaneously at discrete time steps. The input mode for cellular automata is called parallel. One can suppose that all cells fetch their input symbol during a pre-initial step (see Figure 2).

Fig. 2. A two-way cellular automaton

By providing a set of communication symbols the definition is more general than really needed in the following, since here we are interested in the number of messages sent only. However, the definition allows also to restrict the bandwidth of the communication links by bounding the set of communication symbols and, thus, fits well to the devices investigated in [15–17, 30–32, 36].

An input w is accepted by a CA M if at some time i during the course of its computation the leftmost cell enters an accepting state. If all $w \in L(M)$ are accepted with at most $|w|$ time steps, then M is a *real-time* CA. The corresponding family of languages is denoted by $\mathscr{L}_{rt}(\text{CA})$.

An important subclass of cellular automata are so-called *one-way cellular automata* (OCA), where the flow of information is restricted to one way from right to left. For a definition it suffices to require that each cell is now connected to its neighbor to the right only (see, for example, [13, 14] for more on cellular automata).

There are natural possibilities to measure the communication in cellular automata computations [20]. On the one hand, the number of different messages that can be sent over the inter-cell links can be bounded, say, by some constant k. Such devices are called *cellular automata with k-message restricted inter-cell communication*, and are abbreviated as CA_k and OCA_k. On the other hand, the communication can be measured by the number of uses of the links between cells. It is understood that whenever a communication symbol is sent, a communication takes place. Here we do not distinguish whether either or both neighboring cells use the link. For computations we now distinguish the maximal number of communications between each two cells and the total number of communications. Let $f : \mathbb{N} \to \mathbb{N}$ be a mapping. If all $w \in L(M)$ are accepted with computations such that any link between two neighboring cells is used at most $f(|w|)$ times, then M is said to be *max communication bounded by* f (MC(f)-CA, MC(f)-OCA). Similarly, if all $w \in L(M)$ are accepted with computations where the total number of all communications does not exceed $f(|w|)$, then M is said to be *sum communication bounded by* f (SC(f)-CA, SC(f)-OCA).

3.1   Cellular Automata with Limited Inter-Cell Bandwidth

The most restricted setting of the bandwidth of the communication links between two cells is to allow one sole message only. However, these devices are still powerful enough to solve problems such as the firing squad synchronization problem in optimal time [26]. Moreover, it is known [31, 32] that one-message CA_1 can accept rather complicated non-semilinear unary languages in real time, for example, words whose lengths are Fibonacci numbers. On the other hand, there are regular languages which cannot be accepted by CA_k, for $k \geq 1$ [22]. So, the family $\mathscr{L}_{rt}(CA_k)$ is incomparable with the regular and context-free languages. However, if the communication channels of the even one-way cellular automata have a sufficient capacity, the regular languages are accepted. Essentially, every regular language over a $(k-1)$-letter alphabet is accepted by a real-time OCA_k. Basically, the idea of the construction is to simulate a deterministic finite automaton accepting the regular language in all and, thus, in the leftmost cell. The input of the OCA_k is continuously shifted to the left in order to feed the deterministic finite automaton. To this end, $(k-1)$ different messages are sufficient. In addition, one message is emitted by the right boundary cell that signals the end of the input.

We conclude that the limitation of the bandwidth is a serious restriction of the devices. So, the question arises whether the limitations yield gradually increasing restrictions of the computational capacities. The question has been answered in the affirmative for OCA in [36] and for CA in [22]. The infinite, dense, and strict hierarchy depending on the number of messages for the two-way case is a consequence of the proper inclusion $\mathscr{L}_{rt}(CA_k) \subset \mathscr{L}_{rt}(CA_{k+1})$.

Comparing one-way with two-way information flow, the situation for the unrestricted case transfers to devices with limitations of the bandwidth. Let $k \geq 2$ be a constant. Then $\mathscr{L}_{rt}(OCA_k) \subset \mathscr{L}_{rt}(CA_k)$. The inclusion follows for structural reasons. For the strictness, it is well known that all unary languages belonging to $\mathscr{L}_{rt}(OCA)$ are regular [29]. Therefore, it suffices to show that the non-regular language $L = \{ a^{2^x+2x} \mid x \geq 1 \}$ belongs to $\mathscr{L}_{rt}(CA_2)$ [22].

It is a long-standing open problem whether linear-time cellular automata are more powerful than real-time cellular automata. Since linear-time CA_k can accept all regular languages whereas real-time CA_k cannot, we can answer this question in the affirmative for the case of restricted communication. Further results on cellular automata and iterative arrays with limited inter-cell communication can be found, for example, in [17–19, 21].

3.2   Max and Sum Communication Bounded Cellular Automata

Here we consider cellular automata where the communication is quantitatively measured by the number of uses of the links between cells. A simple example that shows the principal idea of limiting the maximal number of communications appearing between each two cells to a *constant* deals with signals as follows.

Example 3. The language $\{ a^n b^n \mid n \geq 1 \}$ belongs to $\mathscr{L}_{rt}(\mathrm{MC}(O(1))\text{-OCA})$.

The acceptance of the language is governed by two signals. The rightmost cell sends a signal B with maximum speed to the left whereas the unique cell which has an a in its input and has a right neighbor with a b in its input sends a signal A with speed $1/2$ to the left. When both signals meet in a cell, an accepting state is entered. In this way, $\{ a^n b^n \mid n \geq 1 \}$ is accepted and each cell performs only a finite number of communications.   □

By a straightforward generalization of the above construction with suitable signals having a certain speed it can be shown that $\{ a^n b^n c^n \mid n \geq 1 \}$, $\{ a^n b^m c^n d^m \mid n, m \geq 1 \}$, and $\{ a_1^n a_2^n \cdots a_k^n \mid n \geq 1 \}$, for $k \geq 1$ and different symbols $a_1, a_2, \ldots, a_k$, are accepted by real-time $\mathrm{MC}(O(1))$-OCA. The languages are not context free. Moreover, though this technique is not applicable directly, $\{ a^n b^{n_1} c^m b^{n_2} \mid n, m \geq 1 \wedge n_1, n_2 \geq 0 \wedge n_1 + n_2 = n \}$, and language $L_k = \{ a^n w \mid n \geq 1 \wedge w \in (b^* c^*)^k b^* \wedge |w|_b = n \}$ belong to $\mathscr{L}_{rt}(\mathrm{MC}(O(1))\text{-OCA})$ as well [19].

Next, we turn to identify the computational capacity of communication bounded real-time devices more generally. In [33, 34] two-way cellular automata are considered where the number of proper *state changes* is limited. Similar as in the present paper the sum of all state changes or the maximal number of the state changes of single cells are considered. It is not hard to see that such a device can be simulated by the corresponding communication bounded device. Whether or not state change bounded devices are strictly weaker than communication bounded ones is an open problem. However, the restrictions introduced in [33, 34] have been investigated with respect to communication in cellular automata. Adapting the proofs we obtain the proper inclusions

$$\mathrm{REG} \subset \mathscr{L}_{rt}(\mathrm{MC}(O(1))\text{-CA}) \subset \mathscr{L}_{rt}(\mathrm{MC}(\sqrt{n})\text{-CA}) \subset \mathscr{L}_{rt}(\mathrm{MC}(n)\text{-CA}),$$

$\mathscr{L}_{rt}(\mathrm{MC}(O(1))\text{-CA}) \subset \mathscr{L}_{rt}(\mathrm{SC}(n)\text{-CA})$, and $\mathscr{L}_{rt}(\mathrm{MC}(O(1))\text{-CA}) \subset \mathrm{NL}$. Actually, there exists an infinite proper hierarchy of real-time $\mathrm{SC}(f)$-CA families [19]. The ingredients are the mirror language $\{ wcw^R \mid w \in \{a, b\}^+ \}$ which is not accepted by any real-time $\mathrm{SC}(f)$-CA if $f \in o(n^2/\log(n))$, and for all $i \geq 1$, language $L_i = \{ w\$^{\varphi_i(|w|)-2|w|} w^R \mid w \in \{a, b\}^+ \}$ defined by the function $\varphi_i : \mathbb{N} \to \mathbb{N}$ which in turn is defined by $\varphi_1(n) = 2^n$, and $\varphi_i(n) = 2^{\varphi_{i-1}(n)}$, for $i \geq 2$. Language L_i is accepted by some real-time $\mathrm{SC}(n \log^{[i]}(n))$-CA, but cannot be accepted by any real-time $\mathrm{SC}(f)$-CA if $f \in o((n \log^{[i]}(n))/\log^{[i+1]}(n))$. From these results, the inclusion $\mathscr{L}_{rt}(\mathrm{SC}(n \log^{[i+1]}(n))\text{-CA}) \subset \mathscr{L}_{rt}(\mathrm{SC}(n \log^{[i]}(n))\text{-CA})$ and, thus, the proper hierarchy is derived.

3.3   Cellular Automata with Minimal Communication

Reducing the communication drastically, but keeping still enough to have non-trivial devices, systems were obtained where each two neighboring cells may communicate only constantly often or where the total number of communications during a computation depends linearly on the length of the input, and systems where the bandwidth of the inter-cell links is limited even to one message. However, all these restrictions do not lead to decidable properties. So, it is interesting to study cellular automata whose allowed communication is reduced to its minimum. To this end, we combine both approaches. Clearly, if there is no communication between two cells the array is split into two parts that work independently of each other. By the definition of acceptance, the right part is useless. So, to identify the minimum of communication, each two cells are allowed to communicate constantly often. Moreover, only one possible message is provided and the information flow is one-way. That is, we consider the class $MC(O(1))$-OCA_1. When the constant is set to one, these devices characterize the regular languages for unary alphabets. For non-unary alphabets non-context-free languages can be accepted [18, 21]. Despite this strong restriction we derive again undecidability for almost all questions for such automata. The results are obtained by a lemma which relates languages accepted by real-time one-way cellular automata with a constant number of communications with languages accepted by real-time one-way cellular automata with a constant number of one-message communications. Finally, this technique, in principle, can be applied also to cellular automata obeying all aforementioned restrictions. In particular, emptiness, finiteness, infiniteness, equivalence, inclusion, regularity, and context-freeness are undecidable for arbitrary real-time $MC(O(1))$-OCA_1. Clearly, the undecidability carries over to, for example, $MC(O(1))$-CA_k with two-way communication and the models $SC(n)$-OCA_k and $SC(n)$-CA_k.

In order to explore the borderline at which non-trivial decidability problems become decidable, additional structural and computational restrictions are imposed. An approach often investigated and widely accepted is to consider a given type of device for special purposes only, for example, for the acceptance of languages having a certain structure or form. From this point of view it is natural to start with unary languages. For general real-time OCA it is known that they accept only regular unary languages [29]. Since the proof is constructive, we derive that the borderline in question has been crossed. So, in [21] acceptors for bounded languages are studied. A language L over some alphabet $\{a_1, a_2, \ldots, a_k\}$ is said to be *bounded*, if $L \subseteq a_1^* a_2^* \cdots a_k^*$. For several devices it is known that they accept non-semilinear languages in general, but only semilinear bounded languages. Since for semilinear sets several properties are decidable [6], constructive proofs lead to decidable properties for these devices in connection with bounded languages. However, provided there are at least two different messages available, by reduction of Hilbert's tenth problem undecidability results for cellular automata are derived in which the number of communications allowed between every two neighboring cells is at least logarithmically or the total number of communications during a computation is linearly bounded in the length of the input. That is, emptiness, finiteness, infiniteness, inclusion, equivalence, regularity, and context-freeness are undecidable for arbitrary real-time $SC(n)$-OCA_2 and $MC(\log n)$-OCA_2 accepting bounded languages.

References

1. Bordihn, H., Kutrib, M., Malcher, A.: Undecidability and hierarchy results for parallel communicating finite automata. Int. J. Found. Comput. Sci. 22, 1577–1592 (2011)
2. Bordihn, H., Kutrib, M., Malcher, A.: On the computational capacity of parallel communicating finite automata. Int. J. Found. Comput. Sci. 23, 713–732 (2012)
3. Bordihn, H., Kutrib, M., Malcher, A.: Measuring communication in parallel communicating finite automata. In: Automata and Formal Languages, AFL 2014. EPTCS (to appear, 2014)
4. Buchholz, T., Kutrib, M.: On time computability of functions in one-way cellular automata. Acta Inform. 35, 329–352 (1998)
5. Choudhary, A., Krithivasan, K., Mitrana, V.: Returning and non-returning parallel communicating finite automata are equivalent. RAIRO Inform. Théor. 41, 137–145 (2007)
6. Ginsburg, S.: The Mathematical Theory of Context-Free Languages. McGraw Hill (1966)
7. Harrison, M.A.: Introduction to Formal Language Theory. Addison-Wesley (1978)
8. Holzer, M., Kutrib, M., Malcher, A.: Complexity of multi-head finite automata: Origins and directions. Theoret. Comput. Sci. 412, 83–96 (2011)
9. Hopcroft, J.E., Ullman, J.D.: Introduction to Automata Theory, Languages, and Computation. Addison-Wesley (1979)
10. Ibarra, O.H.: Reversal-bounded multicounter machines and their decision problems. J. ACM 25, 116–133 (1978)
11. Jurdziński, T., Kutyłowski, M.: Communication gap for finite memory devices. In: Orejas, F., Spirakis, P.G., van Leeuwen, J. (eds.) ICALP 2001. LNCS, vol. 2076, pp. 1052–1064. Springer, Heidelberg (2001)
12. Jurdziński, T., Kutyłowski, M., Loryś, K.: Multi-party finite computations. In: Asano, T., Imai, H., Lee, D.T., Nakano, S.-I., Tokuyama, T. (eds.) COCOON 1999. LNCS, vol. 1627, pp. 318–329. Springer, Heidelberg (1999)
13. Kutrib, M.: Cellular automata – a computational point of view. In: Bel-Enguix, G., Jiménez-López, M.D., Martín-Vide, C. (eds.) New Developments in Formal Languages and Applications. SCI, vol. 113, pp. 183–227. Springer, Heidelberg (2008)
14. Kutrib, M.: Cellular automata and language theory. In: Encyclopedia of Complexity and System Science, pp. 800–823. Springer (2009)
15. Kutrib, M., Malcher, A.: Fast cellular automata with restricted inter-cell communication: Computational capacity. In: Theoretical Computer Science (IFIP TCS2006). IFIP, vol. 209, pp. 151–164. Springer, Boston (2006)
16. Kutrib, M., Malcher, A.: Fast iterative arrays with restricted inter-cell communication: Constructions and decidability. In: Královič, R., Urzyczyn, P. (eds.) MFCS 2006. LNCS, vol. 4162, pp. 634–645. Springer, Heidelberg (2006)
17. Kutrib, M., Malcher, A.: Computations and decidability of iterative arrays with restricted communication. Parallel Process. Lett. 19, 247–264 (2009)
18. Kutrib, M., Malcher, A.: On one-way one-bit O(one)-message cellular automata. Electron. Notes Theor. Comput. Sci. 252, 77–91 (2009)

19. Kutrib, M., Malcher, A.: Cellular automata with sparse communication. Theoret. Comput. Sci. 411, 3516–3526 (2010)
20. Kutrib, M., Malcher, A.: Measuring communication in cellular automata. In: Symposium on Cellular Automata – Journées Automates Cellulaires (JAC 2010), pp. 13–30. Turku Center for Computer Science (2010)
21. Kutrib, M., Malcher, A.: One-way cellular automata, bounded languages, and minimal communication. J. Autom., Lang. Comb. 15, 135–153 (2010)
22. Kutrib, M., Malcher, A.: Cellular automata with limited inter-cell bandwidth. Theoret. Comput. Sci. 412, 3917–3931 (2011)
23. Kutrib, M., Malcher, A.: Two-party Watson-Crick computations. In: Domaratzki, M., Salomaa, K. (eds.) CIAA 2010. LNCS, vol. 6482, pp. 191–200. Springer, Heidelberg (2011)
24. Malcher, A.: Descriptional complexity of cellular automata and decidability questions. J. Autom., Lang. Comb. 7, 549–560 (2002)
25. Martín-Vide, C., Mateescu, A., Mitrana, V.: Parallel finite automata systems communicating by states. Int. J. Found. Comput. Sci. 13(5), 733–749 (2002)
26. Mazoyer, J.: A minimal time solution to the firing squad synchronization problem with only one bit of information exchanged. Technical Report TR 89-03, Ecole Normale Supérieure de Lyon (1989)
27. Mitrana, V.: On the degree of communication in parallel communicating finite automata systems. J. Autom., Lang. Comb. 5, 301–314 (2000)
28. Otto, F.: Asynchronous PC systems of pushdown automata. In: Dediu, A.-H., Martín-Vide, C., Truthe, B. (eds.) LATA 2013. LNCS, vol. 7810, pp. 456–467. Springer, Heidelberg (2013)
29. Seidel, S.R.: Language recognition and the synchronization of cellular automata. Tech. Rep., 79-02. Department of Computer Science, University of Iowa (1979)
30. Umeo, H.: Linear-time recognition of connectivity of binary images on 1-bit inter-cell communication cellular automaton. Parallel Comput. 27, 587–599 (2001)
31. Umeo, H., Kamikawa, N.: A design of real-time non-regular sequence generation algorithms and their implementations on cellular automata with 1-bit inter-cell communications. Fund. Inform. 52, 257–275 (2002)
32. Umeo, H., Kamikawa, N.: Real-time generation of primes by a 1-bit-communication cellular automaton. Fund. Inform. 58, 421–435 (2003)
33. Vollmar, R.: On cellular automata with a finite number of state changes. Computing 3, 181–191 (1981)
34. Vollmar, R.: Some remarks about the 'efficiency' of polyautomata. Int. J. Theoret. Phys. 21, 1007–1015 (1982)
35. Vollweiler, M.: Asynchronous systems of parallel communicating finite automata. In: Non-Classical Models for Automata and Applications (NCMA 2013). books@ocg.at, vol. 294, pp. 243–257. Austrian Computer Society (2013)
36. Worsch, T.: Linear time language recognition on cellular automata with restricted communication. In: Gonnet, G., Panario, D., Viola, A. (eds.) LATIN 2000. LNCS, vol. 1776, pp. 417–426. Springer, Heidelberg (2000)

From Algebra to Logic: There and Back Again

The Story of a Hierarchy[*]

(Invited Paper)

Pascal Weil

CNRS, LaBRI, UMR 5800, F-33400 Talence, France
pascal.weil@labri.fr
Univ. Bordeaux, LaBRI, UMR 5800, F-33400 Talence, France

Abstract. This is a survey about a collection of results about a (double) hierarchy of classes of regular languages, which occurs in a natural fashion in a number of contexts. One of these occurrences is given by an alternated sequence of deterministic and co-deterministic closure operations, starting with the piecewise testable languages. Since these closure operations preserve varieties of languages, this defines a hierarchy of varieties, and through Eilenberg's variety theorem, a hierarchy of pseudo-varieties (classes of finite monoids that are defined by pesudo-identities). The point of this excursion through algebra is that it provides reasonably simple decision algorithms for the membership problem in the corresponding varieties of languages. Another interesting point is that the hierarchy of pseudo-varieties bears a formal resemblance with another hierarchy, the hierarchy of varieties of idempotent monoids, which was much studied in the 1970s and 1980s and is by now well understood. This resemblance provides keys to a combinatorial characterization of the different levels of our hierarchies, which turn out to be closely related with the so-called rankers, a specification mechanism which was introduced to investigate the two-variable fragment of the first-order theory of the linear order. And indeed the union of the varieties of languages which we consider coincides with the languages that can be defined in that fragment. Moreover, the quantifier alternation hierarchy within that logical fragment is exactly captured by our hierarchy of languages, thus establishing the decidability of the alternation hierarchy.

There are other combinatorial and algebraic approaches of the same logical hierarchy, and one recently introduced by Krebs and Straubing also establishes decidability. Yet the algebraic operations involved are seemingly very different, an intriguing problem...

Formal language theory historically arose from the definition of models of computation (automata, grammars, etc) and relied for its first step on combinatorial reasoning, especially combinatorics on words. Very quickly however, algebra and

* This work was partially supported by the ANR through ANR-2010-BLAN-0204.

A.M. Shur and M.V. Volkov (Eds.): DLT 2014, LNCS 8633, pp. 275–278, 2014.

logic were identified as powerful tools for the classification of rational languages, e.g. with the definition of the syntactic monoid of a language and Büchi's theorem on monadic second-order logic. It did not take much time after that to observe that, conversely, formal language theory is itself a tool for algebra and logic.

The results which we will present are an illustration of this back-and-forth movement between languages, algebra and logic. They deal with a hierarchy of classes of rational languages which arises in different contexts and turned out to solve a problem in logic, namely the decidability of the quantifier alternation hierarchy within the two-variable fragment of first-order logic $FO^2[<]$.

The full picture uses a collection of results in logic, combinatorics on words and algebra which were obtained independently of the quantifier alternation hierarchy by various authors over several decades.

Let $\mathcal{R}_0$ be the class of piecewise testable languages, which is natural from a combinatorial and automata-theoretic point of view, and corresponds to the first level of the quantifier alternation hierarchy within $FO^2[<]$ (and within $FO[<]$) as well). This class is rather simple and reasonably well understood, see [1, 10]. We first consider the hierarchies of classes of languages obtained from $\mathcal{R}_0$ by alternatingly closing it under deterministic and co-deterministic closure: we let $\mathcal{L}_0 = \mathcal{R}_0$, $\mathcal{R}_{k+1}$ (resp. $\mathcal{L}_{k+1}$) be the deterministic (resp. co-deterministic) closure of $\mathcal{L}_k$ (resp. $\mathcal{R}_k$).

Results from the 1970s and 1980s [9, 12] show that the classes $\mathcal{R}_k$ and $\mathcal{L}_k$ are varieties (whether a language L belongs to one of these classes depends only on its syntactic monoid) and describe the corresponding varieties of finite monoids $\mathbf{R}_k$ and $\mathbf{L}_k$. Results from the 1960s [5] (see also [6, 11, 17]) shows that their membership problems are decidable and they form an infinite hierarchy.

A first view of the structure of the lattice formed by these varieties can be obtained by using purely algebraic results from the 1970s on a seemingly different hierarchy, that of varieties of idempotent monoids [2]. The theory of the latter varieties is particularly well understood, and one can exhibit for each of them structurally elegant identities and solutions of the word problem (of the corresponding relatively free object) [3].

To completely elucidate the structure of the lattice generated by the $\mathcal{R}_k$ and $\mathcal{L}_k$, Kufleitner and Weil introduced the notion of condensed rankers [8]. These are a rather natural extension of the algorithm to solve the word problem in the relatively free idempotent monoids and have natural connections with deterministic and codeterministic products. But they are also – and foremost – a variant of the rankers introduced by Weiss and Immerman [18] (following the turtle programs of Schwentick, Thérien and Vollmer [13]) to characterize the levels of the quantifier alternation hierarchy of $FO^2[<]$. As a result one can show that the k-th level of this hierarchy coincides with the intersection $\mathcal{R}_{k+1} \cap \mathcal{L}_{k+1}$, thus proving the decidability of each level of the hierarchy [7].

The story does not end there: using algebraic methods similar to those described in his book [15], Straubing showed [16] that the k-th level of the quantifier alternation hierarchy of $FO^2[<]$ is the variety of languages whose syntactic monoid is in the k-th term of the sequence given by $\mathbf{V}_1 = \mathbf{J}$ and $\mathbf{V}_{n+1} = \mathbf{V}_n \square \mathbf{J}$. Here $\mathbf{J}$ is the class of $\mathcal{J}$-trivial monoids, which characterizes piecewise testable languages by Simon's theorem [14] and $\square$ denotes the two-sided block product, the bilateral version of the more classical wreath product. Then Straubing and Krebs showed that every one of these classes of finite monoids is decidable [4], thus providing an alternate proof of the decidability of the quantifier alternation hierarchy, but also giving an alternative characterization of the classes $\mathbf{V}_k$: a finite monoid M is in $\mathbf{V}_k$ if and only if it sits in both $\mathbf{R}_{k+1}$ and $\mathbf{L}_{k+1}$.

The coincidence of these two very differently defined hierarchies raises an intriguing question: what connects the block product with the alternate operation of deterministic and co-deterministic closure?...

References

1. Almeida, J.: Finite Semigroups and Universal Algebra. World Scientific, Singapore (1994)
2. Gerhard, J.: The lattice of equational classes of idempotent semigroups. Journal of Algebra 15, 195–224 (1970)
3. Gerhard, J., Petrich, M.: Varieties of bands revisited. Proceedings of the London Mathematical Society 58(3), 323–350 (1989)
4. Krebs, A., Straubing, H.: An effective characterization of the alternation hierarchy in two-variable logic. In: FSTTCS 2012. LIPIcs, vol. 18, pp. 86–98. Schloss Dagstuhl–Leibniz-Zentrum fuer Informatik, Dagstuhl (2012), doi:10.4230/LIPIcs.FSTTCS.2012.86
5. Krohn, K., Rhodes, J., Tilson, B.: Homomorphisms and semilocal theory. In: Arbib, M. (ed.) The Algebraic Theory of Machines, Languages and Semigroups. Academic Press (1965)
6. Kufleitner, M., Weil, P.: On the lattice of sub-pseudovarieties of DA. Semigroup Forum 81, 243–254 (2010)
7. Kufleitner, M., Weil, P.: The FO^2 alternation hierarchy is decidable. In: CSL 2012. LIPIcs, vol. 16, pp. 426–439. Schloss Dagstuhl–Leibniz-Zentrum fuer Informatik, Dagstuhl (2012), doi:10.4230/LIPIcs.CSL.2012.426
8. Kufleitner, M., Weil, P.: On logical hierarchies within FO^2-definable languages. Logical Methods in Computer Science 8(3:11), 1–30 (2012)
9. Pin, J.E.: Propriétés syntactiques du produit non ambigu. In: de Bakker, J.W., van Leeuwen, J. (eds.) ICALP 1980. LNCS, vol. 85, pp. 483–499. Springer, Heidelberg (1980)
10. Pin, J.É.: Varieties of Formal Languages. North Oxford Academic, London (1986)
11. Rhodes, J., Steinberg, B.: The q-theory of finite semigroups. Springer Monographs in Mathematics. Springer, New York (2009)
12. Schützenberger, M.P.: Sur le produit de concaténation non ambigu. Semigroup Forum 13, 47–75 (1976)
13. Schwentick, T., Thérien, D., Vollmer, H.: Partially-ordered two-way automata: A new characterization of DA. In: Kuich, W., Rozenberg, G., Salomaa, A. (eds.) DLT 2001. LNCS, vol. 2295, pp. 239–250. Springer, Heidelberg (2002)
14. Simon, I.: Piecewise testable events. In: Brakhage, H. (ed.) GI-Fachtagung 1975. LNCS, vol. 33, pp. 214–222. Springer, Heidelberg (1975)
15. Straubing, H.: Finite automata, formal logic, and circuit complexity. Birkhäuser, Boston (1994)

16. Straubing, H.: Algebraic Characterization of the Alternation Hierarchy in $FO^2[<]$ on Finite Words. In: CSL 2011. LIPIcs, vol. 12, pp. 525–537. Schloss Dagstuhl–Leibniz-Zentrum fuer Informatik, Dagstuhl (2011), doi:10.4230/LIPIcs.CSL.2011.525

17. Weil, P.: Some results on the dot-depth hierarchy. Semigroup Forum 46, 352–370 (1993)

18. Weis, P., Immerman, N.: Structure theorem and strict alternation hierarchy for FO^2 on words. Logical Methods in Computer Science 5, 1–23 (2009)

Closure Properties of Pattern Languages

Joel D. Day[1], Daniel Reidenbach[1], and Markus L. Schmid[2]

[1] Department of Computer Science, Loughborough University,
Loughborough, Leicestershire, LE11 3TU, UK
J.Day@lboro.ac.uk,
D.Reidenbach@lboro.ac.uk
[2] Universität Trier, FB IV–Abteilung Informatikwissenschaften,
D-54286 Trier, Germany
MSchmid@uni-trier.de

Abstract. Pattern languages are a well-established class of languages that is particularly popular in algorithmic learning theory, but very little is known about their closure properties. In the present paper we establish a large number of closure properties of the terminal-free pattern languages, and we characterise when the union of two terminal-free pattern languages is again a terminal-free pattern language. We demonstrate that the equivalent question for general pattern languages is characterised differently, and that it is linked to some of the most prominent open problems for pattern languages. We also provide fundamental insights into a well-known construction of E-pattern languages as unions of NE-pattern languages, and vice versa.

Keywords: Pattern languages, Closure properties.

1 Introduction

Pattern languages were introduced by Dana Angluin [1] in order to model the algorithmic inferrability of patterns that are common to a set of words. In this context, a pattern is a sequence of variables and terminal symbols, and its language is the set of all words that can be generated from the pattern by a substitution that replaces all variables in the pattern by words of terminal symbols. Hence, more formally, a substitution is a terminal-preserving morphism, i. e., a morphism that maps every terminal symbol to itself. For example, the pattern language of the pattern $\alpha := x_1 x_1 a x_2 b$, where x_1, x_2 are variables and a, b are terminal symbols, is the set of all words that have a square as a prefix, followed by an arbitrary suffix that begins with the letter a and ends with the letter b. Thus, e.g., abbabbaab is contained in the language of α, whereas bbbaa is not. It is a direct consequence of these definitions that a pattern language is either a singleton or infinite. Furthermore, it is worth noting that two basic types of pattern languages are considered in the literature, depending on whether the variables must stand for nonempty words (referred to as non erasing or NE-pattern languages) or whether they may represent the empty word (so-called extended, erasing or simply E-pattern languages).

A.M. Shur and M.V. Volkov (Eds.): DLT 2014, LNCS 8633, pp. 279–290, 2014.

While the definition of pattern languages is simple, many of their properties are known to be related to complex phenomena in combinatorics on words, such as pattern avoidability (see Jiang et al. [7]) and ambiguity of morphisms (see Reidenbach [12]). Hence, the knowledge on pattern languages is still patchy, despite recent progress mainly regarding decision problems (see, e. g., Freydenberger, Reidenbach [5], Fernau, Schmid [3], Fernau et al. [4] and Reidenbach, Schmid [13]) and the relation to the Chomsky hierarchy (see Jain et al. [6] and Reidenbach, Schmid [14]).

Establishing the closure properties of a class of formal languages is one of the most classical and fundamental research tasks in formal language theory and any respective progress normally leads to insights and techniques that yield a better understanding of the class. In the case of pattern languages, it is known since Angluin's initial work that they are not closed under most of the usual operations, including union, intersection and complement. However, these non-closure properties can be shown by using very basic example patterns and exploiting peculiarities of the definition of pattern languages. For example, if a pattern does not contain a variable, then its language is a singleton; hence the union of any two distinct singleton pattern languages contains two elements, and therefore it cannot be a pattern language. Furthermore, the intersection of two pattern languages given by patterns that start with different terminal symbols is empty and the empty set, although a trivial language, is not a pattern language as well. Since, apart from a strong result by Shinohara [15] on the union of NE-pattern languages, hardly anything is known beyond such immediate facts, we can observe that in the case of pattern languages the existing closure properties fail to contribute to our understanding of their intrinsic properties.

It is the main purpose of this paper to investigate the closure properties of pattern languages more thoroughly. To this end, in Section 3, we consider the closure properties of two important subclasses of pattern languages, namely the classes of terminal-free NE- and E-pattern languages, i. e., pattern languages that are generated by patterns that do not contain any terminal symbols. This choice is motivated by the fact that terminal-free patterns have been a recent focus of interest in the research on pattern languages and, furthermore, most existing examples for non-closure of pattern languages (including the two examples for union and intersection given in the previous paragraph) do not translate to the terminal-free case. In Section 3.1, we completely characterise when the union of two terminal-free pattern languages is again a terminal-free pattern language and, in Section 3.2, we prove their non-closure under intersection, for which the situation is much more complicated compared to the operation of union.

We consider general pattern languages in Section 4, and we provide complex examples demonstrating that it is probably a very hard task to obtain full characterisations of those pairs of pattern languages whose unions or intersections are again a pattern language. In Section 4.3, we also study the question whether an E-pattern language can be expressed by the union of *nonerasing* pattern languages and, likewise, whether an NE-pattern language can be expressed by the union of *erasing* pattern languages. This question is slightly at odds with the classical investigation of closure properties, since we apply a language operation to members of one class and ask whether the resulting language is a member of another class. However, in the case of pattern languages, this makes sense, since every NE-pattern language is a finite union of E-pattern languages and every E-pattern language is a finite union of NE-pattern languages (see Jiang et al. [7]), a phenomenon that has been widely utilised in the context of inductive inference of pattern languages (see, e.g., Wright [17], Shinohara, Arimura [16]).

Due to space constraints, all proofs have been omitted from this paper.

2 Definitions and Preliminary Results

The symbols $\cup$, $\cap$ and $\setminus$ denote the set operations of *union, intersection* and *set difference*, respectively. For sets U and B with $B \subseteq U$, $\overline{B} := U \setminus B$ is the *complement* of B.

Let $\mathbb{N} := \{1, 2, 3, \ldots\}$ and let $\mathbb{N}_0 := \mathbb{N} \cup \{0\}$. For an arbitrary alphabet A, a *word (over A)* is a finite sequence of symbols from A, and ε stands for the *empty word*. The notation A^+ denotes the set of all nonempty words over A, and $A^* := A^+ \cup \{\varepsilon\}$. For the *concatenation* of two words w_1, w_2 we write $w_1 \cdot w_2$ or simply $w_1 w_2$, and w^n stands for the n-fold concatenation of the word w. We say that a word $v \in A^*$ is a *factor* of a word $w \in A^*$ if there are $u_1, u_2 \in A^*$ such that $w = u_1 \cdot v \cdot u_2$. If u_1 (or u_2) is the empty word, then v is a *prefix* (or a *suffix*, respectively) of w. The notation $|K|$ stands for the size of a set K or the length of a word K. A word w is *primitive* if, for any u such that $w = u^k$, $k = 1$. The *primitive root* of a word w is the primitive word u such that $w = u^k$, $k \in \mathbb{N}$.

For any alphabets A, B, a *morphism* is a function $h : A^* \to B^*$ that satisfies $h(vw) = h(v)h(w)$ for all $v, w \in A^*$; h is said to be *nonerasing* if, for every $a \in A$, $h(a) \neq \varepsilon$. A morphism h is *ambiguous (with respect to a word w)* if there exists a morphism g satisfying $g(w) = h(w)$ and, for a letter a in w, $g(a) \neq h(a)$. If such a morphism g does not exist, then h is called *unambiguous (with respect to w)*. A morphism $\sigma : A^* \to B^*$ is *periodic* if for some (primitive) word $w \in B^*$, $\sigma(x) \in \{w\}^*$ for every $x \in A$. The word w will be referred to as the *primitive root* of σ. If $|\sigma(x)| = 1$ for every $x \in A$, then σ is *1-uniform*.

Let Σ be a finite alphabet of so-called *terminal symbols* and X a countably infinite set of *variables* with $\Sigma \cap X = \emptyset$. We normally assume $X := \{x_1, x_2, x_3, \ldots\}$. A *pattern* is a nonempty word over $\Sigma \cup X$, a *terminal-free pattern* is a nonempty word over X; if a word contains symbols from Σ only, then we occasionally call it a *terminal word*. For any pattern α, we refer to the set of variables in α as var(α). If the variables in a pattern α are labelled in the natural way, then it is said to be in *canonical form*, i.e., α is in canonical form if, for some $n \in \mathbb{N}$, var(α) = $\{x_1, x_2, \ldots, x_n\}$ and, for any $x_i, x_j \in$ var(α) with $i < j$, there is a prefix β of α such that $x_i \in$ var(β) and $x_j \notin$ var(β). A pattern α is a *one-variable* pattern if $|\text{var}(\alpha)| = 1$. A morphism $h : (\Sigma \cup X)^* \to (\Sigma \cup X)^*$ is *terminal-preserving* if $h(a) = a$ for every $a \in \Sigma$. The *residual* of a pattern α is the word $h_\varepsilon(\alpha)$, where $h_\varepsilon : (\Sigma \cup X)^* \to (\Sigma \cup X)^*$ is a terminal preserving morphism with $h_\varepsilon(x) := \varepsilon$ for every $x \in$ var(α). A terminal-preserving morphism $h : (\Sigma \cup X)^* \to \Sigma^*$ is called a *substitution*.

Definition 1. *Let Σ be an alphabet, and let $\alpha \in (\Sigma \cup X)^*$ be a pattern. The E-pattern language of α is defined by $L_{E,\Sigma}(\alpha) := \{h(\alpha) \mid h : (\Sigma \cup X)^* \to \Sigma^*$ is a substitution\}. The NE-pattern language of α is defined by $L_{NE,\Sigma}(\alpha) := \{h(\alpha) \mid h : (\Sigma \cup X)^* \to \Sigma^*$ is a nonerasing substitution\}.*

Note that we call a pattern language terminal-free if there exists a terminal-free pattern that generates it.

Some parts of our reasoning in the subsequent sections is based on word equations, which are defined as follows. For a set of unknowns Y, a terminal alphabet Σ, and two words $\alpha, \beta \in (Y \cup \Sigma)^+$, the expression $\alpha = \beta$ is called a *word equation*. The solutions are terminal-preserving morphisms $\sigma : (Y \cup \Sigma)^* \to \Sigma^*$ such that $\sigma(\alpha) = \sigma(\beta)$. The words $\sigma(\alpha) (= \sigma(\beta))$ will be referred to as *solution-words*. It is often convenient to interpret variables from patterns as unknowns, and so word equations will often be formulated from two patterns.

This concludes the basic definitions of this paper. We now begin our investigation of the closure properties of the class of pattern languages. As a starting point, we refer to the corresponding result in the initial paper on pattern languages:

Theorem 1 (Angluin [1]). *NE-pattern languages are not closed under union, intersection, complement, Kleene plus, homomorphism and inverse homomorphism. NE-pattern languages are closed under concatenation and reversal.*

3   Terminal-Free Patterns

As briefly explained in Section 1, the proof of Theorem 1 heavily relies on the fact that patterns can contain terminal symbols. In the present section, we therefore wish to study whether the situation changes if we consider the classes of terminal-free E-pattern languages and terminal-free NE-pattern languages.

3.1   Union

Simple examples show that neither the terminal-free NE-pattern languages nor the terminal-free E-pattern languages are closed under union:

Proposition 1. *Let Σ be an alphabet with $\{a, b\} \subseteq \Sigma$. For every $Z, Z' \in \{E, NE\}$, there does not exist a pattern γ, such that $L_{Z,\Sigma}(\gamma) = L_{Z',\Sigma}(x_1 x_1) \cup L_{Z',\Sigma}(x_1 x_1 x_1)$.*

It is worth noting that the above statement also provides a first minor insight into the topic of expressing E-pattern languages as unions of NE-pattern languages and vice versa. We shall study this subject in Section 4.3 for patterns with terminal symbols in much more detail. In the present section, we merely want to point out that the union of two terminal-free E-pattern languages is indeed never a terminal-free NE-pattern language, and the union of two terminal-free NE-pattern languages cannot be a terminal-free E-pattern language:

Proposition 2. *Let Σ be an arbitrary alphabet, and let α and β be terminal-free patterns. Then there does not exist a terminal-free pattern γ with $L_{E,\Sigma}(\alpha) \cup L_{E,\Sigma}(\beta) = L_{NE,\Sigma}(\gamma)$ or $L_{NE,\Sigma}(\alpha) \cup L_{NE,\Sigma}(\beta) = L_{E,\Sigma}(\gamma)$.*

In the remainder of this section we wish to prove a similarly strong result for the actual closure of the class of terminal-free E- or NE-pattern languages. Hence, we wish to characterise those pairs of terminal-free (NE-/E-)pattern languages where the union again is a terminal-free (NE-/E-)pattern language. Our results shall demonstrate that the union of two terminal-free E-pattern languages can only be a terminal-free E-pattern language if there is an inclusion relation between the two languages, and that the same holds for the NE-pattern languages.

Our reasoning on the E case is based on a result on the inclusion problem for E-pattern languages. In [8], Jiang et al. provide a construction for a morphism τ_k such that, for two patterns α and β, the word $\tau_{|\beta|}(\alpha)$ is contained in $L_{E,\Sigma}(\beta)$ if and only if there exists a morphism φ from β to α. This in turn implies that the erasing languages of two terminal free patterns satisfy a subset relation if and only if there exists a morphism from one pattern to the other. It is not difficult to see that this construction can be further used to satisfy, for patterns α, β, γ, and $k = \max(|\alpha|, |\beta|)$, that $\tau_k(\gamma) \in L_{E,\Sigma}(\alpha) \cup L_{E,\Sigma}(\beta)$ if and only if γ is a morphic image of α or β. Thus if the relation $L_{E,\Sigma}(\alpha) \cup L_{E,\Sigma}(\beta) = L_{E,\Sigma}(\gamma)$ is satisfied, then $L_{E,\Sigma}(\gamma)$ is a subset of (and therefore also equal to) $L_{E,\Sigma}(\alpha)$ or $L_{E,\Sigma}(\beta)$ and we have the following situation.

Lemma 1. *Let Σ be an alphabet, $|\Sigma| \geq 2$, and let α and β be terminal-free patterns. There exists a terminal-free pattern γ with $L_{E,\Sigma}(\alpha) \cup L_{E,\Sigma}(\beta) = L_{E,\Sigma}(\gamma)$ if and only if $L_{E,\Sigma}(\alpha) \subseteq L_{E,\Sigma}(\beta)$ or $L_{E,\Sigma}(\beta) \subseteq L_{E,\Sigma}(\alpha)$.*

It can be observed from simple examples that, in the nonerasing case, inclusion cannot be characterised by the existence of a morphism between the generating patterns. Thus, no equivalent argument can be derived for the nonerasing case. However, a corresponding result can be obtained by looking at the shortest words in the nonerasing languages of α, β and γ. To this end, we define, for a pattern α, the set M_α to be $\{\sigma(\alpha) \mid \sigma : \mathrm{var}(\alpha)^* \to \Sigma^*$ is 1-uniform$\}$.

The set M_α has been used to positive effect in existing literature (see, e.g., [9]). It is particularly useful when considering nonerasing pattern languages because it encodes exactly the original pattern α (up to a renaming of variables). Moreover, it has a number of convenient properties when considering the union of two NE-pattern languages. One such example is that if α is strictly shorter than β, then the set of shortest words in $L_{NE,\Sigma}(\alpha) \cup L_{NE,\Sigma}(\beta)$ will be exactly M_α. Thus, if the union is itself the nonerasing language of some pattern γ, we have that $\gamma = \alpha$ up to a renaming of variables. A similar result can be obtained for the case that $|\alpha| = |\beta|$ by considering $|M_\alpha \cup M_\beta|$.

Lemma 2. *Let Σ be an alphabet, $|\Sigma| \geq 2$, and let α, β be terminal free patterns in canonical form with $|\alpha| = |\beta|$. Suppose that γ is a terminal free pattern (again in canonical form) with $M_\alpha \cup M_\beta = M_\gamma$. Then $\gamma \in \{\alpha, \beta\}$.*

Consequently, we can verify the same statement for nonerasing languages as we have for erasing languages.

Lemma 3. *Let Σ be an alphabet, $|\Sigma| \geq 2$, and let α and β be terminal-free patterns. There exists a terminal-free pattern γ with $L_{\mathrm{NE},\Sigma}(\alpha) \cup L_{\mathrm{NE},\Sigma}(\beta) = L_{\mathrm{NE},\Sigma}(\gamma)$ if and only if $L_{\mathrm{NE},\Sigma}(\alpha) \subseteq L_{\mathrm{NE},\Sigma}(\beta)$ or $L_{\mathrm{NE},\Sigma}(\beta) \subseteq L_{\mathrm{NE},\Sigma}(\alpha)$.*

Note that Lemma 3 extends an equivalent result by Shinohara [15] that holds for alphabets with at least 3 letters.

Thus, in general, the languages of two terminal-free patterns only union together to produce a third in the trivial case.

Theorem 2. *Let $Z, Z' \in \{\mathrm{E}, \mathrm{NE}\}$. Let Σ be an alphabet, $|\Sigma| \geq 2$, and let α, β, γ be terminal-free patterns. Then $L_{Z,\Sigma}(\alpha) \cup L_{Z,\Sigma}(\beta) = L_{Z',\Sigma}(\gamma)$ if and only if $L_{Z,\Sigma}(\alpha) = L_{Z',\Sigma}(\gamma)$ and $L_{Z,\Sigma}(\beta) \subseteq L_{Z,\Sigma}(\alpha)$ or $L_{Z,\Sigma}(\beta) = L_{Z',\Sigma}(\gamma)$ and $L_{Z,\Sigma}(\alpha) \subseteq L_{Z,\Sigma}(\beta)$.*

It is worth noting that, for terminal-free patterns, the inclusion problem – and therefore the question of closure under union – is decidable in the E case (see Jiang et al. [8], as explained above), but still open in the NE case.

3.2   Intersection

In the present section, we wish to investigate if the terminal-free NE- or E-pattern languages are closed under intersection. For the NE case, simple counterexamples such as $\alpha := xyx$ and $\beta := xxy$ can be used to prove the following observation:

Proposition 3. *The terminal-free NE-pattern languages are not closed under intersection.*

We can obtain an equivalent result for the terminal-free E-pattern languages, but our reasoning is significantly more complex and requires the analysis of certain word equations. Moreover, we are able to provide a characterisation for a restricted class of pairs of patterns, and show that, for this class, the situation is non-trivial (i.e., there exist both positive and negative examples). We proceed by considering the link between word equations and intersections of pattern-languages.

If, for a word equation $\alpha = \beta$, the words α and β are over disjoint alphabets, then the set of solutions $\sigma : (\mathrm{var}(\alpha) \cup \mathrm{var}(\beta))^* \to \Sigma^*$ corresponds exactly to the set of pairs of morphisms $\tau_1 : \mathrm{var}(\alpha)^* \to \Sigma^*$, $\tau_2 : \mathrm{var}(\beta)^* \to \Sigma^*$ such that $\tau_1(\alpha) = \tau_2(\beta)$. Thus, it also exactly describes the intersection $L_{\mathrm{E},\Sigma}(\alpha) \cap L_{\mathrm{E},\Sigma}(\beta)$. Furthermore, such an intersection is invariant under renamings of α and of β, so any intersection of E-pattern languages can be described in this way. The next proposition gives a characterisation of when the intersection of two terminal-free E-pattern languages is again a terminal-free E-pattern language in the restricted case that the corresponding word equation permits only periodic solutions. Note that, for α and β over disjoint alphabets, such solutions always exist.

Proposition 4. *Let Σ be an arbitrary alphabet. Let α, β be terminal-free patterns over disjoint alphabets and suppose that the word equation $\alpha = \beta$ permits only periodic solutions. Let w be the shortest non-empty solution-word. Let*

$$\mu := \mathrm{lcm}(\gcd\{|\alpha|_{x_i} \mid x_i \in \mathrm{var}(\alpha)\},\ \gcd\{|\beta|_{y_j} \mid y_j \in \mathrm{var}(\beta)\}).$$

Then $L_{\mathrm{E},\Sigma}(\alpha) \cap L_{\mathrm{E},\Sigma}(\beta)$ is a terminal-free E-pattern language if and only if $\mu = |w|$.

Despite Proposition 4, it is still a non-trivial task to find two terminal-free E-pattern languages whose intersection is not a terminal-free E-pattern language. In particular, it remains to find appropriate patterns α and β such that the word equation $\alpha = \beta$ has only periodic solutions. The following proposition provides such an example, and hence we have the analogous result to Proposition 3.

Proposition 5. *Let Σ be an arbitrary alphabet, and let $\alpha := x_1 x_2 x_1^2 x_2 x_1^3 x_2^2$ and $\beta := x_3 x_4^2 x_3^2 x_4^6 x_3^3$. Then $L_{\mathrm{E},\Sigma}(\alpha) \cap L_{\mathrm{E},\Sigma}(\beta)$ cannot be expressed as a terminal-free E-pattern language.*

It is even possible to give a much stronger statement, showing the extent to which the 'pattern-language mechanism' is incapable of handling this seemingly uncomplicated set of solutions.

Corollary 1. *For any alphabet Σ, $L_{\mathrm{E},\Sigma}(x_1 x_2 x_1^2 x_2 x_1^3 x_2^2) \cap L_{\mathrm{E},\Sigma}(x_3 x_4^2 x_3^2 x_4^6 x_3^3)$ cannot be expressed as a finite union of terminal-free E-pattern languages.*

It is worth noting that the approach above can be used to show that for $\alpha' := x_1 x_2 x_1^2 x_2^2 x_1^3 x_2^3$ and $\beta' := x_3 x_4^2 x_3^2 x_4^7 x_3^3$, one has that $L_{\mathrm{E},\Sigma}(\alpha') \cap L_{\mathrm{E},\Sigma}(\beta') = L_{\mathrm{E},\Sigma}(x_1^6)$. This demonstrates that the intersection of two E-pattern languages can in some cases be expressed as an E-pattern language, and therefore that the problem of whether the intersection of two E-pattern languages form an E-pattern language is nontrivial. However it is worth pointing out that a characterisation of this situation is probably very difficult to acquire due to the challenging nature of finding solution-sets of word equations.

3.3   Other Closure Properties

In this Section, we show that regarding the closure under the operations of complementation, morphisms, inverse morphisms, Kleene plus and Kleene star, terminal-free pattern languages behave similarly to the full class of pattern languages.

Proposition 6. *For every terminal-free pattern α, $\overline{L_{\mathrm{E},\Sigma}(\alpha)}$ is not a terminal-free E-pattern language and $\overline{L_{\mathrm{NE},\Sigma}(\alpha)}$ is not a terminal-free NE-pattern language.*

Proposition 6 does not only prove the non-closure of terminal-free E- and NE-pattern languages under complementation, but also characterises in a trivial way the terminal-free pattern languages whose complement is also a terminal-free pattern language.

Proposition 7. *Let Σ be a terminal alphabet with $|\Sigma| \geq 2$. The terminal-free NE- and E-pattern languages, with respect to Σ, are not closed under morphisms, inverse morphisms, Kleene plus and Kleene star.*

4   General Patterns

As explained in Section 1 and formally stated in Theorem 1, the closure properties of the full classes of NE-pattern languages and of E-pattern languages are understood. In the present section, we therefore wish to expand the more specific insights into the terminal-free pattern languages gained in Section 3 to the full classes. More precisely, with respect to the operations of complementation, intersection and union, we investigate those patterns that exhibit the property that their complement, intersection or union is again a pattern language and we try to characterise these patterns. Our strongest results are with respect to the operation of union.

4.1   Complement

With respect to the full class of E- and NE-pattern language, an analogue of Proposition 6 exists:

Proposition 8 (Bayer [2]). *Let Σ be a terminal alphabet with $\Sigma \geq 2$. For every pattern α, $\overline{L_{E,\Sigma}(\alpha)}$ is not an E-pattern language and $\overline{L_{NE,\Sigma}(\alpha)}$ is not an NE-pattern language.*

In the same way as Proposition 6 does for terminal-free patterns, this proposition yields a trivial characterisation of pattern languages with a complement that again is a pattern language.

4.2   Intersection

It is straightforward to construct patterns α and β such that $L_{E,\Sigma}(\alpha) \cap L_{E,\Sigma}(\beta)$ is not an E-pattern language or $L_{NE,\Sigma}(\alpha) \cap L_{NE,\Sigma}(\beta)$ is not an NE-pattern language. Furthermore, any two terminal-free patterns α and β are an example for the situation that $L_{E,\Sigma}(\alpha) \cap L_{E,\Sigma}(\beta)$ is not an NE-pattern language and, as long as there are at least two symbols in Σ, also for the situation that $L_{NE,\Sigma}(\alpha) \cap L_{NE,\Sigma}(\beta)$ is not an E-pattern language. Moreover, there are non-trivial examples of patterns α, β and γ, such that $L_{NE,\Sigma}(\alpha) \cap L_{NE,\Sigma}(\beta) = L_{E,\Sigma}(\gamma)$:

- $L_{NE,\Sigma}(ax) \cap L_{NE,\Sigma}(xx) = L_{E,\Sigma}(axax)$.
- $L_{NE,\Sigma}(xay) \cap L_{NE,\Sigma}(xxx) = L_{E,\Sigma}(xayxayxay)$.
- $L_{NE,\Sigma}(axa) \cap L_{NE,\Sigma}(xx) = L_{E,\Sigma}(axaaxa)$.
- $L_{NE,\Sigma}(axax) \cap L_{NE,\Sigma}(xbxb) = L_{E,\Sigma}(axbaxb)$.
- $L_{NE,\Sigma}(axy) \cap L_{NE,\Sigma}(xxx) = L_{E,\Sigma}(axaxax)$.

However, it is not known whether or not there are patterns α and β, such that $L_{E,\Sigma}(\alpha) \cap L_{E,\Sigma}(\beta)$ is an NE-pattern language. Moreover, we do not have any characterisations for the situation that the intersection of two pattern languages is again a pattern language.

4.3 Union

Examples of patterns α and β such that $L_{Z,\Sigma}(\alpha) \cup L_{Z,\Sigma}(\beta)$ is not a Z'-pattern language, for all $Z, Z' \in \{E, NE\}$, are provided by Proposition 1.

Theorem 2 is our strongest result in Section 3, as it shows that the union of terminal-free pattern languages can only be a terminal-free pattern language if one of the languages is contained in the other. At first glance it seems a reasonable hypothesis that a similar result might hold for the full class of pattern languages, but in the present section we show that this is not true.

For all but the union of pairs of E-pattern languages and the question of whether they can form an E-pattern language, suitable examples are not too hard to find:

Proposition 9. *Let Σ be a terminal alphabet.*

- $L_{E,\{a,b\}}(aax) \cup L_{E,\{a,b\}}(abx) = L_{NE,\{a,b\}}(ax)$.
- $L_{NE,\Sigma}(abc) \cup L_{NE,\Sigma}(axbxcx) = L_{E,\Sigma}(axbxcx)$.
- $L_{NE,\{a,b\}}(ax_1) \cup L_{NE,\{a,b\}}(bx_1) = L_{NE,\{a,b\}}(x_1x_2)$.

Regarding the question of whether $L_{E,\Sigma}(\alpha) \cup L_{E,\Sigma}(\beta) = L_{E,\Sigma}(\gamma)$ for patterns α, β, γ implies that there is an inclusion relation between $L_{E,\Sigma}(\alpha)$ and $L_{E,\Sigma}(\beta)$, the following three propositions provide increasingly complex counterexamples for alphabet sizes 2, 3, and 4.

Proposition 10. *Let $\Sigma = \{a, b\}$, $\alpha := x_1 a x_2 b x_2 a x_3$, $\beta := x_1 a x_2 b b x_2 a x_3$ and $\gamma := x_1 a x_2 b x_3 a x_4$. Then $L_{E,\Sigma}(\alpha) \nsubseteq L_{E,\Sigma}(\beta)$, $L_{E,\Sigma}(\beta) \nsubseteq L_{E,\Sigma}(\alpha)$ and $L_{E,\Sigma}(\alpha) \cup L_{E,\Sigma}(\beta) = L_{E,\Sigma}(\gamma)$.*

Proposition 11. *Let $\Sigma := \{a, b, c\}$,*

$$\alpha := x_1 a x_2 x_3^6 x_4^3 x_5^6 x_6 b x_7 a x_2 x_8^{12} x_4^6 x_9^{12} x_6 b x_{10},$$
$$\beta := x_1 a x_2 x_3^6 x_4^2 x_5^5 x_6^6 x_7 b x_8 a x_2 x_9^{12} x_4^4 x_5^{10} x_{10}^{12} x_7 b x_{11} \text{ and}$$
$$\gamma := x_1 a x_2 x_3^6 x_4^2 x_5^3 x_6^6 x_7 b x_8 a x_2 x_9^{12} x_4^4 x_5^6 x_{10}^{12} x_7 b x_{11}.$$

Then

$$L_{E,\Sigma}(\alpha) \nsubseteq L_{E,\Sigma}(\beta), \; L_{E,\Sigma}(\beta) \nsubseteq L_{E,\Sigma}(\alpha) \text{ and } L_{E,\Sigma}(\alpha) \cup L_{E,\Sigma}(\beta) = L_{E,\Sigma}(\gamma).$$

Proposition 12. *Let $\Sigma := \{a, b, c, d\}$,*

$$\alpha := x_1 a x_2 x_3^2 x_4^2 x_5^2 x_6 b x_7 a x_2 x_8^2 x_4^2 x_9^2 x_6 b x_{10} c x_{11} x_{12}^2 x_{13}^2 x_{14}^2 x_{15}^2 x_{16} d$$
$$x_{17} c x_{11} x_{18}^2 x_{13}^2 x_{14}^2 x_{19}^2 x_{16} d x_{20} x_{13}^2 x_{14}^2 x_{13}^2 x_{14}^2 x_{13}^2 x_{14}^2 x_{21} x_4^6,$$
$$\beta := x_1 a x_2 x_3^2 x_4^2 x_5^2 x_6^2 x_7 b x_8 a x_2 x_9^2 x_4^2 x_5^2 x_{10}^2 x_7 b x_{11} c x_{12} x_{13}^2 x_{14}^2 x_{15}^2 x_{16} d$$
$$x_{17} c x_{12} x_{18}^2 x_{14}^2 x_{19}^2 x_{16} d x_{20} x_{14}^6 x_{21} x_4^2 x_5^2 x_4^2 x_5^2 x_4^2 x_5^2 \text{ and}$$
$$\gamma := x_1 a x_2 x_3^2 x_4^2 x_5^2 x_6^2 x_7 b x_8 a x_2 x_9^2 x_4^2 x_5^2 x_{10}^2 x_7 b x_{11} c x_{12} x_{13}^2 x_{14}^2 x_{15}^2 x_{16}^2 x_{17} d$$
$$x_{18} c x_{12} x_{19}^2 x_{14}^2 x_{15}^2 x_{20}^2 x_{17} d x_{21} x_{14}^2 x_{15}^2 x_{14}^2 x_{15}^2 x_{14}^2 x_{15}^2 x_{22} x_4^2 x_5^2 x_4^2 x_5^2 x_4^2 x_5^2.$$

Then

$$L_{E,\Sigma}(\alpha) \nsubseteq L_{E,\Sigma}(\beta), \; L_{E,\Sigma}(\beta) \nsubseteq L_{E,\Sigma}(\alpha) \text{ and } L_{E,\Sigma}(\alpha) \cup L_{E,\Sigma}(\beta) = L_{E,\Sigma}(\gamma).$$

We are not able to give equivalent examples for larger alphabets, and we expect the question of their existence to be a complex and important problem. This is because the above examples depend on the ambiguity of terminal-preserving morphisms, which is a phenomenon that underpins many properties of pattern languages. Similar constructions to those in Propositions 10, 11, and 12 have been used to disprove longstanding conjectures on inductive inference (see Reidenbach [10,12]) of and the equivalence problem (see Reidenbach [11]) for E-pattern languages over alphabets of up to 4 letters and, similarly, it has so far not been possible to expand those techniques to arbitrary alphabets. Our examples, thus, suggest a close link between the problem in the current section and the two most important open problems for E-pattern languages over alphabets with at least 5 letters, and we expect that substantial progress on any one of them will require combinatorial insights that will allow the others to be solved as well.

For all $Z, Z' \in \{E, NE\}$, we have seen example patterns α and β such that $L_{Z,\Sigma}(\alpha) \cup L_{Z,\Sigma}(\beta)$ is a Z'-pattern language. We shall now try to generalise these examples in order to obtain characterisations of such pairs of patterns.

For the case $Z = Z' = E$, we are only able to state a necessary condition for $L_{E,\Sigma}(\alpha) \cup L_{E,\Sigma}(\beta) = L_{E,\Sigma}(\gamma)$ that, unfortunately, is not very strong:

Theorem 3. *Let Σ be an alphabet, and let α, β and γ be patterns with $L_{E,\Sigma}(\alpha) \cup L_{E,\Sigma}(\beta) = L_{E,\Sigma}(\gamma)$. Furthermore, let w_α, w_β and w_γ be the residuals of α, β and γ, respectively. Then $w_\gamma = w_\alpha$ and w_γ is a subsequence of w_β or $w_\gamma = w_\beta$ and w_γ is a subsequence of w_α.*

In view of the fact that the examples of Propositions 10, 11 and 12 are rather complicated, we expect that a full characterisation for the case $Z = Z' = E$ is difficult to obtain.

For the case $Z = Z' = NE$, we can present a strong necessary condition that, similarly to Lemma 3, strengthens a result by Shinohara [15]:

Theorem 4. *Let Σ be an alphabet with $\{a, b\} \subseteq \Sigma$ and let α, β and γ be patterns. If $L_{NE,\Sigma}(\alpha) \cup L_{NE,\Sigma}(\beta) = L_{NE,\Sigma}(\gamma)$, then one of the following three statements is true:*

- *$L_{NE,\Sigma}(\alpha) \subseteq L_{NE,\Sigma}(\beta)$ and $\beta = \gamma$.*
- *$L_{NE,\Sigma}(\beta) \subseteq L_{NE,\Sigma}(\alpha)$ and $\alpha = \gamma$.*
- *$|\Sigma| = 2$ and*

$$\alpha = \delta_0 \, a \, \delta_1 \, a \, \delta_2 \ldots \delta_{m-1} \, a \, \delta_m \, ,$$
$$\beta = \delta_0 \, b \, \delta_1 \, b \, \delta_2 \ldots \delta_{m-1} \, b \, \delta_m \, ,$$
$$\gamma = \delta_0 \, x \, \delta_1 \, x \, \delta_2 \ldots \delta_{m-1} \, x \, \delta_m \, ,$$

where $m \geq 1$, $\delta_i \in (X \cup \Sigma)^*$, $0 \leq i \leq m$.

It remains to consider the cases $Z = NE$, $Z' = E$ and $Z = E$, $Z' = NE$, for which we have full characterisations. Before we prove these characterisations, we recall that Jiang et al. show in [7] that, for every pattern α, we can construct finite sets of patterns Γ and Δ such that $L_{E,\Sigma}(\alpha) = \bigcup_{\beta \in \Gamma} L_{NE,\Sigma}(\beta)$ and $L_{NE,\Sigma}(\alpha) = \bigcup_{\beta \in \Delta} L_{E,\Sigma}(\beta)$. More precisely, Γ is the set of all patterns that can be obtained from α by erasing some (possibly none) of the variables and Δ contains all pattern that can be obtained from α by substituting each $x \in \text{var}(\alpha)$ by bx, for some $b \in \Sigma$. We note that the examples $L_{NE,\Sigma}(\text{abc}) \cup L_{NE,\Sigma}(\text{axbxcx}) = L_{E,\Sigma}(\text{axbxcx})$ and $L_{E,\{a,b\}}(\text{aax}) \cup L_{E,\{a,b\}}(\text{abx}) = L_{NE,\{a,b\}}(\text{ax})$ of Proposition 9 are applications of exactly this construction.

The characterisation for the case $Z = \text{NE}$, $Z' = \text{E}$ follows from the fact that we can prove that if we restrict ourselves to unions of only two pattern languages, then $L_{E,\Sigma}(\alpha) = \bigcup_{\beta \in \Gamma} L_{NE,\Sigma}(\beta)$ is the only possible way to describe an E-pattern language by NE-pattern languages.

Theorem 5. *Let Σ be an alphabet with $|\Sigma| \geq 2$ and let α, β and γ be patterns. Then $L_{NE,\Sigma}(\alpha) \cup L_{NE,\Sigma}(\beta) = L_{E,\Sigma}(\gamma)$ if and only if $\alpha \in \Sigma^+$ and $\beta = \gamma = u_1 x^{j_1} u_2 x^{j_2} \ldots x^{j_m} u_{m+1}$, $j_i \in \mathbb{N}_0$, $1 \leq i \leq m$, such that $u_1 u_2 \ldots u_{m+1} = \alpha$.*

With respect to the case $Z = \text{E}$, $Z' = \text{NE}$, we can even present a characterisation for the situation $L_{NE,\Sigma}(\alpha) = \bigcup_{i=1}^{k} L_{E,\Sigma}(\beta_i)$ with $k \leq |\Sigma|$. It shall be explained later on that this characterisation is a generalisation of the construction given by Jiang et al.

Theorem 6. *Let $\ell \geq 2$ and let Σ be an alphabet with $\{a_1, a_2, \ldots, a_\ell\} \subseteq \Sigma$. Furthermore, let $\alpha_1, \alpha_2, \ldots, \alpha_\ell$ and γ be patterns with $L_{E,\Sigma}(\alpha_i) \neq L_{E,\Sigma}(\alpha_j)$, $1 \leq i < j \leq \ell$. Then $\bigcup_{i=1}^{\ell} L_{E,\Sigma}(\alpha_i) = L_{NE,\Sigma}(\gamma)$ if and only if, for some permutation π of $(1, 2, \ldots, \ell)$,*

- *$\Sigma = \{a_1, a_2, \ldots, a_\ell\}$,*
- *$\gamma = u_1 x u_2 x u_3 \ldots u_k x u_{k+1}$, $k \geq 1$, $u_i \in \Sigma^*$, $1 \leq i \leq k+1$, and,*
- *for every i, $1 \leq i \leq \ell$,*

$$\alpha_i = u_1 \alpha_i' a_{\pi(i)} \alpha_i'' u_2 \alpha_i' a_{\pi(i)} \alpha_i'' u_3 \ldots u_k \alpha_i' a_{\pi(i)} \alpha_i'' u_{k+1},$$

where $\alpha_i', \alpha_i'' \in X^*$,
- *for every i, $1 \leq i \leq \ell$, there exists a $y_i \in \text{var}(\alpha_i)$ with $|\alpha_i|_{y_i} = k$ and*
  - *$|\alpha_i'|_{y_i} = 1$ for all i, $1 \leq i \leq \ell$, or*
  - *$|\alpha_i''|_{y_i} = 1$ for all i, $1 \leq i \leq \ell$.*

If we apply the construction of Jiang et al. to a one-variable pattern γ, then we obtain patterns α_i, $1 \leq i \leq |\Sigma|$, that satisfy the conditions of the patterns in the statement of Theorem 6. More precisely, this corresponds to the special case where $\alpha_i' \alpha_i'' = y_i$, $1 \leq i \leq |\Sigma|$. Moreover, it can be easily verified that if γ and patterns α_i, $1 \leq i \leq |\Sigma|$, satisfy the conditions of the statement of Theorem 6, then, depending on whether $|\alpha_i'|_{y_i} = 1$ for all i, $1 \leq i \leq |\Sigma|$, or $|\alpha_i''|_{y_i} = 1$ for all i, $1 \leq i \leq |\Sigma|$, we can obtain patterns β_i from the patterns α_i by replacing $\alpha_i' a_i \alpha_i''$ by $y_i a_i$ or by $a_i y_i$, respectively, and $\bigcup_{i=1}^{|\Sigma|} L_{E,\Sigma}(\beta_i) = L_{NE,\Sigma}(\gamma)$ still holds. Furthermore, the patterns β_i are exactly the patterns that are obtained if we apply the construction of Jiang et al.

Acknowledgments. The authors wish to thank the anonymous referees for their helpful suggestions, which have yielded a stronger version of Proposition 7.

References

1. Angluin, D.: Finding patterns common to a set of strings. Journal of Computer and System Sciences 21, 46–62 (1980)
2. Bayer, H.: Allgemeine Eigenschaften von Patternsprachen. Projektarbeit, Fachbereich Informatik, Universität Kaiserslautern (2007) (in German)
3. Fernau, H., Schmid, M.L.: Pattern matching with variables: A multivariate complexity analysis. In: Fischer, J., Sanders, P. (eds.) CPM 2013. LNCS, vol. 7922, pp. 83–94. Springer, Heidelberg (2013)
4. Fernau, H., Schmid, M.L., Villanger, Y.: On the parameterised complexity of string morphism problems. In: Proc. 33rd IARCS Annual Conference on Foundations of Software Technology and Theoretical Computer Science, FSTTCS 2013. Leibniz International Proceedings in Informatics (LIPIcs), vol. 24, pp. 55–66 (2013)
5. Freydenberger, D., Reidenbach, D.: Bad news on decision problems for patterns. Information and Computation 208, 83–96 (2010)
6. Jain, S., Ong, Y., Stephan, F.: Regular patterns, regular languages and context-free languages. Information Processing Letters 110, 1114–1119 (2010)
7. Jiang, T., Kinber, E., Salomaa, A., Salomaa, K., Yu, S.: Pattern languages with and without erasing. International Journal of Computer Mathematics 50, 147–163 (1994)
8. Jiang, T., Salomaa, A., Salomaa, K., Yu, S.: Decision problems for patterns. Journal of Computer and System Sciences 50, 53–63 (1995)
9. Lange, S., Wiehagen, R.: Polynomial-time inference of arbitrary pattern languages. New Generation Computing 8, 361–370 (1991)
10. Reidenbach, D.: A non-learnable class of E-pattern languages. Theoretical Computer Science 350, 91–102 (2006)
11. Reidenbach, D.: An examination of Ohlebusch and Ukkonen's conjecture on the equivalence problem for E-pattern languages. Journal of Automata, Languages and Combinatorics 12, 407–426 (2007)
12. Reidenbach, D.: Discontinuities in pattern inference. Theoretical Computer Science 397, 166–193 (2008)
13. Reidenbach, D., Schmid, M.L.: Patterns with bounded treewidth. Information and Computation (to appear)
14. Reidenbach, D., Schmid, M.L.: Regular and context-free pattern languages over small alphabets. Theoretical Computer Science 518, 80–95 (2014)
15. Shinohara, T.: Inferring unions of two pattern languages. Bulletin of Informatics and Cybernetics 20, 83–88 (1983)
16. Shinohara, T., Arimura, H.: Inductive inference of unbounded unions of pattern languages from positive data. Theoretical Computer Science 241, 191–209 (2000)
17. Wright, K.: Identification of unions of languages drawn from an identifiable class. In: Proc. 2nd Annual Workshop on Computational Learning Theory, COLT 1989, pp. 328–333 (1989)

Minimal and Hyper-Minimal Biautomata[*]
(Extended Abstract)

Markus Holzer and Sebastian Jakobi

Institut für Informatik, Universität Giessen,
Arndtstr. 2, 35392 Giessen, Germany
{holzer,sebastian.jakobi}@informatik.uni-giessen.de

Abstract. We compare deterministic finite automata (DFAs) and biautomata under the following two aspects: structural similarities between minimal and hyper-minimal automata, and computational complexity of the minimization and hyper-minimization problem. Concerning classical minimality, the known results such as isomorphism between minimal DFAs, and NL-completeness of the DFA minimization problem carry over to the biautomaton case. But surprisingly this is not the case for hyper-minimization: the similarity between almost-equivalent hyper-minimal biautomata is not as strong as it is between almost-equivalent hyper-minimal DFAs. Moreover, while hyper-minimization is NL-complete for DFAs, we prove that this problem turns out to be computationally intractable, i.e., NP-complete, for biautomata.

1 Introduction

The minimization problem for finite automata is well studied in the literature, see, e.g., [8] for a recent overview on some automata related problems. The problem asks for the smallest possible finite automaton that is equivalent to a given one. Because regular languages are used in many applications and one may like to represent the languages succinctly, this problem is also of practical relevance. It is well known that for a given n-state deterministic finite automaton (DFA) one can efficiently compute an equivalent minimal automaton in $O(n \log n)$ time [10]. More precisely, the DFA minimization problem is complete for NL, even for DFAs without inaccessible states [3]. On the other hand, minimization of nondeterministic finite automata (NFAs) is highly intractable, namely PSPACE-complete [12]. These results go along with the structural properties of minimal finite automata. While minimal DFAs are unique up to isomorphism, this is not the case for minimal nondeterministic state devices anymore [1]. In fact, the characterization of minimal DFAs is one of the basic building blocks for efficient DFA minimization algorithms.

When changing from minimization to hyper-minimization a quite similar picture as mentioned above emerges. Hyper-minimization asks for the smallest

---

[*] The first author dedicates this paper to the memory of Professor Dr. Dr. h.c. mult. Wilfried Brauer (1937–2014) in gratitude for his guidance and continuous support.

A.M. Shur and M.V. Volkov (Eds.): DLT 2014, LNCS 8633, pp. 291–302, 2014.

automaton that is equivalent to a given one up to a finite number of exceptions—this form of "equivalence" is referred to as almost-equivalence in the literature. Let us discuss the situation for hyper-minimal DFAs and NFAs in more detail. First, all of the above mentioned computational complexity results remain valid for hyper-minimization. Thus, computing a hyper-minimal DFA can be done in $O(n \log n)$ time [9] and the hyper-minimization problem is NL-complete [5]. In fact it is known that minimization for DFAs linearly reduces to hyper-minimization [9]. Moreover, the intractability result for NFAs remains, that is, hyper-minimization for NFAs is PSPACE-complete [5], just as it is for ordinary NFA minimization. What can be said about the structural properties of hyper-minimal finite state machines? Neither hyper-minimal DFAs nor hyper-minimal NFAs are unique up to isomorphism. Nevertheless, hyper-minimal DFAs obey a structural characterization as shown in [2]. Almost-equivalent hyper-minimal DFAs have isomorphic kernels and isomorphic preambles up to state acceptance. Here the kernel of an automaton consists of the states that are reachable from the start state by an *infinite* number of inputs; all other states belong to the preamble of the automaton.

Recently, an alternative automaton model to deterministic finite automata, the so called *biautomaton* (DBiA) [17] was introduced. Roughly speaking, a biautomaton consists of a *deterministic* finite control, a read-only input tape, and two reading heads, one reading the input from left to right (forward transitions), and the other head reading the input from the opposite direction, i.e., from right to left (backward transitions). An input word is accepted by a biautomaton, if there is an accepting computation starting the heads on the two ends of the word meeting somewhere in an accepting state. Although the choice of reading a symbol by either head is nondeterministic, a deterministic outcome of the computation of the biautomaton is enforced by two properties. In [17] and a series of forthcoming papers [6,7,13,16] it was shown that biautomata share a lot of properties with ordinary finite automata. For instance, as minimal DFAs, also minimal DBiAs are unique up to isomorphism [17]. Moreover, in [6] it was shown that classical DFA minimization algorithms can be adapted to biautomata as well. As a first result we show that biautomaton minimization is NL-complete as for ordinary DFAs.

Now the question arises, which of the structural similarities between almost-equivalent or hyper-minimal DFAs similarly hold for biautomata as well? Moreover, what can be said about the computational complexity of biautomaton hyper-minimization? We give answers to both questions in the forthcoming. Some of the structural similarities found for almost-equivalent and hyper-minimal DFAs carry over to the case of biautomata, but there are subtle differences. On the one hand we show that the kernel isomorphism for almost-equivalent DFAs carries over to almost-equivalent biautomata, but on the other hand, the isomorphism for the preambles of almost-equivalent hyper-minimal DFAs does not transfer to the biautomaton case. In fact, we present an example of two almost-equivalent hyper-minimal biautomata the preambles of which are not isomorphic at all. The observed phenomenon is later used to prove the main result of this paper, namely that hyper-minimizing biautomata is *not* as easy as for DFAs. More precisely, we show that hyper-minimization of biautomata is NP-complete. This is in sharp contrast to the case of hyper-minimal DFAs. Due to space constraints all proofs can be found in the Appendix.

2 Preliminaries

A *deterministic biautomaton* (DBiA) is a sixtuple $A = (Q, \Sigma, \cdot, \circ, q_0, F)$, where Q, Σ, q_0, and F are defined as for deterministic finite automata (DFAs), and where $\cdot$ and $\circ$ are mappings from $Q \times \Sigma$ to Q, called the *forward* and *backward transition function*, respectively. It is common in the literature on biautomata to use an infix notation for these functions, i.e., writing $q \cdot a$ and $q \circ a$ instead of $\cdot(q, a)$ and $\circ(q, a)$. Similar as for the transition function of a DFA, the forward transition function $\cdot$ can be extended to $\cdot \colon Q \times \Sigma^* \to Q$ by $q \cdot \lambda = q$, and $q \cdot av = (q \cdot a) \cdot v$, for all states $q \in Q$, symbols $a \in \Sigma$, and words $v \in \Sigma^*$. The extension of the backward transition function $\circ$ to $\circ \colon Q \times \Sigma^* \to Q$ is defined as follows: $q \circ \lambda = q$ and $q \circ va = (q \circ a) \circ v$, for all states $q \in Q$, symbols $a \in \Sigma$, and words $v \in \Sigma^*$. Notice that $\circ$ consumes the input from right to left, hence the name backward transition function. The DBiA A *accepts* a word $w \in \Sigma^*$ if there are words $u_i, v_i \in \Sigma^*$, for $1 \leq i \leq k$, such that w can be written as $w = u_1 u_2 \ldots u_k v_k \ldots v_2 v_1$, and

$$((\ldots ((((q_0 \cdot u_1) \circ v_1) \cdot u_2) \circ v_2) \ldots) \cdot u_k) \circ v_k \in F.$$

The language accepted by A is $L(A) = \{\, w \in \Sigma^* \mid A \text{ accepts } w \,\}$.

The DBiA A has the $\diamond$-*property*, if $(q \cdot a) \circ b = (q \circ b) \cdot a$, for all $q \in Q$ and $a, b \in \Sigma$, and it has the F-*property*, if we have $q \cdot a \in F$ if and only if $q \circ a \in F$, for all $q \in Q$ and $a \in \Sigma$. The biautomata as introduced in [17] always had to satisfy *both* these properties, while in [6,7] also biautomata that lack one or both of these properties, as well as nondeterministic biautomata were studied. Throughout the current paper, when writing of biautomata, or DBiAs, we always mean deterministic biautomata that satisfy both the $\diamond$-property, and the F-property, i.e., the model as introduced in [17]. For such biautomata the following is known from the literature [6,17]:

- $(q \cdot u) \circ v = (q \circ v) \cdot u$, for all states $q \in Q$ and words $u, v \in \Sigma^*$,
- $(q \cdot u) \circ vw \in F$ if and only if $(q \cdot uv) \circ w \in F$, for all states $q \in Q$ and words $u, v, w \in \Sigma^*$.

From this one can conclude that for all words $u_i, v_i \in \Sigma^*$, with $1 \leq i \leq k$, we have

$$((\ldots ((((q_0 \cdot u_1) \circ v_1) \cdot u_2) \circ v_2) \ldots) \cdot u_k) \circ v_k \in F$$

if and only if

$$q_0 \cdot u_1 u_2 \ldots u_k v_k \ldots v_2 v_1 \in F.$$

Therefore, the language accepted by a biautomaton A can as well be defined as $L(A) = \{\, w \in \Sigma^* \mid q_0 \cdot w \in F \,\}$.

For a state q of an automaton A (DFA or DBiA), the *right language* of q is the language $L_A(q)$ accepted by the automaton that is obtained from A by making q its initial state. Notice that the right language of the initial state q_0 of A is $L_A(q_0) = L(A)$. We say that two automata A and A' are *equivalent*, denoted by $A \equiv A'$, if $L(A) = L(A')$. Similarly, if q is a state of A and q' a state of A', then q and q' are *equivalent*, for short $q \equiv q'$, if $L_A(q) = L_{A'}(q')$. An automaton A is *minimal* if there is no automaton B of the same type, that has fewer states than A and satisfies $A \equiv B$.

We often use regular expressions to describe languages—see, e.g., [11]. As usual we identify an expression with the language it describes, and by abuse of notation we also use regular expressions as names for states.

Recently, the notions of almost-equivalence and hyper-minimality were introduced [2]. Two languages L and L' are *almost-equivalent*, denoted by $L \sim L'$, if their symmetric difference $L \triangle L' := (L \setminus L') \cup (L' \setminus L)$ is finite. This notion naturally carries over to automata and states: two automata A and A' are *almost-equivalent*, for short $A \sim A'$, if $L(A) \sim L(A')$, and two states q and q' of A and, respectively, A' are *almost-equivalent*, for short $q \sim q'$, if $L_A(q) \sim L_{A'}(q')$. An automaton A is *hyper-minimal* if there is no automaton B of the same type, that has fewer states than A and satisfies $A \sim B$. A useful concept for the study of almost-equivalent automata is the partitioning of the state set into preamble and kernel states. A state q of an automaton A is a *kernel state* if it is reachable from the initial state of A by an infinite number of inputs, otherwise q is a *preamble state*. The set of all preamble states of A is denoted by $\mathrm{Pre}(A)$, and the set of kernel states is $\mathrm{Ker}(A)$.

We assume familiarity with the basic concepts of complexity theory [11,18] such as reductions, completeness, and the inclusion chain $\mathsf{NL} \subseteq \mathsf{P} \subseteq \mathsf{NP}$. Here NL is the set of problems accepted by nondeterministic logarithmic space bounded Turing machines. Moreover, let P (NP, respectively) denote the set of problems accepted by deterministic (nondeterministic, respectively) polynomial time bounded Turing machines.

3   Structural Similarity between Hyper-Minimal Automata

Let us first recall some important facts on classically minimal and equivalent DFAs and biautomata. It is well known that two equivalent minimal DFAs must be isomorphic, i.e., there exists a bijective mapping between the state sets of the automata that preserves acceptance of states and is compatible with the transition functions. Moreover, we know that a DFA is minimal if and only if all its states are reachable, and no two distinct states are equivalent. These results carry over to biautomata: the isomorphism between two equivalent minimal biautomata was shown in [17], and the characterization of minimal biautomata in terms of reachable and non-equivalent states was given in [6]. Moreover, one can show the following connection between (minimal) DBiAs and DFAs: if A is a (minimal) DBiA, then the automaton A_{fwd} obtained from A by considering only forward transitions is a (minimal) DFA that accepts the language $L(A)$.

What is known for almost-equivalent and hyper-minimal automata? The notions of almost-equivalence and hyper-minimality were introduced in [2]. There it was shown that two almost-equivalent hyper-minimal DFAs are isomorphic in their kernels, and isomorphic in their preambles (up to acceptance values of preamble states). The following theorem, which summarizes results from [2], formalizes this isomorphism.

Theorem 1. *Let $A = (Q, \Sigma, \delta, q_0, F)$ and $A' = (Q', \Sigma, \delta', q'_0, F')$ be two minimal deterministic finite automata with $A \sim A'$. There exists a mapping $h\colon Q \to Q'$ satisfying the following conditions.*

1. *If $q \in \text{Pre}(A)$ then $q \sim h(q)$, and if $q \in \text{Ker}(A)$ then $q \equiv h(q)$.*
2. *If $q_0 \in \text{Pre}(A)$ then $h(q_0) = q'_0$, and if $q_0 \in \text{Ker}(A)$ then $h(q_0) \sim q'_0$.*
3. *The restriction of h to $\text{Ker}(A)$ is a bijection between the kernels of A and A', that is compatible with taking transitions:*
   3.a *We have $h(\text{Ker}(A)) = \text{Ker}(A')$, and if $q_1, q_2 \in \text{Ker}(A)$ with $h(q_1) = h(q_2)$ then $q_1 = q_2$.*
   3.b *We have $h(\delta(q, a)) = \delta'(h(q), a)$, for all $q \in \text{Ker}(A)$ and all $a \in \Sigma$.*

Further, if A and A' are hyper-minimal then also the following condition holds.

4. *The restriction of h to $\text{Pre}(A)$ is a bijection between the preambles of A and A', that is compatible with taking transitions, except for transitions from preamble to kernel:*
   4.a *We have $h(\text{Pre}(A)) = \text{Pre}(A')$, and if $q_1, q_2 \in \text{Pre}(A)$ with $h(q_1) = h(q_2)$ then $q_1 = q_2$.*
   4.b *We have $h(\delta(q, a)) = \delta'(h(q), a)$, for all $q \in \text{Pre}(A)$ and all $a \in \Sigma$, that satisfy $\delta(q, a) \in \text{Pre}(A)$.*

Notice that the bijection between the preamble states does not preserve finality of states. Further, the mapping h does not necessarily respect the transitions from preamble states to kernel states—see Condition 4.b of Theorem 1. Thus, two almost-equivalent hyper-minimal DFAs can differ in the following:

- acceptance values of preamble states,
- transitions leading from preamble to kernel states,
- the initial state, if the preamble is empty.

However, the transitions connecting preamble and kernel of almost-equivalent DFAs cannot differ arbitrarily. Assume that for some DFA A we have a state $q \in \text{Pre}(A)$, and some symbol $a \in \Sigma$, such that $\delta(q, a) \in \text{Ker}(A)$. Then it could be that the two states $h(\delta(q, a))$ and $\delta'(h(q), a)$ are different, but they must at least be almost-equivalent. This follows from a result from [2]. Also a characterization of hyper-minimal DFAs, which is similar to the characterization of minimal DFAs, was shown in [2]: a deterministic finite automaton is hyper-minimal if and only if it is minimal, and there is no pair of distinct but almost-equivalent states such that one of them is in the preamble.

Now let us investigate, which of the structural similarity results for almost-equivalent hyper-minimal DFAs carry over to biautomata. We first show the following result.

Lemma 2. *Let $A = (Q, \Sigma, \cdot, \circ, q_0, F)$ and $A' = (Q', \Sigma, \cdot', \circ', q_0', F')$ be two biautomata. Let $q \in Q$ and $q' \in Q'$, then $q \sim q'$ if and only if $(q \cdot u) \circ v \sim (q' \cdot' u) \circ' v$, for all words $u, v \in \Sigma^*$. Moreover, $q \sim q'$ implies $(q \cdot u) \circ v \equiv (q' \cdot' u) \circ' v$, for all words $u, v \in \Sigma^*$ with $|uv| \geq k = |Q \times Q'|$.* □

Now we come to a mapping between the states of two almost-equivalent biautomata. As in the case of finite automata, we can find an isomorphism between the kernels of the two automata. However, we cannot find a similar isomorphism between their preambles. Of course, two almost-equivalent hyper-minimal biautomata must have the same number of states, and if their kernels are isomorphic, then also their preambles must be of same size. But still we cannot always find a bijective mapping that preserves almost-equivalence, as in the case of finite automata. We will later see an example for this phenomenon, but first we present our result on the structural similarity between almost-equivalent minimal biautomata.

Theorem 3. *Let $A = (Q, \Sigma, \cdot, \circ, q_0, F)$ and $A' = (Q', \Sigma, \cdot', \circ', q_0', F')$ be two minimal biautomata with $A \sim A'$. There exists a mapping $h\colon Q \to Q'$ that satisfies the following conditions.*

1. *If $q \in \mathrm{Pre}(A)$ then $q \sim h(q)$, and if $q \in \mathrm{Ker}(A)$ then $q \equiv h(q)$.*
2. *If $q_0 \in \mathrm{Pre}(A)$ then $h(q_0) = q_0'$, and if $q_0 \in \mathrm{Ker}(A)$ then $h(q_0) \sim q_0'$.*
3. *The restriction of h to $\mathrm{Ker}(A)$ is a bijection between the kernels of A and A', that is compatible with taking transitions:*
   3.a *We have $h(\mathrm{Ker}(A)) = \mathrm{Ker}(A')$, and if $q_1, q_2 \in \mathrm{Ker}(A)$ with $h(q_1) = h(q_2)$ then $q_1 = q_2$.*
   3.b *We have $h(q \cdot a) = h(q) \cdot' a$ and $h(q \circ a) = h(q) \circ' a$, for all $q \in \mathrm{Ker}(A)$ and all $a \in \Sigma$.* □

Notice that Theorem 3 requires the almost-equivalent DBiAs A and A' to be minimal, but not necessarily hyper-minimal. Of course, the theorem also holds for hyper-minimal automata, since these are always minimal. However, the question is whether we can find more structural similarities—like Statement 4.b from Theorem 1 on DFAs—if both DBiAs are hyper-minimal. Unfortunately the answer is no, as the following example demonstrates.

Example 4. Consider the biautomaton A which is depicted in Figure 1—as usual, transitions which are not shown lead to a non-accepting sink state, which is also not shown. The state labels of the eight states in the lower two rows of the automaton denote the right languages of the respective states. The kernel of A consists of those states, and the sink state. The right languages of the preamble states q_0, q_1, and q_2, are as follows: $L_A(q_0) = L(A) = (a+b)ba^*b + c^*a$, $L_A(q_1) = ba^*b + \lambda$, and $L_A(q_2) = ba^*b$. One can verify that A satisfies the $\diamond$- and the F-property. Let us first show that A is hyper-minimal.

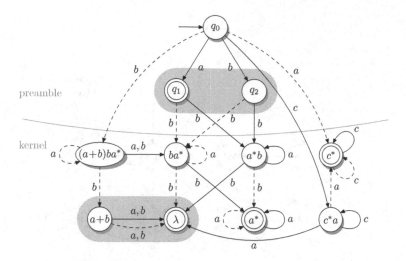

Fig. 1. The hyper-minimal biautomaton A accepting $L(A) = (a+b)ba^*b + c^*a$—solid arrows denote forward transitions by $\cdot$, and dashed arrows denote backward transitions by $\circ$. The gray shading of state pairs denotes almost-equivalence, i.e., we have $q_1 \sim q_2$ and $a + b \sim \lambda$.

Claim. The biautomaton A depicted in Figure 1 is hyper-minimal.   □

Now consider the DBiA A', depicted in Figure 2. This biautomaton accepts the language $L(A') = (a + b)ba^*b + cc^*a$, so it is almost-equivalent to A. Since both A and A' have the same number of states and A is hyper-minimal, the device A' is hyper-minimal, too. Consider a mapping h from the states of A to the states of A', that satisfies the conditions of Theorem 3. Between the kernels of the automata, the mapping is clear. Moreover, since q_0 and q_0' are preamble states, it must be $h(q_0) = q_0'$. This can even be concluded if q_0 and q_0' were not the initial states, because state q_0' of A' is the only state that is almost-equivalent to q_0, and h must satisfy $q \sim h(q)$ for all states q. With the same argumentation we obtain $h(q_1) = h(q_2) = q_2'$. The mapping h is now fully defined, so in this example, there is no other possible mapping from the states of A to the states of A' that preserves almost-equivalence.

Notice that mapping h is not a bijection between the preambles: because we have $h(q_1) = h(q_2) = q_2'$, it is not injective, and it neither is surjective, since no state of A is mapped to state q_1' of B. This shows that the bijection Condition 4.a of Theorem 1 for preambles of deterministic finite automata does not hold for biautomata.

Similarly, also Condition 4.b of Theorem 1 cannot be satisfied here, which is witnessed by the following. We have $h(q_0 \circ a) = h(c^*) = c^*$—here c^* in $h(c^*)$ denotes the kernel state of A, and c^* after the equation symbol denotes the kernel state of A'—but it is $h(q_0) \circ_B a = q_0' \circ_B a = q_1'$, so $h(q_0 \circ a) \neq h(q_0) \circ_B a$.

Of course there exist bijective mappings between the state sets of the two automata A and A', but none of these can preserve almost-equivalence because

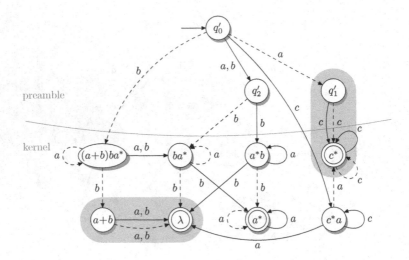

Fig. 2. The hyper-minimal biautomaton A' with $L(A') = (a + b)ba^*b + cc^*a$. The preamble states of A' are q'_0, q'_1, and q'_2 and their right languages are $L_{A'}(q'_0) = L(A')$, $L_{A'}(q'_1) = cc^*$, and $L_{A'}(q'_2) = ba^*b$. The gray shading of state pairs denotes almost-equivalence.

the corresponding almost-equivalence classes in the state sets are not always of same size. For example, there are two states in A that are almost-equivalent to q_1, namely q_1 itself and q_2, but in A' there is only state q'_2 in its equivalence class.   □

Recall that if a biautomaton is minimal, then also the DFA obtained from that biautomaton by considering only forward transitions is minimal, too. Notice that this relation does not hold if we consider hyper-minimal automata: the biautomaton A from Example 4 is hyper-minimal. But the contained DFA A_{fwd} is not hyper-minimal because the two preamble states q_1 and q_2 are almost-equivalent, which contradicts the characterization of hyper-minimal DFAs.

Due to the lack of structural similarity in the preambles of almost-equivalent hyper-minimal biautomata, we do not hope for a nice characterization of hyper-minimal biautomaton, as in the case of hyper-minimal DFAs, or as it is the case for classically minimal automata. Another effect related to these unsatisfying structural properties of hyper-minimal biautomata will show up in the following section, where we show that hyper-minimizing biautomata is computationally hard.

4   Computational Complexity of (Hyper)-Minimization

Given a deterministic finite automaton, it is an easy task to construct an equivalent minimal automaton. A lot of minimization algorithms for DFAs are known, the most efficient of them being Hopcroft's algorithm [10] with a running time of $O(n \log n)$, where n is the number of states of the input DFA. In fact, the decision version of the DFA minimization problem—given a DFA A and an integer n, decide whether there exists an n-state DFA B with $A \equiv B$—is NL-complete [3].

Concerning minimization of biautomata, it was discussed in [6] how classical DFA minimization techniques can also be applied to DBiAs. In the following we investigate the computational complexity of the minimization problem for biautomata, and show that it is NL-complete, too. For proving NL-hardness we use the following NL-complete variant of the graph reachability problem [14,15].

> **Reachability:** given a directed graph $G = (V, E)$ with $V = \{v_1, \ldots, v_n\}$, where every vertex has at most two successors and at most two predecessors, decide whether v_n is reachable from v_1.

We use this variant in order to make sure that the DFA that is obtained from G by a straight-forward construction is also backward deterministic, so that the corresponding biautomaton is not too big. The next theorem reads as follows:

Theorem 5 (DBiA Minimization Problem). *The problem of deciding for a given biautomaton A, and an integer n, whether there exists an n-state biautomaton B with $A \equiv B$, is NL-complete.*   □

Now we turn to hyper-minimization. For deterministic finite automata the situation is similar as in the case of classical minimization: efficient hyper-minimization algorithms with running time $O(n \log n)$ are known [4,9], and it was shown in [5] that the hyper-minimization problem for DFAs is NL-complete. On the one hand, since classical DFA minimization methods also work well for DBiAs, one could expect that hyper-minimization of DBiAs is as easy as for ordinary DFAs. On the other hand, the problems related to the structure of hyper-minimal biautomata, which we discussed in Section 3, already give hints that hyper-minimization of DBiAs may not be so easy. In fact, we show in the following that the hyper-minimization problem for biautomata is NP-complete. To prove NP-hardness we give a reduction from the NP-complete MAX-2-SAT problem [18] which is defined as follows.

> **MAX-2-SAT:** given a Boolean formula φ in conjunctive normal form, where each clause has exactly two literals, and an integer k, decide whether there exists an assignment that satisfies at least k clauses of φ.

Let us describe the key idea of the reduction. Given as instance of MAX-2-SAT a formula φ and number k, we construct a DBiA A_φ such that for every clause that can be satisfied in φ, we can save one state of A_φ, obtaining an almost-equivalent DBiA. Every clause of φ will be translated to a part of the biautomaton using a separate alphabet, so that the clause gadgets in A_φ are mostly independent from each other. Assume that $\varphi_i = (\ell_{i_1} \vee \ell_{i_2})$ is a clause of φ, and the first literal is $\ell_{i_1} = x_u$, and the second is $\ell_{i_2} = \overline{x}_v$, for some variables x_u and x_v. Then the DBiA A_φ contains the structure which is depicted in Figure 3.

The states q_1 and q_1' correspond to literal ℓ_{i_1}, and states q_2 and q_2' to the literal ℓ_{i_2}. States p_u and p_v correspond to the variables x_u and x_v, respectively,

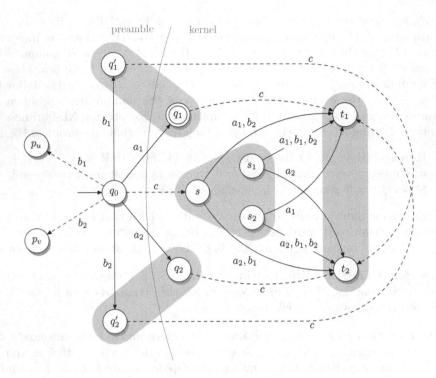

Fig. 3. Simplified structure of A_φ corresponding to the clause $\varphi_i = (x_u \vee \overline{x}_v)$. The gray shading denotes almost-equivalence of states.

and are shared by all clause gadgets related to these variables. Now assume that there is a truth assignment ξ to the variables that satisfies clause φ_i, say by $\xi(x_u) = 1$. Then we make the preamble state p_u accepting, and merge the preamble state q_1' to the almost-equivalent state q_1. To preserve the $\diamond$-property of the biautomaton, we further re-route the backward c transition of the initial state, making state s_1 the target of the transition. The case where the clause φ_i is satisfied by the second literal corresponds to the similar situation, where state p_v stays non-accepting, state q_2' is merged to q_2, and the target of the backward c transition from q_0 is state s_2. The changing of acceptance of preamble states only introduces a finite number of errors. Further, the merging of preamble states to almost-equivalent kernel states also yields an almost-equivalent automaton.[1] Therefore, if k clauses of φ can be satisfied, then k states of A_φ can be saved.

---

[1] In general, this needs some more argumentation. Here the described changes in the biautomaton preserved the $\diamond$-property, and the F-property. Therefore the languages accepted by the original and the modified biautomata are the same as the languages accepted by the contained DFAs (using only forward transitions). Now almost-equivalence of these DFAs, and thus, of the biautomata, follows from the fact that merging preamble states to almost-equivalent kernel states in a DFA preserves almost-equivalence.

The other direction, i.e., the deduction of a truth assignment ξ from a given biautomaton B that is almost-equivalent to A_φ, is similar: let $\xi(x_u) = 1$ if and only if state p_u of automaton B is accepting. Now assume that state B has k states less than A_φ. The reduction will make sure that only the states q_1' and q_2' can be saved. If for example state q_1' is not present in B, then the initial state of B must enter state q_1 on reading symbol b_1 with a forward transition. Due to the F-property of B, the state p_u reached from the initial state by taking a backward b_1 transition must be accepting. Since the variable states are shared by all clause gadgets, the information that p_u is accepting—i.e., that variable x_u should be assigned truth value 1—is transported to all other clause gadgets that use variable x_u. Therefore, no state corresponding to the negative literal $\overline{x}_u$ can be saved, i.e., no clause can be satisfied by a literal $\overline{x}_u$.[2] It may be the case that both states q_1' and q_2' are merged to their almost-equivalent kernel states q_1 and q_2, respectively. But then, due to the $\diamond$-property, the initial state must go to some state s' on reading a c symbol with a backward transition, and this state s' must go to state t_1 on a forward b_1 transition, and it must go to state t_2 on a forward b_2 transition. Such a state is not present in the biautomaton A_φ, so this state s' is an additional state in the preamble of B. Hence, even if both states q_1' and q_2' are merged into the kernel, the clause gadget in B cannot save more than one state compared to the clause gadget in A_φ. Altogether, for every state that B has less than A, there is a clause of φ that is satisfied by ξ.

Containment of the hyper-minimization problem in NP can be seen by an easy guess-and-check-algorithm, therefore, we obtain the following theorem.

Theorem 6 (DBiA Hyper-Minimization Problem). *The problem of deciding for a given biautomaton A, and an integer n, whether there exists an n-state biautomaton B, with $A \sim B$, is NP-complete.* ☐

In [5] another form of equivalence of languages and automata was considered, namely E-equivalence. Given an *error language* $E \subseteq \Sigma^*$, two languages L and L' over Σ are *E-equivalent*, for short $L \sim_E L'$, if $L \triangle L' \subseteq E$, and two automata A and A' are E-equivalent, if $L(A) \sim_E L(A')$. It is shown in [5] that the following E-minimization problem for DFAs is already NP-complete: given two DFAs A and A_E, and an integer n, decide whether there exists an n-state DFA B, with $A \sim_E B$, for $E = L(A_E)$. The E-minimization problem for DBiAs turns out to be NP-complete, too.

Theorem 7 (DBiA E-Minimization Problem). *The problem of deciding for a given biautomaton A, a deterministic finite automaton A_E, and an integer n, whether there exists an n-state biautomaton B, such that $A \sim_E B$, for $E =$*

---

[2] The reader may have noticed that there is still a possibility to "cheat:" one could use accepting and non-accepting copies of variable states in the preamble in order to satisfy a lot more clauses than possible. We take care of this problem in the detailed proof. (The problem can be solved by using many copies $b_{1,j}$ and $b_{2,j}$ of the b_1 and b_2 symbols, each connected to a different copy $p_{u,j}$ of variable states. If the number of these copies is larger than the number of clauses, then the "cheat" turns out to be a bad trade-off.)

$L(A_E)$, is NP-*complete. This also holds, if A_E is a biautomaton instead of a finite automaton.*   □

References

1. Arnold, A., Dicky, A., Nivat, M.: A note about minimal non-deterministic automata. Bull. EATCS 47, 166–169 (1992)
2. Badr, A., Geffert, V., Shipman, I.: Hyper-minimizing minimized deterministic finite state automata. RAIRO–Informatique théorique et Applications/Theoretical Informatics and Applications 43(1), 69–94 (2009)
3. Cho, S., Huynh, D.T.: The parallel complexity of finite-state automata problems. Inform. Comput. 97, 1–22 (1992)
4. Gawrychowski, P., Jeż, A.: Hyper-minimization made efficient. In: Královič, R., Niwiński, D. (eds.) MFCS 2009. LNCS, vol. 5734, pp. 356–368. Springer, Heidelberg (2009)
5. Holzer, M., Jakobi, S.: From equivalence to almost-equivalence, and beyond: Minimizing automata with errors. Internat. J. Found. Comput. Sci. 24(7), 1083–1134 (2013)
6. Holzer, M., Jakobi, S.: Minimization and characterizations for biautomata. In: 5th NCMA. books@ocg.at, vol. 294, pp. 179–193. Österreichische Computer Gesellschaft, Umeå (2013)
7. Holzer, M., Jakobi, S.: Nondeterministic biautomata and their descriptional complexity. In: Jurgensen, H., Reis, R. (eds.) DCFS 2013. LNCS, vol. 8031, pp. 112–123. Springer, Heidelberg (2013)
8. Holzer, M., Kutrib, M.: Descriptional and computational complexity of finite automata—a survey. Inform. Comput. 209(3), 456–470 (2011)
9. Holzer, M., Maletti, A.: An $n \log n$ algorithm for hyper-minimizing a (minimized) deterministic automaton. Theoret. Comput. Sci. 411(38–39), 3404–3413 (2010)
10. Hopcroft, J.: An $n \log n$ algorithm for minimizing the state in a finite automaton. In: Kohavi, Z. (ed.) The Theory of Machines and Computations, pp. 189–196. Academic Press, New York (1971)
11. Hopcroft, J.E., Ullman, J.D.: Introduction to Automata Theory, Languages and Computation. Addison-Wesley (1979)
12. Jiang, T., Ravikumar, B.: Minimal NFA problems are hard. In: Leach Albert, J., Monien, B., Rodríguez-Artalejo, M. (eds.) ICALP 1991. LNCS, vol. 510, pp. 629–640. Springer, Heidelberg (1991)
13. Jirásková, G., Klíma, O.: Descriptional complexity of biautomata. In: Kutrib, M., Moreira, N., Reis, R. (eds.) DCFS 2012. LNCS, vol. 7386, pp. 196–208. Springer, Heidelberg (2012)
14. Jones, N.: Space-bounded reducibility among combinatorial problems. J. Comput. System Sci. 11, 68–85 (1975)
15. Jones, N.D., Lien, Y.E., Laaser, W.T.: New problems complete for nondeterministic log space. Math. Systems Theory 10, 1–17 (1976)
16. Klíma, O., Polák, L.: Biautomata for k-piecewise testable languages. In: Yen, H.-C., Ibarra, O.H. (eds.) DLT 2012. LNCS, vol. 7410, pp. 344–355. Springer, Heidelberg (2012)
17. Klíma, O., Polák, L.: On biautomata. RAIRO–Informatique théorique et Applications / Theoretical Informatics and Applications 46(4), 573–592 (2012)
18. Papadimitriou, C.H.: Computational Complexity. Addison-Wesley (1994)

Deterministic Set Automata

Martin Kutrib, Andreas Malcher, and Matthias Wendlandt

Institut für Informatik, Universität Giessen
Arndtstr. 2, 35392 Giessen, Germany
{kutrib,malcher,matthias.wendlandt}@informatik.uni-giessen.de

Abstract. We consider the model of deterministic set automata which
are basically deterministic finite automata equipped with a set as an additional storage medium. The basic operations on the set are the insertion
of elements, the removing of elements, and the test whether an element
is in the set. We investigate the computational power of deterministic
set automata and compare the language class accepted with the context-
free languages and classes of languages accepted by queue automata. As
results the incomparability to all classes considered is obtained. In the
second part of the paper, we examine the closure properties of the class
of DSA languages under Boolean operations. Finally, we show that deterministic set automata may be an interesting model from a practical
point of view by proving that their emptiness problem is decidable.

1 Introduction

The regular languages and their corresponding automata model, deterministic
and nondeterministic finite automata, are well investigated [6]. It is well known
that this family of languages has many desirable properties. For example, all
commonly studied decidability questions are decidable and the regular languages
are closed under almost all commonly studied operations such as, for example,
the Boolean operation, concatenation, (inverse) homomorphism, and substitution. From a practical point of view, finite automata are in particular interesting,
since many of the decidability questions are decidable in polynomial time and,
in addition, an effective minimization algorithm is known for deterministic finite
automata.

But with respect to the computational power, this model is quite weak since
it builds the lowermost level of the Chomsky hierarchy. Hence, much efforts have
been made to find models that extend the computational power of regular languages by adding storage media, but keep as many 'good' properties as possible.
Consider, for example, the extension by a stack [4] or by a pushdown store [2],
which leads to the context-free languages. For both models nondeterministic
variants are more powerful than deterministic variants, which is in contrast to
finite automata. Moreover, some positive closure properties are lost. On the other
hand, the decidability of emptiness and finiteness is preserved [4,6,9]. In addition, equivalence is decidable for deterministic pushdown automata [10], but not
for the nondeterministic variant [6].

A.M. Shur and M.V. Volkov (Eds.): DLT 2014, LNCS 8633, pp. 303–314, 2014.

Another extension studied is that of a queue. In their general definition queue automata can accept the class of recursively enumerable languages for which all non-trivial decidability questions are undecidable. A meaningful restriction for queue automata is considered in [1] where quasi-real-time computations are studied. The restriction of quasi-real-time means that in any computation the number of subsequent λ-moves is bounded by some constant. Another restriction is investigated in [7] where the number of turns, that is, the changes between an enqueuing and a dequeuing phase, is bounded by some constant. This restriction is similar to finite turn pushdown automata where the number of changes between increasing and decreasing the height of the pushdown store is bounded by some constant. Both restrictions lead to language classes less powerful than the class of recursively enumerable languages. With the latter restriction it is possible to decide the emptiness problem [7].

The paper [3] introduces bag automata which are basically finite automata equipped with a finite number of bags in which the automaton can put symbols and also multiple versions of symbols. The symbols are stored as multisets and, therefore, the order in which they are added to the bags is not remembered. This model is quite powerful, because it is possible to simulate certain counter machines. If the model is restricted to so-called well-formed bag automata, a language class in between the (deterministic) one-counter and the (deterministic) context-free languages is obtained.

In this paper, we consider *deterministic set automata* (DSA) that extend deterministic finite automata by adding the storage medium of a set which, in contrast to bag automata, allows words to be stored and is not a multiset. As operations on the set it is possible to add elements, to remove elements, and to test whether some element is in the set. To prepare a set operation the DSA can write on a one-way write-only tape. For the set operation the contents of that tape are taken and added to the set, removed from the set, or tested. After the set operation, the writing tape is reset to the empty tape and a new set operation may be prepared. A similar model has been introduced by Lange and Reinhardt in [8]. In contrast to DSA, their model may work nondeterministically, allows no remove operations, and a test operation implicitly adds the word tested to the set. The main results in [8] are the decidability of emptiness for the model considered and the closure of the corresponding language class under the operations homomorphism, inverse homomorphism, and intersection with regular languages.

This paper is organized as follows. After the definition of the model and some examples in Section 2, we compare DSA with pushdown automata, quasi-real-time queue automata, and queue automata with finite turns with regard to their computational power. As result we obtain the incomparability with all classes investigated. This shows that DSA can accept languages which are not accepted by the other models. In Section 4 we consider closure properties. It turns out that the language class accepted by DSA is closed under complement and union with regular languages as well as intersection with regular languages, but is not closed under general union and general intersection. Finally, we show that emptiness is decidable for DSA which is a pleasant property from a theoretical as well as from a practical point of view. We would like to note that missing proofs have been omitted due to space constraints.

2 Preliminaries

We write Σ^* for the set of all words over the finite alphabet Σ. The empty word is denoted by λ, and we set $\Sigma^+ = \Sigma^* \setminus \{\lambda\}$. The reversal of a word w is denoted by w^R, and for the length of w we write $|w|$. We use $\subseteq$ for inclusions and $\subset$ for strict inclusions.

A set automaton is a system consisting of a finite state control, a one-way writing tape where transductions of parts of the input can be temporarily stored, and a data structure *set* where words of arbitrary length can be stored. At each time step, it is possible to either write a transduction of the current input letter to the end of the writing tape, to insert or remove the word written on the tape to or from the set, or to test whether the word written on the tape belongs to the set. Each time a set operation {in, out, or test} is done, the content of the writing tape is erased and its head is reset to the left end.

Definition 1. *A* deterministic set automaton, *abbreviated as* DSA, *is a system* $M = \langle S, \Sigma, \Gamma, \lhd, \delta, s_0, F \rangle$, *where*

1. *S is the finite set of* internal states,
2. *Σ is the finite set of* input symbols,
3. *Γ is the finite set of* tape symbols,
4. *$\lhd \notin \Sigma$ is the* right endmarker,
5. *$s_0 \in S$ is the* initial state,
6. *$F \subseteq S$ is the set of* accepting states, *and*
7. *$\delta : S \times (\Sigma \cup \{\lambda, \lhd\}) \to (S \times (\Gamma^* \cup \{\textbf{in}, \textbf{out}\})) \cup (S \times \{\textbf{test}\} \times S)$ is the* partial transition function, *where* **in** *is the instruction to add the content of the tape to the set,* **out** *is the instruction to remove the content of the tape from the set, and* **test** *is the instruction to test whether the content of the tape is in the set. If the transition function is defined for some pair (s, λ) with $s \in S$, then it is not defined for any pair (s, a) with $a \in \Sigma \cup \{\lhd\}$.*

A *configuration* of a DSA $M = \langle S, \Sigma, \Gamma, \lhd, \delta, s_0, F \rangle$ is a quadruple $(s, v, z, \mathbb{S})$, where $s \in S$ is the current state, $v \in \{\Sigma^* \lhd\} \cup \{\lambda\}$ is unread part of the input, $z \in \Gamma^*$ is the content of the tape, and $\mathbb{S} \subseteq \Gamma^*$ is the finite set of stored words. The *initial configuration* for an input string w is set to $(s_0, w\lhd, \lambda, \emptyset)$. During the course of its computation, M runs through a sequence of configurations. One step from a configuration to its successor configuration is denoted by $\vdash$. Let $s, s', s'' \in S$, $x \in \Sigma \cup \{\lambda, \lhd\}$, $v \in \{\Sigma^* \lhd\} \cup \{\lambda\}$, $z, z' \in \Gamma^*$, and $\mathbb{S} \subseteq \Gamma^*$. We set

1. $(s, xv, z, \mathbb{S}) \vdash (s', v, zz', \mathbb{S})$, if $\delta(s, x) = (s', z')$,
2. $(s, xv, z, \mathbb{S}) \vdash (s', v, \lambda, \mathbb{S} \cup \{z\})$, if $\delta(s, x) = (s', \textbf{in})$,
3. $(s, xv, z, \mathbb{S}) \vdash (s', v, \lambda, \mathbb{S} \setminus \{z\})$, if $\delta(s, x) = (s', \textbf{out})$,

4. $(s, xv, z, \mathbb{S}) \vdash (s', v, \lambda, \mathbb{S})$, if $\delta(s, x) = (s', \texttt{test}, s'')$ and $z \in \mathbb{S}$,
5. $(s, xv, z, \mathbb{S}) \vdash (s'', v, \lambda, \mathbb{S})$, if $\delta(s, x) = (s', \texttt{test}, s'')$ and $z \notin \mathbb{S}$.

We denote the reflexive and transitive closure of $\vdash$ by $\vdash^*$. It should be noted that an instruction to remove some z from $\mathbb{S}$ does not test whether $z \in \mathbb{S}$; it only ensures that $z \notin \mathbb{S}$ after the operation. The language accepted by the DSA M is the set $L(M)$ of words for which there exists a computation beginning in the initial configuration and ending in a configuration in which the whole input is read and an accepting state is entered. Formally:

$$L(M) = \{ w \in \Sigma^* \mid (s_0, w \lhd, \lambda, \emptyset) \vdash^* (s_f, \lambda, z, \mathbb{S}) \text{ with } s_f \in F, z \in \Gamma^*, \mathbb{S} \subseteq \Gamma^* \}.$$

The family of all languages accepted by DSA is denoted by $\mathscr{L}(\text{DSA})$.

Example 2. Language $L_1 = \{ wcw \mid w \in \{a, b\}^* \}$ is accepted by a DSA. The idea is to read the whole sequence up to the letter c and to copy it to the tape. When the input head arrives at the c, it stores the word w on the tape in its set. Then it reads the second subword consisting of a's and b's and copies it also to the tape. When the input head arrives at the right endmarker, it tests whether the content on the tape is in the set. If this is the case, then the input is accepted and otherwise rejected.   □

Example 3. Language $L_2 = \{ a^n b^m \$_0 c^n \mid m, n \geq 1 \} \cup \{ a^n b^m \$_1 c^m \mid m, n \geq 1 \}$ is accepted by a DSA. First, the automaton writes for every a in the input an a on the tape. When the first b is read, it adds the word on the tape to the set. Then the automaton writes for every b in the input an b on the tape until the dollar is in the input. Subsequently, the word on the tape, consisting of b's, is added to the set. Depending on whether there has been a $\$_0$ or a $\$_1$ in the input the automaton writes an a or a b for each c in the input on the tape. In the last step, the automaton checks whether the word on the tape is in the set. If the test is successful, the input is accepted and otherwise rejected.   □

Example 4. Language $L_3 = \{ a^n b^n c^n \mid n \geq 1 \}$ is accepted by a DSA in such a way that it writes for every a in the input an a on the tape. When it reads the first b, it adds the content of the tape to the set. Then, it writes for every b in the input an a on the tape and when it reads the first c, it tests whether the word on the tape is in the set. If this is not the case, the input is rejected. Otherwise, for every c in the input an a is written on the tape. If at the end again the word on the tape is in the set, then the input is accepted and otherwise rejected.   □

3   Computational Power of Deterministic Set Automata

In this section, we study the computational power of deterministic set automata. Hence, we compare the model with the known models of pushdown automata and queue automata. Since in general queue automata characterize the recursively enumerable languages, we compare our model with the restricted versions of quasi-real-time queue automata and queue automata with a finite number of turns, that is, the number of changes between enqueuing and dequeuing periods is bounded by a fixed number.

Let us first consider unary languages. It is known that pushdown automata accept only semilinear unary languages, hence regular languages, whereas even quasi-real-time queue automata may accept non-semilinear unary languages [1].

Theorem 5. *Every unary language accepted by a DSA is semilinear.*

Proof. Let $M = \langle S, \{a\}, \Gamma, \lhd, \delta, s_0, F \rangle$ be a DSA accepting a unary language and k be the length of a longest word that M can write in one step on the tape. We may assume that M accepts an infinite language, since finite unary languages are semilinear. Let w be an input such that $|w| > |S|$. When processing this input, the automaton necessarily has to enter a loop. We consider two cases.

First, we assume that there is no situation occurring in which the automaton performs an operation $\{\text{in}, \text{out}, \text{test}\}$ on the set. Then, there will never be a word in the set and M can be easily transformed into an equivalent deterministic finite automaton. Second, we assume that M performs an in-, out, or test-operation after which the content of the tape is deleted. In each computation step, M can write at most k symbols on the tape. Due to the unary input, M can distinguish between at most $|S|$ different situations. Thus, the words written on the tape and possibly added to the set are at most of length $k \cdot |S|$. Hence, we can construct a deterministic finite automaton that simulates M by storing the content on the tape as well as the finite number of words in the set in its state. Since the languages accepted by finite automata are semilinear, the theorem follows. □

With this result we are able to show that the family of languages accepted by deterministic set automata is incomparable with the family of languages accepted by quasi-real-time queue automata.

Theorem 6. *The family of languages accepted by DSA is incomparable with the family of languages accepted by quasi-real-time queue automata.*

Proof. The non-semilinear unary language $\{\, a^n \mid n$ is a Fibonacci number $\}$ is accepted by some quasi-real-time queue automaton [1]. Since by Theorem 5, DSA do not accept non-semilinear unary languages, it remains to show the other direction. The witness language used is language L_2 of Example 3. □

In the next proof as well as in the section for the decidability we need that set automata are in a special form where each state carries the information whether the last action on the set was a **test**-, **in**-, or **out**-operation or was a write operation on the tape. Additionally, it is distinguished between successful and unsuccessful **test**-operations.

Definition 7. *A DSA M is in* action normal form, *if the initial state of M is only visited once at the beginning of the computation and each other state indicates uniquely which action the automaton M did in the last computation step. The states are marked with a corresponding subscript* test+, test-, in, *or* out. *Non-marked states are interpreted as states where the last action was a write operation on the tape.*

Lemma 8. *Any DSA $M = \langle S, \Sigma, \Gamma, \lhd, \delta, s_0, F \rangle$ can be converted into an equivalent DSA $M' = \langle S', \Sigma, \Gamma, \lhd, \delta', s_0', F' \rangle$ in action normal form.*

Proof. In a first step, we construct a DSA $M'' = \langle S'', \Sigma, \Gamma, \lhd, \delta'', s_0'', F'' \rangle$ with a new initial state s_0'' which is visited at most once. Let $S'' = S \cup \{s_0''\}$ with $s_0'' \notin S$. If $s_0 \in F$ we define $F'' = F \cup \{s_0''\}$, and set $F'' = F$ otherwise. The transition function δ'' is defined as follows for $s_i, s_j, s_k \in S$, $a \in \Sigma \cup \{\lambda, \lhd\}$, and $z \in \Gamma^* \cup \{\text{in}, \text{out}\}$.

1. $\delta''(s_0'', \lambda) = (s_0, \lambda)$,
2. $\delta''(s_i, a) = (s_j, z)$, if $\delta(s_i, a) = (s_j, z)$,
3. $\delta''(s_i, a) = (s_j, \text{test}, s_k)$, if $\delta(s_i, a) = (s_j, \text{test}, s_k)$.

In the next step, we define a state set consisting of five pairwise disjoint sets S_{in}, S_{out}, $S_{\text{test+}}$, $S_{\text{test-}}$, and S_{write}. The idea is that we introduce a new state for every state connected with a non-writing operation, whereas the states of S'' indicate writing operations. So, let $S_{\text{write}} = S''$, $S_{\text{test}} = S_{\text{test+}} \cup S_{\text{test-}}$ and $S' = S_{\text{in}} \cup S_{\text{out}} \cup S_{\text{test}} \cup S_{\text{write}}$. We set $s_0' = s_0''$ and $F' = F''$. For the definition of δ', consider $s_i, s_j, s_k \in S''$, $a \in \Sigma \cup \{\lambda, \lhd\}$, and $z \in \Gamma^*$.

1. $\delta'(s_i, a) = (s_{j_{\text{in}}}, \text{in})$ and $\delta'(s_{j_{\text{in}}}, \lambda) = (s_j, \lambda)$, if $\delta''(s_i, a) = (s_j, \text{in})$,
2. $\delta'(s_i, a) = (s_{j_{\text{out}}}, \text{out})$ and $\delta'(s_{j_{\text{out}}}, \lambda) = (s_j, \lambda)$, if $\delta''(s_i, a) = (s_j, \text{out})$,
3. $\delta'(s_i, a) = (s_{j_{\text{test+}}}, \text{test}, s_{k_{\text{test-}}})$, $\delta'(s_{j_{\text{test+}}}, \lambda) = (s_j, \lambda)$, and $\delta'(s_{k_{\text{test-}}}, \lambda) = (s_k, \lambda)$, if $\delta''(s_i, a) = (s_j, \text{test}, s_k)$,
4. $\delta'(s_i, a) = (s_j, z)$, if $\delta''(s_i, a) = (s_j, z)$.

The DSA M' is still a deterministic automaton, since newly introduced λ-transitions start only from newly introduced states. Moreover, M' is in action normal form and is equivalent to M, since the same transitions as in M are performed and the additional λ-transitions do not affect the language accepted.   □

Our next goal is to achieve another normal form for DSA which plays a crucial role in further proofs. Let $M = \langle S, \Sigma, \Gamma, \lhd, \delta, s_0, F \rangle$ be a DSA in action normal form. Thus, $S = S_{\text{in}} \cup S_{\text{out}} \cup S_{\text{test}} \cup S_{\text{write}}$. We note that the tape is empty at the beginning of the computation as well as after each operation on the set. Now we build sets of the form L_{s_i, s_j} with $s_i \in \{s_0\} \cup S_{\text{in}} \cup S_{\text{out}} \cup S_{\text{test}}$ and $s_j \in S_{\text{in}} \cup S_{\text{out}} \cup S_{\text{test}}$ that describe all words that can be written on the tape when the computation starts in state s_i with empty tape and ends in state s_j, and in between no other state performing an operation on the set is entered. Formally we define

$$L_{s_i, s_j} = \{ w_n \in \Gamma^* \mid \text{there is } u \in \Sigma^* \text{ such that } (s_i, u, \lambda, \mathbb{S}) \vdash (s_{i+1}, u_1, w_1, \mathbb{S})$$
$$\vdash^* (s_{i+(n-1)}, u_{n-1}, w_{n-1}, \mathbb{S}) \vdash (s_{i+n}, u_n, w_n, \mathbb{S}) \vdash (s_j, \lambda, \lambda, \mathbb{S}'),$$
$$\text{and } s_{i+1}, s_{i+2}, \dots, s_{i+n} \notin S_{\text{in}} \cup S_{\text{out}} \cup S_{\text{test}} \}.$$

All these sets L_{s_i, s_j} are regular, since an equivalent finite automaton M_{s_i, s_j} can be built from M: M_{s_i, s_j} has S as state set, s_i as initial state, and s_j as only accepting state. We consider all transitions in M from some state $s \in \{s_i\} \cup S_{\text{write}}$ to some state $s' \in \{s_j\} \cup S_{\text{write}}$ writing some word $w \in \Gamma^*$ on the tape. For every such transition we add to M_{s_i,s_j} a transition from s to s' on input w.

We say that a DSA M is in *infinite action normal form* if M is in action normal form and all sets L_{s_i,s_j} are infinite. The next lemma says that we always may assume that a DSA is in infinite action normal form.

Lemma 9. *Any DSA M can be converted into an equivalent DSA M' in infinite action normal form.*

The context-free languages and their important subclass of deterministic context-free languages are one of the best studied families of languages.

Theorem 10. *The family of languages accepted by DSA is incomparable with the (deterministic) context-free languages.*

Proof. By Example 2, the non-context-free language $L_1 = \{ wcw \mid w \in \{a,b\}^* \}$ is accepted by some DSA. So, it suffices to show that the deterministic context-free language $L = \{ wcw^R \mid w \in \{a,b\}^* \}$ is not accepted by any deterministic set automaton. In the following, we are often arguing with two parts of a word in L. So, we call the sequence up to the middle marker c the first part of the word, and the remaining sequence the second part of the word.

The proof is by contradiction. Assuming that L is accepted by some DSA M, we will show as a first step that L is accepted by M in such a way that all possible set operations performed on the first part of the input are a finite number of **in**-operations, and on the second part are a finite number **test**-operations. In a second step, based on M an equivalent one-way multi-head finite automaton accepting L is constructed. This leads to a contradiction.

Let L be accepted by a DSA $M = \langle S, \Sigma, \Gamma, \lhd, \delta, s_0, F \rangle$ which is in infinite action normal form. Thus, $S = S_{\text{in}} \cup S_{\text{out}} \cup S_{\text{test}} \cup S_{\text{write}}$ and we know that all sets L_{s_i,s_j} with $s_i \in \{s_0\} \cup S_{\text{in}} \cup S_{\text{out}} \cup S_{\text{test}}$ and $s_j \in S_{\text{in}} \cup S_{\text{out}} \cup S_{\text{test}}$ are infinite.

Next, we will show that there are no **test**-operations in the first part of an accepted input. Let us first discuss the case when some test in the first part is negative. We consider the subcomputation $(s_i, uvcw^R, \lambda, \mathbb{S}) \vdash^* (s_j, vcw^R, \lambda, \mathbb{S}')$ on input $u'uvcw^R \in L$ with $w = u'uv$, $s_j \in S_{\text{test}}$, $s_i \in \{s_0\} \cup S_{\text{in}} \cup S_{\text{out}} \cup S_{\text{test}}$, and assume that the test has a negative result. Since L_{s_i,s_j} is infinite, there are infinitely many input sequences whose transductions belong to L_{s_i,s_j}. Therefore, there exists some $u'' \in \{a,b\}^+$ with $u \neq u''$, so that the DSA also accepts the input $u'u''vcw^R$, which is a contradiction. We conclude that every test in the first part has to be successful. Now assume as before that the DSA is in state s_i after the processing of the input prefix u'. For any input $\tilde{u}v$ there is a word $u'\tilde{u}vc(u'\tilde{u}v)^R \in L$. Since there are only finitely many words in $\mathbb{S}$, the test has to be successful, and L_{s_i,s_j} is infinite, we can conclude that there are two different words $\tilde{u}$ and $\hat{u}$ whose transduction on the tape is the same. This implies that M is in the same configuration after reading $u'\tilde{u}$ and after reading $u'\hat{u}$. Thus, both words $u'\tilde{u}vc(u'\tilde{u}v)^R$ and $u'\hat{u}vc(u'\tilde{u}v)^R$ are accepted, a contradiction. Hence, we may assume that there is never a test-operation in the first part of accepted inputs.

In a similar way it can be proved that for accepting computations there are never out-operations in the first or in the second part of the input, and that there are never in-operations in the second part of the input.

Next, we turn to show that M can perform only a constant number of in-operations in the first part of accepting computations. Assume contrarily that M performs $k > |S|^2$ input operations in the first part of some input. Then there are two states $s_i \in S_{\text{in}} \cup S_{\text{out}} \cup S_{\text{test}}$ and $s_j \in S_{\text{in}}$ such that M runs from s_i to s_j twice. Consider such a computation on an input $w = zcz^R$ with $z = vuv'u'v''$, where M is in state s_i when it reads the first symbol of u, and is in state s_j when it reads the first symbol of v', and is again in state s_i when it reads the first symbol of u', and is again in state s_j when it reads the first symbol of v''. Then we can conclude out of the fact that M does not perform any test- or out-operations while computing the subword v', that M is in the same configuration after reading the subwords $vuv'u'v''$ and $vu'v'u'v''$. Choosing $u \neq u'$, which is always possible since L_{s_i, s_j} is infinite, we obtain a contradiction, because both words $zcz^R \in L$ with $z = vuv'u'v'$ and $vu'v'u'v''cz^R \notin L$ would be accepted.

Next, we turn to prove that there are only a constant number of tests in the second part of accepted inputs. First, we show that M never performs a negative test-operation in the second part of an accepted input. Assuming the contrary, there is a word $wcuvu' \in L$ such that M is in state $s_i \in S_{\text{in}} \cup S_{\text{out}} \cup S_{\text{test}}$ with empty tape after reading wcu. Now, M reads v, writes some z on the tape, tests z, and enters some state $s_j \in S_{\text{test-}}$ as result of a negative test. Thus, M is in configuration $(s_j, u', \lambda, \mathbb{S})$. Since L_{s_i, s_j} is infinite and the content of the set $\mathbb{S}$ is finite, there is a another word $wcuv'u'$ with $v \neq v'$ such that M is also in configuration $(s_j, u', \lambda, \mathbb{S})$ after reading v'. This is a contradiction, since then $wcuv'u' \notin L$ would be accepted. So, we can conclude that all test-operations performed in the second part of accepted inputs are positive. Assume now that the number of tests is greater than $|S|$. Then one test-state is entered at least twice. We may assume that in the computation on an accepted word $w = zcz^R$ with $z^R = vuv'$ M reaches some state $s_j \in S_{\text{test}}$ when it reads the first symbol of u and again when reading the first symbol of v'. Therefore all words $zcvu^iv'$ with $i \geq 1$ are accepted as well, since we know that there are no out-operations in the second part. Choosing $i = 2$ leads to a contradiction.

Now we know that M never performs test- or out-operations in the first part of accepted inputs and never performs in- or out-operations in the second part of accepted inputs. Furthermore, at most $|S|^2$ in-operations in the first part as well as at most $|S|$ test-operations in the second part are performed. In the following, we describe how M can be simulated by a one-way multi-head finite automaton.

Let uv be an input word and let $u_1, u_2, \ldots, u_n$ be the subwords of u whose transductions are added to the set by in-operations, and $v_1, v_2, \ldots, v_m$ be the subwords of v whose transductions are tested. We construct a one-way multi-head finite automaton M' that leaves $|S|$ many heads at position $p_0 = 1$, that is, at the beginning of the input word, reads the input using some head h, starts to simulate M omitting the simulation of the tape, and leaves $|S|$ many heads in the first part of the input at every moment when M empties its tape and adds some u_i to its set, except for the last in-operation. In the following, these positions are denoted by $p_1, p_2, \ldots, p_{n-1}$. Moreover, for $0 \le i \le |S| - 1$, $h_{i,j}$ denotes the jth head (out of $|S|$ heads) that has been left at position p_i. The states $s_1, s_2, \ldots, s_{n-1}$ the DSA M is in at these moments are stored in the state set of M'. Let us first assume that exactly $n = |S|^2$ in-operations are performed. By counting the number of in-operations in the state set, we know when the last in-operation has been performed. At that moment, M starts to write v_1 on the tape which is eventually tested with the contents of the set. To simulate this behavior by M', we use the heads $h_{0,1}, h_{1,1}, \ldots h_{|S|^2-1,1}$ to start in states $s_0, s_1, s_2, \ldots, s_{n-1}$ at positions $p_0, p_1, p_2, \ldots, p_{n-1}$ to compare the transductions of the words $u_1, u_2, \ldots, u_n$ with the transduction of v_1 read by head h. If an agreement is found when some state $s_j \in S_{\text{test}}$ is entered, that is, M has added some word to the set which is now positively tested, then the simulation is continued by comparing the transductions of the words $u_1, u_2, \ldots, u_n$ with the transduction of v_2 using the heads $h_{0,2}, h_{1,2}, \ldots h_{|S|^2-1,2}$. This behavior is continued until all m tests have been simulated successfully. Finally, it is checked with head h whether M enters an accepting state. In this case M' accepts the input and rejects otherwise. The following two pictures show the situation of dropping all heads in the first part, and the simulation for the test of the transduction of v_1. The rightmost head is head h.

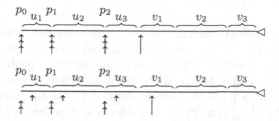

Let us now discuss the case when $n < |S|^2$ in-operations have been performed. In this case, it not clear which in-operation is the last one that starts the comparing phase. To manage this case, we drop another $|S|^2$ heads at every position $p_0, p_1, \ldots, p_{n-1}$ and interpret every in-operation as the last operation which starts the comparing phase. If the next operation is an in-operation, we start a new comparing phase with a new set of heads. If the next operation is a test-operation, we continue the comparing phase with a new set of heads. Altogether, we need at most $|S|^3 + |S|^4 + 1$ heads.

In summary, the simulation shows that L is accepted by a one-way multi-head finite automaton. This is a contradiction, since it is known that L is not accepted by any one-way multi-head finite automaton.   □

Next, we derive the incomparability of the family of languages accepted by deterministic set automata with the family of languages accepted by queue automata with finite turns as follows.

Theorem 11. *The family of languages accepted by DSA is incomparable with the family of languages accepted by queue automata with finite turns.*

Proof. We consider language L_3 of Example 4. Let us assume that L_3 is accepted by some finite-turn deterministic queue automaton. It is shown in [7] that any k-turn deterministic queue automaton can be converted into an equivalent $2k$-flip deterministic flip-pushdown automaton which is basically a deterministic pushdown automaton with the additional ability to reverse the current contents of the pushdown store. Thus, L_3 can be accepted by such an automaton with a finite number of flips. On the other hand, it is shown in [5] that L_3 cannot even be accepted by any nondeterministic flip-pushdown automaton with a finite number of flips. Hence, L_3 is not accepted by any finite-turn deterministic queue automaton.

Let us now consider the union $L = L' \cup L''$ with $L' = \{\, a^n b^m c^n \mid m, n \geq 1 \,\}$ and $L'' = \{\, a^n b^m c^{n+m} \mid m, n \geq 1 \,\}$. It is not difficult to construct a queue automaton with one turn which accepts L. On the other hand, it can be shown that L is not accepted by any DSA. $\qquad\square$

4   Closure Properties

In this section, we investigate the closure properties of DSA with respect to the Boolean operations.

Lemma 12. *The family of languages accepted by DSA is closed under complementation.*

Proof. The closure under complementation for deterministic finite automata can be easily proved by interchanging accepting and rejecting states. We cannot translate this idea directly to DSA, because mainly three problems may occur. First, the given DSA may not read its input completely by either entering a configuration in which no next move is defined (1) or by entering an infinite λ-loop (2). Second, the given DSA may perform λ-steps leading from an accepting state to a rejecting state and back (3).

Now, let M be a DSA for which we want to construct a DSA accepting its complement. To overcome problem (1), we introduce a new non-accepting state s_{rej} to which all undefined transitions of M are directed. Additionally, we define further moves from s_{rej} which shift the input head to the end of the input. For problem (3), we note that M can accept at the earliest after reading the endmarker. Thereafter additional λ-steps may be possible. We now want to achieve that in this case M enters an accepting state as soon as possible which cannot be left. To this end, a new accepting state s_{acc} is added for with the transition function is undefined. Moreover, we double the state set of M and store in every state the information whether the endmarker has been passed. If we now have a transition entering an accepting state with the knowledge that the endmarker has been passed, we redirect such a transition to enter state s_{acc}. By these modifications we have obtained an equivalent DSA M' in which problems (1) and (3) do no longer occur. However, M' may enter infinite λ-loops.

Next, we transform M' into infinite action normal form and note that by the construction problems of type (1) and (3) are not occurring. Let us distinguish two cases: first, we assume that on infinite λ-loops only states from S_{write} can be visited. By an inspection of the transition function we can check in advance which states from S_{write} will end in an infinite λ-loop. Then, any transition ending in such a state will be redirected to s_{rej}. Second, we assume that we have an infinite λ-loop in which some state $s_1 \in S_{\text{in}} \cup S_{\text{out}} \cup S_{\text{test}}$ is entered. Let $s_2 \in S_{\text{in}} \cup S_{\text{out}} \cup S_{\text{test}}$ be the next, not necessarily different, non-writing state along the λ-loop. Then, we consider the language L_{s_1,s_2} which has to be infinite due to the infinite action normal form. On the other hand, M' is deterministic, no input is read while moving from s_1 to s_2, and the tape is empty when starting in s_1 and when having reached s_2. Hence, it is only possible to write one word on the tape while moving from s_1 to s_2. This implies that L_{s_1,s_2} is finite. Thus, this case cannot occur.

Now, we have obtained that any computation ends when the whole input and the endmarker is read either in an accepting or non-accepting state which cannot be left once entered. Thus, the standard technique for constructing an automaton that accepts the complement can be used: all accepting states become non-accepting states and all non-accepting states become accepting states.   □

Lemma 13. *The family of languages accepted by DSA is not closed under union and intersection.*

Proof. In the proof of Theorem 11, we have shown that the language $L = L' \cup L''$ with $L' = \{\, a^n b^m c^n \mid m, n \geq 1 \,\}$ and $L'' = \{\, a^n b^m c^{n+m} \mid m, n \geq 1 \,\}$ is not accepted by any DSA. Similar to the construction in Example 3, we can construct DSA accepting L' as well as L''. Thus, we obtain that $\mathscr{L}(\text{DSA})$ is not closed under union. Moreover, since $\mathscr{L}(\text{DSA})$ is closed under complementation by Lemma 12, it cannot be closed under intersection.   □

Lemma 14. *The family of languages accepted by DSA is closed under intersection with regular languages and under union with regular languages.*

Proof. A DSA can simulate a given deterministic finite automaton in parallel to its computation by using the standard cross product construction. Thus, family $\mathscr{L}(\text{DSA})$ is closed under intersection with regular languages. Since $\mathscr{L}(\text{DSA})$ is closed under complementation by Lemma 12, it is closed under union with regular languages as well.   □

5   Decidability of Emptiness

Here we turn to show that emptiness is decidable for deterministic set automata. This is of interest both from a theoretical and practical point of view. For example, it is known that emptiness is decidable for pushdown automata and stack automata [4], but is undecidable for deterministic quasi-real-time queue automata [1] and deterministic one-way multi-head finite automata.

Theorem 15. *It is decidable whether a given deterministic set automaton accepts the empty language.*

Proof. The proof here will only be sketched. Given a DSA M, the basic idea is to construct a meta automaton M' and to explore all possible paths up to a certain length in its state graph to find a path from the initial state to some accepting state. If such a path does not exist, the accepted language is empty.

In a first step, the DSA M is transformed into the meta automaton M' whose states are the initial state and the in- out-, and test-states of M. The edges of M' are labeled with regular languages. The language of an edge connecting state s_i with s_j represents all strings that can be written on the tape when a computation of M passes from s_i to s_j.

The next step is to elaborate several properties of accepting paths of M and M'. In particular, it can be shown that there exists an accepting path of bounded length if there is an accepting path at all.

Finally, it can be analyzed which of the paths of M' can be expanded to paths of M, that is, how a path can be evaluated to represent an accepting computation of M. If there is no such path, then the language accepted by M is empty and non-empty otherwise.   □

References

1. Cherubini, A., Citrini, C., Crespi-Reghizzi, S., Mandrioli, D.: QRT FIFO automata, breadth-first grammars and their relations. Theoret. Comput. Sci. 85, 171–203 (1991)
2. Chomsky, N.: On certain formal properties of grammars. Inform. Control 2, 137–167 (1959)
3. Daley, M., Eramian, M.G., McQuillan, I.: The bag automaton: A model of nondeterministic storage. J. Autom., Lang. Comb. 13, 185–206 (2008)
4. Ginsburg, S., Greibach, S.A., Harrison, M.A.: One-way stack automata. J. ACM 14, 389–418 (1967)
5. Holzer, M., Kutrib, M.: Flip-pushdown automata: $k + 1$ pushdown reversals are better than k. In: Baeten, J.C.M., Lenstra, J.K., Parrow, J., Woeginger, G.J. (eds.) ICALP 2003. LNCS, vol. 2719, pp. 490–501. Springer, Heidelberg (2003)
6. Hopcroft, J.E., Ullman, J.D.: Introduction to Automata Theory, Languages, and Computation. Addison-Wesley (1979)
7. Kutrib, M., Malcher, A., Mereghetti, C., Palano, B., Wendlandt, M.: Input-driven queue automata: Finite turns, decidability, and closure properties. In: Konstantinidis, S. (ed.) CIAA 2013. LNCS, vol. 7982, pp. 232–243. Springer, Heidelberg (2013)
8. Lange, K.-J., Reinhardt, K.: Automaten mit der Datenstruktur Menge. In: Kutrib, M., Worsch, T. (eds.) 5. Theorietag Automaten und Formale Sprachen, pp. 159–167. Universität Giessen, Giessen (1995)
9. Ogden, W.F.: Intercalation theorems for stack languages. In: Proceedings of the First Annual ACM Symposium on Theory of Computing (STOC 1969), pp. 31–42. ACM Press, New York (1969)
10. Sénizergues, G.: $L(A) = L(B)$? decidability results from complete formal systems. Theoret. Comput. Sci. 251, 1–166 (2001)

The Minimum Amount of Useful Space:
New Results and New Directions[*]

Klaus Reinhardt[1] and Abuzer Yakaryılmaz[2,3,**]

[1] Wilhelm-Schickard-Institut für Informatik, University of Tübingen, Germany
Currently Humboldt University of Berlin, Germany
[2] University of Latvia, Faculty of Computing, Raina bulv. 19, Rīga, LV-1586, Latvia
[3] National Laboratory for Scientific Computing, Petrópolis, RJ, 25651-075, Brazil
`klaus.reinhardt@uni-tuebingen.de, abuzer@lncc.br`

Abstract. We consider minimal space requirements when using memory with restricted access policy (pushdown - hence giving pushdown automata (PDAs), and counter - hence giving counter automata (CAs)) in connection with two-way and realtime head motion. The main results are that: (i) $\log \log n$ is a tight space lower bound for accepting general nonregular languages on weak realtime PDAs, (ii) there exist unary nonregular languages accepted by realtime alternating CAs within weak $\log n$ space, (iii) there exist nonregular languages accepted by two-way DPADs within strong $\log \log n$ space, and, (iv) there exist unary nonregular languages accepted by two-way CAs with quantum and classical states within middle $\log n$ space and bounded error.

1 Introduction

It is a fundamental research direction to determine the minimum amount of useful "resources" which are necessary adding to a realtime deterministic finite automaton to recognize a nonregular language. There have been introduced many different "resources" such as the access way to the input (realtime, one-way, or two-way), computation mode or model (deterministic, nondeterministic, alternating, probabilistic or quantum), type of the working memory (counter, stack, or tape), etc. Moreover, unary languages needs a special attention since they may have resource requirements different from those used for languages built on general (binary) alphabets. We focus on the *minimum amount of useful space* and present some new results.

Firstly, we show that realtime nondeterministic pushdown automata (PDAs) can recognize a nonregular language with $\log \log n$ weak space. Then, we show that their two-way deterministic counterparts can recognize the same language with $\log \log n$ strong space. These bounds are tight since even two-way alternating Turing machines (TM) cannot recognize any nonregular language with less space. In the case of unary languages, it is a well-known fact that one-way

---

[*] The related technical reports are arXiv:1309.4767 and arXiv:1405.2892.
[**] Yakaryılmaz was partially supported by CAPES, ERC Advanced Grant MQC, and FP7 FET project QALGO.

A.M. Shur and M.V. Volkov (Eds.): DLT 2014, LNCS 8633, pp. 315–326, 2014.

nondeterministic PDAs can recognize only regular languages. Their alternating counterparts, on the other hand, were shown to recognize any unary language in deterministic exponential time with linear exponents by giving a simulation of the computation of linear-space alternating TMs [3]. However, this simulation is very space inefficient and it does not seem applicable if we replace the stack with a counter. Thirdly, we show that realtime alternating one-counter automata can recognize some nonregular unary languages with $\log n$ weak space[1]. Here we also present a trade-off to alternation depth. Note that two-way deterministic one-counter automata (2DCAs) cannot recognize any unary nonregular language using a sublinear space [5]. Bounded-error two-way quantum models can recognize some nonregular languages in constant-space [2]. But, we do not know whether constant-space is sufficient for unary nonregular languages. Lastly, we show that 2DCAs having two–qubits can recognize some nonregular unary languages by using middle logarithmic space on its counters for the members.

We assume that the reader is familiar with the definitions of classical computational models and so we provide only the definition for 2DCAs using a fixed-size quantum memory (in the next section). We present our results in Section 3 with a discussion of the known results. We also identify some new directions and formulate a few open questions. We put the proofs in Section 4 which also includes our trade-off results regarding alternation depth (Section 4.3). We refer the reader to [16] for a complete reference of quantum computation.

2   Preliminaries

We use three different modes of space usage [20]: (i) *Strong space* refers to the space used by the machine on all possible inputs, (ii) *middle space* refers to the space used by the machine on the inputs it gives the decision of "acceptance", and, (iii) *weak space* refers to the minimum space used by the machine on an accepting path. The length of the input is denoted by n in space bounds throughout the paper.

A two-way one-counter automaton with quantum and classical states (2QCCA) [22] is such an automaton endowed with a constant-size quantum register. If we remove the counter we obtain a two-way finite automaton with quantum and classical states (2QCFA) [2]. In the original definition of 2QCFA, the automaton can apply unitary and measurement operators to its quantum part. Here we allow our quantum models to apply a superoperator (see Fig. 1), a generalization of classical and unitary operators including measurement. This does not change the computational power of 2QCFAs and 2QCCAs in general [2]. The only remaining open case is when the operators are defined using rational numbers. We present our quantum algorithms using rational superoperators.

A 2QCCA $\mathcal{M}$ is a 8-tuple $\mathcal{M} = (S, Q, \Sigma, \delta, s_1, s_a, s_r, q_1)$, where S is the set of classical states, $s_1 \in S$, $s_a \in S$, and $s_r \in S$ $(s_a \neq s_r)$ are the initial, accepting, and rejecting states, respectively, Q is the set of quantum states, q_1 is the

---

[1] Space means here the value of the counter and not the space needed to represent the counter value.

A superoperator $\mathcal{E}$ is composed of a finite number of operation elements, $\mathcal{E} = \{E_1, \ldots, E_k\}$, satisfying

$$\sum_{i=1}^{k} E_i^\dagger E_i = I, \tag{1}$$

where $k > 0$ and the indices are the measurement outcomes. When a superoperator, say $\mathcal{E}$, is applied to the quantum register in state $|\psi\rangle$, i.e. $\mathcal{E}(|\psi\rangle)$, we obtain the measurement outcome i with probability $p_i = \langle\widetilde{\psi}_i|\widetilde{\psi}_i\rangle$, where $|\widetilde{\psi}_i\rangle$, the unconditional state vector, is calculated as $|\widetilde{\psi}_i\rangle = E_i|\psi\rangle$ and $1 \le i \le k$. Note that using unconditional state vector simplifies calculations in many cases. If the outcome i is observed ($p_i > 0$), the new state of the system, which is obtained by normalizing $|\widetilde{\psi}_i\rangle$, is given by $|\psi_i\rangle = \frac{|\widetilde{\psi}_i\rangle}{\sqrt{p_i}}$. Moreover, as a special operator, the quantum register can be initialized to a predefined quantum state. This initialize operator has only one outcome.

Fig. 1. The details of superoperators [23]

initial quantum state, Σ not containing the left and right end-markers (¢ and $, respectively) is the input alphabet, and, δ is the transition function composed of δ_q and δ_c that governs the quantum and classical part, respectively.

The given input $w \in \Sigma^*$ is placed on the input tape as ¢$w$$. At the beginning of the computation, the input head is on symbol ¢, the automaton in state s_1 and $|q_1\rangle$ in the classical and quantum parts, respectively, and the counter value is zero. Assume that the automaton is in state $s \in S$, the tape head is on symbol $\sigma \in \Sigma \cup \{¢, \$\}$, the status of the counter is $\theta \in \{0, \pm\}$, and the quantum state is $|\psi\rangle$, where $\pm$ means the value of the counter is nonzero. Each step is composed by a quantum and then a classical transition. In the quantum part, $\delta_q(s, \sigma, \theta)$ determines a superoperator which is applied to the quantum register and a classical outcome, say τ, is observed. The quantum state is updated to $|\psi_\tau\rangle$. Then, the following transition is implemented in the classical part: $\delta_c(s, \sigma, \theta, \tau) = (s', d_i, c)$ such that the automaton enters state $s' \in S$, the input head and the value of the counter are updated with respect to $d_i \in \{\leftarrow, \downarrow, \rightarrow\}$ and $c \in \{-1, 0, 1\}$, respectively. When the automaton enters s_a (s_r), the input is accepted (rejected).

3 Our Results and New Directions

3.1 Deterministic, Nondeterministic, and Alternating Machines

An almost complete picture is known for TMs. Thus, no weak $o(\log\log(n))$-space alternating TM can recognize a nonregular language and there exists a unary nonregular language recognized by strong $O(\log\log(n))$-space deterministic TM [20]. For one-way TMs, the tight bounds are given in Table 1, taken from a recent paper by Yakaryılmaz and Say [26] in which it was shown that all these bounds are tight for almost all realtime TMs. (One-way head is a restricted two-way head which is not allowed to move to left and realtime head is a restricted one-way head which stays on the same symbol at most a fixed number of steps.)

Table 1. Minimum space used by one-way TMs for recognizing nonregular languages

	General input alphabet			Unary input alphabet		
	Strong	Middle	Weak	Strong	Middle	Weak
Deterministic TM	$\log n$	$\log n$	$\log n$	$\log n$	$\log n$	$\log n$
Nondeterministic TM	$\log n$	$\log n$	$\log \log n$	$\log n$	$\log n$	$\log \log n$
Alternating TM	$\log n$	$\log \log n$	$\log \log n$	$\log n$	$\log n$	$\log \log n$

Open Problem 1. *[26] Are the double logarithmic lower bounds for the recognition of the nonregular unary languages by real-time nondeterministic and alternating TMs tight?*

If a TM has a stack as memory, then we obtain a pushdown automaton (PDA). It is known that no weak $o(n)$-space bounded one-way deterministic PDA can recognize any nonregular language [10] and it is a well-known fact that a real-time deterministic PDA can recognize the nonregular language $\{a^n b^n \mid n \geq 0\}$ in strong linear space. For one-way nondeterministic PDAs, a weak logarithmic space algorithm was given for a nonregular language [19]. We improve this bound to weak $\log \log n$ space. We denote the reverse of string c by c^R. Our language is REI composed by the non-prefixes of the following infinite word $bc_0 ac_1^R bc_1 a \cdots bc_k ac_{k+1}^R bc_{k+1} ac_{k+2}^R b \cdots$, where

- $c_k = eb_1 db_{k,1} db_1^R eb_2 db_{k,2} db_2^R eb_3 db_{k,3} db_3^R e \cdots eb_{k,\lceil logk \rceil} db_{\lceil logk \rceil} db_{k,\lceil logk \rceil}^R e$ is a counter representation for k augmented with subcounters,
- $b_i \in \{0,1\}^*$ is the binary representation of i, and
- $b_{k,i} \in \{0,1\}$ is the i-th last bit (value 2^i) in the binary representation of k.

Theorem 1. *Realtime nondeterministic PDAs can recognize nonregular language* REI *with weak* $\log \log n$ *space.*

This bound is also tight for one-way/realtime alternating PDAs since alternating TMs cannot recognize any nonregular language in $o(\log \log n)$ weak space [20].

Open Problem 2. *What are the tight strong/middle space bounds for one-way/ realtime nondeterministic and alternating PDAs for the recognition of nonregular languages?*

In case of unary language, we know that one-way nondeterministic PDAs cannot recognize any nonregular language [11]. Realtime alternating one-counter automata (CAs), on the other hand, can recognize some unary nonregular languages even in weak logarithmic space on the counter. We define two unary languages: UPOWER $= \{a^{2^n} \mid n \geq 0\}$ and UPOWER+ $= \{a^{2^n + 2n} \mid n \geq 0\}$.

Theorem 2. *Realtime alternating CAs can recognize nonregular* UPOWER+ *in weak logarithmic space.*

Open Problem 3. *What are the tight space bounds for realtime/one-way alternating CAs for the recognition of nonregular unary and binary languages?*

In Section 4.2, we first present a one-way algorithm for UPOWER and then our realtime algorithm for UPOWER+. Both algorithms have a linear alternation depth (for the members). We also investigate (in Section 4.3) whether we can have a shorter alternation depth. We present a realtime algorithm for UPOWER with logarithmic alternation depth but it uses a linear counter for the members.[2] Moreover, we show that if we replace the counter with a stack, then we can have only a single alternation using a linear space on the stack.

In the case of two-way PDAs, we have tight bounds due to the following

Theorem 3. *Two-way deterministic PDAs can recognize* REI *in strong* $\log \log n$*-space.*

In [5], it was shown that any unary language recognized by a two-way deterministic PDA using sublinear space on its stack is regular. Moreover, two-way deterministic CAs can recognize nonregular unary language UPOWER with linear space. Therefore, linear-space is a tight bound for both two-way deterministic PDAs and CAs. Currently, we do not know whether nondeterminism or using random choices can help for unary languages.

Another interesting direction is to identify the tight space bounds for one-way/realtime multi-counter/pushdown automata. Yakaryılmaz and Say [26] showed that realtime deterministic automata with k-counter can recognize some nonregular languages in middle $O(n^{\frac{1}{k}})$ space, where $k > 1$. They also remark that the same result can be followed by bounded-error probabilistic 1-counter automata but the error bound increases depending on the value of k.

3.2 Probabilistic and Quantum Machines

We start by observing that the probabilistic models are special cases of their quantum counterparts. In the unbounded error case, realtime probabilistic finite automata (PFAs) can recognize unary nonregular languages [17]. Therefore, it is interesting to consider the bounded error case. One-way PFAs can recognize only regular languages with bounded-error [18]. Two-way PFAs can recognize some nonregular languages but only with exponential expected time [7,6]. With an arbitrary small space, two-way probabilistic TMs can recognize nonregular languages [9] in polynomial time. One-way probabilistic TMs, on the other hand, cannot recognize any nonregular language in space $o(\log \log n)$ [8,13].

Two-way quantum finite automata (QFAs), on the other hand, can recognize some nonregular languages in polynomial time [2]. If the input head is quantum, i.e. the head can be in a superposition of more than one place on the input tape, then one-way QFAs can recognize some nonregular languages in linear time [15,1,24,21]. But, it is still not known whether two-way QFAs can recognize any nonregular unary language with bounded-error. Note that 2PFAs cannot recognize a nonregular unary language with bounded-error [12].

---

[2] Our algorithm is a slightly modified version of the algorithm provided by Ďuriš [4].

Here we show that using a fixed-size quantum memory can save some space for bounded-error 2QCCA on unary nonregular languages when considering middle-space.

Theorem 4. *The nonregular language* UPOWER *can be recognized by a 2QCCA with bounded-error and the automaton uses middle logarithmic space in its counter.*

One-way probabilistic PDAs cannot recognize any nonregular unary language with bounded-error [14] but the question is open for its quantum counterpart. On the other hand, realtime bounded-error probabilistic PDAs can recognize the following binary language by using middle logarithmic space:

$$\{b_1ab_2^Rab_3^Rab_4a\cdots ab_{2k-1}ab_{2k}^R \mid k > 0 \text{ and } b_i \text{ is the binary represenation of } i\}.$$

Currently, we do not know any better result and whether quantumness helps.

4   The Details of the Proofs

4.1   Proof of Theorem 1

A realtime non-deterministic PDA accepts words which are non-prefixes of the infinite word by guessing and verifying errors of the following kind:

- Some error in the format which means there is a part between a b and the following a respectively between an a and the following b which is not in $(e\{0,1\}^*d\{0,1\}d\{0,1\}^*)^*e$ and thus can not be the representation of a counter c_k or c_k^R. This can be recognized already using the finite control. (We also check if the part after the last a respectively b can not be a prefix.)
- One of the counter representations is not starting correctly with a 1 in the first sub-counter. This means some be is not followed by $1d$ or some eb is not following $d1$. Again, this can be recognized using the finite control.
- One of the sub-counters b_i is not correct or not correctly incrementing. This means one of the counter representations c_k would contain a defective part $eb_idb_{k,i}dv^Re$ or db_{i-1}^Revd with $b_i \neq v \in \{0,1\}^*$. This can be recognized using $|b_i|$ space on the push-down store. Assuming this i-th sub-counter is the first (and thus smallest) where this error occurs, the space is bounded by $|b_i| \leq \log i \leq \log k$.
- The part between two sequential a's is not a correct palindrome of the form $ac_k^Rbc_ka$. One possibility would be that the highest sub-counter is not correct. This means the outer part $ae\{0,1\}^*d\cdots b\cdots d\{0,1\}^*ea$ is already not palindromic. Again, this can be recognized using $|b_k| \leq \log k$ space on the push-down store (assuming that the sub-counters before b are correct). Assuming now that the sub-counters are correct, an error in the main counter can be recognized by guessing the position of the wrong bit $b_{k,i}$, pushing the following sub-counter b_i on the push-down store, then guessing the corresponding position in the second part, verifying the sub-counter value there and checking that the bit $b_{k,i}$ really differs.

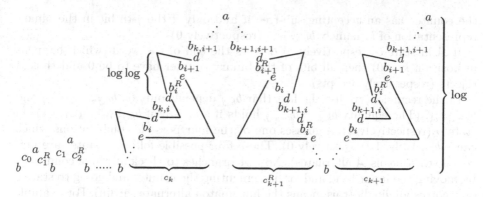

Fig. 2. The structure of a prefix of the infinite word, where each counter representation c_k consists of a logarithmic bounded number of bits, and each of this bits is controlled by a subcounter represented by at most $\log\log$ bits

- The part between two sequential b's does not have the correct form $bc_k ac_{k+1}^R b$. Like in the previous case, the last sub-counter might not be correct. This means they are either not identical (the part $d\{0,1\}^* ebe\{0,1\}^* d$ around the b is not palindromic) although there is no length increment in b_k, or the highest sub-counter (written reverse) after the b is not the highest sub-counter before the b incremented by one although there is an increment in the length from $b_k = 1111\cdots 1$ to $b_{k+1} = 1000\cdots 0$. This can be recognized similarly as in the previous case. Assuming now that the sub-counters are correct, an error in the main counter can be recognized again similar to the previous case verifying that either $b_{k,i}$ changed although $b_{k,i-1}$ did not change from 1 to 0, or $b_{k,i}$ did not change although $b_{k,i-1}$ changed from 1 to 0 (or $i = 1$).

In each case, the automaton only needs to store a sub-counter b_i with $i \le \lceil \log k \rceil$. This means the required space is $\le \lceil \log i \rceil \le \lceil \log(\lceil \log k \rceil) \rceil$. We may assume that everything was correct before the found error, which means that indeed k counters are occurring and the input must have size $n > k$. Thus the push-down size $\log(\log n)$ is sufficient to guess and verify the smallest occurring incorrectness. □

4.2 Proof of Theorem 2

First, we give a one-way automaton $\mathcal{A}$ for UPOWER since we need incrementations and decrementations without a movement of the input head.

The idea of the construction is to represent decrementing binary counters along the length of the input. The counter is only used to address a single bit of this binary counter identifying an assertion of this bit with the existence of an accepting sub-tree. Here we need to distinguish two (existential) states o and z and construct $\mathcal{A}$ in a way such that the configuration in state o (respectively z) with the input head on the k-th last position in the word and value j in the counter has an accepting sub-tree if and only if the j-th bit in the binary representation of k, namely $b_{k,j}$, is 1 (respectively 0).

If $\mathcal{A}$ in state o (respectively z) detects the end of the word (which happens in position $k = 0$) then all bits of the binary counter have to be 0 and thus $\mathcal{A}$ rejects (respectively accepts).

In the recursion, we use the fact that $b_{k,j}$ depends only on $b_{k,j-1}$, $b_{k-1,j}$ and $b_{k-1,j-1}$. (The equality $b_{k,j} = b_{k-1,j}$ holds if and only if $b_{k,j-1} \geq b_{k-1,j-1}$). In state o (respectively z) $\mathcal{A}$ guesses one of the four possible combinations which cause $b_{k,j}$ to be 1 (respectively 0). These four possible following states are universal (this means $\mathcal{A}$ alternates). Now $\mathcal{A}$ branches to check all three conditions by moving the input head and/or decrementing the counter and going to state o or z correspondingly (this means the automaton alternates again). For example $\mathcal{A}$ might guess in state o that the reason for $b_{k,j} = 1$ was $b_{k,j-1} = 0$, $b_{k-1,j} = 0$, and $b_{k-1,j-1} = 1$, then the corresponding universal branch is to decrement the counter going to z, to move the input head going to z, and to decrement the counter while at the same time move the input head going to o.

Now given an input consisting of a's, $\mathcal{A}$ wants to have an accepting sub-tree if the length of the input is 2^n, so it starts with looping in an existential state while incrementing the counter to guess n on the counter. Then, to verify that $k = 2^n$ which means $b_{k,n} = 1$ and $b_{k,j} = 0$ for all $j \neq n$, it suffices to check $b_{k,n} = 1$ and $b_{k,j} = 0$ for all $j < n$ and $b_{j,n} = 0$ for all $j < k$. So $\mathcal{A}$ alternates and branches universally to state o (to check $b_{k,n} = 1$), to a universal loop decrementing the counter with branches to z (to check $b_{k,j} = 0$ for all $j \neq n$) and to a universal loop moving the input head with branches to z (to check $b_{j,n} = 0$ for all $j < k$).

This completes the construction for UPOWER. The detection of the end of the input word can be replaced by first guessing that the end of the input word is reached, then decrementing the counter to zero and then finally checking that the end of the input word is reached. Once n is guessed, the remaining computation will then make exactly n decrements of the counter one each path. Changing the automaton to read an additional input symbol for each of the n increments and decrements makes the automaton realtime (i.e. at most two steps per symbol) and changes the accepted language from UPOWER to UPOWER+.   □

4.3   A Trade-Off to Alternation Depth

The proof of Theorem 2 requires linear alternation depth. On the other hand, it is possible to recognize UPOWER with only logarithmic alternation depth but this requires using linear counter values. The details are given below.

Let $\mathcal{A}$ be our realtime alternating one-counter automaton. We assume that $\mathcal{A}$ updates the value of the counter from the set $\{i \in \mathbb{Z} \mid -3 \leq i \leq 3\}$ instead of $\{-1, 0, 1\}$.[3] Let a^m be the input, where $m \geq 0$. The automaton $\mathcal{A}$ non-deterministically picks a position on the input, say j_1, by reading j_1 symbols. The value of counter is set to j_1 meanwhile. Then, $\mathcal{A}$ makes a universal choice:

---

[3] $\mathcal{A}$ can be easily modified without changing its program to use the latter set for updates on the counter, and so, this assumption is not essential but makes our algorithm easier to follow.

- In the first branch, $\mathcal{A}$ reads j_1 more symbols on the input tape by using the counter and accepts the input only if there is exactly j_1 symbols in the remaining part of the input. In other words, this branch returns "true" only if j_1 is the exact half of m, i.e. $2j_1 = m$. Otherwise, this branch returns "false" and so the parent universal node never returns "true".

- In the second branch, we assume that $j_1 = \frac{m}{2}$ since the result in this branch is insignificant for any other value of j_1 due to the first branch. The automaton $\mathcal{A}$ nondeterministically picks a new position, say j_2, by reading j_2 symbols. For each reading symbol, $\mathcal{A}$ decrements the value of counter by 3, and so the new counter value becomes $j_1 - 3j_2$. Then, $\mathcal{A}$ makes another universal choice very similar to previous one. In the *new* first branch, $\mathcal{A}$ adds 1 to the counter for each reading symbol. The input is accepted if the counter hits to zero when finishing the input. Otherwise, the input is rejected. So, this branch returns "true" only if $j_1 - 2j_2 = 0$, i.e. j_2 is the exact half of j_1. We can again assume that $j_2 = \frac{j_1}{2}$ in the *new* second branch and so this branch starts with the counter value $-j_2$. The details are the same as the previous universal branch except that $\mathcal{A}$ increments the value of counter by 3 for the new nondeterministically picked position j_3, and later $\mathcal{A}$ adds -1 to the counter for each reading symbol in order to check the counter hitting to zero when finishing the input which happens only if $2j_3 = j_2$. By assuming $j_3 = \frac{j_2}{2}$, we can start the newest universal branch with the counter value of $+j_3$. This procedure is repeated in this way (the updates on the counter, in the second branches, alternates between -3 and 3, followed by adding respectively $+1$ and -1 in the first child branch) until there remains only a single symbol after a universal choice where the input is accepted.

If m is a power of 2, $\mathcal{A}$ has an accepting computation tree where the values of j's are set as

$$j_1 = \frac{m}{2}, j_2 = \frac{m}{4}, \ldots, j_i = \frac{m}{2^i}, \ldots, j_{\log_2 m - 1} = 1 \tag{2}$$

and the counter value is set to $+j_1, -j_2, +j_3, -j_4, \cdots$ at the beginning of each universal choice. Note that this is the unique setting of j's, the next one is the exact half of the previous one, leading to ending up with an accepting tree. Therefore, there is a single accepting tree for each member of UPOWER. If m is not a power of 2, then there is no such setting as given in (2), and so there is no accepting tree, i.e., $\mathcal{A}$ fails at least once to find the exact half of the remaining input in an iterative step. It is clear that the depth of alternation is logarithmic for members.

In a similar way, UPOWER can be recognized with only one alternation but requiring a linear pushdown store instead of a counter as follows: The automaton guesses some word $\in \{a, b\}^* c \{a, b\}^*$ onto the pushdown store then alternates and verifies if the word has the form $w^R c w$ where $|w|$ is the input length, $w_{2^i} = b$ for all i (other positions in w are a) and $w_n = b$. This verfication starts with branching to check if the part after c has the input length and a loop on a universal state in which one symbol is popped from the push-down store and at the same time the input head is moved. Each time the automation branches to do the following two checks:

1. The same letter $\in \{a, b\}$ must occur at the position with the same distance to c. To find this position, the automaton stops to move the input head until c is reached and then continues until the end of the input is reached.
2. The same letter $\in \{a, b\}$ must also occur at the position with the double distance to c and the following symbol must be an a. This time, to find the position, two of the symbols after c are popped for each remaining position on the input.

4.4   Proof of Theorem 3

Here we can follow the same construction as in the proof of Theorem 1. (We might even use an easier version abandoning the reverse written parts.) Instead of guessing the kind of incorrectness, we have to check it one by one in an appropriate order to make sure that we find the smallest occurring incorrectness first:

Assume the automaton already checked that the sequence is correct until c_k then it first checks that the last sub-counter is consistent in the next counter, then that the sub-counters in the next counter are correctly incrementing and then that b_k is incremented correctly by going back and forth for each bit using the sub-counter content in the pushdown store to find the corresponding position in the next counter representation (crossing exactly one a respectively b).

Checking that sub-counters are correctly incrementing can be done by repeatedly pushing a binary sub-counter representation to the push-down store and comparing it with the next sub-counter representation. Finding the identical sub-counter representation in the next counter representation can be done by comparing the push-down contents with each sub-counter representation on the way; if a comparison fails, the automaton can reverse and on the way back to the beginning of the counter restore the push-down contents using the prefix which had so far been identical.   □

4.5   Proof of Theorem 4

Recently, Yakaryılmaz [22] introduced a new programming technique for 2QC-CAs and it was shown that $\texttt{USQUARE} = \{a^{n^2} \mid n \geq 1\}$ can be recognized by them for any error bound by using $O(\sqrt{n})$-space on its counter for the members. Based on this technique, we show that logarithmic space can also be useful.

2QCFAs can recognize $\texttt{POWER} = \{a^n b^{2^n} \mid n \geq 1\}$ such that any member is accepted with probability 1 and any non-member is rejected a probability arbitrarily close to 1 [25]. Let $\mathcal{P}$ be such a 2QCFA rejecting any member with a probability at least $\frac{4}{5}$ (see Appendix A of arxiv:1405.2892v1). One important property of $\mathcal{P}$ is that it reads the input from left to right in an infinite loop and another one is that it uses three quantum states.[4]

We present a 2QCCA $\mathcal{UP}$ for $\texttt{UPOWER}$ calling $\mathcal{P}$ as a subroutine such that any member is accepted with probability 1 and any non-member is rejected with probability at least $\frac{4}{5}$. Moreover, the counter value of $\mathcal{UP}$ never exceeds the

---

[4] If we use computable complex numbers then a single qubit is sufficient [2].

logarithm of the input length for the members. The pseudo-code of $\mathcal{UP}$ is given below. Let $w = a^m$ be the input.

> **REJECT (ACCEPT)** w if it is a^0 (a^1)
> FOR $i = 1$ TO m
>   RUN $\mathcal{P}$ on $w' = a^i b^m$
>   IF $\mathcal{P}$ accepts w' THEN **TERMINATE** FOR-LOOP
>   IF $\mathcal{P}$ rejects w' AND $i = m$ THEN **REJECT** the input
> END FOR
> **ACCEPT** w with a nonzero probability at most $\left(\frac{1}{9}\right)^m$
> **RESTART** the algorithm (from the second line)

In order to implement the FOR-LOOP, we use the counter. Its value is set to i, and then $\mathcal{P}$ can be simulated on $w' = a^i b^m$. It is clear that, if a^m is a member, $\mathcal{P}$ never rejects $w' = a^{\log_2 m} b^m$ and so the counter value is never set to $\log_2 m + 1$. That is, the FOR-LOOP is terminated with certainty and i is never set to m which is the only case the input might be rejected. Therefore, for the members of UPOWER, the decision of "rejection" is never given in FOR-LOOP. Therefore, they are accepted exactly. For the non-members, the input is rejected with a probability at least $\left(\frac{8}{9}\right)^m$ at the end of the FOR-LOOP. Since the input can be accepted with a probability at most $\left(\frac{1}{9}\right)^m$ after the FOR-LOOP, the rejecting probability is at least 8^m times greater than the accepting probability after the FOR-LOOP. Therefore, any non-member is rejected with a probability at least $\frac{8}{9}$. It is clear that for the members, the counter value never exceeds $\log(|w|)$, so the space complexity is logarithmic for the members. Note that, the rejecting probability can be arbitrary close to 1 by running $\mathcal{UP}$ sufficient times. □

Acknowledgements. We thank Pavol Ďuriš for kindly providing a one-way alternating one-counter automaton for UPOWER using a linear counter value for the members, Alexander Okhotin and Holger Petersen for their answers to our questions on the subject matter of this paper, and, the anonymous reviewers for their very helpful comments (based on which we rewrote the abstract and made many corrections within the text).

References

1. Amano, M., Iwama, K.: Undecidability on quantum finite automata. In: STOC 1999: Proceedings of the Thirty-first Annual ACM Symposium on Theory of Computing, pp. 368–375 (1999)
2. Ambainis, A., Watrous, J.: Two–way finite automata with quantum and classical states. Theoretical Computer Science 287(1), 299–311 (2002)
3. Chandra, A.K., Kozen, D.C., Stockmeyer, L.J.: Alternation. Journal of the ACM 28(1), 114–133 (1981)
4. Ďuriš, P.: Private communication (October 2013)
5. Ďuriš, P., Galil, Z.: On reversal-bounded counter machines and on pushdown automata with a bound on the size of their pushdown store. Information and Control 54(3), 217–227 (1982)

6. Dwork, C., Stockmeyer, L.: A time complexity gap for two-way probabilistic finite-state automata. SIAM Journal on Computing 19(6), 1011–1123 (1990)
7. Freivalds, R.: Probabilistic two-way machines. In: Gruska, J., Chytil, M.P. (eds.) MFCS 1981. LNCS, vol. 118, pp. 33–45. Springer, Heidelberg (1981)
8. Freivalds, R.: Space and reversal complexity of probabilistic one-way Turing machines. In: Karpinski, M. (ed.) FCT 1983. LNCS, vol. 158, pp. 159–170. Springer, Heidelberg (1983)
9. Freivalds, R., Karpinski, M.: Lower space bounds for randomized computation. In: Shamir, E., Abiteboul, S. (eds.) ICALP 1994. LNCS, vol. 820, pp. 580–592. Springer, Heidelberg (1994)
10. Gabarró, J.: Pushdown space complexity and related full-A.F.L.s. In: Fontet, M., Mehlhorn, K. (eds.) STACS 1984. LNCS, vol. 166, pp. 250–259. Springer, Heidelberg (1984)
11. Ginsburg, S., Rice, H.G.: Two families of languages related to ALGOL. Journal of the ACM 9(3), 350–371 (1962)
12. Kaņeps, J.: Regularity of one-letter languages acceptable by 2-way finite probabilistic automata. In: Budach, L. (ed.) FCT 1991. LNCS, vol. 529, pp. 287–296. Springer, Heidelberg (1991)
13. Kaņeps, J., Freivalds, R.: Minimal nontrivial space complexity of probabilistic one-way Turing machines. In: Rovan, B. (ed.) MFCS 1990. LNCS, vol. 452, pp. 355–361. Springer, Heidelberg (1990)
14. Kaņeps, J., Geidmanis, D., Freivalds, R.: Tally languages accepted by Monte Carlo pushdown automata. In: Rolim, J.D.P. (ed.) RANDOM 1997. LNCS, vol. 1269, pp. 187–195. Springer, Heidelberg (1997)
15. Kondacs, A., Watrous, J.: On the power of quantum finite state automata. In: FOCS 1997, pp. 66–75 (1997)
16. Nielsen, M.A., Chuang, I.L.: Quantum Computation and Quantum Information. Cambridge University Press (2000)
17. Paz, A.: Introduction to Probabilistic Automata. Academic Press, New York (1971)
18. Rabin, M.O.: Probabilistic automata. Information and Control 6, 230–243 (1963)
19. Reinhardt, K.: A tree-height hierarchy of context-free languages. International Journal of Foundations of Computer Science 18(6), 1383–1394 (2007)
20. Szepietowski, A.: Turing Machines with Sublogarithmic Space. LNCS, vol. 843. Springer, Heidelberg (1994)
21. Yakaryılmaz, A.: Superiority of one-way and realtime quantum machines. RAIRO - Theoretical Informatics and Applications 46(4), 615–641 (2012)
22. Yakaryılmaz, A.: One-counter verifiers for decidable languages. In: Bulatov, A.A., Shur, A.M. (eds.) CSR 2013. LNCS, vol. 7913, pp. 366–377. Springer, Heidelberg (2013)
23. Yakaryılmaz, A.: Public qubits versus private coins. In: The Proceedings of Workshop on Quantum and Classical Complexity, pp. 45–60. Univeristy of Latvia Press (2013), ECCC:TR12-130
24. Yakaryılmaz, A., Say, A.C.C.: Efficient probability amplification in two-way quantum finite automata. Theoretical Computer Science 410(20), 1932–1941 (2009)
25. Yakaryılmaz, A., Say, A.C.C.: Succinctness of two-way probabilistic and quantum finite automata. Discrete Mathematics and Theoretical Computer Science 12(2), 19–40 (2010)
26. Yakaryılmaz, A., Say, A.C.C.: Tight bounds for the space complexity of nonregular language recognition by real-time machines. International Journal of Foundations of Computer Science 24(8), 1243–1253 (2013)

Debates with Small Transparent Quantum Verifiers

Abuzer Yakaryılmaz[1,2,*], A.C. Cem Say[3], and H. Gökalp Demirci[4]

[1] University of Latvia, Faculty of Computing,
Raina bulv. 19, Rīga, LV-1586, Latvia
[2] National Laboratory for Scientific Computing,
Petrópolis, RJ, 25651-075, Brazil
[3] Boğaziçi University, Department of Computer Engineering,
Bebek 34342 İstanbul, Turkey
[4] University of Chicago, Department of Computer Science,
Chicago, IL 60637, USA
abuzer@lncc.br, say@boun.edu.tr, demirci@cs.uchicago.edu

Abstract. We study a model where two opposing provers debate over the membership status of a given string in a language, trying to convince a weak verifier whose coins are visible to all. We show that the incorporation of just two qubits to an otherwise classical constant-space verifier raises the class of debatable languages from at most NP to the collection of all Turing-decidable languages (recursive languages). When the verifier is further constrained to make the correct decision with probability 1, the corresponding class goes up from the regular languages up to at least E.

Keywords: quantum finite automata, quantum computing, probabilistic finite automata, Arthur-Merlin games, debate systems, zero-error.

1 Introduction

It is well known that the model of alternating computation is equivalent to a setup where two opposing debaters try to convince a resource-bounded deterministic verifier about whether a given input string is in the language under consideration or not [4]. Variants of this model where the verifier is probabilistic, and the communications between the debaters are restricted in several different ways, have been studied [6,11,9]. Quantum refereed games, where the messages exchanged between the debaters are quantum states, were examined by Gutoski and Watrous [13].

Most of the work cited above model the verifier as opaque, in the sense that the outcomes of its coin throws are not visible to the debaters, who have a correspondingly incomplete picture of its internal state during the debate. These models can therefore be classified as generalizations of private-coin interactive

* Yakaryılmaz was partially supported by CAPES, ERC Advanced Grant MQC, FP7 FET project QALGO, and BÜVAK.

A.M. Shur and M.V. Volkov (Eds.): DLT 2014, LNCS 8633, pp. 327–338, 2014.

proof systems [12] to the competing multiple provers case. In this paper, we focus on models where all of the verifier's coins, as well as all communications, are publicly visible to all parties, making them generalizations of Arthur-Merlin games [3]. A recent result [17] established that a very small quantum component is sufficient to expand the computational power of classical proof systems of this kind considerably, by studying a setup where an otherwise classical constant-space verifier is augmented by a quantum register of just two qubits. We modify that protocol to show that the addition of a two-qubit quantum register to the classical finite state verifier raises the class of debatable languages from at most NP to that of all Turing-decidable languages. We also study the case where the verifier is required to take the correct decision with probability 1. We show that small quantum verifiers outperform their classical counterparts in this respect as well, exhibiting an increase from the class of regular languages to at least $E = DTIME(2^{O(n)})$.[1]

The rest of this paper is structured as follows: Section 2 describes our model and reviews previous work. Our result on the computational power of the model with a two-qubit constant-space verifier in the two-sided bounded error case is presented in Section 3. Section 4 contains an examination of the more restricted zero-error case. Section 5 concludes the paper with some remarks on the possible usage of multihead automata as verifiers.

2   Preliminaries

Consider an interactive system consisting of three actors: two debaters named Player 1 (P1) and Player 0 (P0), respectively, and a Verifier (V). All actors have access to a common input string w. P1 tries to convince V that w is a member of the language L under consideration, whereas P0 wants to make V reject w as a non-member. The debaters communicate with each other through a communication cell which is seen by every actor. Each debater writes a symbol in the communication cell when its turn comes, and V executes a further step of its computation, taking this communication and the outcomes of its coin into account. The debate continues in this way until the computation of V is terminated as it reaches a decision. We assume that both debaters see the coin outcomes of V as they occur, and thereby have complete information about the state of the verifier at any point.

In such a setup, we say that language L has a *debate checkable by a machine V with error bound* $\epsilon \in [0, \frac{1}{2})$ if

- for each $w \in L$, P1 is able to make V accept w with probability at least $1 - \epsilon$, no matter what P0 says in return,
- for each $w \notin L$, P0 is able to make V reject w with probability at least $1 - \epsilon$, no matter what P1 says in return.

A language is said to be *debatable* if it has a debate checkable by some verifier.

---

[1] Note that E is a proper subset of $EXP = DTIME(2^{poly(n)})$.

Note that the class of debatable languages is closed under complementation.

We focus on verifiers which are only allowed to operate under constant space bounds. When V is set to be a deterministic two-way finite automaton, the system described above is equivalent to an alternating two-way finite automaton, and the class of debatable languages coincides with the regular languages [15]. When one replaces V with a two-way probabilistic finite automaton, the class in question becomes one that should be denoted ∀BC-SPACE(1) in the terminology of [5], and is known to contain some nonregular languages [10], and to be contained in NP. We will show that the addition of a small amount of quantum memory to the probabilistic model increases the power hugely, all the way to the class of decidable languages.

The public-coin quantum verifier model that we will use is the two-way finite automaton with quantum and classical states (2QCFA) [2], in which the quantum and classical memories are nicely separated, allowing a precise quantification of the amount of "quantumness" required for our task. Such automata execute quantum and classical moves alternately at each step:

- First, a superoperator[2] (Figure 1), determined by the current classical state and the symbols being scanned on the input tape and in the communication cell, is applied to the quantum register (the quantum memory of the machine), with the outcome of the operator being automatically sent to the debaters. All entries of quantum operators are rational numbers, meaning that the probabilities of the outcomes are always rational,[3] and the debaters can easily keep track of the superposition in the quantum register.
- Then, the next classical state and tape head movement is determined by the current classical state and the observed outcome.

Execution halts when an outcome associated with "acceptance" or "rejection" is observed.

One obtains the definition of the quantum Arthur-Merlin (qAM) systems of [17] when one removes P0 from the picture described above. Our results on small quantum verifiers for debates are based on the following result:

Fact 1. *For any error bound $\epsilon > 0$, every Turing-recognizable language (recursively enumerable language) has an Arthur-Merlin system where the verifier uses just two quantum bits, (i.e. four quantum states,) members of the language are accepted with probability 1, nonmembers are accepted with a probability not greater than ϵ, and a dishonest P1 can cause the machine to run forever without reaching a decision.*

Proof. We outline the basic idea, referring the reader to [17] for a detailed exposition of this proof. Let T be the single-tape Turing machine recognizing the

---

[2] The usage of superoperators generalizes and simplifies the quantum transition setup of the 2QCFA's of [2]; see [20].

[3] The classical probabilistic finite automata, to which we compare our quantum model, can only flip fair coins. It is known that this is sufficient for two-way automata to realize any rational transition probability.

For a 2QCFA with j quantum states, each superoperator $\mathcal{E}$ is composed of a finite number of $j \times j$ matrices called *operation elements*, $\mathcal{E} = \{E_1, \ldots, E_k\}$, satisfying

$$\sum_{i=1}^{k} E_i^{\dagger} E_i = I, \tag{1}$$

where $k \in \mathbb{Z}^+$, and the indices are the measurement outcomes. When a superoperator $\mathcal{E}$ is applied to a quantum register in state $|\psi\rangle$, then we obtain the measurement outcome i with probability $p_i = \langle \widetilde{\psi}_i | \widetilde{\psi}_i \rangle$, where $|\widetilde{\psi}_i\rangle$ is calculated as $|\widetilde{\psi}_i\rangle = E_i |\psi\rangle$, and $1 \leq i \leq k$. If the outcome i is observed ($p_i > 0$), the new state of the system is obtained by normalizing $|\widetilde{\psi}_i\rangle$ which is $|\psi_i\rangle = \frac{|\widetilde{\psi}_i\rangle}{\sqrt{p_i}}$. Moreover, the quantum register can be set to a predefined quantum state by an initialize operator with a single outcome.

Fig. 1. Superoperators (adapted from [17])

language L under consideration. For any input string w, P1 (the only debater in this restricted scenario) is supposed to send the computation history (i.e. the sequence of configurations) of T on input w to the verifier V. Some of the possible outcomes of V's observations of its quantum register will be interpreted as "restart" commands to P1. At any point, V may interrupt P1 and ask it to restart sending the computation history from the beginning in this manner. (In fact, the verifier is highly likely to require a restart at each step.)

Whenever the verifier catches P1 lying (i.e. giving an incorrect configuration description), it rejects the input. If the verifier reads a computation history sent by P1 all the way to its completion by a halting configuration without detecting an incorrect configuration, it halts with a decision paralleling the one described in that history with a certain non-zero probability, and requests a restart with the remaining probability.

A classical public-coin finite automaton faced with this task would not be able to compare two consecutive configuration descriptions c_i and c_{i+1} (which may be very long).[4] A 2QCFA verifier handles this problem by encoding the substrings in question into the amplitudes of its quantum states.[5] Let $next(c)$ denote the description of the configuration that is the legitimate successor, according to the transition function of T, of configuration c, and let $e(x)$ denote an integer that encodes string x according to a technique to be described later. After the description c_{i+1} has been read, the amplitudes of the quantum states of V form the vector $\alpha \left(1 \quad e(next(c_i)) \quad e(c_{i+1}) \quad e(next(c_{i+1}))\right)^{\mathsf{T}}$, where α is a small rational number. (The amplitude of the first state is used as an auxiliary value during the encoding [17], as will also be seen in the next section.)

---

[4] The associated complexity class is known [7] to be included in P. When the verifier is allowed to hide its coins, its power increases [8].

[5] Actually, this encoding can also be performed by a classical probabilistic machine [16]. It is the subsequent subtraction that is impossible for classical automata.

When P1 concludes the presentation of c_{i+1}, V executes a move that has the effect of subtracting $\alpha e(next(c_i))$ from $\alpha e(c_{i+1})$, rejecting with a probability equal to the square of the difference, continuing with some little probability after placing the encoding of $next(c_{i+1})$ into the second state's amplitude and resetting the third and fourth amplitudes to zero for beginning the next encode-compare stage, and requesting a restart with the remaining probability. If c_{i+1}'s description is indeed equal to the valid successor of c_i, the subtraction mentioned above yields zero probability of rejection. Otherwise, the rejection probability arising from a transition error within a computation history is guaranteed to be a big multiple of the acceptance probability that may arise due to that spurious history ending with an accepting configuration.

If $w \in L$, P1 need only obey the protocol, sending the accepting computation history, restarting each time V tells it to do so. In each try, P1 has a small but nonzero probability of sending the full history without being interrupted, leading to a nonzero probability of halting with acceptance. Since V will detect no transition errors between configurations, the probability of rejection is zero.

If $w \notin L$, any attempt of P1 to trick V to accept w with high probability by sneaking a transition error to the history and ending it with an accepting config-uration will be foiled, since the rejection probability associated with the defect in the history can be guaranteed to be as big a multiple of the final acceptance prob-ability as one desires. There is, however, one annoyance that P1 can cause V in this case: If P1 sends an infinite-length "configuration description" at any point[6] during its presentation, V will never reach the point where it compares the two amplitudes it uses for encoding, and it will therefore fail to halt.   □

3   Small Transparent Verifiers for All Decidable Languages

Our first result is a generalization of the proof of Fact 1 to the setup with two debaters described in the previous section.[7]

Theorem 1. *For every error bound $\epsilon > 0$, every Turing-decidable language has a debate checkable by a 2QCFA with four quantum states, and with error bounded by ϵ.*

Proof. We modify the verifier V described in the proof of Fact 1 in Section 2 to obtain a new verifier V_1 as follows: V_1 listens to both P0 and P1 in parallel. In the protocol imposed by V_1, both debaters are expected to behave exactly as P1 was supposed to behave in that earlier proof; transmitting the computation

---

[6] Except at the beginning, since V can check the first configuration itself by matching it with the input.

[7] In separate work, the techniques of [17] were used to define a model called q-alternation [18]. This model is distinct from debate checking in the same sense that the two equiva-lent definitions of classical nondeterminism (the "probabilistic machine with zero cut-point" and the "verifier-certificate" views) lead to quantum counterparts ([1] and [14], respectively) which are remarkably different from each other.

history of the single-tape Turing machine T for language L on input string w, interrupting and restarting transmissions whenever V_1 observes an outcome associated with the "restart" action in its quantum register.

The strategy of V_1 is based on the fact that the two debaters are bound to disagree at some point about the computation history of T on w. As long as the same description is coming in from both debaters, V_1 uses the same technique mentioned in the proof of Fact 1, to be described in more detail shortly, for encoding the successive configurations. At the first point within a history when a mismatch between the two debaters is detected, V_1 uses its register to flip a fair coin to choose to trace one or the other debater's transmission from that time. The chosen debater's description of what it purports to be the computation history is then checked exactly as in the earlier proof, and the other debater is ignored until a restart is issued by V_1 to both players during (or at the end of) that check. The truthful debater always obeys the protocol. In the case that the other debater's transmission is identical to that of the truthful one, V_1 parallels the decision of T depicted by both debaters.

If it sees the debater it is tracing violating the protocol, for instance, making a transition error, V_1 rules in favor of the other player. When it sees a debater announcing the end of a computation history, V_1 decides in that debater's favor with some probability, and demands a restart with the remaining probability. Like the program described in the proof of Fact 1, V_1 is constructed so that the probability of the decision caused by the detection of a transition error in a computation history is guaranteed to be much greater than the probability of the decision caused by mimicking the result described at the end of that history.

A full description of V_1 would involve the complete presentation of its classical transition function, as well as all the operation elements of every superoperator associated with every triple of state, input symbol, and debate symbol. We will give a higher-level description of the program and its execution at a level that will allow the interested reader to construct the full 2QCFA if she wishes to do so.

A segment of computation which begins with a (re)start, and ends with a halting or restarting configuration will be called a "round" [19]. In each such round, each debater is supposed to transmit a string of the form

$$c_1\$\$c_2\$\$\cdots c_{h-1}\$\$,$$

where c_1 is the description of the start configuration of T on w, each c_{i+1} is the legal successor of the corresponding c_i, and c_{h-1} is the last configuration in the computation history before the halting configuration. (V_1 will be able to understand whether the successor of c_{h-1} is an accepting or rejecting configuration by focusing on the symbols around the tape head in c_{h-1}.) We assume that each configuration description ends with the blank symbol $\#$, and that the alphabet Γ used to write the configurations does not include the $\$ symbol. Fix an ordering of the symbols in Γ, and let $e(\sigma)$ denote the position of any symbol $\sigma \in \Gamma$ in this ordering. Let m be an integer greater than the cardinality of Γ, we will fix its value later.

The state of the quantum register is set to $|\psi_{1,0}\rangle = (1\ 0\ 0\ 0)^{\mathsf{T}}$ at the beginning of each round.

Let l_i be the length of $c_1\$\$c_2\$\$\cdots c_i\$\$$ $(i > 0)$.

As it reads the string $w_1 = c_1\$\$$ from the debaters, V_1 both compares it with the input to catch a debater that may lie at this point, and also applies a superoperator corresponding to each symbol of w_1 to the register in order to encode $next(c_1)$ as a number in base m (times a factor that will be described later) into the amplitude of the second quantum state. One operation element of the superoperator $\mathcal{E}_{1,j}$ applied when reading the jth symbol, say, σ, of w_1 is

$$E_{1,j,1} = \frac{1}{d}\begin{pmatrix} 1 & 0 & 0 & 0 \\ e(\sigma) & m & 0 & 0 \\ 0 & 0 & 0 & 0 \\ 0 & 0 & 0 & 0 \end{pmatrix},$$

where d is an integer which has the properties to be described now.[8] Since $\mathcal{E}_{1,j}$ would not obey the wellformedness criterion (Equation 1 in Figure 1) if its only operation element were $E_{1,j,1}$, we add as many 4×4 rational matrices as necessary as *auxiliary operation elements* of $\mathcal{E}_{1,j}$ to complement its single *main operation element* $E_{1,j,1}$ to ensure that Equation 1 is satisfied. Furthermore, we do this for all superoperators to be described in the rest of the program in such a way that each of their main operation elements can be written with the same factor $\frac{1}{d}$ in front, as we just did for $E_{1,j,1}$. This is the property that d must satisfy, and such a d can be found easily [17,20].

The observation outcome associated with all auxiliary operation elements will be interpreted as a "restart" command to the debaters. Some operation elements to be described below are associated with halting (acceptance or rejection). The outcomes of all remaining operation elements, including the $E_{1,j,1}$, are "continue" commands.

Depending on whether the length of T's configuration description increases as a result of its first move or not, we have the following cases:

- If $|next(c_1)| = |c_1|$, the main operation elements of $\mathcal{E}_{1,|c_1|}$ and $\mathcal{E}_{1,|c_1\$|}$ are

$$\frac{1}{d}\begin{pmatrix} 1 & 0 & 0 & 0 \\ e(\#) & m & 0 & 0 \\ 0 & 0 & 0 & 0 \\ 0 & 0 & 0 & 0 \end{pmatrix} \text{ and } \frac{1}{d}\begin{pmatrix} 1 & 0 & 0 & 0 \\ 0 & 1 & 0 & 0 \\ 0 & 0 & 0 & 0 \\ 0 & 0 & 0 & 0 \end{pmatrix},$$

respectively, since the encoding of $next(c_1)$ is finished by superoperator $\mathcal{E}_{1,|c_1|}$.
- If $|next(c_1)| = |c_1| + 1$, and the $|c_1|$th symbol of $next(c_1)$ is σ, the main operation elements of $\mathcal{E}_{1,|c_1|}$ and $\mathcal{E}_{1,|c_1\$|}$ are

$$\frac{1}{d}\begin{pmatrix} 1 & 0 & 0 & 0 \\ e(\sigma) & m & 0 & 0 \\ 0 & 0 & 0 & 0 \\ 0 & 0 & 0 & 0 \end{pmatrix} \text{ and } \frac{1}{d}\begin{pmatrix} 1 & 0 & 0 & 0 \\ e(\#) & m & 0 & 0 \\ 0 & 0 & 0 & 0 \\ 0 & 0 & 0 & 0 \end{pmatrix},$$

---

[8] Note that the "names" we are using for the superoperators are based on their application position on the debater transmissions; this same superoperator would be applied again (but would have a different index in our exposition) if another σ comes up elsewhere in the transmission of c_1.

respectively, since the encoding of $next(c_1)$ is finished by superoperator $\mathcal{E}_{1,|c_1|+1}$.

The main operation element of $\mathcal{E}_{1,|c_1\$\$|}$ just multiplies the state vector by $\frac{1}{d}$.

As long as the debaters are in agreement, and a halting configuration has not been detected, each configuration description block $w_i = c_i\$\$$ $(i \geq 2)$ is processed in the following manner. The state vector is

$$|\widetilde{\psi_{i,0}}\rangle = \left(\frac{1}{d}\right)^{l_i-1} (1 \quad e(next(c_{i-1})) \quad 0 \quad 0)^\mathsf{T}$$

at the beginning of the processing. The tasks are:

1. To encode c_i and $next(c_i)$ into the amplitudes of the third and fourth quantum states, respectively, during the processing of the substring $c_i\$$, and
2. To accept (resp. reject) the input if $next(c_i)$ is an accepting (resp. rejecting) configuration, or to prepare for the $(i+1)^{st}$ configuration description block if $next(c_i)$ is not a halting configuration, during the processing of the final $\$$ symbol.

The details of superoperators to encode c_i and $next(c_i)$ are similar to the ones given above. For each $j \in \{1, \ldots, |c_i|-1\}$, the main operation element of $\mathcal{E}_{i,j}$ is

$$\frac{1}{d}\begin{pmatrix} 1 & 0 & 0 & 0 \\ 0 & 1 & 0 & 0 \\ e(\sigma) & 0 & m & 0 \\ e(\gamma) & 0 & 0 & m \end{pmatrix},$$

where σ and γ are the j'th symbols of c_i and $next(c_i)$, respectively. $\mathcal{E}_{i,|c_i|}$ and $\mathcal{E}_{i,|c_i\$|}$ handle the two cases where $e(next(c_i))$ may or may not be longer than $e(c_i)$, similarly to the superoperators seen for the processing of c_1 [17]. Thus, before applying $\mathcal{E}_{i,|c_i\$\$|}$, the state vector becomes

$$|\widetilde{\psi_{i,|c_i\$\$|}}\rangle = \left(\frac{1}{d}\right)^{l_i-1} (1 \quad e(next(c_{i-1})) \quad e(c_i) \quad e(next(c_i)))^\mathsf{T}. \tag{2}$$

Task (2) described above is to be realized by operator $\mathcal{E}_{i,|c_i\$\$|}$, which has one main operation element, as described in Figure 2.

After a disagreement between the debaters is noticed, the verifier picks a debater with probability $\frac{1}{2}$. (A fair coin can be implemented in this setup by the superoperator $\mathcal{E} = \{E_{h_1} = \frac{1}{2}I, E_{h_2} = \frac{1}{2}I, E_{t_1} = \frac{1}{2}I, E_{t_2} = \frac{1}{2}I\}$ with the outcomes for the first two operation elements interpreted as heads and the other ones as tails, for instance.) The processing of the transmission of the chosen debater is the same as the processing of the common stream, except for the last superoperator dealing with the final $\$$ symbol of each description block. That superoperator has two main operation elements. The first one realizes the first actual transition correctness check:

$$\frac{1}{d}\begin{pmatrix} 0 & 0 & 0 & 0 \\ 0 & 1 & -1 & 0 \\ 0 & 0 & 0 & 0 \\ 0 & 0 & 0 & 0 \end{pmatrix}.$$

DESCRIPTION	OPERATOR
If $next(c_i)$ is a halting configuration, then this operator is applied with the action of acceptance or rejection, as indicated by $next(c_i)$, associated with the outcome. The input is thereby accepted or rejected with probability $p_1 = \left(\frac{1}{d}\right)^{2l_i}$. The round is terminated in this case.	$\frac{1}{d}\begin{pmatrix} 1 & 0 & 0 & 0 \\ 0 & 0 & 0 & 0 \\ 0 & 0 & 0 & 0 \\ 0 & 0 & 0 & 0 \end{pmatrix}$
If $next(c_i)$ is not a halting configuration, then this operator is applied. The state vector becomes $\widetilde{\lvert\psi_{i+1,0}\rangle} = \left(\frac{1}{d}\right)^{l_i}(1 \; e(next(c_i)) \; 0 \; 0)^{\mathsf{T}}$.	$\frac{1}{d}\begin{pmatrix} 1 & 0 & 0 & 0 \\ 0 & 0 & 0 & 1 \\ 0 & 0 & 0 & 0 \\ 0 & 0 & 0 & 0 \end{pmatrix}$

Fig. 2. Operation element for preparing for the next configuration in the debater stream

The associated action of this operation element is to reject the claim of this debater. Therefore, when talking to P0 (resp., P1), the input is accepted (resp., rejected) with probability $\left(\frac{1}{d}\right)^{2l_i}(e(next(c_{i-1})) - e(c_i))^2$, which is zero if the check succeeds $(next(c_{i-1}) = c_i)$, and is at least $p_2 = \left(\frac{1}{d}\right)^{2l_i} m^2$ if the check fails $(next(c_{i-1}) \neq c_i)$. Since the last symbols of $next(c_{i-1})$ and c_i are identical, the value of $\lvert e(next(c_{i-1})) - e(c_i)\rvert$ can not be less than m in this case.

The second main operation element is the one already described in Figure 2, which either halts and decides, or readies the state vector for scanning the next configuration (with small probability) depending on whether that next configuration is a halting one or not.

Note that if the chosen debater is cheating and never sends any $'s, then the communication with it terminates with probability 1 without any decision.

The overall acceptance probability of such a "program with restart" equals the ratio of the acceptance probability to the halting probability in a single round [19]. The probability that the truthful debater will be selected after the disagreement is $\frac{1}{2}$. If this happens, V_1 will reach a halting state with the correct decision with some small probability, and restart with the remaining probability. In case the other debater is selected, there are two different possibilities of deception. If that debater presents an infinite "configuration", V_1 will restart sooner or later. Otherwise, if a finite but spurious history with one or more incorrect transitions is presented, V_1 may make the wrong decision with some small probability p_1, but this is more than compensated by the much greater probability p_2 of its making the correct decision earlier on, when the transition error(s) in this history were detected. Overall, the error rate of ϵ of V_1 is bounded by $\frac{p_1}{p_1+p_2} = \frac{1}{m^2+1}$, and can be tuned down to any desired positive value by choosing m, the base of the encoding used, to be a sufficiently large integer. □

4   Debates with Zero Error

In classical computation, the benefits of using random bits come at the cost of incurring some nonzero probability of error; and "zero-error" probabilistic finite automata can be shown trivially to be no more powerful than their deterministic counterparts. We will now show that randomness without some tolerance of error is not useful for classical finite-state verifiers of debates, and then prove that things change in the quantum case.

Theorem 2. *The computational power of a public-coin probabilistic debate checking system is reduced to the level of its deterministic counterpart when the verifier is not allowed to make any error in its final decision.*

Proof. Assume that a language L has a debate checkable by a probabilistic verifier V_p with zero error. We construct a deterministic verifier V_d with the same space and time bounds as V_p. V_d mimics V_p, except that whenever it needs to simulate a coin throw of V_p, it reads the corresponding bit from P0.

If a string $w \in L$, then P1 is able to convince V_p to accept w, no matter what P0, or the coins of V_p, can "say." Note that in such a case, P1 will be able to convince V_d to accept w, no matter what P0 can say. If $w \notin L$, then P0 is able to convince V_p to reject w, no matter what P1, or the coins of V_p, can say. In this case, P0 would of course be able to convince V_d to reject w, regardless of what P1 might say.   □

As mentioned in Section 2, languages with debates checkable by deterministic finite state verifiers are regular, whereas probabilistic verifiers can handle some nonregular languages when some error is allowed. We will now see that our small quantum verifiers can do much more with zero error.

Theorem 3. *Every language in the class* E *has a debate checkable by a 2QCFA with four quantum states, and with zero error.*

Proof. Since E = ASPACE(n) (the class of languages recognized by alternating Turing machines (ATMs) using linear space) [4], it is sufficient to show how to trace the execution of a linear-space alternating Turing machine (ATM). Let A be an ATM that decides a language L, using at most kn tape squares for its computation on any string of length n, for a positive integer k. Assume, without loss of generality, that A alternates between existential and universal states at each step, and that the start state is an existential state.

We construct a 2QCFA V_0 that checks debates on membership in L. V_0 is a variant of the verifier V_1 described in the proof of Theorem 1. In this version, the debaters play a game to produce a computation history of A on the input w of length n. The protocol dictates that P1 starts by announcing the first existential choice to be made. Both debaters then transmit the start configuration of A on w parallelly. P0 then announces the first universal choice as a response to the first move of P1, followed by both debaters transmitting the configuration that A would reach by executing the choice announced by P1 in the beginning. In general, the choice that determines configuration c_{i+1} is announced by the corresponding debater before the transmission of configuration c_i. As usual, the verifier may order the debaters to restart the whole thing at any step.

After using it to check that the first configuration description is accurate, V_0 starts moving its reading head on the input tape back and forth at the appropriate speed to make sure that neither debater sends a configuration description longer than nk symbols in the rest of the transmission, deciding against any debater seen to violate this rule. As described for the verifiers in our earlier proofs, V_0 scans the parallel transmissions, encoding the last configuration descriptions it has seen, as well as their legal successors according to the choices that have already been announced by the debaters. If the debaters send the same complete history, V_0 halts and announces the result in that history. If the debaters disagree, V_0 flips a coin and picks one debater's transmission to trace, just like V_1. Unlike V_1, however, V_0 does not trace this debater until it sends a halting configuration. Instead, V_0 just performs the transition check between the previously sent configuration and the presently sent one,[9] and then issues a restart command. V_0 does not imitate any decision of A that it may see in the transmission of the chosen debater; the only way that V_0 can halt without any restarts after choosing a debater is by detecting a transition error, and deciding in favor of the other debater.

If both debaters obey the protocol, then P1 will always be able to demonstrate an accepting computation history of A on w if $w \in L$, and P0 will always be able to demonstrate a rejecting computation history of A on w if $w \notin L$. So let us examine the case where one debater is lying.

If V_0 chooses the truthful debater to trace, it will detect no error, and so will restart with certainty. If it chooses the other debater, it will detect a transition error and announce the correct decision with some probability, and restart with the remaining probability. There is no possibility that V_0 can make an error.   □

5   Concluding Remarks

It is well known that finite automata with k classical input heads can use them as one can use logarithmic space; for instance, to count up to $O(n^k)$. One can therefore extend the argument of Theorem 3 to APSPACE (the class of languages recognized by ATMs using polynomial space), which equals EXPTIME [4], concluding that every language in the class EXPTIME has a zero-error (public-coin) debate checkable by a multiple-head 2QCFA with four quantum states

Debate systems with deterministic logarithmic-space (or equivalently, multi-head finite-state) verifiers which have the additional property that P0 can hide some of its messages to the verifier from P1 are known to correspond to the class EXPTIME. If one upgrades the verifier in this model to a probabilistic version, but demands that it should still make zero error, the computational power does not change, since zero-error probabilistic machines can be derandomized easily. We can therefore also state that every language in the class EXPTIME has such a "partial-information" debate checkable by a private-coin multiple-head two-way probabilistic finite automaton with zero error.

Acknowledgements.   We thank the anonymous reviewers for their helpful comments.

---

[9] If the chosen debater attempts to send an exceedingly long configuration at this point, it will be caught by the control implemented by the input head.

References

1. Adleman, L.M., DeMarrais, J., Huang, M.D.A.: Quantum computability. SIAM Journal on Computing 26(5), 1524–1540 (1997)
2. Ambainis, A., Watrous, J.: Two–way finite automata with quantum and classical states. Theoretical Computer Science 287(1), 299–311 (2002)
3. Babai, L.: Trading group theory for randomness. In: STOC 1985, pp. 421–429 (1985)
4. Chandra, A.K., Kozen, D.C., Stockmeyer, L.J.: Alternation. Journal of the ACM 28(1), 114–133 (1981)
5. Condon, A.: Computational Models of Games. MIT Press (1989)
6. Condon, A., Feigenbaum, J., Lund, C., Shor, P.: Probabilistically checkable debate systems and approximation algorithms for PSPACE-hard functions (extended abstract). In: STOC 1993, pp. 305–314. ACM (1993)
7. Condon, A., Ladner, R.E.: Probabilistic game automata. Journal of Computer and System Sciences 36(3), 452–489 (1988)
8. Condon, A., Lipton, R.J.: On the complexity of space bounded interactive proofs (extended abstract). In: FOCS 1989, pp. 462–467 (1989)
9. Demirci, H.G., Say, A.C.C., Yakaryılmaz, A.: The complexity of debate checking. Theory of Computing Systems (2014), doi:10.1007/s00224-014-9547-7
10. Dwork, C., Stockmeyer, L.: Finite state verifiers I: The power of interaction. Journal of the ACM 39(4), 800–828 (1992)
11. Feige, U., Kilian, J.: Making games short (extended abstract). In: STOC 1997, pp. 506–516. ACM (1997)
12. Goldwasser, S., Micali, S., Rackoff, C.: The knowledge complexity of interactive proof systems. SIAM Journal on Computing 18(1), 186–208 (1989)
13. Gutoski, G., Watrous, J.: Toward a general theory of quantum games. In: STOC 2007, pp. 565–574 (2007)
14. Kitaev, A.Y., Shen, A., Vyalyi, M.N.: Classical and Quantum Computation. American Mathematical Society (2002)
15. Ladner, R.E., Lipton, R.J., Stockmeyer, L.J.: Alternating pushdown automata. In: FOCS 1978, pp. 92–106 (1978)
16. Rabin, M.O.: Probabilistic automata. Information and Control 6, 230–243 (1963)
17. Yakaryılmaz, A.: Public-qubits versus private-coins. Tech. rep. (2012), ECCC:TR12-130
18. Yakaryılmaz, A.: Quantum alternation. In: Bulatov, A.A., Shur, A.M. (eds.) CSR 2013. LNCS, vol. 7913, pp. 334–346. Springer, Heidelberg (2013)
19. Yakaryılmaz, A., Say, A.C.C.: Succinctness of two-way probabilistic and quantum finite automata. Discrete Mathematics and Theoretical Computer Science 12(2), 19–40 (2010)
20. Yakaryılmaz, A., Say, A.C.C.: Unbounded-error quantum computation with small space bounds. Information and Computation 279(6), 873–892 (2011)

Embedding Finite and Infinite Words into Overlapping Tiles

(Short Paper)

Anne Dicky and David Janin

LaBRI, IPB, Université de Bordeaux,
351, cours de la Libération,
F-33405 Talence, France
{dicky,janin}@labri.fr

Abstract. In this paper, we study languages of finite and infinite bi-rooted words. We show how the embedding of free ω-semigroups of finite and infinite words into the monoid of birooted words can be generalized to the embedding of two-sorted ω-semigroups into (some notion of) one-sorted ordered ω-monoids. This leads to an algebraic characterization of regular languages of finite and infinite birooted words that generalizes and unifies the known algebraic characterizations of regular languages of finite and infinite words[1].

1 Introduction

Infinite strings naturally arise in Software Engineering as models of (potentially) non-terminating system behaviors. From an abstract point of view, infinite strings are defined as infinite concatenations of non-empty finite strings, with an infinite associativity law ensuring that this infinite product is compatible with the standard concatenation of finite strings. This leads to the notions of ω-semigroups and morphisms that provide an algebraic characterization of regular languages of infinite words (see [25]). Programming languages such as Haskell [8] allow for effectively defining infinite streams of values by means of lazy evaluation mechanisms. In view of application to temporal media programming, that is, finite and infinite sequences of media type values such as images, sounds or control events, the abstract data type implicitly induced by ω-semigroups is enriched with the parallel product of finite or infinite strings. This leads to effective tools for handling temporal media types [7].

It has recently been advocated that there are benefits in embedding finite strings as well as infinite streams into (some notion of) *tiled temporal media* [9]. Typical temporal media synchronization constructs such as musical pickups (anacruses) lead to distinguishing the effective starts of temporal media – the first note of a melody, from their logical starts – the first strong beat of that melody [1]. Then, the notion of tiled temporal media allows for a fairly simple modeling of these multi-level synchronization constructs. This comes from the

---

[1] See http://hal.archives-ouvertes.fr/hal-00910002 for a complete version.

A.M. Shur and M.V. Volkov (Eds.): DLT 2014, LNCS 8633, pp. 339–347, 2014.

fact that they are equipped with a *tiled product* that is neither a sequential nor a parallel product but both [9, 17].

In the general setting of higher-dimensional strings and tiling semigroups [16, 18, 19], the tiled temporal media and the related tiled product lay in the mathematical framework of inverse semigroup theory [21] and semigroups with local units [5, 6, 20]. Projecting tiled temporal media onto semantical tags, we obtain sorts of tiled words. In the finite case, the induced algebra is the inverse semigroup of McAlister whose elements are birooted words [22, 23].

Aiming at providing a robust mathematical framework for handling languages of tagged tiled temporal media, we extend here the language-theoretical tools available for languages of finite birooted words (see [3, 11, 13, 14]) to the case of infinite birooted words.

For such a purpose, we show that the obvious embedding of the free ω-semigroups of finite and infinite words into the monoid of birooted words can be generalized to the embedding of *two-sorted* ω-semigroups into *one-sorted* ordered ω-monoids.

This leads to an algebraic characterization of regular languages of finite and infinite birooted words that generalizes and unifies the algebraic characterizations of regular languages of finite and infinite words.

2   From Finite or Infinite Words to Birooted Words

Let A be a finite alphabet. Let $(A^*, \cdot)$ be the free monoid of finite strings on the alphabet A, and let $(A^+, A^\omega, \cdot, *, \pi)$ be the associated free ω-semigroup (see [25]) with finite product $\cdot$, mixed product $*$ and infinite product π.

In the sequel, both the finite product $u_1 \cdot u_2$ when $u_1, u_2 \in A^*$ or the mixed product $u_1 * u_2$ when $u_1 \in A^*$ and $u_2 \in A^\omega$ may simply be denoted by $u_1 u_2$.

The set $A^\infty = A^* \cup A^\omega$ of finite and infinite strings is ordered by the *prefix order* $\leq_p$, defined, for every u and $v \in A^\infty$, by $u \leq_p v$ when either $u = v$, or u is finite and there is $w \in A^\infty$ such that $v = uw$. Extended with a maximum element denoted by 0, the set $A^\infty + 0$ ordered by the prefix order $\leq_p$ is a complete lattice. The *prefix join* $u \vee_p v$ of two words u and $v \in A^\infty$ is then the least word $w \in A^\infty$, if it exists, such that we have both $u \leq_p w$ and $v \leq_p w$, or 0 otherwise. Then, for every $u \in A^*$ and $v \in A^\infty$, the *right residual* $u^{-1}(v)$ of v by u, is defined as the word $w \in A^\infty$, unique if it exists, such that $v = uw$. We take $u^{-1}(v) = 0$ otherwise. By definition, $u^{-1}(v) \neq 0$ if and only if $u \leq_p v$.

Definition 1. *A positive (right) birooted word u is a pair $u = (u_1, u_2)$ where $u_1 \in A^*$ and $u_2 \in A^\infty$. The word $u_1 u_2 \in A^\infty$ is called the domain of the birooted word u, and the word u_1, its root path. The birooted word (u_1, u_2) is finite when u_2 is finite. The set of positive birooted words on the alphabet A is denoted by $T^\infty(A)$. The set of finite positive birooted words on A is denoted by $T^+(A)$.*

The product $u \cdot v$ of two birooted words $u = (u_1, u_2)$ and $v = (v_1, v_2)$ is defined as the birooted word $w = (w_1, w_2)$ with $w_1 = u_1 v_1$ and $w_2 = v_1^{-1}(u_2) \vee_p v_2$ when $v_1^{-1}(u_2) \vee_p v_2 \neq 0$. Otherwise, we take $u \cdot v = 0$ for some new birooted word 0, with $0 \cdot u = u \cdot 0 = 0$ for every $u \in T^\infty(A)$, and $0 \cdot 0 = 0$.

Examples. The following examples illustrate the definition of the product:

$$(ab, ab) \cdot (a, bc) = (aba, bc) \qquad (ab, (ab)^\omega) \cdot (a, bc) = 0$$
$$(ab, (ab)^\omega) \cdot (a, ba) = (aba, (ba)^\omega) \qquad (ab, ab) \cdot (a, (ba)^\omega) = (aba, (ba)^\omega)$$
$$(ab, ab) \cdot (a, (bc)^\omega) = (aba, (bc)^\omega) \qquad (ab, ac) \cdot (a, (ba)^\omega) = 0$$
$$(1, ab) \cdot (1, abc) = (1, abc) \qquad (1, ab) \cdot (1, ac) = 0$$

Let $T_0^\infty(A)$ (resp. $T_0^+(A)$) be the set of positive birooted words (resp. finite positive birooted words) extended with 0.

Theorem 1. *The set $T_0^\infty(A)$ equipped with the above product is a partially ordered monoid with unit $1 = (1, 1)$.*

Let $u \mapsto u^R$ and, resp., $u \mapsto u^L$ be the right projection (or reset) and, resp., the left projection (or co-reset) defined on the monoid $T_0^\infty(A)$ by

$$0^R = 0^L = 0, \quad u^R = (1, u_1 u_2) \quad \text{and} \quad u^L = (1, u_2)$$

for every $u = (u_1, u_2) \in T_0^\infty(0)$. Then the *natural order relation* (see [10, 20]) is defined by

$$u \leq v \quad \text{when} \quad u = u^R \cdot v \cdot u^L, \quad \text{or, equivalently,} \quad u = u^R \cdot v$$

for every u and $v \in T_0^\infty(A)$. Elements u such that $u \leq 1$ are called *subunits* and the set of subunits is denoted by $U(T_0^\infty(A))$.

Theorem 2. *The natural order relation on $T_0^\infty(A)$ is a partial order relation stable under product, that is if $u \leq v$ then $uw \leq vw$ and $wu \leq wv$ for all $u, v, w \in T_0^\infty(A)$. Subunits and idempotents coincide, that is, we have $u \leq 1$ if and only if $u = uu$ for all $u \in T_0^\infty(A)$. Moreover, the set $U(T_0^\infty(A))$ ordered by the natural order is a complete lattice with product as meet.*

An immediate interest of the notion of finite and infinite birooted words is that the (two sorted) free omega monoid (A^+, A^ω) can be embedded into the (one sorted) monoid $T_0^\infty(A)$. More precisely, let $u * v$ be the *mixed product* defined by $u * v = (u \cdot v)^R$ for all $u \in T_0^\infty(A)$ and $v \in U(T_0^\infty(A))$, and let $\pi(u_i)_{i \in \omega}$ be the *infinite product* defined by $\pi(u_i)_{i \in \omega} = \bigwedge_{n \in \omega}(u_0 \cdot u_1 \cdots u_{n-1})^R$ for all infinite sequences of birooted words $(u_i)_{i \in \omega}$. Then we have:

Theorem 3. *The pair $(T_0^+(A), U(T_0^\infty(A)))$ equipped with the finite product, the mixed product and the infinite product is a well-defined ω-semigroup.*

Moreover, let $\theta_f : A^* \to T_0^+(A)$ be the mapping defined by $\theta_f(v) = (v, 1)$ for every finite word $v \in A^*$ and let $\theta_\omega : A^\omega \to U(T_0^\infty(A))$ be the mapping defined by $\theta_\omega(w) = (1, w)$ for every infinite word $w \in A^\omega$.

Then, the pair of mappings $(\theta_f, \theta_\omega) : (A^+, A^\omega) \to (T_0^+(A), U(T_0^\infty(A)))$ is an ω-semigroup embedding, that is, a one-to-one mapping that preserves the finite, the mixed and the infinite products.

In the particular case where the infinite sequence $(u_i)_{i \in \omega}$ is *constant*, that is, when there is some $v \in T_0^\infty(A)$ such that $u_i = v$ for every $i \in \omega$, then we write v^ω for $\pi(u_i)_{i \in \omega}$. One can easily check that for every idempotent (or subunit) $u \leq 1$ we have $u^\omega = u$.

3   Embedding ω-semigroups into Ordered ω-monoids

We show here that the above embedding of the free ω-semigroup (A^+, A^ω) into the monoid of birooted words $T^\infty(A)$ can be generalized to an embedding of any ω-semigroup S into (some notion of) ordered ω-monoid $M(S)$.

Let M be a monoid partially ordered by a relation $\leq$. We assume that the order relation $\leq$ is stable under product, i.e. if $x \leq y$ then $xz \leq yz$ and $zx \leq zy$ for every x, y and $z \in M$. The set $U(M)$ of *subunits* of the partially ordered monoid M is defined by $U(M) = \{y \in M : y \leq 1\}$.

The following definition is adapted from [14] and then, following [20], refined with the congruence property [4, 10].

Definition 2 (Adequately ordered and E-ordered monoids). A (stable) partially ordered monoid M is an *adequately ordered monoid* when:

(A1) idempotent subunits: for every $x \in M$, if $x \leq 1$ then $xx = x$,
(A2) left and right projection: for every $x \in M$, both the left projection defined by $x^L = \min\{y \in U(M) : xy = x\}$ and the right projection defined by $x^R = \min\{y \in U(M) : yx = x\}$ exist in $U(M)$,

It is an Ehresmann-ordered monoid (or E-ordered monoid) when, moreover:

(A3) congruence property: for every $x, y, z \in M$, if $x^L = y^L$ then $(xz)^L = (yz)^L$ and if $x^R = y^R$ then $(zx)^R = (zy)^R$,

Examples. Every monoid trivially ordered is an adequately ordered monoid. Every inverse monoid ordered by the natural order [21] is also an adequately ordered monoid with left and right projections defined by $x^R = x \cdot x^{-1}$ and $x^L = x^{-1} \cdot x$ for every element x.

The notion of adequately ordered monoid is extended here with an infinite (right) product as follows.

Definition 3 (E-ordered ω-monoid). An *E-ordered ω-monoid* is an E-ordered monoid M equipped with an infinite product operator $\pi : M^\omega \to U(M)$ satisfying the following properties:

(I1) *subunit preservation:* for every $(x_i)_{i \in \omega}$ such that for every $i \in \omega$ we have $x_i = x$ for some $x \in U(M)$, then $\pi(x_i)_{i \in \omega} = x$,
(I2) *monotonicity:* for every infinite sequences $(x_i)_{i \in \omega} \in M^\omega$ and $(y_i)_{i\omega}$, if $x_i \leq y_i$ for every $i \in \omega$ then $\pi(x_i)_{i \in \omega} \leq \pi(y_i)_{i \in \omega}$,
(I3) *mixed associativity:* for every infinite sequence $(x_i)_{i \in \omega} \in M^\omega$, for every $x \in M$, given $x_i' = x$ when $i = 0$ and $x_i' = x_{i-1}$ when $i > 0$, we have $(x \cdot (\pi(x_i)_{i \in \omega}))^R = \pi(x_i')_{i \in \omega}$
(I4) *infinite associativity:* for every infinite sequence $(x_i)_{i \in \omega} \in M^\omega$, for every strictly increasing sequence $(k_i)_{i \in \omega}$ of positive integers with $k_0 = 0$, given $y_i = x_{k_i} \cdot x_{k_i+1} \cdots x_{k_{i+1}-1}$ defined for every $i \in \omega$, we have $\pi(y_i)_{i \in \omega} = \pi(x_i)_{i \in \omega}$.

Definition 4 (Monoid completion). Let $S = (S_f, S_\omega)$ be an ω-semigroup with finite product $\cdot : S_f \times S_f \to S_f$, mixed product $* : S_f \times S_\omega \to S_\omega$ and infinite product $\pi : (S_f)^\omega \to S_\omega$. By definition (see [25]) the finite, mixed and infinite product are related by mixed and infinite associativity laws. Let S_f^1 be the semigroup S_f extended with a unit and let $\mathcal{P}^*(S_\omega)$ be the set of non-empty subsets of S_ω.

The *monoid completion* $M(S)$ of the ω-semigroup S is defined to be the $M(S) = S_f^1 \times \mathcal{P}^*(S_\omega) + 0$ equipped with the product $\cdot$ defined, for every non zero element (x, X) and $(y, Y) \in M(S)$ by $(x, X) \cdot (y, Y) = (xy, y^{-1}(X) \cap Y)$ when $y^{-1}(X) \cap Y \neq 0$ with $y^{-1}(X) = \{z \in S_\omega : y * z \in X\}$ and is defined to be 0 all other cases, with $y^{-1}(X) = \{z \in S_\omega : y * z \in X\}$. Element of $M(S)$ are also ordered by the relation $\leq$ defined over $M(S)$ by taking 0 to be the smallest element and by $(x, X) \leq (y, Y)$ when $x = y$ and $X \subseteq Y$ for every (x, X) and $(y, Y) \in M(S)$.

Theorem 4. *The set $M(S)$ with the above product $\cdot$ and the natural order $\leq$ is an E-ordered monoid with unit $1 = (1, S_\omega)$.*

We define the infinite product π almost by iteration although, as well known in ω-language theory, the infinite product itself cannot be defined as a limit since many regular languages are not closed in prefix topology.

Among subunits of $M(S)$, the meet $\wedge$ in the order correspond to the product. Indeed, a subunit in $M(S)$ is either zero (with just behave like $(1, \emptyset)$) or of the form $(1, X)$ for some non empty $X \subseteq S_\omega$ with $(1, X) \cdot (1, Y) = (1, X \cap Y)$. The monoid of subunits $U(S)$ is thus isomorphic to the power set $\mathcal{P}(S_\omega)$ with intersection as product. In the next definition, when appropriate, we thus may use at will the meet operator $\wedge$ in place of the product.

Definition 5 (Infinite product). Let $(x_i)_{i \in \omega} \in (M(S))^\omega$. Let $\pi^0() = 1$ and for every $n \in \omega$, let $\pi^{n+1}(x_i)_{i \leq n} = (x_0 \cdot \pi^n(x_{i+1})_{i < n})^R$. Then, the infinite product $\pi(x_i)_{i \in \omega}$ is defined by

$$\pi(x_i)_{i \in \omega} = (1, X_\omega) \wedge \bigwedge_{n \in \omega} \pi^n(x_i)_{i \leq n}$$

with X_ω defined by $X_\omega = S_\omega$ when $J = \{i \in \omega : x_i \leq 1\}$ is finite and X_ω defined by $X_\omega = \{x_\omega\}$ when J is infinite, with $x_\omega = \pi(s_{j_i})_{i < \omega}$ where $(j_i)_{i \in \omega}$ is the increasing enumeration of the elements of J and, for very $j \in J$, the (non zero) element x_j is of the form $x_j = (s_j, X_j)$.

Theorem 5. *The E-ordered monoid $M(S)$ equipped with the above infinite product is an adequately ordered ω-monoid.*

Last, we aim at showing that the ω-monoid $S = (S_f, S_\omega)$ can be embedded as an ω-monoid into $M(S)$. This means defining over $M(S)$ an ω-monoid structure $(M_f(M), M_\omega(S))$, which can be done by taking $M_f(S) = M(S)$ for the "finitary" part and $M_\omega(S) = U(M(S))$ for the "infinitary" part. The infinite product is the one already defined, and the *mixed product* of two elements $x \in M(S)$ and $y \in U(M(S))$ is defined to be $x * y = (x \cdot y)^R$.

Theorem 6. *The mapping $\theta = (\theta_f, \theta_\omega) : (S_f, S_\omega) \to (M(S), U(M(S)))$, defined by $\theta_f(x) = (x, S_\omega)$ for every $x \in S_f$ and by $\theta_\omega(y) = (1, \{y\})$ for every $y \in S_\omega$, is a one-to-one ω-monoid morphism.*

4 Application to Language Theory

A language of birooted words is a set $X \subseteq T^\infty(A)$ of *non-zero* birooted words. The class of such languages is equipped with the boolean operators union (also called sum), intersection and complement, plus operators derived from the structure of $T_0^\infty(A)$: for all X and $Y \subseteq T^\infty(A)$, the product $X \cdot X$, the star X^*, as well as X^+ and the omega X^ω, are defined by extending the corresponding operators on $T_0^\infty(A)$ in a point-wise manner, *always omitting* the birooted word zero possibly resulting from these products. Additionally, we define the left and right projections X^L and X^R of the language X as the sets $X^L = \{x^L \in T^\infty(A) : x \in X\}$ and $X^R = \{x^R \in T^\infty(A) : x \in X\}$.

Theorem 7. *The class of languages of (positive) birooted words definable in MSO is closed under all the operators defined above. Moreover, it is finitely generated from the finite languages of (positive) birooted words, sum, product, star, omega and left and right projections.*

Proof. Easily follows from standard relativization techniques of mathematical logic.

Combining the notions of Muller ω-word automata [24] (see also [25]) and of tile automata [14], we define birooted words ω-automata as follows.

Definition 6. A (finite) *birooted word ω-automaton* is a tuple $\mathcal{A} = \langle Q, \delta, K, W \rangle$ with a (finite) *set of states* Q, a *transition function* $\delta : A \to \mathcal{P}(Q \times Q)$, a *finitary acceptance condition* $K \subseteq Q \times Q$ and an *infinitary acceptance condition* $W \subseteq \mathcal{P}(Q)$. For technical reasons, we always assume that $\emptyset \in W$.

A *run* of the automaton $\mathcal{A}$ on a birooted word $u = (u_1, u_2)$ is a labeling mapping $\rho : [0, |u_1 * u_2| + 1[\to Q$ that satisfies the *local consistency* property : $(\rho(k), \rho(k+1)) \in \delta((u_1 * u_2)[k])$ for every $0 \leq k \leq |u_1| + |u_2|$.

The run ρ of the automaton $\mathcal{A}$ on the birooted word $u = (u_1, u_2)$ is *locally accepting* when the pair of states $(\rho(0), \rho(|u_1|))$ that marks the input and output roots belongs to K.

The run ρ is *globally accepting* when the set of states that occur infinitely often belongs to W, i.e. $\{q \in Q : |\varphi^{-1}(q)| = \infty\} \in W\}$. Observe that when (u_1, u_2) is finite then the global acceptance constraint is always satisfied since we assume that $\emptyset \in W$.

The language $L(\mathcal{A})$ is then defined as the set of birooted words $(u_1, u_2) \in T^\infty(A)$ for which there exists a locally and globally accepting run of $\mathcal{A}$ on (u_1, u_2).

Theorem 8. *A language $L \subseteq T^\infty(A)$ of non nul birooted words is recognized by a finite-state birooted word ω-automaton if and only if L is definable in MSO and upward closed.*

Corollary 1. *The class of languages recognized by a finite-state birooted word ω-automaton is closed under the sum, intersection, product, star and omega operations. Moreover, it is finitely generated from finite languages of (positive) birooted words, sum, product, star and omega, and the* upward closure *of left and right projections.*

The next definitions and theorem extend a similar result proved in [14] for languages of finite birooted words.

A mapping $\varphi : T_0^\infty(A) \to T$ from birooted words to an adequately ordered monoid T is a *premorphism* when $\varphi(1) = 1$ and $\varphi(xy) \leq \varphi(x)\varphi(y)$ and if $x \leq y$ then $\varphi(x) \leq \varphi(y)$ for every x and $y \in T_0^\infty(A)$.

The premorphism φ is *adequate* when, moreover, $\varphi(x^L) = (\varphi(x))^L$ and $\varphi(x^R) = (\varphi(x))^R$ for every $x \in S$, and if $x^L y^R \neq 0$ and $x^L \vee y^R = 1$ then $\varphi(xy) = \varphi(x)\varphi(y)$ for every x and $y \in S$. In the latter case we say that the product xy is disjoint.

The adequate premorphism φ is *ω-adequate* when T is an adequately ordered ω-monoid and, for every infinite sequence $(u_i)_{0 \leq i} \in (T_\infty^0(A))^\omega$, we have $\varphi(\pi(u_i)_{0 \leq i}) \leq \pi(\varphi(u_i))_{0 \leq i}$ and, moreover, if the product $\pi(u_i)_{0 \leq i}$ is disjoint (i.e. for every $0 < k$, given $x = u_0 \cdot u_1 \cdots u_{k-1}$ and $y = \pi(u_{i+k})_{0 \leq i}$ we have $x^L y^R \neq 0$ and $x^L \vee y^R = 1$), then we also have $\varphi(\pi(u_i)_{0 \leq i}) = \pi(\varphi(u_i))_{0 \leq i}$.

A language $L \subseteq T^\infty(A)$ is *quasi-recognizable* when there exists a finite adequately ordered ω-monoid S and an ω-adequate premorphism $\varphi : T_0^\infty(A) \to S$ such that $L = \varphi^{-1}(\varphi(T))$.

Theorem 9. *Let $L \subseteq T_0^\infty(A)$ be a language of non-zero finite or infinite birooted words. The language L is quasi-recognizable if and only if it is a finite boolean combination of upward-closed (for the natural order) languages definable in monadic second order logic (MSO).*

Remark 1. As a concluding remark, following [12,15], we could also aim at generalizing the present approach to trees, possibly leading to further developments in the very subtle and difficult emerging algebraic theory of languages of infinite trees [2]. However, there is no evidence yet that such a generalization could lead to a successful algebraic characterization of infinite tree languages.

References

1. Berthaut, F., Janin, D., Martin, B.: Advanced synchronization of audio or symbolic musical patterns: an algebraic approach. International Journal of Semantic Computing 6(4), 409–427 (2012), http://hal.archives-ouvertes.fr/hal-00794196
2. Blumensath, A.: Recognisability for algebras of infinite trees. Theoretical Comp. Science 412(29), 3463–3486 (2011)
3. Dicky, A., Janin, D.: Two-way automata and regular languages of overlapping tiles. Research report RR-1463-12, LaBRI, Université de Bordeaux (2013), http://hal.archives-ouvertes.fr/hal-00717572

4. Dubourg, E., Janin, D.: Algebraic tools for the overlapping tile product. In: Dediu, A.-H., Martín-Vide, C., Sierra-Rodríguez, J.-L., Truthe, B. (eds.) LATA 2014. LNCS, vol. 8370, pp. 335–346. Springer, Heidelberg (2014), http://hal.archives-ouvertes.fr/hal-00879465

5. Fountain, J., Gomes, G., Gould, V.: A Munn type representation for a class of E-semiadequate semigroups. Journal of Algebra 218, 693–714 (1999)

6. Fountain, J., Gomes, G., Gould, V.: The free ample monoid. Int. Jour. of Algebra and Computation 19, 527–554 (2009)

7. Hudak, P.: A sound and complete axiomatization of polymorphic temporal media. Tech. Rep. RR-1259, Department of Computer Science, Yale University (2008)

8. Hudak, P., Hugues, J., Peyton Jones, S., Wadler, P.: A history of Haskell: Being lazy with class. In: Third ACM SIGPLAN History of Programming Languages (HOPL). ACM Press (2007)

9. Hudak, P., Janin, D.: Tiled polymorphic temporal media. Research report RR-1478-14, LaBRI, Université de Bordeaux (2014), http://hal.archives-ouvertes.fr/hal-00955113

10. Janin, D.: Quasi-inverse monoids (and premorphisms). Research report RR-1459-12, LaBRI, Université de Bordeaux (2012), http://hal.archives-ouvertes.fr/hal-00673123

11. Janin, D.: Quasi-recognizable vs MSO definable languages of one-dimensional overlapping tiles. In: Rovan, B., Sassone, V., Widmayer, P. (eds.) MFCS 2012. LNCS, vol. 7464, pp. 516–528. Springer, Heidelberg (2012), http://hal.archives-ouvertes.fr/hal-00671917

12. Janin, D.: Algebras, automata and logic for languages of labeled birooted trees. In: Fomin, F.V., Freivalds, R., Kwiatkowska, M., Peleg, D. (eds.) ICALP 2013, Part II. LNCS, vol. 7966, pp. 312–323. Springer, Heidelberg (2013), http://hal.archives-ouvertes.fr/hal-00784898

13. Janin, D.: On languages of one-dimensional overlapping tiles. In: van Emde Boas, P., Groen, F.C.A., Italiano, G.F., Nawrocki, J., Sack, H. (eds.) SOFSEM 2013. LNCS, vol. 7741, pp. 244–256. Springer, Heidelberg (2013), http://hal.archives-ouvertes.fr/hal-00659202

14. Janin, D.: Overlapping tile automata. In: Bulatov, A.A., Shur, A.M. (eds.) CSR 2013. LNCS, vol. 7913, pp. 431–443. Springer, Heidelberg (2013)

15. Janin, D.: On languages of labeled birooted trees: Algebras, automata and logic. Information and Computation (in print, 2014), http://hal.archives-ouvertes.fr/hal-00982538

16. Janin, D.: Towards a higher dimensional string theory for the modeling of computerized systems. In: Geffert, V., Preneel, B., Rovan, B., Štuller, J., Tjoa, A.M. (eds.) SOFSEM 2014. LNCS, vol. 8327, pp. 7–20. Springer, Heidelberg (2014)

17. Janin, D., Berthaut, F., DeSainte-Catherine, M., Orlarey, Y., Salvati, S.: The T-calculus : towards a structured programming of (musical) time and space. In: ACM Workshop on Functional Art, Music, Modeling and Design (FARM), pp. 23–34. ACM Press (2013), http://hal.archives-ouvertes.fr/hal-00789189

18. Kellendonk, J.: The local structure of tilings and their integer group of coinvariants. Comm. Math. Phys. 187, 115–157 (1997)

19. Kellendonk, J., Lawson, M.V.: Tiling semigroups. Journal of Algebra 224(1), 140–150 (2000)

20. Lawson, M.V.: Semigroups and ordered categories. I. the reduced case. Journal of Algebra 141(2), 422–462 (1991)
21. Lawson, M.V.: Inverse Semigroups: The theory of partial symmetries. World Scientific (1998)
22. Lawson, M.V.: McAlister semigroups. Journal of Algebra 202(1), 276–294 (1998)
23. McAlister, D.: Inverse semigroups which are separated over a subsemigroups. Trans. Amer. Math. Soc. 182, 85–117 (1973)
24. Muller, D.: Infinite sequences and finite machines. In: Fourth Annual Symp. IEEE, Switching Theory and Logical Design, pp. 3–16 (1963)
25. Perrin, D., Pin, J.E.: Infinite Words: Automata, Semigroups, Logic and Games, Pure and Applied Mathematics, vol. 141. Elsevier (2004)

Author Index